SEDGES OF KARNATAKA (INDIA)
(FAMILY CYPERACEAE)

V.P. PRASAD

Botanical Survey of India,
Western Circle, Pune-411 001

&

N.P. SINGH

Botanical Survey of India,
Kolkata-700 001.

Published by:

SCIENTIFIC PUBLISHERS (INDIA)
5A, New Pali Road, P.O. Box 91
Jodhpur 342 001, India
E-mail: info@scientificpubonline.com
Website: www.scientificpubonline.com

Print, 2022

ISBN: 978-81-72333-03-4
E-ISBN: 978-93-87913-97-4

Printed in India

FOREWORD

The family Cyperaceae is a difficult and challenging group, as far as its taxonomy is concerned. In India though the family is common and forms an important component of wetlands everywhere, there are only very few works on its taxonomy. The minute flowers enclosed in glumes and arranged in complicated inflorescences of spikelets, along with the inadequate herbarium collections (often with immature spikelets and without the important underground parts), cause great difficulties in identifying the species, and very few have ventured to study the taxonomy and relationships of Indian sedges. Intensive and extensive explorations in India during the recent years have added a large number of additions as new taxa, new records and new introductions into the country and these have only increased this difficulty. Further the generic concepts and delimitations have been considerably altered and nomenclature of most species have been changed during recent years. Therefore, the old national and regional floras have become grossly inadequate to identify Indian species and a complete up-to-date revision of the family is long overdue.

In this background the present book, **Sedges of Karnataka** is most welcome. The authors have made a very commendable contribution after years of extensive field work in the state followed by critical study of the species in the field as well as in the herbarium and good attempt has been made to put the results into an up-to-date taxonomic revision of the family for the State of Karnataka. The work deals with 163 species, 6 subspecies and 9 varieties belonging to 23 genera. While most of the species are cosmopolitan and common to other states in India and particularly in Peninsular India, there are 9 species endemic to Karnataka State. Two new species have been discovered from the state during this work. As a result of critical studies, 7 species known earlier from the Karnataka are excluded for various reasons and another 7 species reported for the first time from the state.

The detailed introductory part deals with the geography, topography, soil and climate of the state followed by a brief history of the family, its morphological peculiarities, distribution, ecological and habitat patterns, dispersal systems and classification within the family. Discussions on the new findings, endemism, distribution and uses of sedges are also included. The systematic part provides a comprehensive description of the family, the literature relevant to the family in India, followed by an identification key to the genera. For each genus, after the description a simple utilitarian key to the species helps one to easily identify the species. Detailed nomenclatural citations and descriptions are provided for each species along with notes on habitats, ecology, distribution, phenology etc. Lastly citation of the herbarium specimens studied along with their respective herbarium acronyms are included. Line illustrations depicting habit and floral parts are provided for 60 species, which would greatly help in identification. Photographs are also provided showing habits and habitats.

It is hoped that the present work would greatly help Botany students, teachers and research workers, naturalists and conservationists particularly dealing with wetlands, for identifying the sedges not only of Karnataka but also of other states of India. I am sure that this work would generate interest in the study of the sedges in India and would contribute to the early completion of the family for Flora of India series.

Dr. N.P. Balakrishnan
Ex-Joint Director,
Botanical Survey of India
Southern Circle,
Coimbatore 641 003

PREFACE

In the present study an effort is made to provide a detailed up-to-date account of the family Cyperaceae in the state of Karnataka. To serve this purpose many field trips were undertaken to collect plant specimens and visited different herbaria to study the sedge specimens from Karnataka.

A general account of the family is provided in which a brief history of the family, morphology, distribution, habitat, dispersal and classification are discussed. During the past six years hundreds of specimens were studied to make detailed descriptions of most of the species. Sixty illustrations covering almost all the genera in the state are included which make identification of the concerned taxa easy. Detailed descriptions of a total number of 163 species, 6 subspecies and 9 varieties coming under 23 genera are provided along with information on flowering and fruiting season, habitat, distribution, etc. Keys to the species and infraspecific taxa were prepared by sorting out characters from the descriptions made and these keys will be useful for identification of sedges.

We hope this book will be useful, especially to students, botanists, environmentalists, agriculturists and foresters. As most species have a wide range of distribution, this can be used to identify sedges of other parts of the country also especially of peninsular India.

V.P. PRASAD
&
N.P. SINGH

ACKNOWLEDGEMENTS

The present work is an off-shoot of a long term project of Botanical Survey of India on the State Flora of Karnataka and in this connection the authors are thankful to:

The Director, Botanical Survey of India (BSI), Kolkata for his kind permission to work on the family Cyperaceae of Karnataka and for providing all the facilities.

Shri M.K. Vasudeva Rao, Joint Director, Dr. P. Lakshminarasimhan, Scientist 'C', other scientists and staff of BSI, Western Circle, Pune for their help in different ways.

Forest Department of Karnataka state for permitting to collect sedge specimens from different parts of the state.

Authorities of different herbaria, like BLAT, BNHS, JCB, MGH, MH and RRCBI for permitting to consult their herbaria.

Authorities of Royal Botanic Gardens, Kew, especially Dr. David Simpson, Higher Scientific Officer, Dr. Sri Krishna Murti, Ex-Indian Botanical Liaison Officer and Dr. Jayasri Bhattacharya, Present Indian Botanical Liaison Officer for the critical study of some herbarium specimens and for providing necessary literature.

Dr. K. Gopalakrishna Bhat, Department of Botany, Poornaprajna College, Udupi and Dr. A. Jaffar Hussain, Department of Botany, Presidency College, Chennai for their help.

Mrs. Mini Prasad for rewriting most part of the manuscript and for labelling all the specimens collected during this study.

Shri K.V. Sasidharan, BSI, Pune who did all the work in computer for the excellent production of the final manuscript.

Shri A.P. Jain, Ex-Library & Information Assistant, BSI, Jodhpur and Shri Pawan Kumar, Scientific Publishers (India), Jodhpur for their keen interest in publishing this work and Shri Rajesh Ojha, Scientific Publishers (India), Jodhpur for the computer type setting.

ABBREVIATIONS

ca	...	*circa* : about
Chrom. No.	...	Chromosome Number
Coll.	...	Collector
comb.nov.	...	*combinatio novo* : new combination
Distrib.	...	Distribution
e.g.	...	*exempli gratia* : for example
et al.	...	and others
Fig./f.	...	Figure
Fls.	...	Flowers
Frts.	...	Fruits
Furth.	...	Further
Ic.	...	*Icon* : Illustration
Illus.	...	Illustration
l.c.	...	*loco citato* : at the place cited
nom. cons.	...	*nomen conservandum* : conserved name
nom. illeg.	...	*nomen illegitimum* : illegitimate name
nom. nud.	...	*nomen nudum* : naked name
p.	...	page
pl.	...	plate
p.p.	...	*pro parte* : partly
R.F.	...	Reserve Forest
S.F.	...	State Forest
s.n.	...	*sine numero* : without number
sect.	...	section
subsp.	...	subspecies
t.	...	*tabula* : plate
var.	...	variety

HERBARIA CONSULTED

BLAT	...	Blatter Herbarium, St. Xavier's College, Mumbai.
BNHS	...	Bombay Natural History Society, Mumbai.
BSI	...	Botanical Survey of India, Western Circle, Pune.
JCB	...	Herbarium, St. Joseph's College, Bangalore (Presently in Centre for Ecological Sciences, Indian Institute of Science, Bangalore).
MACS	...	Maharashtra Association for Cultivation of Science, Pune. (Presently Agharkar Research Institute, Pune).
MH	...	Madras Herbarium (Botanical Survey of India, Southern Circle), Coimbatore.
MGH	...	Manasa Gangotri Herbarium, Botany Department, Mysore University, Mysore.
RRCBI	...	Regional Research Centre (Ayurveda), Central Council for Research in Ayurveda & Siddha, Bangalore.

LIST OF ILLUSTRATIONS

LIST OF PHOTOGRAPHS

CONTENTS

INTRODUCTION

Karnataka is very rich in its flora and fauna due to the diverse topographic and climatic conditions in different parts of the state. There is tremendous variation in the ecological conditions from the western coastal line to the drier eastern plains through the forests of western ghats. Certain parts of western ghats form very good habitat for the dense forests and these areas are of high botanical interest. More than 3,900 species of angiosperms with a large number of infra-specific taxa belonging to 1,370 genera and 199 families were reported so far from the state (Sharma *et al.*, 1984).

Cyperaceae is the 7th largest family in Karnataka, when the number of species is taken into account. But the largest genus in the state is *Fimbristylis* Vahl with 39 species followed by *Cyperus* L. with 38 species both Cyperaceae members.

GEOGRAPHY OF THE STATE

Karnataka state lies almost in the central part of the western peninsular India between 11°40′ and 18°27′ North latitude and 74°5′ and 78°33′ East longitude with a total area of 1,92,204 sq km. The state is bounded in the north by Maharashtra, Andhra Pradesh in the east, Tamil Nadu in the South, Kerala in the south-west and the small state of Goa and the long stretch of Arabian sea in the west. The state is divided into 19 administrative districts. Though 7 more new districts were created recently, the old district names are used in the present study covering the whole state. Scattered, small, uninhabited or sparsely populated islands are found not far away from the coast of Uttara Kannada and Dakshina Kannada.

TOPOGRAPHY

Physiographically Karnataka is divided in different ways. Sharma *et al.* (1984) considered two regions, the hilly 'Malnad' region and the plain 'Maidan'. The 'Malnad' region comprises mostly the hilly western ghats

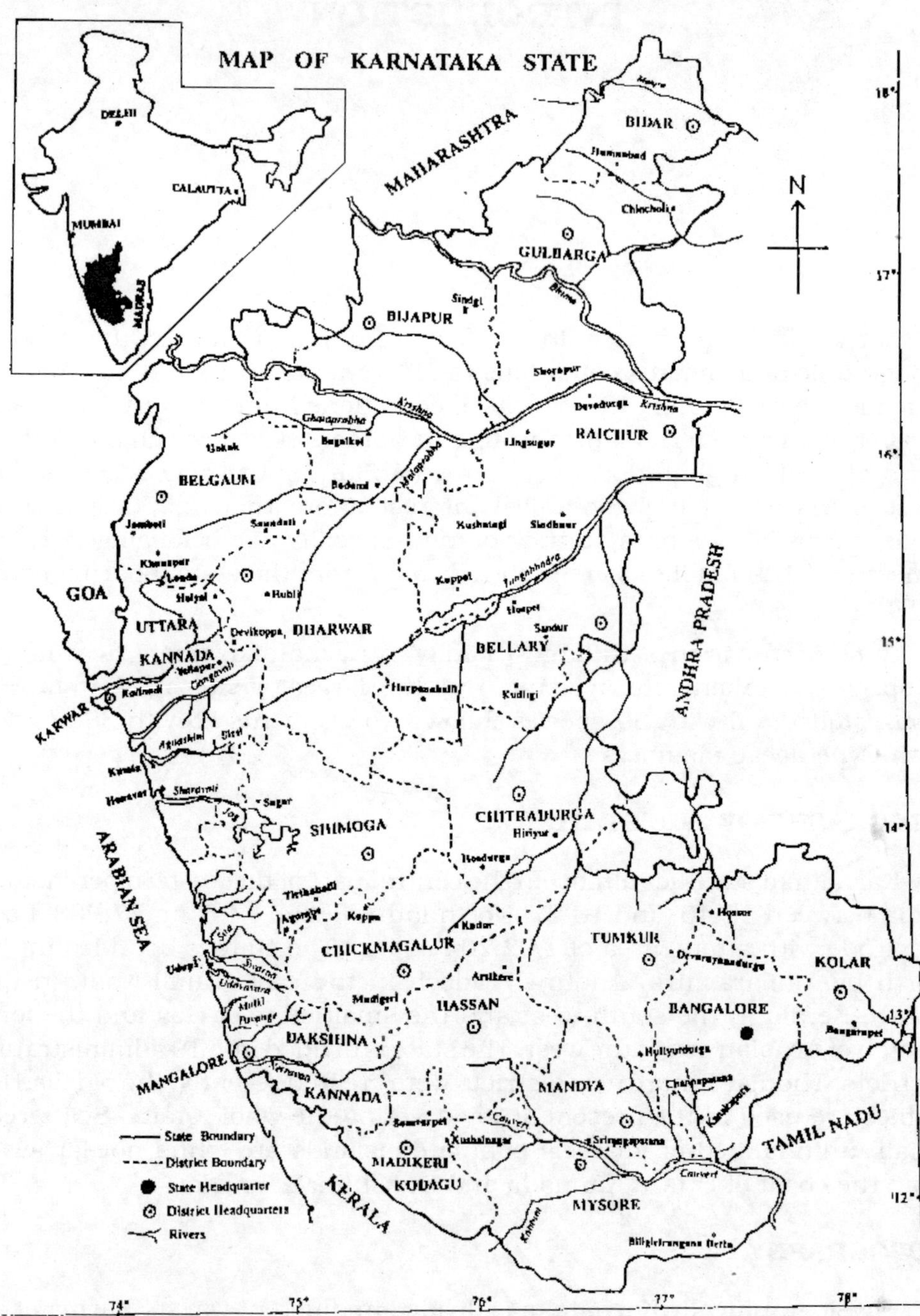

Map 1. Map of the Karnataka State

Photo 1. *Cyperus compressues* L. Inflorescence

Photo 2. *Fimbristylis cymosa* R.Br.-Habit

Photo 3. *Mariscus javanicus* (Houtt.) Merr. & Metcalfe - Habit

Photo 4. *Pycreus macrostachyog* (Lamk.) J. Raynal - Inflorescence

Photo 5. *Lipocarpha chinensis* (Osb.) Kern - Habit

Photo 6. *Rhynchospora wightiana* (Nees) Steud. - Habit

Photo 7. *Cyperus arenarius* Retz. - in sandy sea shore

Photo 8. Cyperus malaccensis Lamk. - forming large patches in brackish water

Photo 9. Brackish water in mangroves - Habitat for some halophytic sedges

Photo 10. Wetland - Habitat for many sedges

Photo 11. Schoenoplectus Mucronatus (L.) Palla - in a shallow wetland, along with Nymphaea Nouchali etc.

Photo 12. Schoenoplectus lateriflorus (Gmel.) Lye - as a weed in paddy field

Photo 14. Cyperus rotundus L. - as a weed in garden

Photo 13. Courtoisina cyperoides (Roxb.) J. Sojal and Pycreus macrostachyos (Lamk.) J. Raynal - as weeds in paddy field.

which falls in the districts of Uttara Kannada, Dakshina Kannada, Shimoga, Chikmagalur, Hassan, Kodagu and to some extend in Belgaum and Dharwar. The 'Maidan' is the plain country found in all the other districts and known as the Karnataka plateau.

According to Saldanha (1984) the state has got 5 physiographic regions as, 1. A narrow coastal belt from Mangalore to Karwar, 2. A mountain range formed of western ghats, 3. Interior Karnataka known as the *Maidan* which is again divided into Northern *Maidan* and Southern *Maidan*, 4. Southern interior Karnataka or the Southern *Maidan*, and 5. the two mountain ranges on the Deccan plateau, the Bababudan Range in Chikmagalur district and the Biligirirangan Range at the Southern tip of the state.

It seems to be more convenient to divide the state into three physiographic regions as shown below.

The Western Coastal belt : It is a narrow coast line of about 320 km from Karwar in the north to Mangalore in the south passing through the districts of Uttara Kannada and Dakshina Kannada. This coastal belt is often very narrow, even less than 5 km wide in certain areas in the north, but extending up to 65 km in the south. All the west flowing rivers originating in the western ghats cross the coastal belt to drain in the Arabian sea.

The mountain range of Western ghats or *Malnad* : This region is formed by the hilly range of western ghats passing through the state from north to south. It passes through all the districts except those of Eastern Karnataka. Many rivers, both west flowing and east flowing arise in the Western ghats. The important west flowing rivers are Sharavati, Agnashini, Gangavali, Kali, all passing through Uttara Kannada and Netravati, Pavanje, Mulki, Udyavara, Swarnanadi and Sitanadi passing through Dakshina Kannada district.

The plain or *Maidan* : The interior Karnataka formed by an inland plateau of varying elevations between 300 and 800 m can be called as the *Maidan* or plain country. It can be divided into the northern and southern plateau. The Northern plateau is undulating, dotted with low hills and the elevation is between 300 and 450 m. The Southern plateau is also undulating and the higher elevation is up to 800 m. All the east flowing rivers of the state are running through the *maidan*. The main east flowing rivers are Krishna, Tungabhadra and Cauvery.

GEOLOGY :

Geology of the state is discussed in detail by Saldanha (1984) in the introductory Chapters of Flora of Karnataka. He has discussed about four main types of geological formations in Karnataka as follows :

The Archaean complex made up of Dharwar Schists and granitic gneisses. The southern portion comprising nearly three-fourth of the state is occupied by this ancient complex of highly metamorphosed rocks. The Dharwar schists have the oldest rocks in the area. The Dharwar Schists are found in 7-8 bands running in a south easterly direction. These are rich in mineral ores. There are gold mines in Kolar and Hutti; while Castle Rock, Chiknaikanahalli, Kudremuk and Sandur have iron ores and there is copper in Chitradurga.

The Granite gneisses cover large area of the state from the southern boundary up to Belgaum and Raichur in the north. These have been classified into Champion, Peninsular, Nilgiri and Bellary Gneisses. Champion Gneisses are the oldest in this series and occurs in island like intrusion in the Dharwar Schists near Bhadravi, Channagiri, Chitradurga and Kolar. Peninsular Gneisses occupy the largest area. It is a mixture of several types of granite rocks and appear as greyish, smooth hillocks with strong master joints. Nilgiri gneisses also known as Charnockite is a plutonic intrusion into the Dharwar Schists. It occurs in several patches in the southern districts.

The northern districts of Karnataka have 2 more pre-cambrian systems of non fossiliferous, sedimentary formations. The Kaladgi series has horizontal rocks that run from Belgaum to Bijapur. Excellent exposures of this can be seen in Belgaum district along the Ghataprabha river at Gokak and along Malaprabha river at Peacock Gorge. The Bhima series occur north-east of Kaladgi series on either side of the Bhima river. It conists of non-fossiliferous sand stones, lime stones and shales.

The Deccan trap occur in the four northern most districts of the state. It was formed by a series of horizontal lava flows that may have occurred during the Tertiary period. Inter trappean beds are fossiliferous and found intercalated with different flows of trappean lava.

Tertiary and more recent laterite is found on the coast and in several districts in the Deccan Plateau. The high level laterite found in Belgaum, Bellary, Chikmagalur and Hassan districts are fossil laterites formed by subareal weathering and leaching in different rocks. Mining for Bauxite, Kaolin and Manganese ores is being done in the lateritic areas which disturb the natural vegetation of the area.

Coastal alluvium is of recent origin and lies just behind the western seaboard.

SOIL

Soil in Karnataka has been divided into the following groups.

1. **Coastal Alluvium :** It is limited to the narrow coastal belt but sometimes extending inwards along the rivers. These soils are sandy with very low clay or organic content.

2. **Lateritic soils :** It is found mainly in the eastern side of coastal alluvium, extending from Belgaum in the north to Kodagu in the south. It is found in the districts of Bangalore, Bidar and Kolar also. These soils are with clayey or loamy surface and are acidic. Laterite soils in forest areas are rich in humus, with high organic matter.

3. **Dark brown clay :** The adjoining areas of Kodagu and Mysore districts have a narrow strip of dark brown clayey soils which support a good forest belt.

4. **Red soils :** Red soils along with Red loam cover major portion of the state, particularly in the south. Red soil is found in the districts of Tumkur, Kolar, Bangalore, Mandya and Mysore.

5. **Red loams :** Red loam is rich in organic matter and is acidic especially in forest areas. It is found in parts of Shimoga, Chikmagalur, Hassan and Mysore districts.

6. **Red and black soils :** The districts of Belgaum, Dharwar, Bellary and parts of Raichur have mixed red and black soils.

7. **Deep black soils :** Deep black soil with a mixture of alluvial gravel and lime stone are found in river valleys formed by the Bhima, Krishna and their tributaries.

8. **Trap soil :** Black soils of trappean origin occur in northern districts of Bidar, Gulbarga, Bijapur and part of Belgaum. The black soils have good water retaining capacity.

CLIMATE :

Climate in Karnataka is of tropical monsoon type. Three seasons viz., Winter, Summer and Monsoon are distinct. Temperature varies from place to place in the same season depending mainly on altitude. Winter is from December to February and is usually without or with meager rainfall. Average minimum temperature in winter drops to 10°C or even below at high ranges of Western Ghats. On extreme cold days of winter the minimum temperature may drop down to as low as 2°C at high altitudes. But average minimum temperature in coastal areas is about 22°C in winter.

Summer season is from March to May and the average maximum temperature shoots up to 39°C or more in the drier areas of northern interior Karnataka. Maximum summer temperature can reach up to 46°C in drier areas. But average maximum summer temperature at high altitude and in coastal areas is less being around 28°C and 33°C respectively.

Monsoon season is from June to November, of which maximum precipitation is through south-west monsoon from June to September. But north-east monsoon from October to November is also distinct. The total annual rainfall is much low in the drier plains of Eastern Karnataka ranging from 45 to 90 cm. The average annual rainfall in the northern plains is only 71 cm. The area between Bellary and Chitradurga experience a subdesertic condition with an average annual rainfall of 55 cm only. But in the southern interior Karnataka rain fall is higher with the mean annual rainfall of 114 cm. But it is as high as 150-750 cm in the heavy rainfall areas of western ghats extending from Uttara Kannada in the north to Kodagu in the south. Agumbe in Shimoga district is the wettest place in the state with an average rainfall of 827 cm.

The relative humidity is much higher in western ghat areas and much low in the comparatively drier plains of Eastern Karnataka.

PAST AND PRESENT WORK

The history of floristics in Karnataka has been discussed in detail by Razi (1979), Sharma *et al.* (1984) and Saldanha (1984). Singh (1988) also discussed about the past explorations and taxonomical studies, particularly of Eastern Karnataka. Hence, it is preferred here to avoid a detailed account of the past floristic works in Karnataka. However, it is worth mentioning the historical workers like Buchnan Hamilton (1807), Someran (1879), Lovery (1888), Camaron (1894), Talbot (1909 & 1911) and Barnes (1944). Prof. B.A. Razi has studied the Flora and vegetation of Karnataka more than 30 years and trained many students in the field of Taxonomy and Floristics. Saldanha (1984 & 1996) published 2 volumes of *Flora Karnataka* and subsequent volumes are awaited. Singh (*l.c.*) did extensive exploration of the eastern districts and published the *Flora of Eastern Karnataka* in 2 volumes.

As far as family Cyperaceae of the state is concerned, no detailed study was undertaken in the past. Information on sedges of Karnataka is scattered in different district Floras and papers dealing with flora of small areas. To mention a few are Satyanarayan & Shankarnarayan (1964), Ramaswamy & Razi (1973), Hooper (1976), Arora *et al.* (1981), Rao & Razi (1981), Yoganarasimhan *et al.* (1981), Singh (*l.c.*) and Keshvamurthy & Yoganarasimhan (1990).

Clarke (1893-1894) reported only 9 species specifically from the state of Karnataka in the *Flora of British India.* But the 48 species mentioned as distributed throughout India, 27 species in Deccan peninsula and 3 species in the Malabar region also may be considered distributed in Karnataka, as these are actually found in the state. Hence, the total number of sedge species reported from Karnataka in the *Flora of British India* may be considered as 87. Cooke (1908) reported 28 species from the areas coming under the present political boundary of Karnataka in the *Flora of the*

Presidency of Bombay. It includes the South Maratha country, often mentioned by him. This number seems to be too small for the northern parts of Karnataka. But few species reported by him from Konkan and Deccan also must be from the present state of Karnataka. 80 sedge speceis reported by Fischer (1931) in *Flora of the Presidency of Madras* seems comparatively a good number. It may be mentioned that southern parts of the state provide more suitable habitats for the sedge family.

As far as recent studies are concerned, not many people have worked on the family in Karnataka. Gopalakrishna Bhat of Udupi has made good collections of Cyperaceae specimens from the southern districts which are extant in the herbaria of Mysore University, Mysore and Indian Institute of Science, Bangalore. Bhat & Razi (1971) enumerated 62 speceis of Cyperaceae from Mysore city and surrounding areas. Govindarajalu, one of the well known Cyperologists of India (1973, 1974, 1975, 1990, 1994, 1996 & 1998) described a few new species from the state. Hooper (1976) has provided a good account of the sedges of Hassan district. Sharma *et al.* (1984) provided a list of 155 species of the family in the State Flora- Analysis. But a detailed study of individual species has not been done so far. Hence, the present study was undertaken to get detailed up-to-date information on each species of the family.

MATERIALS AND METHODS

The present Flora was prepared in a modified pattern of Flora of India by studying thousands of herbarium specimens and by gathering data on different aspects of the species from various already published articles besides some observations made in the field. Hundreds of Cyperaceae specimens were collected by conducting exploration tours to different parts of the state. This is in addition to thousands of specimens already available in various herbaria like BSI, MH, BLAT, JCB, MGH, etc.

First of all a list of all the species reported from Karnataka was prepared by referring all the available Floras and floristic accounts published in various journals. All the Cyperaceae specimens available in BSI and other herbaria were studied, to make a detailed description of each species. For this purpose other herbaria like MH at Coimbatore; BLAT and BNHS at Mumbai; JCB, and RRCBI at Bangalore and MGH at Mysore were visited. During the course of this study thousands of spikelets were dissected to make detailed descriptions of each species and the infraspecific taxa wherever necessary. Data on phenology, habitat, distribution, etc. of each species were collected and included after the species description. Chromosome number and uses, if any, are also provided wherever available. Nomenclature of each species was also checked and updated. Full citation of the correct name and synonyms are given along with reference to relevant regional Floras and district Floras, besides the Flora of British India. Vernacular names are also provided, wherever available. Special observations if any are given as a note after the species description. Specimens examined at various herbaria for each species are also included. Illustrations of 55 species, 4 subspecies and 1 variety, covering most of the genera, are provided in the text and the figure numbers are furnished after the citations. Similarly photographs of some species are provided depicting their habit mostly in their natural habitat. Key to the species was made by sorting out the characters from the description made for each species during the present study.

VEGETATION

Flora of the state is very rich and diverse due to the favourable topographic and climatic conditions. Gradual change in vegetation can be observed from littoral vegetation of the western coastal belt to the thorny scrubs of the drier eastern plains through the dense evergreen or moist deciduous forests of western ghats. Vegetation types of Karnataka were discussed extensively by earlier workers. It started with Eners (1907) who wrote about the evergreen forests of Manjarabad forest range. Kadambi (1939- 1950) has published many papers on the montane and Ghat forests. Later on Satyanarayan (1958-1961) studied the vegetation of western ghats and Arora (1961-1964) studied the forest types of North Kanara in detail. Razi (1950-1972) has discussed the forest types of Karnataka in many of his papers.

Vegetation of the state varies from tropical wet evergreen forests to scrub jungles depending on the altitude and rainfall of the areas. Champion and Seth (1968) in *A Revised Survey of the Forest types of India* divided Karnataka forests into 3 main categories as southern tropical moist forests, southern tropical dry forests and southern montane forests. Each of these forest types are further divided as follows:

I. *Southern tropical moist forests.*

   A. Southern tropical wet evergreen forests.

      1. West coast tropical evergreen.

      2. Southern hill top tropical evergreen.

      3. Wet bamboo brakes.

   B. Southern tropical semi-evergreen forests.

      1. West coast semi-evergreen.

      2. Lateritic semi-evergreen.

      3. Moist bamboo brakes.

C. Southern moist deciduous forests.

D. Littoral mangroves.

II. *Southern tropical dry forests.*

A. Southern tropical dry deciduous forests.

1. Mixed deciduous forests.

2. Lateritic scrub.

3. Bamboo brakes.

B. Southern tropical thorn forests.

III. *Southern montane forests.*

A. Southern sub-tropical broad leaved forests.

B. Southern sub-tropical hill savanna.

C. Southern montane wet temperate forests.

D. Southern montane wet grasslands.

Considering the topography, bioclimate and soil, the vegetation of Karnataka was grouped into 4 categories by Saldanha (1984), viz., 1) Littoral vegetation, 2) West coast tropical vegetation, 3) Upland deciduous vegetation and 4) Southern tropical montane forests. Each of these categories are further divided into smaller groups.

Based on the frequency of rainfall, Sampathkumaran (1924) divided the vegetation of the then Mysore state into 3 parallel north-south zones or belts, as 1) The evergreen belt, 2) the deciduous forests and 3) the dry deciduous fuel tract and scrub. This system was followed by subsequent workers like Razi (1950a) and Sharma *et al.* (1984). It seems to be convenient even today to follow this classification for the present day state also with slight modification by including the coastal and mangrove vegetation. Hence vegetation in Karnataka can be divided as follows.

1. Littoral vegetation and mangroves :

The sandy beaches and adjacent areas of west coast are characterised by many species like *Spinifex littoreus* (Burm.) Merr., *Remirea maritima* Abul., *Ipomoea pes-caprae* (L.) R. Br. and *Canavalia rosea* (Sw.) DC. which are adapted to grow in sandy saline soils of coastal areas. Other common species found here are herbs like *Hydrophylax maritima* L.f. and *Launaea sarmentosa* (Willd.) Alston. *Pandanus fascicularis* Lamk. and *Calophyllum inophyllum* L. are also found here.

Mangrove vegetation is found in some pockets only, especially in the estuaries of the coastal districts. Common species found in the mangroves are *Avicennia marina* Vierh., *A. officinalis* L., *Sonneratia caseolaris* (L.f.)

Engl., *Kandelia candal* (L.) Druce and *Rhizophora mucronata* Lamk. These species are characterised by stilt roots and vivipary. Patches of *Acanthus ilicifolius* L. can be seen in shallow brackish water of these areas.

2. The evergreen forests :

The evergreen forests are at the altitude ranging from 600- 1000 m, in the heavy rainfall regions of the western ghats, which receive a mean annual rainfall of 350 cm or more. Width of this belt vary from 10 km to 64 km. At higher altitudes of 1200 to 1800 m or more, the hill tops are exposed to high velocity winds and hence trees of stunted growth are found on grassy meadows. But dense Shola type forests are found in the valleys protected from the winds. The evergreen forests can be divided into 1) the moist evergreen belt and 2) the mixed belt of evergreen and deciduous forests.

The moist evergreen forest is in a very narrow belt along the western ghats with ravines and narrow valleys. In this region the rainfall is between 250 and 600 cm. If Champion and Seth (*l.c.*) is followed, this belt falls under Southern tropical wet evergreen forests. Common species found in this region are *Dipterocarpus indicus* Bedd., *Calophyllum apetalum* Willd., *Hopea ponga* (Dennst.) Mabb., *Mesua nagassarium* (Burm.) Kosterm., *Vateria indica* L., *Artocarpus hirsutus* Lamk., etc. Herbaceous species are rare in evergreen forests. Species composition of the evergreen forests changes from place to place.

The mixed belt of evergreen and deciduous forests are dominated by deciduous trees, more herbs and less number of evergreen tree species. According to Champion & Seth (*l.c.*) these forests come under Southern tropical semi-evergreen forests. Here rainfall varies from 150-250 cm. The main deciduous trees found here are *Terminalia paniculata* Roth, *Lagerstroemia microcarpa* Wight, *Eleocarpus serratus* L. and *Diospyros* spp.

3. The deciduous forest belt :

The deciduous forests are found on the eastern side of the evergreen belt. Rainfall is in an intermediate range of 75-150 cm. Width of this belt varies from 32 to 48 km. According to Champion & Seth (*l.c.*) it falls under Southern tropical moist deciduous forests. The main species found here are *Tectona grandis* L.f. along with *Grewia tillifolia* Vahl, *Lagerstroemia microcarpa* Wight, *Dillenia pentagyna* Roxb., *Kydia calycina* Roxb., *Dalbergia latifolia* Roxb., *Terminalia crenulata* Roth, *T. paniculata* Roth, *T. alata* Heyne ex Roth, *T. bellirica* (Gaertn.) Roxb., etc.

4. The Dry deciduous Fuel Tract and Scrub :

This belt is situated on the eastern parts of the state adjacent to the deciduous forest belt, occupying the remaining area of the state. Here the rainfall is between 35 and 75 cm. Depending upon the availability of rainfall this can be divided into the superior and inferior types. In the superior type the main species found are *Tectona grandis* L.f., *Boswellia serrata* Roxb. ex Coleb., *Anogeissus latifolia* (DC.) Wall. ex Guill. & Perr., *Sterculia urens* Roxb., *Acacia chundra* (Rottl.) Willd., *Dalbergia latifolia* Roxb., *Hardwickia binata* Roxb., *Dillenia pentagyna* Roxb., *Kydia calycina* Roxb., species of *Terminalia, Lagerstroemia microcarpa* Wight, *Haldina cordifolia* Ridsdale, *Grewia tillifolia* Vahl, *Santalum album* L., etc. According to Champion & Seth (*l.c.*) this falls under the Southern tropical dry deciduous forests. There are some xerophytic species and Bamboos also here along with the deciduous species.

The inferior type is a dryland tract with more xerophytic species and plants of stunted growth. This can be considered as the Southern tropical thorn and scrub forests of Champion & Seth (*l.c.*). *Hardwickia binata* Roxb. is the main tree species found here along with mostly thorny plants like species of *Acacia* and *Zizyphus, Capparis divaricata* Lamk., *Aegle marmelos* (L.) Corr., *Dichrostachys cinerea* (L.) Wight & Arn., *Catunaregum spinosa* (Thunb.) Tiruv., etc. Lands of very low quality have scrub forests which are usually composed of *Euphorbia antiquorum* L., *E. tirucalli* L., *Dodonaea viscosa* Jacq., *Cassia auriculata* L., etc.

FAMILY CYPERACEAE - A GENERAL ACCOUNT

Cyperaceae or the sedge family is one of the largest families with *ca* 5,000 species (Bruhl, 1995) coming under *ca* 80 genera. It is one of the most widely distributed plant groups and hence attracted the attention of most Phytogeographers. Cyperaceae members are always herbs except the African genus *Microdracoides* which is tree-like. Sedges are characterised by the grass-like or rush-like habit with or without rhizomes or stolons, the minute, inconspicuous flowers enclosed by the distichously or spirally arranged glumes on a spikelet and the indehiscent fruits known as nuts or achenes.

A BRIEF HISTORY

Though there is a tendency to avoid the sedges in the field, even by the present day botanists, mainly due to the difficulty in identifying them, there are pre-Linnaean contributions to the family by Tournefort (1719) and Micheli (1729). Linnaeus (1753 & 1754) in his historical *Species Plantarum* and *Genera Plantarum* described 5 genera and 81 species of the family. Subsequent workers went on describing new taxa of this family also, while doing the floristics of different regions of the world. Other historical works on the family are of Burman (1768), Linnaeus (1767 & 1771), Rottboell (1773), Retzius (1786-1791), Willdenow (1797-1830), Vahl (1805-1806), Brown (1810), Roxburgh (1820, 1824, 1832) and Miquel (1855-1859). However, workers like Nees (1834), Kunth (1837), Steudel (1854-1855), Boott (1858-1862), Boeckeler (1868-1877), Bentham (1881, 1883) and Pax (1888) have done more detailed study on the family. Kukenthal (1909, 1935-1936, 1938-1944 and 1949-1951) has done monographic studies on different groups of Cyperaceae. Kern (1954-1974) and Koyama (1955-1988) are the recent world authorities of the sedge family.

Though Indian Cyperaceae is included in many of the above mentioned works, Clarke (1893, 1894) was the first one to do extensive study on Indian Cyperaceae, whose work is included in J.D. Hooker's *Flora of British India.* A total number of 449 Cyperaceae species coming under 28 genera are enumerated in this Flora. In the regional Floras published in the first half of the twentieth century also, the family is well represented. Most of the species reported by Cooke (1908) in *Flora of the Presidency of Bombay* and by Fischer (1931) in Gamble's *Flora of the Presidency of Madras* are found in Karnataka, which was part of both the Presidencies earlier.

Not many Indian botanists have worked on family Cyperaceae. The late Prof. E. Govindarajalu (1966-1998) was the most important Indian Cyperologist of the recent past who devoted more than 30 years in studying the sedges of the country. He has described many new species especially from South India. Other important Indian workers on the family are Verma and others like Veena Chandra and Rao (1979-1982), Shah (1963-1967) and others. The late G.L. Shah along with co-workers like Suryanarayana, Deshpande and Bhatt (1967-1974) and Rao (1981-1982) also had done good work on this family. In Karnataka very limited work have been done on this family. Govindu & Thirumalachar (1960) and Bhat (1971) did the Cyperaceae of Mysore and surroundings. Bhat (1983) studied the sedges of Kodagu and Dakshina Kannada districts also.

MORPHOLOGY

Underground parts

Sedges are usually annual or perennial herbs. Annuals are with fibrous roots and perennials with short or long-creeping rhizomes. But the annual or perennial habit is often difficult to make out in the herbarium specimens especially when the underground portion is missing and also when the annual or perennial habit is variable. Annual or perennial habit is often used to distinguish Cyperaceae members at different taxonomic levels especially in combination with other characters. The rhizomes are usually small, woody, but some times are long-creeping or emitting stolons which often bear tubers as in *Bolboschoenus maritimus* (L.) Palla and *Cyperus rotundus* L. The rhizomes and stolons are clothed with scales which usually disintegrate leaving fibrous remains.

Stems

Stems are usually tufted but occasionally solitary when the rhizomes or stolons are long-creeping which usually produce a solitary stem from each node. Although stems are erect in most cases, there are species especially of certain *Scleria* which straggle over other herbs and shrubs. Stems can be capillary or slender to very stout, few millimetres (*Eleocharis* sp. and *Cyperus* sp.) to several metres long (in some *Scleria* sp.), without nodes or nodose as in *Scleria* sp. and *Fuirena* sp., solid or hollow, at times septate as in *Eleocharis dulcis* (Burm.f.) Trin. ex Hensch. and

Schoenoplectus articulatus (L.) Palla, trigonous, triquetrous or even almost 3- winged, at times terete, multiangular or strongly compressed, very rarely ancipitous as in *Fimbristylis complanata* (Retz.) Link., usually smooth, at times scabrid especially on the angles. Some of the stem characters given above are helpful to identify the sedges at species level.

Leaves :

Leaves are usually 3-ranked when the stems are trigonous or triquetrous, often distichous or polystichous, usually in a basal cluster, but at times a few or almost all cauline (*Fuirena* sp., *Rhynchospora* sp., *Scleria* sp., *Diplacrum* sp.). In genus *Eleocharis* and some species of *Fimbristylis* leaves are reduced to bladeless sheaths. Leaves are sessile, herbaceous or coriaceous, shorter to much longer than the stem, flat, plicate, revolute, conduplicate or involute. In *Fimbristylis littoralis* Gaudich. leaves are laterally flattened and equitant. Leaf sheaths vary from few millimetres to several centimetres, membranous or herbaceous, usually closed, loosely or tightly enclosing the stem and the colour varies from reddish, purplish, blackish, yellowish to greenish. Mouth of the sheath is concave, truncate or convex. Ligule (a projection at the top of the sheath) is generally absent but sometimes represented by a fringe of hairs or a membrane. Many species of *Fimbristylis* can be grouped on the basis of presence or absence of ligule. In species of *Scleria* mouth of the sheath is produced beyond the base of the leafblade on its opposite side which is known as contraligule. Leaf characters especially its presence or absence, basal or cauline arrangement, presence or absence of ligule, etc. are useful for identifying certain genera as mentioned above. These characters can be used to distinguish many species also as in the case of *Fimbristylis*.

Inflorescence :

As the flowers in sedges are very minute and inconspicuous, inflorescence in this family is constituted by the arrangement of spikelets. Inflorescence is always terminal, but some times appears to be lateral (in fact pseudolateral as the main bract elongates and looks like continuation of the stem as in *Schoenoplectus* sp.). In the genus *Eleocharis* and in certain species of *Fimbristylis* inflorescence is reduced to a single spikelet. Otherwise inflorescence is normally anthelate or capitate. Anthelate or umbellate inflorescence can be simple, compound, decompound or supradecompound. Ultimate branches of the anthela terminate in single to several spikelets. Sometimes spikelets are arranged in short to long spikes or in clusters. Inflorescence are usually subtended by foliaceous or glumaceous involucral bracts. Number of the bracts vary from one to several. Normally the lowest bract is the longest and gradually becoming shorter upwards. Inflorescence characters are variable in most genera and hence not a very reliable taxonomic character. But it is constant in the genus *Eleocharis* and in several species of *Fimbristylis*, *Lipocarpha*, *Schoenoplectus*, *Kyllinga*, etc.

Prophylls :

Prophylls are small scale-like organs, varying in form and size, attached to different parts in the inflorescence. Taxonomically these are important structures. In *Cyperus* base of the rays are covered by a tubular prophyll. In *Carex* such tubular prophyll is found at the base of peduncles of the spikes.

The pistillate flowers in the species of *Carex* are enclosed in a sac-like structure known as utricle or perigynia which are also prophylls. A typical utricle in the genus *Carex* has a lower inflated portion and a beak above, which is usually bidentate. As there is a lot of variation in the structure of utricles of caricoid genera it is a good character for species differentiation.

Prophylls are also present as small scales in spikelets. In *Cyperus* there is one prophyll in each spikelet, i.e. the upper one of the lowest two empty glumes which are smaller than the upper flowering glumes.

Spikelets :

Spikelets in Cyperaceae vary in their size, shape and colour and are found either solitary or in inflorescence. Each spikelet is generally subtended by a prophyll and below that a bract. It consists of a rachilla bearing glumes and flowers. Morphology of the spikelets is very important in the taxonomy of Cyperaceae. Koyama (1961b) differentiated different kinds of spikelets, of which Scirpoid, Cyperoid, Rhynchosporoid, Scleroid, and Caricoid types are found in the genera found in Karnataka.

Spikelets of *Scirpus*, *Eleocharis*, *Fuirena*, *Eriophorum* and most species of *Fimbristylis* and *Bulbostylis* have tightly and spirally arranged glumes. Flowers are hermaphrodite and sometimes hypogynous bristles or scales are present which are homologous to perianth units. In *Cyperus* and allied genera spikelets differ from the above type in always having distichous glumes. Also the second empty glume at base becomes a prophyll. In *Kyllinga* and *Mariscus* the rachilla are jointed at the base.

Spikelets in *Scleria* are usually unisexual. The male spikelets are with spirally or more or less distichously arranged glumes. But the female spikelets consist of a single pistillate flower, subtended by a few glumes. The pistillate flower is usually considered as terminal, but some times it is difficult to make out whether it is terminal or axillary.

In *Carex* the flowers are unisexual and borne on a spike-like structure. The male flowers are borne laterally on the axis of the spike-like structure and each consists of stamens subtended by a glume. The female flowers are also arranged similarly and each consists of a solitary pistil enclosed by a sac-like utricle formed of a prophyll. The utricle is subtended by a glume. Here each staminate flower and pistilate flower, subtended by a glume is often considered as a spikelet. In *Carex* the male and female flowers (or spikelets) are arranged in different parts on the same spike. The female flowers are often arranged in the basal part and the male flowers in the upper part or vice-versa. Some times the male flowers are arranged in the basal as well as upper part with the female flowers in the middle or the female flowers in the basal and upper parts with male flowers in the middle. At times both the male and female flowers may be intermixed alternately. Rarely the male and female flowers may be found on separate spikes as in *C. maculata* Boott.

Glumes :

Glumes in Cyperaceae members exibit important taxonomic characters because of the variation in their arrangement, number, size, shape, colour, nervation, apex, etc. As explained earlier glumes are arranged distichously in *Cyperus* and allied genera or spirally in genera like *Eleocharis, Fuirena, Schoenoplectus*, etc. But in some species of *Fimbristylis* the lower glumes of the spikelets are distichous and the upper ones spiral. Shape of the glumes can be linear, lanceolate, oblong, ovate, orbicular, obovate or spathulate, the apex can be acute, obtuse, sinuate, mucronate, aristate, etc. The size varies with in the range of a few millimetres and the colour varies from stramineous to different shades of red and brown. It can be faintly to strongly striate and often keeled with a faint to prominent midnerve.

Persistence or deciduous nature of the rachilla and presence or absence of wings on the rachilla are also important taxonomic characters.

Flowers :

Flowers are minute and inconspicuous in the family Cyperaceae. But there are many very useful floral characters like presence or absence of perianth bristles or scales, number of stamens, size of the anthers, nature of the connective appendages of the anthers, continuous or articulated nature of the style on the ovary, number and length of the stigmas, etc. Perianth is absent in most of the genera like *Cyperus, Fimbristylis, Bulbostylis, Scleria, Carex*, etc. But it is found in the form of hypogynous bristles in genera like *Eleocharis, Schoenoplectus, Bolboschoenus*, etc. and in *Eriophorum* sp. (not found in Karnataka) the bristles are divided to the base into many hair-like segments. Perianth is modified to scales in *Fuirena* sp., *Lipocarpha* and *Hypolytrum*. In *Fuirena* often both the bristles and scales are found together. Number of stamens vary from 1 to 3, free except in some species of *Carex* and the filaments usually elongating while fruiting. The style can be uniformly slender, thickened towards the base and hence with a pyramidally thickened base in most tri-stigmatic species.

Fruits :

Fruits in Cyperaceae are one seeded, indehiscent and usually known as nuts or achenes. Most of the nut characters in Cyperaceae are stable and hence reliable in the taxonomy of the family. Nuts are sessile or subsessile or sometimes stipitate. It can be lenticular, planoconvex or trigonal and the shape vary as oblong, obovoid, ovoid or globose; apex can be obtuse, acute, apiculate or beaked. Difference in the size of the nut is often useful to differentiate the species within a genus. Colour of the nuts vary as different shades of brown, black and yellow in most of the genera. In *Scleria* nuts are usually pure white and rarely bluish. Gradual change in the colour is usually found when the nut matures. Surface of the nut varies according to the species. It can be smooth or with different kinds of ornamentation as pitted, trabeculate, zonate, tessellate, verrucose, puncticulate or with some prominent glandular outgrowth as in *Fimbristylis dipsacea* (Rottb.) Clarke. In many species, apex of the nut is crowned with the persistent style base which is very prominent on the nuts of *Bulbostylis* as a bulbiform, button-like structure. The shape and size of the hypogynium or disc is very important in the identification of species of *Scleria*.

DISTRIBUTION

Most genera found in Karnataka have got a wide range of world distribution. Some of them possess a world wide range of distribution. Some are restricted to the tropical and subtropical regions only. These are *Bulbostylis, Fuirena, Hypolytrum, Lipocarpha* etc. Some are extending to the temperate zones, e.g. *Cyperus, Eleocharis, Fimbristylis, Rhynchospora, Scleria, Carex,* etc. Of these, most of the *Carex* species are found in temperate regions and in tropical zones mostly confined to hill stations only. But in the case of *Scleria* most species are found in tropical countries and only few are found in temperate regions. Similarly *Cyperus* and *Fimbristylis* are distributed mostly in the lowlands and hills of tropics and the number of species decreases towards the temperate region. A detailed account of species distribution is discussed in the next Chapter.

HABITAT

In general, Cyperaceae members are found in varied habitats. Sometimes habitat preference of different species of a genus vary. But there are certain genera confined to particular habitats. For example, species of *Carex* and *Hypolytrum* are generally found in hilly forest areas. Similarly

Scleria is also usually found in forests. The monotypic genus *Remirea* and a few species of *Cyperus* and *Fimbristylis* are found in the saline sandy sea coasts only. *Cyperus*, one of the most dominant genera in Karnataka is found in different habitats but mostly in lowland, wet or marshy areas. Similarly *Fimbristylis* is also found in different habitats, but many in hilly areas. A few species of *Fimbristylis* and *Mariscus* are found in saline muddy areas or among the mangroves.

Many species, especially of *Cyperus, Fimbristylis, Mariscus, Eleocharis*, etc. are found as weeds in agricultural fields. Habitat of different species found in Karnataka are discussed in next Chapter also.

DISPERSAL

Dispersal of sedges very much depends on the habitat of the species. Species growing in and arround water is mostly dispersed through water only. At times there are some kind of adaptations for specific kind of dispersal by the morphological modification of the nut.

Nuts of the species growing along the sea coast are usually dispersed through sea water, e.g. in the nuts of *Remirea maritima* there is a swollen, corky rachilla internode overtopping the nut which helps to float in sea water. Species found in and around fresh water bodies like marshes, swampy areas, rice fields and other wetlands or along river banks, lakes, etc. are also dispersed through water. Nuts of most of the fresh water wetland species of sedges do not have any specific modification or device for their dispersal. Most of these nuts sink in water and are often carried by flowing water or may survive till the water dries up in seasonal water bodies. But there are few species in which some devices are found for their dispersal. In *Carex* the nut is enclosed by a sac-like utricle which floats on water surface and is dispersed by water current. Nuts of many sedges found as weeds are brought to agricultural lands, especially rice fields through irrigation water or along with the crop seeds. Tubers and stolons of some species like *Cyperus rotundus* are also dispersed through irrigation water.

Aquatic birds, especially the wild ducks play an active role in dispersal of sedge nuts. Mc Atee (1939) reported presence of 5,228 sedge seeds in a single bird stomach. Metcalf (1931) consider 14 species of sedges as most important food for wild ducks in North Dakota. Ridley (1930) accounts for many distributions in the genus *Scirpus* on the basis of dispersal by ducks either through sticking of the nuts to mud on the birds feet or passing through the stomach unharmed. He found *Mallard* and *Pintail* ducks (which winter in many parts of India also) feeding on *Scirpus* sp. Nuts of some species may withstand the activity of digestive enzymes and are dispersed through the excreta. A key for the identification of sedge seeds found in bird stomachs has been compiled for the North American species by Beetle (1943). The hypogynous bristles often found persistent on the nuts in the species of *Schoenoplectus, Eleocharis, Bolboschoenus, Fuirena*, etc. also may be helpful for their dispersal. These bristles are usually antrorsely or retrorsely scabrous and hence can cling to the feathers of birds. Like birds, herbivorous animals also may be dispersing the nuts of certain sedges in the same way. Man also sometimes disperse the nuts, tubers, rhizomes, etc. of sedges unknowingly by the agricultural activities or by transporting soil or mud from one place to another place. In *Eriophorum* sp. the perianth bristles are divided to the base into numerous hair-like structures which help in its dispersal through wind.

CLASSIFICATION

Taxonomy and classification of the family Cyperaceae is interesting, but a bit complicated due to the presence of very large number of species and the highly complex nature of certain floral parts. Hence, a monographic study of the whole family at global level is rather impossible and such studies has been done at regional levels mainly for certain genera like *Carex* only as done by Ohwi (1936-1944) and Akiyama (1935-1955) for Eastern Asia, by Mackenzie (1931-1940), Beetle (1940-1946) and Svensen (1929-1939) for North America, by Kreczetowicz (1935) and others for erstwhile USSR and Nelmes (1946-1955) & Kern (1952-1974) for Malesian region. A monographic study of the family for the Indian subcontinent or for the country as a whole is not yet done. A revision of the family in Ceylon (Sri Lanka) by Koyama (1985) is probably the best work done in South Asian region in the recent past.

Koyama (1961b) has discussed the history of the classification of the family starting with Nees (1835) who was the first one to create tribes under this family, which are validly published. In the beginning he created nine tribes as *Cypereae, Hypolytreae, Chrysitricheae, Scirpeae, Rhynchosporeae, Cladieae, Sclerieae, Elyneae* and *Cariceae* and later on added two more as *Ficinieae* and *Fuireneae* in the *Flora Brasiliensis* vol.2 (1842). Bentham and Hooker (1883) used sex of the flowers for the first time to divide the family into two main groups as **Monoclines** having bisexual flowers and **Diclines** bearing unisexual flowers. These major groups are equivalent to the subfamilies of subsequent workers. There are three tribes under **Monoclines** as *Scirpeae, Hypolytreae* and *Rhynchosporeae* and another three tribes under **Diclines** as *Cryptangieae, Sclerieae* and *Cariceae*. Same classification was followed by Hutchinson (1959, while adding tribe *Cypereae*) and Mattfeld (1936). But the latter treated the tribe *Hypolytreae* under **Diclines** by considering their flowers as unisexual and replaced the tribe *Cryptangieae* with *Lagenocarpeae*. Bentham & Hooker's 'Monoclines' and 'Diclines' were also changed as **Scirpoideae** and **Caricoideae** respectively, the former having bisexual flowers and the latter having unisexual flowers.

Ascherson and Graebner (1902) seperated the tribe *Rhynchosporeae* from the subfamily **Scirpoideae** and raised it to the level of a subfamily as **Rhynchosporoideae** as it have many empty glumes and only a few fruit-bearing glumes, compared to other tribes of **Scirpoideae**. Thus, the family was divided into three subfamilies as **Scirpoideae, Rhynchosporoideae** and **Caricoideae**. Ohwi (1944) mostly agreed with this sytem, but transferred the tribe *Sclerieae* to the subfamily **Rhynchosporoideae** and also added one more tribe *Gahnieae* to it. This system of classification is therefore based on the sex of the flowers as well as the number of empty or fertile glumes in the spikelets. Ohwi (*l.c.*) found *Sclerieae* more close fo *Rhynchosporeae* than to *Cariceae* and hence placed it under **Rhynchosporoideae**.

The systems based on the sex and the number of flowers can be summarized as proposed by Mackenzie for the North American Flora (Vol. 18).

Flowers hermaphrodite :	Spikelets many - flowered.
	Tribe 1. *Scirpeae.*
	Spikelets 1-2 - flowered.
	Tribe 2. *Rhynchosporeae.*
Flowers unisexual :	Achenes (Nuts) naked.
	Tribe 3. *Sclerieae.*
	Achenes in a closed sac.
	Tribe 4. *Cariceae.*

Clarke has made several systems, but the one given in the *Flora of Tropical Africa* Vol. 8 (1902) seems to be his final one. In this system the family is divided into three suborders (subfamilies) *viz.*, **Scirposchoeneae, Mapanieae** and **Cariaceae**, the first one having three tribes and the other two having one tribe each. Later on, the illegitimate names of these subfamilieis were corrected and tribe *Sclerieae* was seperated from *Cariceae* by Marloth (1915) in the *Flora of South Africa*, but fully accepted Clarke's system as follows :

Subfam. I. **Scirpoideae :**	Tribe 1. *Cypereae*
	Tribe 2. *Scirpeae*
	Tribe 3. *Schoeneae*
Subfam. II. **Mapanioideae**	Tribe 4. *Mapanieae*
Subfam. III. **Caricoideae**	Tribe 5. *Sclerieae*
	Tribe 6. *Cariceae*

Here Clarke created a new subfamily **Mapanioideae** to accomodate the genera of the tribe *Hypolytreae.* The Mapanioid genera are distinct from other groups of Cyperaceae by the cymose arrangement of spikelets bearing a terminal flower, while in others spikelets are spicate with axillary flowers only. *Slerieae* and *Cariceae* were kept under the same subfamily **Caricoideae**. But according to Koyama (1961b) *Scleria* is closely related to **Rhynchosporoideae** in its spikelets. The prophyll-modified, sac-like utricles found in *Carex* is not found in *Scleria.* Hence Koyama preferred to separate *Sclerieae* from *Caricoideae* to keep it under the subfamily Rhynchosporoideae.

It is clear from the above discussion that the main characters used for the major divisions of the family have been sex of the flowers, number of fruit-bearing flowers within a spikelet and to an extent presence or absence of the terminal flowers. According to Koyama (*l.c.*) sex of flowers is less important to make major divisions in the family. He found that presence or absence of terminal flowers and the prophyll are the most important characters for major divisions. Depending mainly on these characters he proposed to divide the family into six tribes under four subfamilies as shown below.

Subfam. I. **Mapanioideae**	Tribe 1. *Hypolytreae*
Subfam. II. **Scirpoideae**	Tribe 2. *Scirpeae*
	Tribe 3. *Cypereae*
Subfam. III. **Rhynchosporoideae**	Tribe 4. *Rhynchosporeae*
	Tribe 5. *Sclerieae*
Subfam. IV. **Caricoideae**	Tribe 6. *Cariceae*

According to Koyama (*l.c.*) **Mapanioideae** is most distinct of all other subfamilies because of their cymose spikelets with one terminal flower and several lateral flowers. All the others having spicate spikelets, without any terminal flower.

The subfamilies **Scirpoideae** and **Rhynchosporoideae** are morphologically similar and usually treated together under **Scirpoideae**. But Koyama separated them based on less number of flowers in **Rhynchosporoideae** compared to other *Scirpoid* genera. Usually in all the *Scirpoid* genera all the glumes are flower-bearing but in *Rhynchosporoid* genera some glumes are empty and hence the less number of flowers. According to him *Scleroid* spikelets are not fundamentally different from *Rhynchosporoid* type, but is a reduced condition of the latter type only. Usually *Scleria* is treated under subfamily **Caricoideae**. But Koyama transfered the tribe *Sclerieae* to the subfamily **Rhynchosporoideae**.

Though there are some differences among the above discussed systems, all are based on the supposition that the strictly unisexual flowers of

Caricoid genera have evolved from the primitive bisexual flowers of **Cyperoideae**. But the link between the bisexual flowers of **Cyperoideae** and the unisexual flowers of **Caricoideae** is not yet found in the form of a vestige of the other sex or of the hypogynous bristles or scales in the flowers of *Caricoideae.* But according to Kern (1974) if the so called partial inflorescence of *Hypolytreae* is considered homologous with the bisexual flowers in Cyperaceae, this tribe should be placed at the beginning of the subfamily **Cyperoideae** and not at the end of **Caricoideae**. He classified the family into five tribes under two subfamilies for the *Flora Malesiana* (1974), as shown below.

A. Subfamily : **CYPEROIDEAE**
- I. Tribe - *Hypolytreae*
- II. Tribe - *Cypereae*
- III. Tribe - *Rhynchosporeae*

B. Subfamily : **CARICOIDEAE**
- IV. Tribe - *Sclerieae*
- V. Tribe - *Cariceae*

Recently Bruhl (1995) proposed a new suprageneric classification of the family. As per this classification the family is divided into two subfamilies and 12 tribes. There are four tribes under the subfamily **Cyperoideae** viz., *Cypereae* (17 genera), *Scirpeae* (28 genera), *Abildgaardeae* (7 genera) and *Arthrostylideae* (4 genera). The subfamily **Caricoideae** has eight tribes as *Rhynchosporeae* (4 genera), *Schoeneae* (27 genera), *Cryptangieae* (5 genera), *Trilepideae* (4 genera), *Cariceae* (6 genera), *Sclerieae* (2 genera), *Bisboeckelereae* (4 genera) and *Hypolytreae* (14 genera).

He has prepared a database of the sedge genera of the world and the list of characters utilised by him for this purpose includes morphological observations, anatomical characters, embryo morphology, cytology, photosynthetic pathways, host specificity of pathogenic fungi, etc. He divided many genera based on the characters, like variation in the photosynthetic pathway. Acceptance of such narrow generic concepts resulted in 122 genera under the family. But he agrees that several of these genera should be merged in the interest of acquiring satisfactory generic circumscriptions.

Though different sytstems of classification are discussed above, in the present work all the 23 genera found in Karnataka are arranged in alphabetic order as per the guidelines for the Flora of India Project. Similarly all the species are also arranged alphabetically under the respective genus.

DISCUSSION AND FINDINGS

The present study on family Cyperaceae in Karnataka revealed that there are a total number of 163 species, 6 subspecies and 9 varieties belonging to 23 genera of the family in the state. Earlier Sharma *et al.* (1984) provided a list of 155 species (excluding *Cyperus pedunculatus* (R.Br.) Kern, a synonym of *Remirea maritima* Aubl. reported in the same Flora), 1 subspecies and 11 varieties in the State Flora-Analysis, who missed 2 species reported earlier by Blatter & McCann (1934). Besides these, 10 more species were reported from the state by different workers subsequently. Hence, the total number of species reported prior to the present study may be considered as 167 including the recent report of two new species by Govindarajulu (1997, 1998). But as a result of this study 7 Cyperaceae species given below had to be excluded from the state flora as they are not actually found in Karnataka.

Species excluded:

The following species are excluded from the state flora and reason for excluding these are given in next Chapter under the respective genus.

1. *Cyperus alternifolius* L.
2. *Cyperus fuscus* L.
3. *Diplacrum reticulatum* Holtt.
4. *Fimbristylis glabra* Hochst. ex Steud.
5. *Fuirena pubescens* (Lamk.) Kunth
6. *Fuirena tuwensis* Deshpande et Shah
7. *Pycreus flavescens* (L.) Beauv.

Fimbristylis annua var. *paucispiculata* Blatt. & McC. reported earlier from the state is treated as a synonym of **F. dichotoma** (L.) Vahl in the present work and three species reported in the State Flora Analysis (Sharma *et al., l.c.*) are treated here as infraspecific taxa as shown below.

1. *Carex glaucina* Boeck. is treated as a variety as **C. caricina** var. **glaucina** (Boeck.) Ghildyal & Bhattacharyya
2. *Kyllinga bifolia* Miq. is treated as a subspecies as **K. melanosperma** subsp. **bifolia** (Miq.) Karthik.
3. *Mariscus pictus* Nees is treated as a variety as **M. cyperinus** var. **pictus** (Nees) Karthik.

Hence, as a result of the present study total number of the species would have gone down from 167 to 156 if some species were not reported from the state for the first time as a result of this study. Further, of the 11 varieties reported earlier from the state, 5 are merged with the typical varieties in the present study, 4 are given the same varietal status and one elevated to the subspecies level. The varieties made synonyms of their species are :

1. *Cyperus iria* var. *paniciformis* Clarke
2. *Fimbristylis tenera* var. *oxylepis* (Steud.) Clarke
3. *Pycreus globosus* (syn. of. *P. flavidus*) var. *nilagiricus* (Steud.) Clarke
4. *P. globosus* var. *stricta* Clarke
5. *P. polystachyos* var. *laxiflorus* (Benth.) Clarke

The four varieties given the same status are :

1. *Cyperus digitatus* var. *hookeri* (Boeck.) Clarke (treated as *C. digitatus* var. *khasiana* (Clarke) Kern).
2. *C. distans* var. *pseudonutans* Kuekenth.
3. *C. exaltatus* var. *dives* (Del.) Clarke
4. *Pycreus diaphanus* var. *gracilescens* (Kukenth.) Hooper

Cyperus nutans var. *eleusinoides* (Kunth) Haines is treated as a subspecies in the present study.

Though the total number of species after the exclusions and reductions came down to 156, there is an addition of 7 species as a result of the present study as explained below and hence, the total number of species has actually gone up to 163. Of the 7 species added during the present study, 2 are new to science from Karnataka and the other 5 are reported for the first time from the state.

New species described:

During the course of the present study, a lot of specimens were found to be wrongly identified and some of them were of doubtful identity and hence interesting. Identity of hundreds of wrongly identified specimens was corrected, but most of them were already reported from the state. But a few of these specimens were not matching with any other known species and hence they were sent to the herbarium of Royal Botanic Gardens, Kew (UK) along with detailed descriptions and illustrations for confirming their identity. As a result, two of these were found to be new to science and hence described as new species from the state (Prasad & Singh, 1997c & 1999). These species are **Fimbristylis simpsonii** Prasad & N.P. Singh from Shimoga district and **Pycreus kanarensis** Prasad & N.P. Singh from Kanara (without exact locality).

Other additions:

As mentioned earlier a detailed study of many misidentified old specimens and a few newly collected specimens lead to certain very interesting findings. Besides the two new species mentioned above, a few other species were found to be new additions to the state flora of Karnataka. As per the present study there are 5 species, 1 subspecies and 2 varieties which are being reported for the first time from the state (Prasad & Singh, 1996a, 1997c & 2001). **Cyperus amabilis** Vahl, **Fimbristylis consanguinea** Kunth, **Fimbristylis microcarya** F.v. Muell., **Fimbristylis narayanii** Fischer and **Mariscus javanicus** (Houtt.) Merr. & Metcalf are the species found as new records for Karnataka during the present study, besides the two new species mentioned earlier. Though *Cyperus malaccensis* Lamk. and *Eleocharis spiralis* (Rottb.) Roem. & Schult. were reported long back by Blatter & McCann (1934) these were not noticed by subsequent workers until the final stages of the present study. Further, **Cyperus rotundus** subsp. **tuberosus** (Rottb.) Kukenth., **Scleria levis** var. **pubescens** (Steud.) C.Z. Zheng and **Bulbostylis barbata** var. **pulchella** (Thw.) Clarke are the infraspecific taxa reported for the first time from the state.

Of the above new records, presence of *Scleria levis* var. *pubescens* (Steud.) C.Z. Zheng in Karnataka is interesting. In India this variety is reported from North-Eastern states only. On the other hand *Mariscus javanicus* (Houtt.) Merr. & Metcalf is common along the coastal belt of Peninsular India including Karnataka, especially in Dakshina Kannada, though this is not reported from the state prior to this study.

New combination:

Cyperus laxus Lamk. subsp. **macrostachyus** (Boeck.) Prasad & N.P. Singh is a new combination made by elevating its status from variety level. There are considerable differences between the typical subspecies and the above taxon to treat it as a subspecies [kindly see the description of the subspecies in next Chapter].

Distribution and endemism

Of the 23 genera of the family, *Fimbristylis* Vahl is the largest genus in the state having 39 species followed by *Cyperus* L. with 38 species. But *Cyperus* has got an additional 4 subspecies and 3 varieties in the state. Hence these two genera together form *ca* 47% of the total sedge species in the state. *Scleria* Berg. and *Pycreus* Beauv. are the other major genera having more than 10 species each. There are 8 genera in the state, each having a single species only. Of these, 3 namely *Ascopholis* Fischer, *Queenslandiella* Dom. and *Remirea* Aubl. are monotypic. The number of species and infraspecific taxa (if any) under each genus are given below.

Table - 1. Number of species and infraspecific taxa under each genus according to number of species.

S.No.	Genus	No. of species	No. of infraspecific taxa
1	*Fimbristylis* Vahl	39	–
2	*Cyperus* L.	38	4 subspecies & 3 varieties
3	*Pycreus* Beauv.	13	1 variety
4	*Scleria* Berg.	11	1 variety
5	*Mariscus* Vahl	9	1 variety
6	*Eleocharis* R. Br.	7	–
7	*Kyllinga* Rottb.	7	1 subspecies
8	*Schoenoplectus* (Reichb.) Palla	7	–
9	*Carex* L.	6	1 variety
10	*Fuirena* Rottb.	5	–
11	*Bulbostylis* Kunth	4	1 variety
12	*Rhynchospora* Vahl	3	–
13	*Diplacrum* R. Br.	2	–
14	*Lipocarpha* R. Br.	2	–
15	*Rikliella* J. Raynal	2	–
16	*Ascopholis* Fischer	1	–
17	*Bolboschoenus* Asch. ex Palla	1	1 subspecies
18	*Courtoisina* J. Sojak	1	–
19	*Hemicarpha* Nees	1	–
20	*Hypolytrum* L.C. Rich.	1	–
21	*Kyllingiella* Haines & Lye	1	–
22	*Queenslandiella* Dom.	1	–
23	*Remirea* Aubl.	1	–

Sedge family is one of the most widely distributed plant groups and hence attracted the attention of phytogeographers since long. The distribution pattern of the family is being used as an index for phytogeographical interpretations (Clarke, 1898; Raymond, 1951; Shah, 1967). Clarke (*l.c.*) had based the phytogeographical sub-subareas of British India by taking evidence from the distribution of family Cyperaceae.

Many Cyperaceae species found in Karnataka have got a very wide range of distribution globally as well as within the country. Most of them are disrtibuted in the tropical regions of Asia, Africa and often extending to Australia. Majority of the 23 genera found in Karnataka have very wide range of distribution in the tropics, some extending to subtropics and a few extending to the temperate regions also. *Carex*, the largest genus of the family with about 2,000 species throughout the world is poorly represented in the state having just 6 species and one variety only, because the genus is mostly of temperate species. Though there are *ca* 160 species of *Carex* in India, at least 100 of them are found in the Himalayan region only. The genus *Fimbristylis* distributed mostly in Indomalaysia and Australia is well represented in peninsular India as a whole, with 50 species endemic to this region. In Karnataka a total number of 39 species are found and is the most dominant sedge genus in the state. Of the 3 monotypic genera found in Karnataka, *Ascopholis* Fischer is endemic to South India and the other two namely, *Queenslandiella* Dom. and *Remirea* Aubl. are wides.

Five species namely *Bolboschoenus maritimus* (L.) Palla, *Cyperus compressus* L., *C. rotundus* L., *Fimbristylis dichotoma* (L.) Vahl and *Pycreus polystachyos* (Rottb.) Beauv. found in Karnataka are cosmopolitan in distribution. *Eleocharis atropurpurea* is found in Asia, Australia, America and extending to Europe. Other three species having a very wide range of distribution are *Cyperus difformis* L., *C. haspan* L. and *Rhynchospora rugosa* (Vahl) Gale being found in the tropical and subtropical regions of the world. *Fimbristylis bisumbellata* (Forsk.) Bub. extends from old world tropics up to the mediterannian region of Europe. Another three species extending from the tropics to Europe are *Pycreus flavidus* (Retz.) Koyama, *Schoenoplectus articulatus* (L.) Palla and *S. mucronatus* (L.) Palla.

Nineteen species and one subspecies of Cyperaceae in Karnataka are pantropic in distribution. Similarly 23 species and 2 subspecies found in the state are common to Asia, Africa and Australia, whereas 8 species are common to Asia, Africa and America. There are 21 species and 2 varieties found in Asia and Africa only and another 17 species and 1 subspecies are common to Asia and Australia only. There are 22 species, 2 subspecies and 2 varieties of the sedges in the state common to South and South- East Asia. There are a few taxa restricted to South Asia only of which 3 species and 1 variety namely *Bulbostylis barbata* var. *pulchella* (Thw.) Clarke, *Carex lindleyana* Nees, *Fuirena capitata* (Burm.f.) Koyama and *Mariscus clarkei* (Cooke) Koyama are found only in peninsular India and Sri Lanka. One species namely *Cyperus alulatus* Kern is found only in India and Pakistan and *C. digitatus* var. *khasiana* (Clarke) Kern extends from Southern peninsular India and North-East India to Sri Lanka and Myanmar respectively.

Endemism :

Nayar (1977) estimated about 2,100 endemic species of flowering plants in peninsular India which is about 32% of its total flora. There are about 58 endemic genera in peninsular India, of which *Ascopholis* Fischer is the only one belonging to family Cyperaceae. Peninsular region of India has the maximum degree of endemism next only to Himalayas. It is well-known that endemic taxa are usually restricted to Islands, mountain peaks and peninsular regions. Turrill's (1964) contention that next to Islands, peninsular regions provide favourable conditions for endemism is true in the case of peninsular India also. Characteristic endemic species of the Western Ghats were enumerated by Subramanyam & Nayar (1974), who mentioned that Western Ghat summits are comparable with islands regarding endemic species. In general, Peninsular India has 32% of the endemics, while the rest of the country has only 27% (Nayar, 1980). The western ghats of Karnataka is known for a few endemic genera and many endemic species. Agumbe, Bababudan hills, Coorg, Uttara Kannada, Dakshina Kannada and Shimoga are areas of high endemic contents (Ahmedulla & Nayar, 1987).

According to Ahmedulla & Nayar (*l.c.*) there are 58 species and 7 varieties of Cyperaceae endemic to peninsular India. But at present there must be at least 66 species of Cyperaceae endemic to peninsular India, of which the genus *Fimbristylis* contribute the maximum number. Prasad & Singh (1997a) have reported a total number of 30 species of *Fimbristylis* endemic to Peninsular India. Further studies by them (1997d) corrected the number as 37 for Peninsular India and a total number of 46 species for the whole country. Some more species were described subsequently, thus making the total number as 50 endemic species of *Fimbristylis* for Peninsular India and 60 for the whole country.

As a result of the present study, out of the total number of 163 species of Cyperaceae in Karnataka, only 9 were found to be endemic to the state. This includes the two new species described during the present study.

List of species endemic to Karnataka :

1. *Fimbristylis bispicula* Govind.
2. *F. breviculma* Govind.
3. *F. dimorphonucifera* Govind.

4. *F. simpsonii* Prasad & N.P. Singh
5. *Kyllinga eglandulosa* Govind. & Ramani
6. *Pycreus atroglumosus* (Govind.) P. & V. Singh
7. *P. kanarensis* Prasad & N.P. Singh
8. *P. mahadevanii* Govind.
9. *P. plurinodosus* (Govind.) P. & V. Singh

Besides these, another 11 species and 2 varieties found in Karnataka are endemic to Peninsular India (see list below), which includes species of the monotypic genus *Ascopholis* also.

List of the species and infraspecific taxa in Karnataka which are Endemic to Peninsular or Southern India (In addition to the list of 9 species endemic to Karnataka given above).

1. *Ascopholis gamblei* Fischer
2. *Bulbostylis subspinescens* Clarke
3. *Carex caricina* var. *glaucina* (Boeck.) Ghildyal & Bhattacharyya
4. *Mariscus cyperinus* var. *pictus* (Nees) Karthik.
5. *Fimbristylis kingii* Clarke ex Boeck.
6. *F. lawiana* (Boeck.) Kern
7. *F. monospicula* Govind.
8. *F. narayanii* Fischer
9. *F. pseudomicrocarya* Govind.
10. *F. semidisticha* Govind.
11. *F. virella* Govind.
12. *F. woodrowii* Clarke
13. *Fuirena trilobites* Clarke

PHENOLOGY

Phenology of sedges may vary depending mainly on the climatic conditions. Majority of the sedges in Karnataka were in flower and fruit from August to December and probably this is the best season for collecting Cyperaceae specimens. Number of flowering and fruiting sedge species were found to be low from March to June. Out of the 163 species, 6 subspecies and 9 varieties, just 34 taxa were found in flower and fruit in April and 36 in March. Species found in flower and fruit during this period are mostly perennials and a few annuals growing in and around perennial water bodies. However, most favourable season is during August to December. A

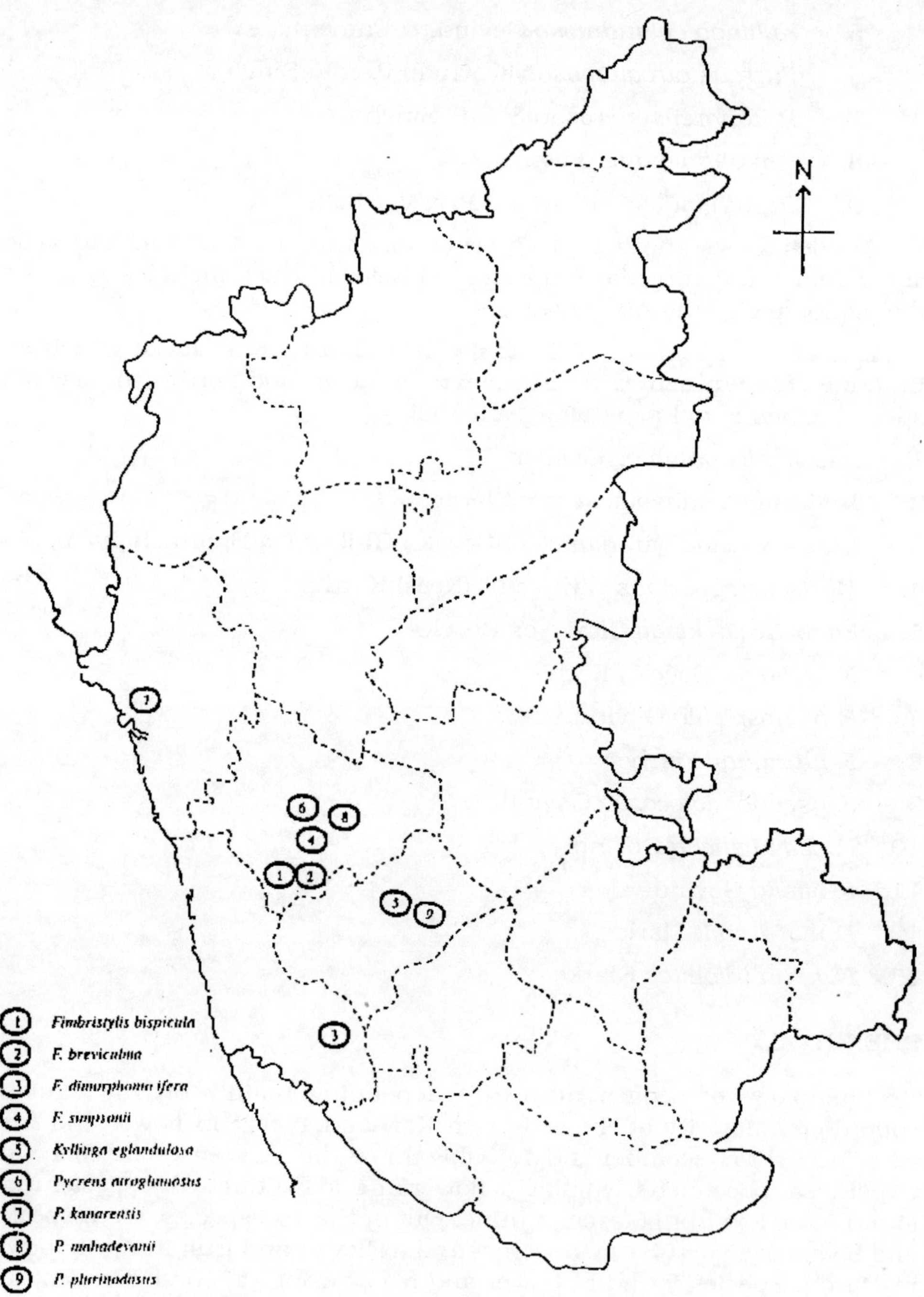

Map 2. Distribution of Endemic species in the state

maximum number of 118 taxa were in flower and fruit in October and 116 during November. Almost all the annual species and many perennials produce flowers and fruits during this period, i.e., during and just after the South-West monsoon as well as North-East monsoon. There is a gradual increase in the number of species in flower from May onwards reaching a peak in October and then gradually decreasing from November onwards reaching the minimum in April (see the graph). But there can be slight change in the phenology depending mainly on the variation in monsoon season.

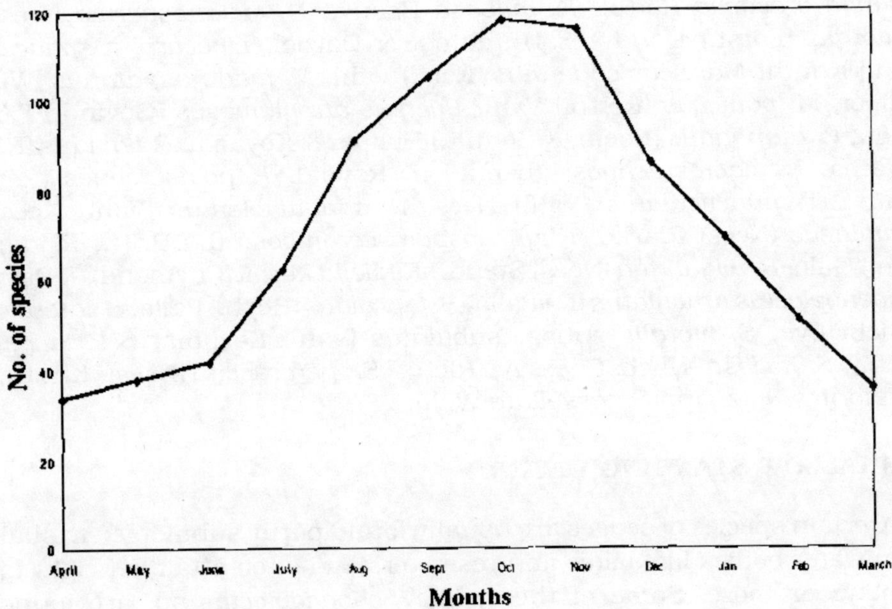

Graph - 1. Monthly variation in number of species in flowers and fruits.

HABITAT

As mentioned in last Chapter sedges are found every where, in very varied habitats. Most species are found in particular habitat, but some can be found in different habitats also. Various kinds of habitats and the sedge species of Karnataka found in such habitats are as follows.

A. LOWLAND WET AREAS :

Most of the species in the state are found in the lowland moist or wet areas like in and around rice fields, lakes, reservoirs, river banks, river beds, wetlands etc. 63 species found in such areas are : *Bolboschoenus maritimus* (L.) Palla, *Cyperus alopecuroides* Rottb., *C. difformis* L., *C. exaltatus* Retz., *C. involucratus* Rottb., *C. haspan* L., *C. iria* L., *C. nutans* Vahl, *C. pangorei*

Rottb., *C. platystylis* R. Br., *C.rotundus* L., *Eleocharis acutangula* (Roxb.) Schult., *E. atropurpurea* (Retz.) Presl., *E. congesta* Don, *E. dulcis* (Burm.f.) Trin. ex Hensch., *E. geniculata* (L.) Roem. & Schult., *E. retroflexa* subsp. *chaetaria* (Roem. & Shult.) Koyama, *Fimbristylis acuminata* Vahl, *F. aestivalis* (Retz.) Vahl, *F. aphylla* Steud., *F. bisumbellata* (Forsk.) Bub., *F. complanata* (Retz.) Link, *F. dichotoma* (L.) Vahl, *F. lawiana* (Boeck.) Kern, *F. littoralis* Gaudich., *F. merrillii* Kern, *F. miliacea* (L.) Vahl, *F. squarrosa* var. *esquarrosa* Makino, *F. tetragona* R.Br., *F. tomentosa* Vahl, *Fuirena ciliaris* (L.) Roxb., *F. umbellata* Rottb., *F. wallichiana* Kunth, *F. trilobites* Clarke, *Kyllinga brevifolia* Rottb., *K. bulbosa* Beauv., *K. melanosperma* Nees, *K. nemoralis* (Forst.) Dandy ex Hutchins. & Dalziel, *Lipocarpha sphacelata* (Vahl) Kunth, *Mariscus cyperinus* (Retz.) Vahl, *M. maderaspatanus* (Willd.) Napper, *M. paniceus* (Rottb.) Vahl, *Pycreus atroglumosus* (Govind.) P.& V. Singh, *P. diaphanus* (Roem. & Schult.) Hooper & Koyama, *P. flavidus* (Retz.) Koyama, *P. macrostachyos* (Lamk.) J. Raynal, *P. polystachyos* (Rottb.) Beauv., *P. puncticulatus* (Vahl) Nees, *P. sanguinolentus* (Vahl) Nees, *P. stramineus* (Nees) Clarke, *Rhynchospora corymbosa* (L.) Britt., *R. rugosa* (Vahl) Gale, *R. wightiana* (Nees) Steud., *Rikliella kernii* (Raymond) J. Raynal, *Schoenoplectus articulatus* (L.) Palla, *S. juncoides* (Roxb.) Palla, *S. lateriflorus* (Gmel.) Lye, *S. litoralis* subsp. *subulatus* (Vahl) Koyama, *Scleria biflora* Roxb., *S. foliosa* Hochst. ex A. Rich., *S. pergracilis* (Nees) Kunth, *S. poaeformis* Retz. and *S. tessellata* Willd.

B. SHALLOW STANDING WATER:

Certain species of sedges are usually found partly submerged in shallow open water bodies like lakes and reservoirs and often attain a height up to 1 metre or more. Some of these, like *Schoenoplectus* sp. are found in association with other floating vegetation also. Height of these plants usually depend on the water depth. Species found in such habitats are : *Cyperus alopecuroides* Rottb., *C. articulatus* L., *Eleocharis acutangula* (Roxb.) Schult., *E. dulcis* (Burm.f.) Trin. ex Hensch., *Lipocarpha chinense* (Osb.) Kern, *Schoenoplectus corymbosus* (Roth ex Roem. & Schult.) J. Raynal, *S. mucronatus* (L.) Palla, *S. litoralis* (Schrad.) Palla.

Among these, *Eleocharis dulcis* often forms large patches in open water areas by means of vegetative propogation.

C. FOREST AREAS :

In Karnataka species of *Carex* and *Scleria* are usually confined to forest areas. Species found in forest areas are : *Carex caricina* (Don) Ghildyal & Bhattacharyya, *C. indica* subsp. *lataebrunea* (Clarke) Koyama, *C. maculata* Boott, *C. speciosa* Kunth, *Cyperus laxus* subsp. *macrostachyus* (Boeck.) Prasad & N.P. Singh, *Diplacrum caricinum* R. Br., *Fimbristylis cinnamometorum* (Vahl) Kunth, *F. eragrostis* (Nees & May.) Hance, *Hypolytrum nemorum* (Vahl) Spreng., *Scleria corymbosa* Roxb., *S. lithosperma* (L.) Sw., and *S. terrestris* (L.) Fassett.

D. GRASSLANDS :

Many species of sedges grow in association with grasses usually in moist soils. Species found mostly in grasslands are *Cyperus tenuiculmis* Boeck., *Eleocharis geniculata* (L.) Roem. & Schult., *Fimbristylis consanguinea* Kunth, *F. ovata* (Burm.f.) Kern., *F. pseudomicrocarya* Govind., *F. schoenoides* (Retz.) Vahl, *F. tenera* Schult., *F. woodrowii* Clarke, *Kyllinga odorata* Vahl, *Mariscus clarkei* (Cooke) Koyama, *M. compactus* (Retz.) Boldingh, *M. dubius* (Rottb.) Kukenth. ex Fischer, *Scleria annularis* Nees ex Steud., *S. biflora* Roxb., *S. levis* Retz., and *S. tessellata* Willd.

E. HIGH ALTITUDE :

Some species prefer high altitude habitats, often at high range grasslands. Species found in such habitats are *Carex baccans* Nees, *C.lindleyana* Nees, *Fimbristylis kingii* Clarke ex Boeck., *F. monospicula* Govind., *F. semidisticha* Govind., *Pycreus plurinodosus* (Govind.) P. & V. Singh, and *Scleria levis* var. *pubescens* (Steud.) C.Z. Zheng.

F. MOIST SANDY SOIL :

Species found in moist sandy soil are the following: *Cyperus alulatus* Kern, *C. castaneus* Willd., *C. cuspidatus* Kunth, *Eleocharis geniculata* (L.) Roem. & Schult., *Fuirena capitata* (Burm.f.) Koyama, *Hemicarpha isolepis* Nees, *Mariscus sumatrensis* (Retz.) J. Raynal, *Rikliella kernii* (Raymond) J. Raynal, *R. squarrosa* (L.) J. Raynal and *Schoenoplectus senegalensis* (Hochst. ex Steud.) Palla ex J. Raynal .

G. COASTAL AREAS :

There are some species characteristic to the vegetation of sandy saline soil of coastal areas. These species are characterised by the well developed root systems or highly developed rhizomes and stolons. Such taxa found along the coast line of Karnataka are *Bulbostylis barbata* var. *pulchella* (Thw.) Clarke, *B. subspinescens* Clarke, *Cyperus arenarius* Retz., *C. bulbosus* Vahl, *C. stoloniferus* Retz., *Queenslandiella hyalina* (Vahl) Ballard and *Remirea maritima* Aubl. Species like *Cyperus esculentus* L. and *C.stoloniferus* Retz. are found along the margins of brackish water bodies also.

H. SALINE MARSHY AREAS :

A few halophytic sedge species are found along the muddy shores or other marshy areas of brackish waters. They are *Cyperus laevigatus* L., *C. malaccensis* Lamk., *Eleocharis spiralis* (Rottb.) Roem. & Schult., *Fimbristylis ferruginea* (L.) Vahl, *F. polytrichoides* (Retz.) R. Br., *F. sieberiana* Kunth and *Mariscus javanicus* (Houtt.) Merr. & Metcalfe .

I. ROCKY SLOPES :

Diplacrum africanum (Benth.) Clarke, *Fimbristylis breviculma* Govind., *F. falcata* (Vahl) Kunth, *F. narayanii* Fischer and *F. simpsonii* Prasad & N.P. Singh are usually found on moist rocky slopes.

J. ROCKY RIVER BEDS :

Cyperus corymbosus Rottb., *C. distans* L.f. and *C. pangorei* Rottb. are often found on rocky river beds. They are also found in somewhat dry muddy areas (e.g. *Cyperus macer* Clarke and *C. pygmaeus* Rottb).

Cyperus cephalotus Vahl, very rarely *C. alopecuroides* Rottb. and *C. platystylis* R. Br. are found as floating vegetation as these species often grow on some floating substances like decaying plant material.

Sedges as Weeds: There are many sedges found as weeds in agricultural fields especially in rice fields, along roadsides, in lawns and gardens. It is very difficult to eradicate weeds belonging to Cyperaceae, because of their well-established underground parts and propogation through the rhizomes, stolons or tubers. *Cyperus rotundus* L. is one of the most troublesome weeds having cosmopolitan distribution and is one of the most serious weeds of the world (Sen, 1981). In a study conducted in the agricultural fields of Jodhpur, Rajasthan this species was found as the most dominant species (Prasad 1989). It propogates very fast by means of the nut-like tubers and is everywhere, in agricultural fields, wastelands, roadsides, gardens, lawns, river banks and so on. Other common weeds belonging to this family found in Karnataka, particularly in cultivated fields are the following :

Cyperus haspan L., *C. iria* L., *C. laevigatus* L., *C. nutans* Vahl, *C. procerus* Rottb., *Bulbostylis barbata* (Rottb.) Clarke, *B. puberula* (Poir.) Clarke, *Eleocharis acutangula* (Roxb.) Schult., *E. atropurpurea* (Retz.) Presl., *E. dulcis* (Burm.f.) Trin. ex Hensch., *E. geniculata* (L.) Roem. & Schult., *Fimbristylis aestivalis* (Retz.) Vahl, *F. argentea* (Rottb.) Vahl, *F. bisumbellata* (Forsk.) Bub., *F. complanata* (Retz.) Link, *F. littoralis* Gaudich., *F. sieberiana* Kunth, *Fuirena umbellata* Rottb., *Lipocarpha chinensis* (Osb.) Kern, *Pycreus diaphanus* (Roem. & Schult.) Hooper & Koyama, *P. macrostachyos* (Lamk.) J. Raynal, *P. pumilus* (L.) Nees, *P. puncticulatus* (Vahl) Nees, *P.*

sanguinolentus (Vahl) Nees, *Rhynchospora corymbosa* (L.) Britt., *R. squarrosa* (L.) J. Raynal and *Schoenoplectus articulatus* (L.) Palla.

USEFUL SEDGES

When the size of the family is taken into consideration, there is no much economical importance to the family Cyperaceae, because only limited number of sedge species are reported to be useful. A perusal of literature revealed that many species of sedges found in Karnataka are used for some purpose in other parts of the world or in other states of India. But it seems only a few species are utilized in Karnataka and no work on these lines has been done during the present studies either. However, this aspect is being discussed here for potential uses of sedges in the state in future. The economically weaker sections of the society, especially the tribals must be using the sedges for different purposes as in medicine, for food, for thatching and making baskets, as fodder, etc. Similarly native medical men or traditional physicians or *Vaidyas* must be using the tubers and rhizomes of certain sedges for their medical preparations. For example, tubers of *Cyperus rotundus* L., is widely used in Karnataka for worms and to cure stomach disorders.

As some potential useful sedge species are available in the state, uses of these species are discussed here as reported elsewhere or as reported from the state. Hence even if all these species are not utilised at present, these can be utilised in future depending upon their availability. Uses of sedges are reported by many earlier workers as Beetle (1943, 1950), Caius (1935), Chopra *et al.* (1958), Kirtikar & Basu (1918), Metcalf (1931), Muller (1915), Pathak (1920) and Rubsy (1909). Datta & Banerjee (1978) compiled the uses of weeds of rice fields in West Bengal, in which a few sedge species were also included. Vartak (1982) found about 28 species of sedges for their utility in nature conservation and various domestic uses. Jain & De Filipps (1991) compiled the Medicinal plants of India. Recently Nayar *et al.* (1989 & 1994) compiled the economic plants of India in which family Cyperaceae is also well represented. The following account is mostly as reported by the above and other workers and also based on some personal observations. There are different kinds of uses of sedges as Medicinal, Food, Mat making and thatching, Fodder, Conservation, etc.

A. **Medicine:** Most important use of sedges is probably medicinal, because tubers of many species have been found to be of medicinal properties. Most of the earlier works cited above are of medicinal uses. There are many potential medicinally important species in Karnataka (e.g. *Cyperus rotundus, C. articulatus, C. esculentus, C. iria, C. stoloniferus, Fimbristylis falcata, Kyllinga nemoralis, Schoenoplectus articulatus*, etc.). Medicinal uses of some species are as follows :

1. **Cyperus articulatus** L. : Tubers can be used for a tonic and as a stimulant. They are bitterish, aromatic and act as sedative in dyspeptic disorders (WI).

2. **C. esculentus** L. : The tubers are used as a stimulant and as aphrodisiac.

3. **C. exaltatus** Retz. : Tubers crushed with seeds of *Cuminum cyminum* and the extract is used for dysentery by Gowde tribals in Karnataka (Karne, 1998).

4. **C. iria** L. : The whole plant is used for stomach ache, as a stimulant and astringent. Munda tribes of Chota Nagpur grind the tuber along with that of *C. rotundus* and drink in fever.

5. **C. rotundus** L. : This is the most widely used sedge for medicinal purpose. Though it is one of the most troublesome weeds, the tubers are used medicinally in different parts of the world as it has a very wide range of global distribution. In Ayurveda it is used for stomachic and other stomach disorders, anthelmintic, loss of apetite, astringent, leprosy, thirst, fever, blood diseases, biliousness, dysentery, intense itching, pain, vomiting, epilepsy, ophthalmia, etc. The tubers are usually used as a worm remedy. In Kerala the tubers are ground, mixed with milk and given to children for worms and other stomach disorders. In Konkan fresh tubers are said to be applied to the breast to stimulate the flow of milk. Dry tubers are often sold in the market for using in medicine and perfumery.

6. **C. stoloniferus** Retz. : Tubers of this species are also good for stomachic and considered as a good stimulant for the heart.

7. **Fimbristylis falcata** (Vahl) Kunth : The plant is used in dysentery. The rhizome can be used for ring worm and other skin diseases.

8. **Kyllinga bulbosa** Beauv. : It is used against infection or poison and useful for healing wounds and regulating the heat of the body. The root extract or a decoction is used in fever and diabetes. Oil boiled with its roots is used to relieve intensive itching of the skin. The roots yield an essential oil which is used to stimulate liver function.

9. **K. nemoralis** (Forst.) Dandy ex Hutchins. & Dalziel : The plant is used for stomach complaints and as a cooling medicine in fevers. It is also used as an antidote in many parts of India. The *Malekudiya* tribes of Coorg district use the extract of the whole plant for *Asthma* (Kshirsagar, 1998).

10. **Schoenoplectus articulatus** (L.) Palla : It is used as a purgative. The tubers are effective against diarrhoea and vomiting.

11. **Scleria levis** Retz. : The nut is used to cure coughing; it is eaten with betel nut.

12. **S. lithosperma** (L.) Sw. : A decoction of the root is drunk after childbirth.

B. **Mat making and Thatching:** Few tall sedges are usually used for making mats and baskets and also for thatching purpose. Though a few such species are available, such uses are not reported from the state. Rusby (1909) stated that bulrushes represent 'one of the most important uncultivated textile material in the world' as many new-world species of *Scirpus* have been used by the Red Indians for thatching, mat making, basket making, etc. In addition to such uses Beetle (1950) reported the use of *Scirpus* sp. called *'totora'* in Peru and Bolivia for making boats by cleverly lashing them together. The potential useful sedge species found in Karnataka which can be used for mat making, basketry and thatching are *Cyperus corymbosus* Rottb., *C. malaccensis* Lamk. and *Schoenoplectus articulatus* (L.) Palla. *C. corymbosus* known as *'Chinese matgrass'* is used extensively for making mats especially in South India and Bengal (Maheshwari & Singh, 1965). Nayar *et al.* (1989) reported *Fimbristylis cymosa* R. Br. also as used for mat making. The stems of *Cyperus iria* L. can also be used for making mats (Datta & Banerjee, 1978).

C. **Food and Fodder:** The bulb-like tubers of *Cyperus bulbosus* Vahl are used as food by villagers in Konkan and other coastal areas. The tubers can be roasted or pounded into flour (WI). The tubers of *Eleocharis dulcis* (Burm.f.) Trin. ex Hensch. known as *Chinese water chestnut* is eaten in Goa and Konkan in extreme drought period. This species is often cultivated for the tubers in China which resulted in the development of a strain having bigger and more sweet tubers (Kern, 1974). Tubers of *Cyperus esculentus* L. is also edible and in Southern Europe it is cultivated (WI). Since the above species are found in Karnataka these can be considered as potential famine plants. Some species like *Cyperus corymbosus* Rottb., *Schoenoplectus articulatus* (L.) Palla, *S. littoralis* (Schrad.) Palla, etc. growing abundantly in marshy areas are used as fodder. In coastal areas *Fimbristylis polytrichoides* (Retz.) R. Br. forms a major part of the herbage. Many species of sedges grow in grasslands, wastelands, wetlands and other habitats along with grass and hence form part of herbage of the grazing cattle.

D. **Conservation of Nature and Wildlife:** The heavy root system of many sedges acts as good soil binder and hence prevent soil erosion. *Cyperus corymbosus, C. malaccensis, Schoenoplectus litoralis*, etc. are examples for this. Coastal species like *Cyperus arenarius* Retz. are also considered as good soil binders. Vartak (*l.c.*) reported some species occuring along slow streams, on lake margins and around isolated ponds as shelter for wildlife and as a source of food for them. *Cyperus digitatus* Roxb., *C. exaltatus* Retz., *Shoenoplectus articulatus* are the prominent species reported by him as liked by aquatic birds. There are some species which form food for animals, especially for wild waterfowls. In the world famous Keoladeo National Park in Rajasthan large patches of *Eleocharis dulcis* are found and here tubers are formed in late winter. It was observed that Sarus Crane (*Grus antigone*) and Siberian Crane (*Grus leucogeranus*), a highly endangered bird species wintering in this sanctuary feed on these tubers (Prasad, 1988). But their most preferred food in the park was found to be the tubers of *Bolboschoenus maritimus* (L.) Palla, which is also very common there. Tubers of *Cyperus rotundus* also form part of their food. Wild boars also feed on the tubers of these species. Purple Moorhen (*Porphyrio porphyrio*), another water bird was also found feeding on the lower part of the stems of *Eleocharis dulcis.*

In the wetlands of Bharatpur it was observed that Purple Moorhen usually nest in the central part of a tuft of partly submerged *Cyperus alopecuroides* Rottb. (Prasad, *l.c.*) so that the nest is encircled by the tall leaves which probably protect the eggs or chicks from raptors.

Other uses of Sedges: Dry tubers of *Cyperus rotundus* are used in perfumery and in making incence sticks. Tubers of *Cyperus articulatus* are used in perfumery (Maheshwari & Singh, 1965). Some common sedges like *Cyperus malaccensis* and *Mariscus javanicus* have been tried in the production of paper pulp and the results seem to be encouraging. Some more species like *Eleocharis dulcis* which form large patches in shallow water bodies may also be tried for making paper pulp.

Cyperus involucratus Rottb. is the only ornamental sedge species grown in gardens, which is common in Karnataka also.

SYSTEMATIC TREATMENT OF THE FAMILY

CYPERACEAE

A.L. Juss., Gen. 26. 1789, nom. cons.

Type genus : *Cyperus* L.

Annual or perennial herbs, grass-like or rush-like (the monotypic African genus *Microdracoides* is tree-like); perennials with short or long-creeping rhizomes, often emitting stolons. Stems solid or hollow, at times septate, usually trigonous, rarely terete or 2-sided, usually nodeless. Leaves usually 3-ranked, rarely distichous or polystichous, usually radical, sometimes cauline, usually sheathing at base; blades sessile, linear or rarely lanceolate, at times much reduced or absent; sheaths closed, rarely open; ligule when present a rim of short hairs or membranous. Inflorescence varied, mostly anthelate, paniculate, racemose, spicate or capitate, bearing few to many spikelets, rarely reduced to a single spikelet, usually subtended by 1- several leafy involucral bracts. Spikelets few to many-flowered. Glumes distichously or spirally imbricated, all or some bearing an axillary flower. Flowers simple, inconspicuous, bisexual or unisexual. Perianth usually absent, often consisting of bristles or scales. Stamens 1-3, free, rarely connate (in few species of *Carex*); filaments usually elongate after anthesis; anthers basifixed, introrse, opening longitudinally. Ovary solitary, superior, 2-3 carpellate; style often thickened at base; stigmas 2 or 3, rarely style undivided. Fruit an indehiscent nut (achene), free or at times surrounded by a modified prophyll (utricle).

A large family having about 80 genera and about 5000 species distributed throughout the world. There are 39 genera and *ca* 580 species in India. In Karnataka there are 163 species, 6 subspecies and 8 varieties coming under 23 genera.

Literature: **BRUHL, J.J.** (1995) Sedge genera of the world: relationships and a new classification of the Cyperaceae, in *Austr. Syst. Bot.* 8:125-305. **KERN, J.H.** (1974) Cyperaceae in: van Steenis, *Fl. Males.* 1,7: Part 3:

435-753. **KERN, J.H.** & **NOOTEBOOM, H.P.** (1974) Cyperaceae II in: van Steenis, *Fl. Males.* 1, 9: part 1: 107-187. **KOYAMA, T.** (1961) Classification of the family Cyperaceae I, in *Journ. Fac. Sci. Univ. Tokyo* Sect. III, Bot. 8: 37-148. **KOYAMA, T.** (1961) Classification of the family Cyperaceae III, in *Quart. Journ. Taiwan Mus.* 14: 159-194. **KOYAMA, T.** (1962) Classification of the family Cyperaceae II, in *Journ. Fac. Sci. Univ. Tokyo* Sect. III, Bot. 8: 149-278. **RAO, A.S. & VERMA, D.M.** (1981) Cyperaceae, in *Rec. Bot. Surv. India* 21 (2): 209-220.

Key to the Genera

1a. Flowers bisexual 2

1b. Flowers unisexual 22

2a. Spikelets 1-2-flowered 3

2b. Spikelets few to many-flowered 6

3a. Rachilla disarticulating above the 2 empty glumes 4

3b. Rachilla not disarticulating 5

4a. Annuals with fibrous roots; nut planoconvex or subtrigonous; enclosed in the scale (utricle) .... **1. Ascopholis**

4b. Perennials with short or horizontally creeping rhizome; nut laterally flattend, not enclosed in the scale .... **13. Kyllinga**

5a. Rhizome long-creeping, with long internodes; plants confined to sea-coasts .... **19. Remirea**

5b. Rhizome short, not with long internodes; plants not confined to sea-coasts .... **14. Kyllingiella**

6a. Glumes distichous 7

6b. Glumes spiral (at times distichous in *Fimbristylis*; see key to the species of *Fimbristylis*) 11

7a. Rachilla deciduous 8

7b. Rachilla persistent 10

8a. Stigmas 2; nut biconvex .... **18. Queenslandiella**

8b. Stigmas 3; nut trigonous 9

9a. Annuals; keel of the nut-bearing glumes winged .... **5. Courtoisina**

9b. Perennials; keel of the nut-bearing glumes not winged .... **16. Mariscus**

10a. Stigmas 2; nut laterally flattened, with one angle facing the rachilla .... **17. Pycreus**

10b. Stigmas 3 (rarely 2 or with a subentire style); nut trigonous, triquetrous, at times lenticular, with one face opposite the rachilla .... **6. Cyperus**

11a. Flowers with a perianth of hypogynous scales or bristles or of both ... 12

11b. Flowers without a perianth of hypogynous scales or bristles 19

12a. Perianth consists of bristles only 13

12b. Perianth consists of scales (in *Fuirena* often bristles also found). 16

13a. Leaves reduced to bladeless sheaths 14

13b. Leaves well developed 15

14a. Inflorescence reduced to a single terminal spike; style articulated with the ovary; nut with a beak of persistent style base .... **8. Eleocharis**

14b. Inflorescence anthelate or a head, pseudolateral; style continuous with the ovary; nut even if beaked not with a constriction ...... **22. Schoenoplectus**

15a. Stems triquetrous; style continuous with the overy, hardly or not dilated at the base, caducous .... **2. Bolboschoenus**

15b. Stems trigonous to almost terete; style articulated with the ovary, dilated at the base, with persistent base .... **20. Rhynchospora**

16a. Inflorescence pseudolateral .... **11. Hemicarpha**

16b. Inflorescence terminal 17

17a. Hypogynous scales folded, sharpely keeled; stigmas 2; nut with a spongy beak .... **12. Hypolytrum**

17b. Hypogynous scales not folded, nor keeled; stigmas 3; nut without a spongy beak 18

18a. Stems noded; leaves cauline, ligulate; inflorescence paniculate; nut not enclosed by hypogynous scales .... **10. Fuirena**

18b. Stems not noded; leaves basal, eligulate; inflorescence capitate; nut enclosed by hypogynous scale .... **15. Lipocarpha**

19a. Style base dilated, constricted or articulated above the nut 20

19b. Style base not dilated, continuous with the nut 21

20a. Nut with a button-like thickening formed by the persistent style-base .... **3. Bulbostylis**

20b. Nut without a button-like thickening of the style-base ... **9. Fimbristylis**

21a. Leaves reduced to sheaths .... **22. Schoenoplectus**

21b. Leaves well developed .... **21. Rikliella**

22a. Nut enlcosed in an utricle .. .. **4. Carex**

22b. Nut not enclosed in an utricle 23

23a. Contraligule present; nut not enveloped by two glumes ......**23. Scleria**

23b. Contraligule absent; nut closely enveloped by 2 glumes which fall off along with the nut .... **7. Diplacrum**

1. ASCOPHOLIS

Fischer in Bull. Misc. Inf. 1931: 104. 1931.

Glabrous herbs with fibrous roots. Stems solitary, erect, swollen at the base and clothed with fleshy scales, leafy near the base only. Inflorescence terminal, bracteate, bearing 3 - several, sessile, globose spikes. Spikelets spicately arranged on short axis, terete, 1-flowered; rachilla disarticulating above the 2 empty glumes, leaving a knob. Glumes 2, opposite; lower empty; upper spathiform, bearing a single hermaphrodite flower enclosed in a long utricular scale which split half way at one side. Bristle 0 or 1. Stamens 3. Style continuous with the ovary, base not enlarged; stigmas 3. Nut planoconvex or subtrigonous, oblong, with a beak continuous with the style.

Monotypic genus, endemic to South India.

Ascopholis gamblei Fischer in Bull. Misc. Inf. 1931: 104. 1931 *et* in Gamble, Fl. Pres. Madras 1679. 1931 (3: 1164. 1957, repr.ed.); Hooper in Saldanha & Nicolson, Fl. Hassan 658. 1976; Sharma *et al*., Fl. Karnataka 303. 1984; Karthik. *et al*., Fl. Ind. Enum. Monocot. 32. 1989.

Type: India, Tamilnadu, Ootacamund (Ooty), *Gamble.*

Perennials with bulbous base. Stems 4-12 cm long, striate, base swollen with papery brown sheaths. Leaves shorter or longer than stem, filiform or very narrowly linear. Spikes oblong, *ca* 10 mm long. Involucral bracts 4, linear, 1-7 cm long, broad towards base. Spikelets linear. Glumes 2.5 - 3 mm long, hyaline and brown- dotted, veined. Hypogynous scale 3.5 - 3.8 mm long, pale brown, often with darker dots; bristles 0 or 1, capillary, much shorter than ovary, white. Style trigonous. Nut planoconvex or subtrigonous, narrowly oblong, *ca* 2.5 mm long, dark brown.

Fls. & Frts. : July - Oct.

Habitat : Dry deciduous forests.

Distrib. : Endemic to South India. KARNATAKA: Bangalore, Hassan.

Specimens examined: Bangalore: Central College, Bangalore, *Saldanha* 19085, 25.8.1976 (JCB). Hassan: 6 miles before Arsikere, on Tiptur-Arsikere Road, *Nicolson et al.* 2236, 22.10.1971 (JCB); Between Harisave and Channaryapattana, *Saldanha et al.* 1940, 27.7.1978 (JCB).

2. BOLBOSCHOENUS

Asch. ex Palla in W.D. J. Koch, Syn. Deutsch. Schweiz. Fl. ed. 3, 2531. 1904. *Scirpus* α *Bolboschoenus* Asch., Fl. Brandenb. 1: 753. 1864 (infrageneric rank not indicated). *Scirpus* sect. *Bolboschoenus* (Asch.) Beetle in Amer. Journ. Bot. 29: 82. 1942.

Perennials with swollen tubers and long stolons. Stems triquetrous. Leaves cauline, laminate; lower sheaths often septate, nodose. Inflorescence terminal, corymb-like or capitate; bracts leafy, several. Spikelets large, with spirally arranged glumes. Flowers hermaphrodite; hypogynous bristles 0-6. Stamens 3. Style base not enlarged, persistent. Nut large, more or less smooth. Embryo schoenoplectoid. Anatomy eucyperoid.

About 8 species, mainly in eastrn Asia and eastern N. America. 2 in India, 1 in Karnataka.

Literature : **BEETLE, A.A.** (1942) Studies in the genus *scirpus* IV. The section *Bolboschoenus* Palla, in *Amer. Journ. Bot.* 29: 82-88. **NORDLINDH, T.** (1972) Notes on the variation and taxonomy in the *Scirpus maritimus* complex, in *Bot. Not.* 125: 397- 405.

Bolboschoenus maritimus (L.) Palla in Koch, Syn. Deutsch. Schweiz. Fl. ed. 3, 2532. 1904; Hooper in Saldanha & Nicolson, Fl. Hassan 658. 1976; Sharma *et al.*, Fl. Karnataka 303. 1984, *p.p.;* Singh, Fl. E. Karnataka 2: 627. 1988, *p.p. Scirpus maritimus* L., Sp. Pl. 1, 51. 1753; Clarke in Hook.f., Fl. Brit. India 6: 658. 1893; Cooke, Fl. Pres. Bombay 2: 893. 1908 (3: 407. 1958, repr.ed.); Fischer in Gamble, Fl. Pres. Madras 1666. 1931 (3: 1156. 1957, repr.ed.); Ramaswamy & Razi, Fl. Bangalore 99. 1973; Kern in van Steenis, Fl. Males. 1, 7: 499. 1974. *Schoenoplectus maritimus* (L.) Lye in Blyttia 29: 145. 1971; Karthik. *et al.*, Fl. Ind. Enum. Monocot. 69. 1989.

Key to the subspecies

1a. Tubers always ovoid, up to 3 cm thick; inflorescence a compound umbel; spikelets oblong to oblong- ovoid, 8-13 x 3-5 mm; glumes 5-7 x 3-4 mm; anthers linear-oblong, *ca* 3 mm long; stigmas 3 (rarely 2), *ca* 3.5 mm long; nut planoconvex to almost trigonous, obovate, rather acute than apiculate at apex; epidermal cells not prominent. .... subsp. **maritimus**

1b. Tubers oblong, ellipsoid or ovoid, up to 1 cm thick; inflorescence a cluster of 2-6 sessile spikelets (rarely 1 or 2 on short rays), often reduced to a single spikelet; spikelets ovoid, 13-23 x 7-11 mm; glumes 7-10 x 2.5-3 mm; anthers oblong, *ca* 2 mm long; stigmas strictly 2, 2-3 mm long; nut strongly compressed, lenticular, prominently apiculate at apex; epidermal cells on the nut very prominent. .... subsp. **affinis**

subsp. **maritimus**

Perennials with long, horizontal, creeping rhizomes producing large, stout tubers at nodes, 16-90 cm high. Tubers ovoid, up to 3 x 2 cm or more, blackish. Stems erect, solitary from a tuberous base, trigonous, smooth, striate, 6-15 mm thick near base (including the leaf-sheaths), 1-3 mm thick just below inflorescence. Leaves well developed, cauline, flat, upper ones usually overtopping the inflorescence, gradually narrowed to a fine triquetrous tip, 4-10 mm wide in middle portion, striate, with a prominent midrib, sheaths tightly closed except basal ones, smooth, truncate or slightly produced at mouth; ligule absent. Inflorescence usually a compound umbel, rarely reduced to a single spikelet, 3-7 cm long, 3-8 cm wide. Bracts usually several, not sheathing; lower ones foliaceous, overtopping the inflorescence. Rays quite unequal, 1-4.5 cm long. Spikelets solitary or in culsters of 3 (-few), terete, oblong to oblong-ovoid, subacute at apex, 8-13 (-32) x 3-5 mm, ferrugineous, densely many-flowered. Glumes membranous, ovate or oblong - ovate, emarginate or lacerate at apex, 5-7 x 3-4 mm, pubescent outside; midrib strong, produced into a long antrorsely scabrid awn at apex. Perianth bristles 3-6, unequal, shorter than nut, retrorsely scabrous, caducous. Stamens 3; filaments elongate up to 5 mm; anthers linear-oblong, *ca* 3 mm long, with a well developed bristly connective appendage. Ovary obovate, *ca* 0.6 mm long; style *ca* 2 mm long; stigmas 2 or 3, longer than style, *ca* 3.5 mm long. Nut planoconvex to trigonous, broadly obovate to suborbicular in outline, 2.8 - 3 x *ca* 2 mm, blackish brown, smooth.

Fls. & Frts.: July - April.

Habitat : Edges of rice fields, swampy areas along with other tall sedges and in other moist localities.

Distrib. : Old world. INDIA: Throughout (except the North-East). KARNATAKA: Bangalore, Dharwar, Hassan, Mysore, Raichur.

Specimens examined : Bangalore: Bangalore, *Camaron* 499, 26.3.1890 (MH). Dharwar: Haveri, *Talbot* 2266, 2.1.1890 (BSI). Hassan: Hassan, *Hooper* & *Gandhi* 2394, 11.11.1991 (JCB). Mysore: Birle Homotal, *Mahajan* 34734 A, 23.4.1958 (BSI); St. Philomina College, *Bhat* 31, 20.7.1970 (MGH); Mandakalli, *Bhat* 20, 12.9.1970 (JCB). Raichur: Shorapur-Lingsugur Road, *Singh* 129550, 13.2.1975 (BSI).

subsp. **affinis** (Roth) Koyama in Brittonia 31: 284. 1979. *Scirpus affinis* Roth in Roem. & Schult. Syst. Veg. 2: 140. 1817; Kern in Reinwardtia 6: 33. 1961; Karthik. *et al.*, Fl. Ind. Enum. Monocot. 70. 1989. *S. maritimus* L. var. *affinis* (Roth) Clarke in Hook.f., Fl. Brit. India 6: 659. 1893; Cooke, Fl. Pres. Bombay 2: 893. 1908 (3: 408. 1958, repr.ed.); Fischer in Gamble, Fl. Pres. Madras 1666. 1931 (3: 1156. 1957, repr.ed.). *S. strobilinus* Roxb. (Hort. Beng. 6. 1814, *nom. nud.*), Fl. Indica, eds. Carey & Wall. 1: 222. 1820. *Bolboschoenus affinis* (Roth) Drobov, Trudy Bot. Muz. Imp. Akad. Nauk. 16: 139. 1916. **Fig. 1.**

Plants 15-50 cm high. Tubers oblong, ellipsoid or ovoid, 10-25 x 6-9 mm. Stems 2.5 - 6 mm thick near the base, 1-2 mm thick below the

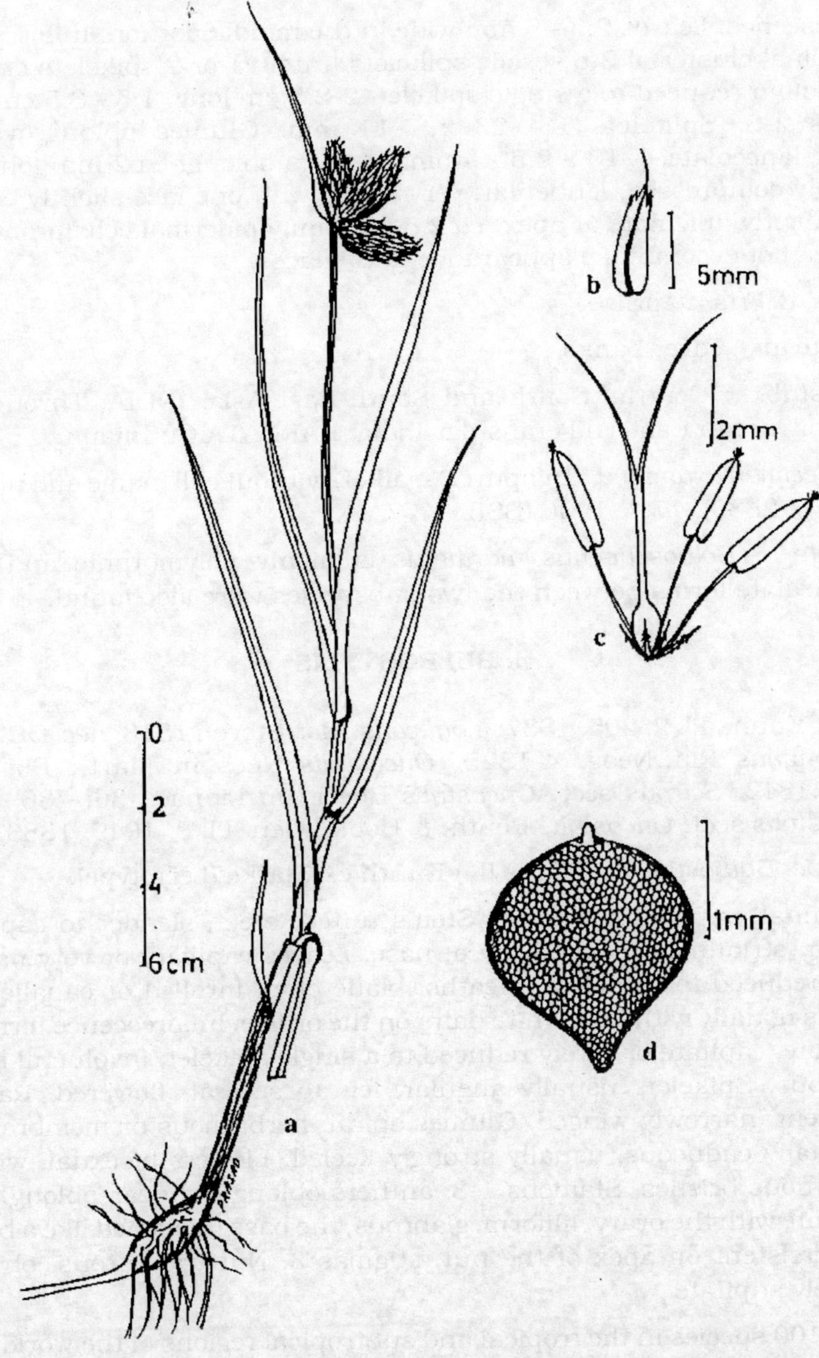

Fig. 1. *Bolboschoenus maritimus* subsp. *affinis* (Roth) Koyama
a. Habit, **b.** Glume, **c.** Flower, **d.** Nut

inflorescence. Leaves 2.5 - 5 mm wide in the middle portion. Inflorescence a terminal cluster of 2-6 sessile spikelets (rarely 1 or 2 spikelets on short rays), often reduced to a single spikelet, 2-2.5 cm long, 1.5 - 3.5 cm wide. Bracts 2-3. Spikelets 13- 23 x 7-11 mm. Glumes oblong-ovate to oblong-lanceolate, 7-10 x 2.5 -3 mm. Anthers oblong, *ca* 2 mm long. Nut strongly compressed, lenticular, very rarely with one face slightly convex, prominently apiculate at apex, *ca* 2.5 x 2 mm; epidermal cells prominent, giving a honeycomb-like appearance to the faces.

Fls. & Frts. : April.

Habitat : River banks.

Distrib. : Central, South and South-East Asia. INDIA: Throughout, probably not in Eastern Peninsular India. KARNATAKA: Bijapur.

Specimen examined : Bijapur: Bagalkot, without coll. name and number (Acc. No. 2839), 17.4. 1906 (BSI).

Note : *Bolboschoenus maritimus* is highly polymorphic and some intermediate forms between the two subspecies were also found.

3. BULBOSTYLIS

Kunth, Enum. Pl. 2: 205. 1837, *nom.cons., non* Steven 1813, *nec.* DC.1836. *Stenophyllus* Raf.,Neog. 4.1825. *Oncostylis* Nees in Mart., Fl. Bras. 2,1:80.1842. *Scirpus* sect. *Oncostylis* Boeck. in Linnaea 36: 736. 1870. *Fimbristylis* sect. *Oncostylis* Benth. & Hook., Gen. Pl. 3: 1049. 1883.

Type: Bulbostylis capillaris (L.) Kunth ex Clarke (Lectotype).

Annual or perennial herbs. Stems tufted, erect, slender to capillary, angular, striate or sulcate, leafy at base. Leaves capillary or very narrow, rarely reduced to bladeless sheaths; blade often inrolled or canaliculate; sheaths usually with long white hairs on the orifice. Inflorescence terminal, anthelate, capitate or rarely reduced to a single spikelet. Involucral bracts foliaceous. Spikelets usually angular, few to several- flowered. Rachilla persistent, narrowly winged. Glumes spiral, herbaceous or membranous, acropetally caducous, usually strongly keeled. Flowers bisexual, without hypogynous bristles. Stamens 1-3; anthers oblong or linear-oblong. Style articulate with the ovary, filiform, glabrous, the base thickened like a button and persistent on apex of the nut; stigmas 3. Nut triquetrous, obovate, minutely stipitate.

ca 100 species in the tropical and subtropical regions of the world, with the highest concentration in the tropical Africa followed by America. In India only 5 species; 4 in Karnataka.

Literature : **BODARD, M.** (1963) Notes pre'liminaires a'la revision du genre *Bulbostylis* (Cyperacees), in *Bull. Soc. Bot. Fr.* 110: 158-160. **GORDON-GRAY, K.D.** (1971) *Fimbristylis* and *Bulbostylis* generic limits as seen by a student of Southern African species, in *Mitt. Bot. Staatssamml. Munchen* 10: 549-574 (Chrom. Nos.). **KRAL, R.** (1971) A treatment of *Abildgaardia, Bulbostylis* and *Fimbristylis* (Cyperaceae) for North America, in *Sida* 4 (2): 57-227 (Generic delimitation). **LYE, K.A.** (1971) The generic concept of *Bulbostylis* Kunth ex C.B. Cl., in *Mitt. Bot. Staatssamml. Munchen* 10: 539-547.

Key to the species

1a. Spikelets in heads; nut smooth.. 2

1b. Spikelets umbellate, rarely reduced to a single spikelet; nut transversely wrinkled 3

2a. Stems very slender, glabrous; head not prickly; glumes strongly keeled .... **1. B. barbata**

2b. Stems thick, hairy; head almost prickly; glumes scarcely keeled .... **4. B. subspinescens**

3a. Inflorescence with loosely arranged spikelets on long rays; glumes muticous at apex, glabrous .... **2. B. densa**

3b. Inflorescence with crowded spikelets on very short rays, often contracted to nearly head-like; glumes mucronate at apex, pubescent .... **3. B. puberula**

1. **Bulbostylis barbata** (Rottb.) Clarke in Hook.f., Fl. Brit. India 6: 651. 1893; Fischer in Gamble, Fl. Pres. Madras 1662. 1931 (3: 1153. 1957, repr. ed.); Kern in van Steenis, Fl. Males. 1, 7: 539. 1974; Datar & Vartak in Journ. Univ. Poona, Sci. Tech. 48: 24, t. 3, f. 1-6. 1976; Hooper in Saldanha & Nicolson, Fl. Hassan 659. 1976; Rao & Razi, Fl. Mysore 557. 1981; Rao & Verma, Cyp. NE India 40, t. 21-21b. 1983; Sharma *et al.*, Fl. Karnataka 303. 1984; Koyama in Dassanayake & Forsberg, Rev. Handb. Fl. Ceylon 5: 327. 1985; Singh, Fl. E. Karnataka 2: 627. 1988; Karthik. *et al.*, Fl. Ind. Enum. Monocot. 33. 1989; Keshava Murthy & Yoganarasimhan, Fl. Coorg 499. 1990. *Scirpus barbatus* Rottb., Progr. 27. 1772 *et* Descr. & Ic. 52, t. 17, f. 4. 1773. *Isolepis barbata* (Rottb.) R. Br., Prodr. Fl. Nov. Holl. 222. 1810; Nees in Wight, Contr. Bot. India 109. 1834. *Fimbristylis barbata* (Rottb.) Benth., Fl. Austr. 7: 321. 1878; Ramaswamy & Razi, Fl. Bangalore 104. 1973. *Stenophyllus barbata* (Rottb.) Cooke, Fl. Pres. Bombay 2: 887. 1908 (3: 401. 1958, repr.ed.). *Chavuri hullu.* **Fig. 2.**

Illus. : Rao & Verma, *l.c.*; Datar & Vartak, *l.c.*; Matthew, Illus. Fl. Tamilnadu Carnatic Pl. 778. 1982.

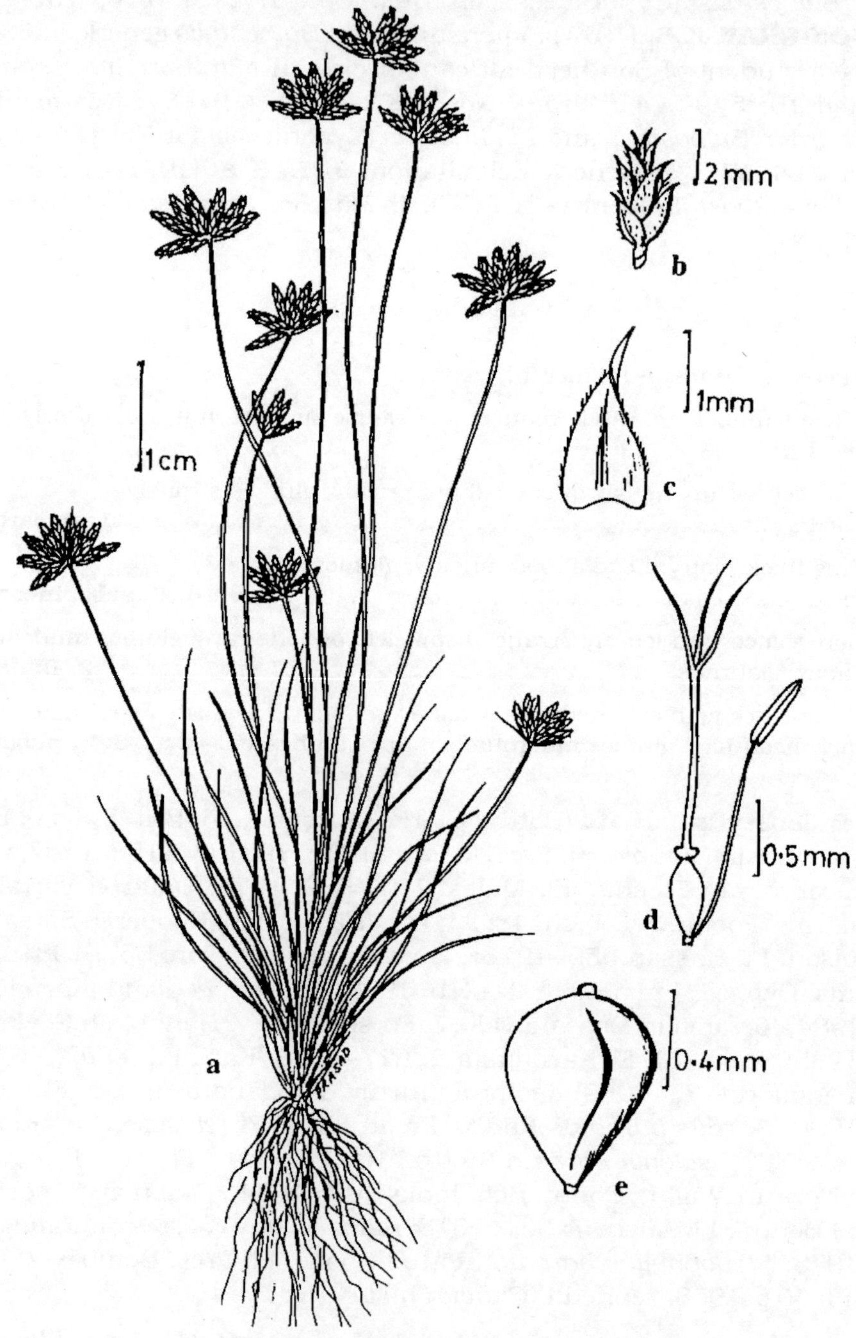

Fig. 2. *Bulbostylis barbata* (Rottb.) Clarke
a. Habit, **b**. Spikelet, **c**. Glume, **d**. Flower, **e**. Nut

Key to the varieties

1a. Stems 0.3 - 0.4 mm thick; leaf sheaths stramineous to brownish; glumes usually puberulous. ....var. **barbata**

1b. Stems 0.5-1 mm thick; leaf sheaths purple brown; glumes usually glabrous or subglabrous. ....var. **pulchella**

var. **barbata**

Annuals, highly tufted, 3-20 cm high. Stems very slender, sulcate, smooth, glabrous, 0.3-0.4 mm thick. Leaves much shorter than stem, capillary to very slender, acute at apex, glabrous or scabrous on lower surface towards apex, 0.2-0.5 mm wide; sheaths membranous, stramineous to brownish when dry; ligule absent; orifice white-hyaline, usually with long white hairs; leaves at times reduced to sheaths. Inflorescence a terminal head of (1-) 2-20 sessile spikelets, sub-spherical in later stage, 3-10 mm across. Involucral bracts 1-3, shorter to much longer than the inflorescence, dilated at base, usually scabrous on lower surface, 5-25 mm long. Spikelets ovoid to ovoid-lanceolate, subacute at apex, angular, 3-7 x 1-1.7 mm, few-flowered. Glumes membranous, ovate, mucronate, 2-2.5 x 1.25-1.5 mm, strongly keeled, brownish, with green keel, usually puberulous on outer surface, often glabrous; margin ciliolate; mucro usually excurved. Stamen 1; filament elongate up to 2.5 mm; anther linear-oblong, *ca* 0.7 mm long. Ovary obovate, *ca* 0.5 mm long; style *ca* 1 mm long; stigmas 3, shorter than style. Nut triquetrous, broadly obovoid, smooth, *ca* 0.8 x 0.5 mm, stramineous, finely reticulate with minute isodiametric cells in many rows.

Fls. & Frts. : May - Jan.

Habitat : Dry open sandy places, dry gravelly soil, roadsides, shallow depressions, on rocks and in open sandy areas in forests and plantations. Also found as a common weed in cultivated fields. Sea level to *ca* 1300 m, but usually at low altitude.

Distrib. : Widely distributed in warm regions of the Old World and Southern U.S.A. INDIA: Throughout. KARNATAKA: Bangalore, Belgaum, Bellary, Bijapur, Dakshina Kannada, Hassan, Kodagu, Kolar, Mandya, Mysore, Raichur, Shimoga, Tumkur, Uttara Kannada.

Specimens examined : Bangalore: Kanakapura to Sangam, *Ravindra* 1559, 19.7.1978 (JCB). Belgaum: Gokak, *Sheodye s.n.* (Acc. No. 2644), 17.8.1909 (BSI). Bellary: Benkal plantation, Moka, *Singh* 132888, 12.9.1974 (BSI); Moka R.F., *Singh* 143096, 19.8.1976 (BSI). Bijapur: Badami, without coll. name, *s.n.* (Acc. Nos. 2645, 2646, 2647, 2648, 2649, 2650, & 2651), 30.8.1892 (BSI); Badami, without coll. name, *s.n.* (Acc. No.

79153), without date (BLAT). Dakshina Kannada: Malpe, near Udupi, *Vartak s.n.* (S. No. 14425), Nov. 1971 (MACS); Malpe, *Bhat* 296, 23.9.1976 (MGH); Surathkal beach, *Saldanha & Prakash* 4038, 12.11.1978 (JCB). Hassan: Bourdalboore, *Nicolson et al.* 2344, 26.10.1971 (JCB). Kodagu: Kushalnagar, *Bhat* 1044, 21.9.1981 (MGH). Kolar: Narasapura, *Ramesh & Ravindra* 1483, 7.7.1978 (JCB). Mandya: Melukote, *Dinesh* 662, 7.11.1982 (MGH). Mysore: Nangangad, *Camaron* 555, 6.10.1890 (MH); Chamundi, *Bhat* 44, 20.7.1970 (JCB); Manasagangotri, *Bhat* 27, 14.7.1970 (MGH). Raichur: Konagi, *Wadhwa* 44865, 29.9.1958 (BSI); Kalluru, *Wadhwa* 45028, 2.10.1958 (BSI); Koppal-Kushtagi Road, 38th km, *Singh* 141523, 9.11.1975 (BSI). Shimoga: Agumbe, *Mohana Mani* 45, Jan. 1964 (MGH). Tumkur: Hiriyur-Tumkur Road, *Rao* 73217, 27.8.1961 (BSI); Ujne State forest, *Yoganarasimhan* 404, 15.8.1974 (RRCBI). Uttara Kannada: Karwar, *Talbot* 555, 10.8.1883 (BSI). Without exact locality, *Vartak s.n.* (S. No. 8542), Nov. 1971 (MACS).

Note: A variable species especially in the colour and texture of the glumes. Specimens were found with stramineous to bright brown glumes. Similarly, glabrous, slightly pubescent and densely puberulous glumes were also found. In certain specimens mucro of the glumes are more excurved giving the inflorescence a prickly appearance. But in many specimens mucro is not or slightly excurved.

var. **pulchella** (Thw.) Clarke in Hook. f., Fl. Brit. India 6: 652. 1893; Karthik. *et al.*, Fl. Ind. Enum. Monocot. 33. 1989. *Isolepis pulchella* Thw. Enum. Pl. Zeyl. 350. 1864. *Scirpus thwaitesii* Boeck. in Linnaea 38: 380. 1874. *Bulbostylis barbata* (Rottb.) Clarke subsp. *pulchella* (Thw.) Koyama in Bot. Mag. (Tokyo) 93: 341. 1980 *et* in Dassanayake & Fosberg, Rev. Handb. Fl. Ceylon 5: 328. 1985.

Plants more rigid than the typical variety. Stems 0.5-1 mm thick, usually light purple towards the base. Leaf sheaths purplish brown. Glumes usually glabrous or subglabrous.

Fls. & Frts. : Aug. - Jan.

Habitat : Restricted to the coastal areas.

Distrib. : Restricted to Sri Lanka and Peninsular India. KARNATAKA: Dakshina Kannada and Uttara Kannada.

Specimens examined : Dakshina Kannada: Mangalore, *Foulker* 4865, Jan. 1902 (MH); Neleshwar, without coll. name, 15323, 9.11.1917 (MH); Ucchila, Mangalore Taluk, *Sheriff & Suresh* 590, 11.9.1985 (MH). Uttara Kannada: Karwar, *Talbot* 1788, 20.8.1885 (MH).

Note : The main difference found in this taxon is the more thick stems. Also the leaf sheaths were found purple-brown (in dry specimens). Lower glumes are longer (up to 3.3 mm) in the typical variety also. Moreover glumes in Karnataka specimens are mostly glabrous or only slightly pubescent, but mostly puberulous in the typical variety. Hence the hairy nature seems to be quite variable. As there is not much difference and as intermediate forms are also found it is preferred here to treat this as a variety only, as done by Clarke, *l.c.*.

2. **Bulbostylis densa** (Wall. ex Roxb.) Hand.-Mazz. in Karst. & Schenk, Vegetationsb. 20, 7: 16. 1930; Kern in van Steenis, Fl. Males. 1, 7: 538. 1974; Hooker in Saldanha & Nicolson, Fl. Hassan 659. 1976; Datar & Vartak in Journ. Univ. Poona, Sci. Tech. 48: 26, t. 4, f. 1-8. 1976; Rao & Verma, Cyp. NE India 40. 1982; Sharma *et al.*, Fl. Karnataka 303. 1984; Koyama in Dassanayake & Fosberg, Rev. Handb. Fl. Ceylon 5: 324. 1985; Karthik. *et al.*, Fl. Ind. Enum. Monocot. 33. 1989; Keshava Murthy & Yoganarasimhan, Fl. Coorg 499. 1990. *Scirpus densus* Wall. in Roxb., Fl. Ind. 1: 231. 1820. *Isolepis trifida* Nees in Wight, Contr. Bot. India 108. 1834. *Bulbostylis capillaris* var. *trifida* Clarke in Hook.f., Fl. Brit. India 6: 652. 1893; Fischer in Gamble, Fl. Pres. Madras 1662. 1931 (3: 1153. 1957, repr.ed.). *Fimbristylis capillaris* (L.) Gray var. *trifida* (Nees) Koyama in Journ. Fac. Sci. Univ. Tokyo 3, 8(3): 103. 1961; Ramaswamy & Razi, Fl. Bangalore 105. 1973. **Fig.3.**

Type : India.

Illus. : Datar & Vartak, *l.c.*; Matthew, Furth. Illus. Fl. Tamilnadu Carnatic Pl. 623, 1988.

Annuals, highly tufted, 3-25 cm high. Stems very slender, sulcate, smooth, glabrous, 0.2-0.5 mm thick. Leaves shorter than stem, very slender, acute at apex, 0.5-12 cm long, 0.2-0.4 mm wide, glabrous, at times scabrous towards apex on the lower surface and margins; sheaths membranous, usually with white hairs on the hyaline orifice. Inflorescence simple to subcompound, rarely contracted and nearly head-like, with (1-) 2-7 spikelets, up to 4 cm broad. Involucral bracts very short, glume-like. Rays 1-5, filiform, up to 2 cm long. Spikelets solitary, ovate to oblong-ovate, subacute at apex, angular, 3-5 x 1.5-3 mm, subdensely few-flowered. Glumes spiral, membranous, broadly ovate, acute at apex, strongly keeled, minutely ciliolate at margins, *ca* 2 x 1.2 mm, brownish, often hyaline towards base. Stamens 2; filaments elongate up to 2 mm; anthers oblong, acute at apex, *ca* 0.4 mm long. Ovary obovate, *ca* 0.4 mm long; style *ca* 1 mm long; stigmas 3, much shorter than style. Nut triquetrous, broadly obovoid, minutely stipitate, *ca* 1 x 0.8 mm, stramineous to greyish-brown, densely granular-puncticulate; epidermal cells minute, isodiametric, not very clear on the corrugate surface.

Fls. & Frts. : Aug. - Jan.

Habitat : Sandy soil, slopes of hills, along roadsides, open wet places, edges of drainage channels and as a weed in rice fields.

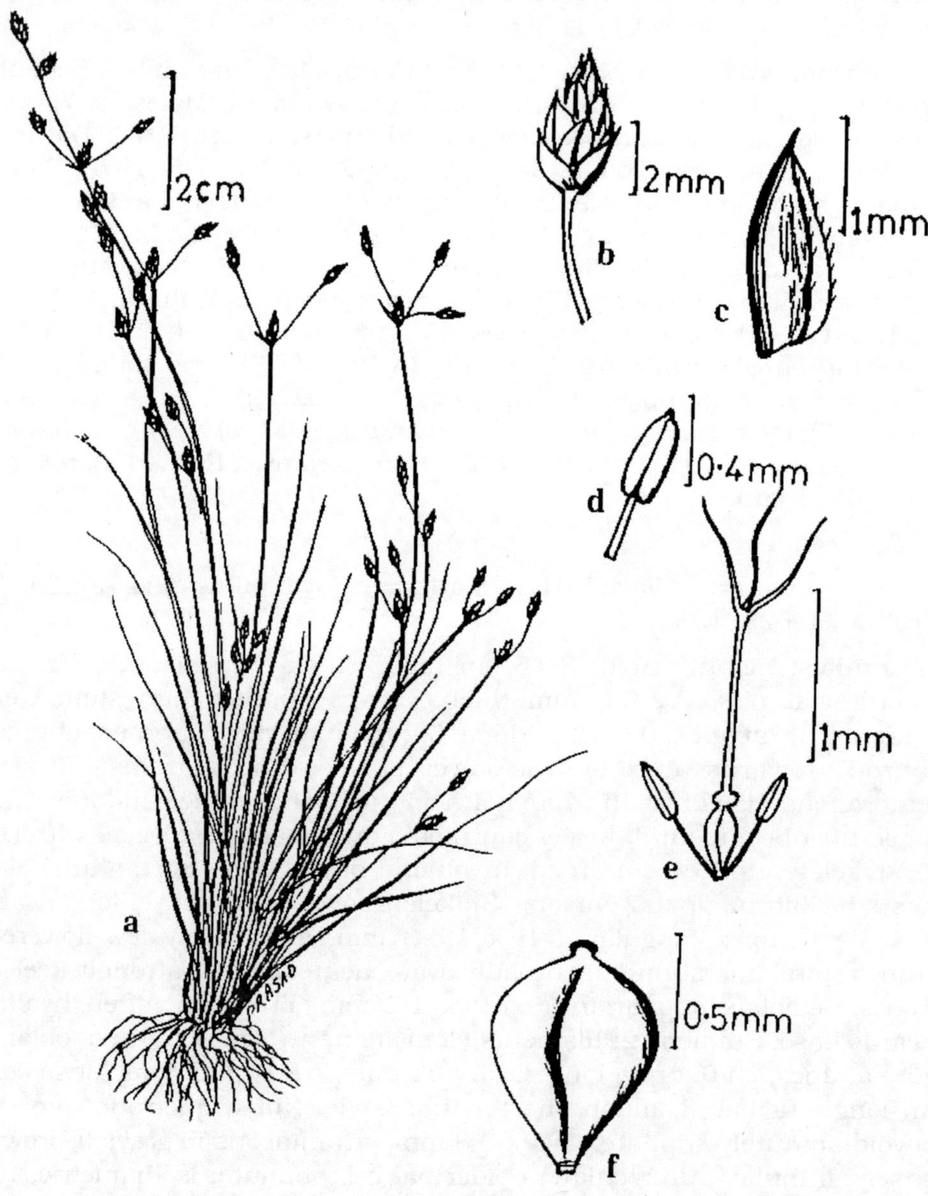

Fig. 3. *Bulbostylis densa* (Wall. ex Roxb.) Hand.-Mazz.
a. Habit, **b.** Spikelet, **c.** Glume, **d.** Stamen, **e.** Flower, **f.** Nut.

Distrib. : Widely distributed in the tropics and subtropics of old world (South & South-East Asia, tropical Africa and Australia). INDIA: Throughout. KARNATAKA: Bangalore, Chikmagalur, Dakshina Kannada, Hassan, Kodagu, Mysore, Shimoga, Uttara Kannada (Sharma *et al., l.c.*).

Specimens examined : Bangalore: Nandi hills, *Hooper & Saldanha* 18094, 4.12.1971 (JCB); Bannergatta, *Saldanha & Rao* 18644, 7.9.1975 (JCB). Chikmagalur: Dattatraya peetha, *Ramesh & Murthy* 4805, 2.12.1978 (JCB). Dakshina Kannada: Indrali, Udupi, *Bhat* 280, 22.8.1976 (MGH); Kapu, *Bhat* 446, 15.1.1977 (MGH). Hassan: Achihalli, *Saldanha* 17980, 16.9.1969 (JCB); Byra, *Nicolson et al.* 2266, 23.10.1971 (JCB). Kodagu: Bridal path of Rajaseat, *Rao* 74408, 20.7.1961 (BSI); Abbe falls, Mercara, *Bhat* 1051, 25.9.1981 (MGH); Talakaveri, *Bhat* 1072, 27.9. 1981 (MGH). Mysore: Locality (?), *Talbot* 3119, 7.9.1893 (BSI). Shimoga: Kanagalgudda, *Raghavan* 900254, 19.8.1960 (BSI); Agumbe, *Bhat* 699, 26.9.1980 (MGH). Without locality (Mysore & Carnatic), *Thomson, s.n.*, without date (MH).

Note : *B. densa* is closely related to *B.capillaris* (L.) Clarke, an American species.

3. **Bulbostylis puberula** (Poir.) Clarke in Hook.f., Fl. Brit. India 6: 652. 1893; Fischer in Gamble, Fl. Pres. Madras 1662. 1931 (3: 1153. 1957, repr.ed.); Ramaswamy & Razi, Fl. Bangalore 110. 1973; Kern in van Steenis, Fl. Males. 1, 7: 540, f. 37.1974; Datar & Vartak in Journ. Univ. Poona. Sci. Tech. 48: 28, t. 5, f. 1-7. 1976; Yoganarasimhan *et al.*, Fl. Chikmagalur 362. 1981; Sharma *et al.*, Fl. Karnataka 303. 1984; Koyama in Dassanayake & Fosberg, Rev. Handb. Fl. Ceylon 5: 326. 1985; Karthik. *et al.*, Fl. Ind. Enum. Monocot. 33. 1989. *Scirpus puberulous* Poir. in Lamk., Enc. 6: 767. 1804. *Isolepis gracilis* Nees in Wight, Contr. Bot. India 109. 1834. **Fig.4.**

Type: Madagascar.

Illus. : Kern, *l.c.*; Datar & Vartak, *l.c.*

Annuals with fibrous roots, densely tufted, 10-35 cm high. Stems very slender, sulcate, puberulous or glabrous, often hispid below the inflorescence, 0.4-0.5 mm thick. Leaves much shorter than stem, acute at apex, 0.3-0.5 mm wide, puberulous on the lower surface and margins, sheaths membranous, puberulous, stramineous; orifice with long, white hairs. Some of the leaves reduced to sheaths. Inflorescence simple, rarely subcompound, often congested to almost head-like inflorescence, rarely reduced to a single spikelet, 1-1.5 cm wide, with 1-few spikelets. Involucral bracts 2-4, filiform, longest usually overtopping the inflorescence, up to 3 cm long; rays up to 5, 1-6 mm long. Spikelets solitary, oblong-ovoid or ovoid, subacute at apex, 3-5 x 1.5-2 mm, few-flowered. Glumes spiral, membranous, broadly ovate, mucronulate at apex, strongly keeled, 2-2.2 x *ca* 1.5 mm, ciliolate on margins, densely pubescent, pale to dark brown; mucro excurved in later stage. Stamen 1; filament elongate up to 2mm;

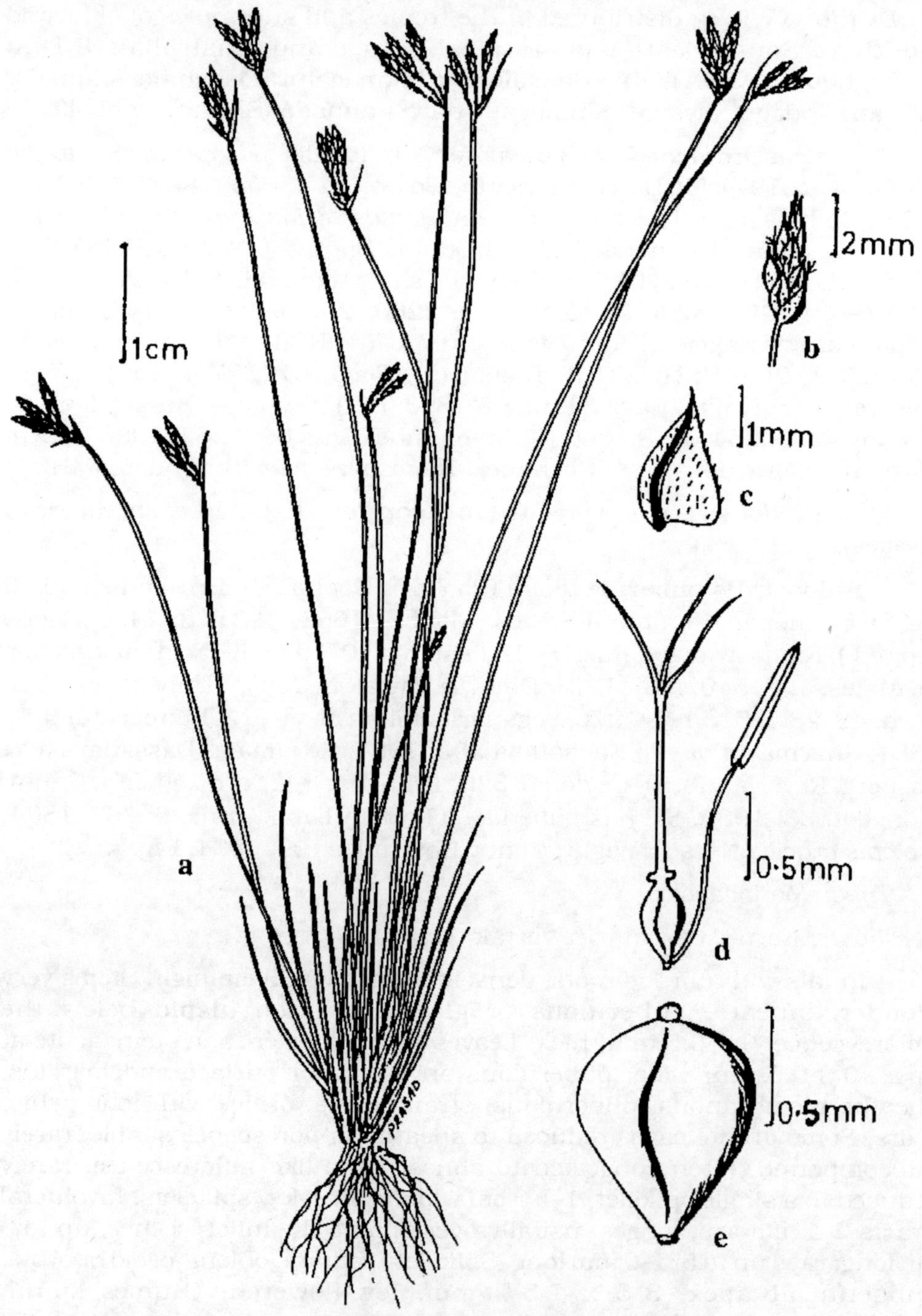

Fig. 4. *Bulbostylis puberula* (Poir.) Clarke
a. Habit, **b**. Spikelet, **c**. Glume, **d**. Flower, **e**. Nut

anther linear-oblong, *ca* 0.8 mm long. Ovary obovate, *ca* 0.4 mm long; style *ca* 1 mm long; stigmas shorter than style. Nut triquetrous, broadly obovoid, *ca* 1 x 0.8 mm, transversely wavy-wrinkled, stramineous, minutely stipitate; epidermal cells longitudinally oblong.

Fls. & Frts. : Aug. - Jan.

Habitat : Sandy areas at low altitudes especially near sea coast and other open dry sandy areas. A common weed in paddy fields.

Distrib. : Widely distributed in tropical Asia and Africa; also in Southern U.S.A. INDIA: Restricted to the peninsular region. KARNATAKA: Bangalore (Ramaswamy & Razi, *l.c.*), Chikmagalur (Yoganarasimhan *et al.*, *l.c.*), Dakshina Kannada, Uttara Kannada.

Specimens examined : Dakshina Kannada: Mangalore, *Foulker* 4869, Jan. 1902 (MH); Pilicode, without coll. name, 16274, 4.12.1919 (MH); Kudlu, without coll. name, 16757, 2.9.1920 (MH); Malpe, *Bhat* 297, 23.9.1976 (MGH); Kapu, *Bhat* 348, 1.11.1978 (MGH). Uttara Kannada: Karwar, *Talbot* 1512, 10.8.1885 and 1314, 1.10.1885 (BSI).

Uses : In China it is used as a diuretic (Kern, *l.c.*)

Note : The plant looks like *B. barbata*; but can be distinguished by the hairy nature of the stem and inflorescence and so also by the stalked spikelets.

4. **Bulbostylis subspinescens** Clarke in Hook.f., Fl. Brit. India 6: 652. 1893; Fischer in Gamble, Fl. Pres. Madras 1662. 1931 (3: 1153. 1957, repr.ed.) Datar & Vartak, Journ. Univ. Poona, Sci. & Tech. 48: 30, t. 6, f. 1-7. 1976; Sharma *et al.*, Fl. Karnataka 304. 1984; Karthik. *et al.*, Fl. Ind. Enum. Monocot. 33. 1989.

Illus. : Datar & Vartak, *l.c.*

Similar to *B. barbata* (Rottb.) Clarke, but can be distinguished by the following characters.

Whole plant pubescent. Stems thick, rigid, curved, often twisted, shallowly grooved, hairy. Leaves up to half the length of the stem, rigid, channelled, hispid- pubescent. Inflorescence an almost prickly head. Bracts 3, about as long as the head. Spikelets almost stellately spreading. Glumes scarsely keeled, densely pubescent. Nut pale brown.

Fls. & Frts. : Not seen.

Habitat : Sandy soil in the sea coast.

Distrib. : Endemic to peninsular India. KARNATAKA: Dakshina Kannada (Datar & Vartak, *l.c.*).

Note: This species is included as reported by Datar & Vartak, *l.c.*

Specimens were not seen. According to Clarke it may be treated as a variety of *B.barbata* growing in sandy areas of sea coast. The description given above is compiled from Clarke, *l.c.* and Fischer, *l.c.*. But Datar & Vartak feel that it is a distinct species which differs from *B.barbata* Kunth by the general appearance of the leaves and inflorescence. Also habitat of both the taxa are quite different.

4. CAREX

L., Gen. Pl. ed. 1, 280. 1737. *Vignea* Beauv. in Lestib., Ess. Fam. Cyp. 22. 1819.

Type : *Carex acuta* L.

Perennials with tufted or creeping rhizomes. Stems usually trigonous and solid, rarely terete and hollow, often clothed at the base by persistent, disintegrated leaf-sheaths or their fibrous remains. Leaves 3-ranked, radical and cauline, linear or rarely lanceolate; sheaths closed. Plants monoecious, rarely dioecious. Inflorescence racemose, spicate, fasciculate or paniculate, with 2 - many spikes or with a single terminal spike. Spikes bisexual or unisexual; when unisexual normally terminal one staminate and lateral ones pistillate; lateral spikes often with an urticuliform or ocreiform cladoprophyll surrounding it. Flowers unisexual, solitary in the axil of glumes; perianth absent. Staminate flowers triandrous, borne at axil of staminate glumes. Pistillate flowers consisting of a single pistil, enclosed in a sac-like prophyll (utricle); style branched above into 2 or 3 stigmas which protrude through the small orifice at the apex of the utricle. Utricle chartaceous, membranaceous or coriaceous, closely enveloping the nut. Nuts trigonous, planoconvex or lenticular, sessile or stipitate.

The largest genus of the family with *ca* 2000 species distributed throughout the world. *Ca* 160 species in India; 6 in Karnataka.

Literature : **AKIYAMA, S**. (1969) Illustrations of the genus *Carex* (Cyperaceae), in *Sci. Rep.Ranazawa univ.* 14.1969:71- 174, 51 Pl. **BOOTT, F.** (1858-1867) *Illustrations of the genus Carex* 4 Vols. **KERN, J.H.** (1974) *Carex* in: van Steenis, *Fl. Males.* 1,7: 107-187. **KUKENTHAL, G.** (1909) *Carex* in: Engler, *Pflanzenr.* 38 (IV.20): 67-824. **NELMES, E**. (1946) A key to the Carices of Malaysia and Polynesia, in *Kew Bull.* 1946: 5-29. **NELMES, E.** (1951) The genus *Carex* in Malaysia, in *Reinwardtia* 1: 221-450. **NELMES, E**. (1955) The genus *Carex* in Indo-China including Thailand and Lower Burma, in *Mem. Mus. Hist. Nat.* (Paris) *n. ser.* B. 4 : 83-182. **ROBERTSON, A.**(1979) History of the classification of the genus *Carex*, in *Taxon* 28 (5/6): 535-548.

Key to the species

1a. Spikes not androgynous (male glumes on a separate terminal male spike) . . . . **5. C. maculata**

1b. Spikes androgynous 2

2a. Inflorescence usually of 1 terminal and 1-2 [according to Nelmes 1-3 (-4)] solitary, axillary spikes . . . . **6. C. speciosa**

2b. Infloresence a compound panicle with secondary panicles bearing many to numerous spikes 3

3a. Spikes with staminate part longer than or almost equal in length to the pistillate part 4

3b. Spikes with staminate part always shorter than pistillate part 5

4a. Secondary panicles congested; spikes with staminate part longer than pistillate part . . . . **4. C. lindleyana**

4b. Secondary panicles not congested; spikes with staminate part almost equal in length to pistillate part . . . . **3. C. indica** subsp. **laete-brunnea**

5a. Leaves 7-15 mm wide; spikes 3-6 cm long; male glumes *ca* 5.5 mm long; female glumes 3-3.5 mm long; nut 3-3.5 mm long . . . . **1. C. baccans**

5b. Leaves 6-7 mm wide (8-12 mm in var. *glaucina*); spikes 5-25 mm long; male glumes up to 3.5 mm long; female glumes 1.5 - 2.5 mm long; nut 2-2.5 mm long . . . . **2. C. caricina**

1.**Carex baccans** Nees in Wight, Contr. Bot. India 122. 1834; Clarke in Hook.f., Fl. Brit. India 6: 722. 1894; Fischer in Gamble, Fl. Pres. Madras 1687. 1931 (3: 1169. 1957, repr.ed.); Nelmes in Reinwardtia 1: 322. 1951; Rao & Razi, Fl. Mysore 557. 1981; Rao & Verma, Cyp. NE India 76. 1982; Sharma *et al.*, Fl. Karnataka 304. 1984; Koyama in Dassanayake & Fosberg, Rev. Handb. Fl. Ceylon 5: 371. 1985; Karthik *et al.*, Fl. Ind. Enum. Monocot. 34. 1989; Keshava Murthy & Yoganarasimhan, Fl. Coorg 499. 1990.

Type : India, *Wight* 1912.

Illus. : Boott, Illus. Carex. 2: 83, tt. 234-236, 238, 239. 1860.

Perennials, loosely tufted, up to 150 cm high. Rhizome stout, woody, oblique. Stems erect, prominently trigonous, 3-5 mm thick, smooth, with fuscous or purple brown sheaths at base which slightly disintegrate into fibres. Leaves cauline, throughout the stem, but more dense in the lower part, upper ones overtopping the stem, widely linear, gradually narrowed to a long acute apex, 7-15 mm wide, coriaceous, scabrid on upper surface. Sheaths 10-15 cm long, tightly enclosing the stem, eventually splitting into fibres on the membranous ventral side; basal ones reddish-brown; upper ones greenish. Inflorescence a compound, interrupted panicle, occupying the upper 1/3 to 1/2 of the stem; secondary panicles 5-8, single at each node, elliptical to oblong, 5-20 cm long, 3-5 cm wide, lower ones more exserted; peduncles smooth or scaberulous. Lower bracts leaf-like, overtopping the inflorescence; sheaths as in cauline leaves; upper ones smaller; bracteoles glume-like, usually aristate. Cladoprophylls utriculiform. Spikes numerous, sessile, suberect, cylindrical, androgynous, 3-6 cm long, subdensely flowered; female part usually longer than male part; male part linear. Male glumes oblong-lanceolate, mucronate to short-awned at apex, *ca* 5.5 mm long. Stamens 3. Female glumes broadly ovate or oblong-ovate, obtuse to acute at apex, basal ones often with a short awn at apex, 3-3.5 x 2-2.2 mm, glabrous or hispidulous. Style short, not thickened at base; stigmas 3. Utricle obscurely trigonous, inflated, loosely enveloping the nut, ellipsoid to subglobose, 3.5-4.5 x 1.7-2.2 mm, subcoriaceous, distinctly nerved, narrowly marginate, glabrous except at margins which are minutely hispidulous, usually curved at apex, ultimately reddish; beaks subterete, bidentate at apex. Nut trigonous, ellipsoid, 3-3.5 x 1-1.2 mm (including *ca* 0.5 mm long beak), short stipitate, ultimately dark brown.

Fls. & Frts. : Nov. - Feb.

Habitat : Common in wet slopes and grasslands at high elevations.

Distrib. : Sri Lanka, Indo-China, South China, Formosa and Malesia. INDIA: Peninsular India, Central & North-East India. KARNATAKA: Chikmagalur, Kodagu, Mysore.

Specimens examined : Chikmagalur: Bababudan hills, *Talbot* 3249, 29.12.1893 (BSI); Kemmangudi, *Ahuja* 65638, 9.12.1959 (BSI); Hulikottai, *Raghavan* 86968, 27.2.1967 (BSI); On way to Bubbabudan hills, Kemmangudi, *Yoganarasimhan* 1802, 21.11.1974 (MGH). Kodagu: Talakaveri, *Bhat* 810, 19.12.1980 (MGH); Pushpagiri, *Bhat* 1102, 6.11.1981 (MGH). Mysore: Gopalaswamy hills, Bandipur, *Naithani* 23223, 29.1.1965 (MH).

2. **Carex caricina** (Don) Ghildyal & Bhattacharyya in Journ. Econ. Tax. Bot. 7: 602. 1985; Karthik. *et al.*, Fl. Ind. Enum. Monocot. 34. 1989. *Cyperus caricinus* Don, Prodr. Fl. Nepal 39. 1825. *Carex filicina* Nees in Wight, Contr. Bot. India 123. 1834; Clarke in Hook.f., Fl. Brit. India 6: 717. 1894; Fischer in Gamble, Fl. Pres. Madras 1686. 1931 (3: 1169. 1957, repr.ed.); Nelmes in Reinwardtia 1: 304. 1951; Rao & Razi, Fl. Mysore 557. 1981; Yoganarasimhan *et al.*, Fl. Chikmagalur 359. 1981; Rao & Verma, Cyp. NE India 74. 1982; Sharma *et al.*, Fl. Karnataka 304. 1984; Koyama in Dassanayake & Fosberg, Rev. Handb. Fl. Ceylon 5: 375. 1985.

Illus. : Boott, Illus. Carex 3: 105, Pls. 311 & 312. 1862; Matthew, Illus. Fl. Tamilnadu Carnatic pl. 779. 1982.

Key to the varieties

1a. Leaves 6-7 mm wide; spikes 5-18 mm long; beak on the utricle not prominently bifid at tip .... var. **caricina**

2b. Leaves 8-12 mm wide; spikes 15-25 mm long; beak on the utricle prominently bifid at tip .... var. **glaucina**

var. **caricina**

Perennials with short woody rhizome, 50-100 cm high. Stems erect, prominently trigonous, with 3-6 nodes, smooth, 2-4 mm thick near the base. Leaves basal and 2-5 cauline, shorter than stems, linear, narrowed to a long acute apex, 6-7 mm wide, striate, smooth, antrorsely scabrous on the margins towards apex; sheaths of cauline leaves 2-8 cm long, loosely surrounding the stem, with membranous mouth; basal sheaths short, reddish brown, disintegrating into fibres; ligule membranous. Inflorescence a compound panicle, 15-50 cm long, 3-5 cm wide; secondary panicles 3-7, single or binate, ovate-lanceolate or pyramidal in outline, 4-15 cm long (including the long, slender, smooth peduncle), 2-6 cm wide. Rachis smooth below, hispidulous above. Bracts foliaceous, shorter than inflorescence, but longer than secondary panicles; sheaths as in upper leaves. Spikes sessile, androgynous, narrow, cylindrical, 5-18 mm long, 2-4 mm broad, loosely to subdensely flowered, male part slender, usually shorter than female part. Cladoprophylls glumiform, broadly ovate, 1-1.5 x *ca* 1 mm; midvein excurrent into 1-4 mm long, hispidulous awn. Male glumes longer, up to 3.5 mm long. Stamens 3; filaments up to 3 mm long. Female glumes ovate to oblong- ovate, obtuse or acute at apex, 1.5 - 2.5 x 1-1.5 mm, pale brown to reddish-brown, glabrous or hispid above; midrib at times ending in a short, hispid awn. Stigmas 3. Utricle trigonous, ellipsoid, 3.5-4 x *ca* 1 mm (including 1-1.5 mm long beak), somewhat recurved above, greenish to stramineous, glabrous, with 4-5 nerves on each face. Nut trigonous, ellipsoid, 2-2.5 x *ca* 1 mm (including the short stipe and beak).

Fls. & Frts. : April-May; Sept-Nov.

Habitat : As an undergrowth in swampy and marshy places in forests at high altitude.

Distrib. : Sri Lanka, Nepal, Myanmar, Vietnam, Indo-China, Malesia and Japan. INDIA: Peninsular India, North and North-East India. KARNATAKA: Chikmagalur, Dakshina Kannada, Kodagu, Mysore, Uttara Kannada.

Specimens examined : Chikmagalur: Bababudan hills, *Govinda s.n.* (Acc. No. 95995), 20.5.1953 (MH); Hosakere village forests, on way to

Bhyrapura forests, *Yoganarasimhan* 1387, 16.11.1972 (RRCBI). Dakshina Kannada: Chittakanam, Sambaji R.F., *Srinivasan* 68095, 31.10.1981 (MH). Kodagu: Nagarhole, *Arora, s.n.* (Acc. Nos. 2882 & 2891), 6.5.1959 (BSI). Mysore: Santaveri, *Talbot* 3076, 6.9.1893 (BSI); Continaribetta, Bedguli section, *Rao* 80080, 18.4.1962 (BSI); Biligiri Rangan hills, *Rao* 980, 4.10.1970 (JCB). Uttara Kannada: Nilkund, *Talbot* 3531, 1.12.1895 (BSI); Maringundi, *Fernandez* 1789, 19.11.1950 (BLAT); Castle Rock, *Jain* 29167, 15.11.1957 (BSI); Castle Rock, *Almeida* MRA-1262, 25.2.1981 (BNHS).

var. **glaucina** (Boeck.) Ghildyal & Bhattacharyya in Journ. Econ. Tax. Bot. 7: 602. 1985; Karthik. *et al.*, Fl. Ind. Enum. Monocot. 35. 1989. *Carex glaucina* Boeck. in Linnaea 40: 353. 1876; Hooper in Saldanha & Nicolson, Fl. Hassan 660. 1976; Sharma *et al.*, Fl. Karnataka 304. 1984; Keshava Murthy & Yoganarasimhan, Fl. Coorg. 500. 1990. *C. mercarensis* Hochst. ex Steud. var. *major* Steud. ex Clarke in Hook.f., Fl. Brit. India 6: 719. 1894; Cooke, Fl. Pres. Bombay 2: 906. 1908 (3: 421. 1958, repr.ed.). *C. filicina* Nees var. *glaucina* (Boeck.) Kuekenth. in Engl., Pflanzenr. 4 (20), Heft 38: 274. 1909. *C. lindleyana* Nees var. *major* (Steud.) Fischer in Gamble, Fl. Pres. Madras 1687. 1931 (3: 1169. 1957, repr.ed.).

Illus. : Matthew, Furth. Illus. Fl. Tamilnadu Carnatic Pl. 624. 1988.

It differs from the typical variety by the slightly broader leaves, longer spikelets and prominently bifid tip on the beak of the utricle.

Fls. & Frts. : Oct.- Feb.

Habitat : Same as in typical variety.

Distrib. : Endemic to Peninsular India. KARNATAKA: Bangalore, Belgaum, Dakshina Kannada, Hassan, Kodagu, Mysore (Sharma *et al., l.c.*), Shimoga, Uttara Kannada.

Specimens examined : Bangalore: Magodi estate, Sableshpur, *Saldanha* 10042, 6.10.1965 (JCB); Shiradi ghat, *Saldanha* 11630, 9.2.1968 (JCB). Belgaum: Jamboti Reserve Forest, *Prasad* 172886, 15.12.1994 (BSI); Dudwawada, R.F., Londa, *Prasad* 172893, 17.12.1994 (BSI). Dakshina Kannada: Naravir R.F., without coll. name, 18141, 21.11.1927 (MH); Charmadi ghat, *Bhat* 544, 24.11.1978 (MGH); Kodachadri, *Bhat* 590, 17.11.1978 (MGH). Hassan: Bannuhalla, *Saldanha* 15300, 10.10.1969 (JCB); Hullahalli, *Jarret & Saldanha* 815, 6.10.1970 (JCB); Below Panorama Point, Shiradi, *Nicolson et al.* 2332, 25.10.1971 (JCB). Kodagu: Talakaveri, *Bhat* 810, 19.12.1980 (MGH); Kushalnagar, *Bhat* 1047, 21.9.1981 (JCB); Pushpagiri, *Bhat* 1122, 6.11.1981 (JCB); Abbe falls, Mercara, *Yoganarasimhan et al.* 4452, 21.11.1983 (RRCBI). Shimoga: Varahi, Hulical, *Raghavan* 83043, 8.10.1962 (BSI); Hulical ghat, *Raghavan* 83077, 9.10.1962 (BSI); Ghat Road, Agumbe, *Raghavan* 83170, 13.10.1962 (BSI); Yedur, *Raghavan* 90481, 17.11.1964 (BSI). Uttara Kannada: without exact locality, *Beddome s.n.* (Acc. No. 74037), 1866 (MH); Yellapur, *Talbot* 667, 21.9.1883 (BSI) and *s.n.* 25.9.1884 (BSI); Ambighat, *Talbot* 1321, 10.10.1885 (BSI); without exact locality, *Talbot* 1564, without date (BSI); Castle Rock, *Gammie* 15887, 24.10.1902 (BSI); 15765, 28.10.1902 (BSI); without exact locality, *Chibber s.n.* Nov. 1910 (BSI); Guddhali peak, Karwar, *Bell* 2117, Dec. 1920 (BLAT); Near Dudhsagar, *Fernandez* 543, 28.11.1949 (BLAT); Kurkeri, *Puri* 2001, 20.12.1957 (BSI); Caramuldi forests, near Castle Rock, *Kanodia* 93030, 14.11.1963 (BSI).

Note: Though it is treated as a distinct species by different workers, it seems there is no much difference between this and the typical *C. caricina*. The differentiating characters given above also seem to be variable. Hence it is best treated as a variety of *C. caricina*, as in this work here.

3. **Carex indica** L. subsp. **laete-brunnea** (Clarke) Koyama in Dassanayake & Fosberg, Rev. Handb. Fl. Ceylon 5: 374. 1985. *C. indica* var. *laete-brunnea* Clarke in Hook.f., Fl. Brit. India 6: 715. 1894. *C. dietrichiae* Boeck. in Flora 58: 122. 1875; Nelmes in Reinwardtia 1: 273. 1951; Karthik. *et al.*, Fl. Ind. Enum. Monocot. 36. 1989. *C. indica sensu* Hooper in Saldanha & Nicolson, Fl. Hassan 660. 1976; Sharma *et al.*, Fl. Karnataka 304. 1984.

Type: Sri Lanka.

Tufted perennials with short, woody rhizome, up to 80 cm high. Stems erect, trigonous, with flat faces, 1-2.5 mm thick below the middle portion, smooth. Leaves radical and subradical, at times one cauline, long-acuminate at apex, 5-10 mm wide. Basal sheaths brownish or with blackish nerves on the back, membranous on ventral side, ultimately disintegrating to fibres. Inflorescence a compound, rather slender panicle, branched into several sessile spikes; secondary panicles 3-8, distantly arranged on long peduncles, single from each node, erect, oblong or oblong-lanceolate, 2-7 cm long, 1-3.5 cm broad; upper ones approximate to rather distant, short-peduncled. Bracts leafy, usually overtopping the inflorescence, with sheathing base; orifice membranous, brown. Spikes numerous, sessile, androgynous, cylindric, 5-15 mm long, male and female parts almost equal in length. Staminate part slender. Male glumes oblong-ovate, obtuse to acute at apex, *ca* 4 x 1.2 mm, membranous. Stamens 3; filaments elongate up to 3.5 mm. Female glumes ovate or oblong-ovate, truncate to obtuse or subacute at apex, 2.5-3 x 1-1.5 mm, brownish, usually glabrous. Utricle trigonous, ellipsoid or obovoid-ellipsoid, beaked at apex, 3.5-5 x 1.2-2 mm. Nut prominently trigonous with concave faces, ellipsoid, beaked at apex, with a short stipe, *ca* 2.5 x 1.2 mm, ultimately dark brown. Stigmas 3.

Fls. & Frts. : Sept., Dec.

Habitat: Shady forest clearings in ghats; rare.

Distrib. : Sri Lanka, Thailand, Malesia, Micronesia, Queensland. INDIA: Probably restricted to Southern India. KARNATAKA: Chikmagalur, Hassan (Hooper, *l.c.*).

Specimen examined : Chikmagalur: Santaveri, *Talbot* 3076, 6.9.1893 (BSI).

Note: From the description provided by Hooper, *l.c.* it is assumed that the specimen mentioned belongs to the subspecies *laete- brunnea*. Probably in South India *C. indica* L. is represented by this subspecies only.

4. **Carex lindleyana** Nees in Wight, Contr. Bot. India 121. 1834; Clarke in Hook.f., Fl. Brit. India 6: 721. 1894; Fischer in Gamble, Fl. Pres. Madras 1686. 1931 (3: 1169. 1957, repr.ed.); Sharma *et al.*, Fl. Karnataka 304. 1984; Koyama in Dassanayake & Fosberg, Rev. Handb. Fl. Ceylon 5: 379, f. 31. 1985; Karthik. *et al.*, Fl. Ind. Enum. Monocot. 38. 1989.

Type : Peninsular India.

Illus. : Koyama, *l.c.*

Perennials, often loosely tufted. Rhizome short, creeping or sub-erect, covered with reddish-brown scales and their fibrous remnants. Stems erect, 20-70 cm high, 0.8-2 mm thick near the middle portion, distantly 1-3 noded. Leaves mostly basal, 1-3 upper ones distantly arranged on the stem, linear, gradually narrowed to a subacute apex, 20-35 cm long, 5-7.5 mm wide; sheaths of cauline leaves 3-8 cm long; basal sheaths reddish- brown. Inflorescence paniculate, bearing up to 5 secondary panicles, up to 20 cm long. Secondary panicles usually solitary, ovoid or oblong-ovoid, 1.5-4 cm long, 1-1.5 cm thick, densely bearing many spikes in spicate or paniculate manner; upper ones subsessile or on short peduncles; lower ones on long hispid peduncles. Bracts leafy, shorter than inflorescence; upper most not sheathing. Spikes androgynous, 5-11 mm long, staminate part longer than pistillate part. Male glumes lanceolate, acuminate at apex, usually short awned, *ca* 4 x 1 mm, membranous. Stamens 3; anthers linear-oblong, *ca* 1.5 mm long. Female glumes ovate-lanceolate, gradually narrowed to an acuminate apex, *ca* 3 x 1.3 mm, membranous, brown in the middle portion, pale towards margins; keel 1-nerved, often ending in a short awn. Utricle obtusely trigonous, ovate-elliptic or elliptic, narrowed to a beak at apex, 3.5-4 x *ca* 1.2 mm; beak bifid at apex. Nut triquetrous, broadly elliptic, obtuse at both ends, 2.2-2.5 x 1-1.3 mm, dark brown when mature. Style-base not thickened; stigmas 3.

Fls. & Frts. : Not known from Karnataka.

Habitat : Wet grasslands in high ranges.

Distrib. : Sri Lanka. INDIA: South India. KARNATAKA: Chikmagalur and Shimoga (Sharma *et al.*, *l.c.*)

Note : This species is included on the authority of Sharma *et al.*, *l.c.* Specimens from Karnataka were not found in BSI. Hence description is based on specimens from Tamilnadu.

5. **Carex maculata** Boott in Trans. Linn. Soc. 20:128. 1846; Clarke in Hook.f., Fl. Brit. India 6: 735. 1894; Kukenth. in Engl., Pflanzenr. 4 (20), Heft 38: 427. 1909; Fischer in Gamble, Fl. Pres. Madras 1686. 1931 (3: 1169. 1957, repr.ed.); Nelmes in Kew Bull. 1950: 205. 1950 *et* in Reinwardtia 1: 414. 1951; Hooper in Saldanha & Nicolson, Fl. Hassan 660. 1976; Rao & Verma, Fl. NE India 80. 1982; Sharma *et al.*, Fl. Karnataka 304. 1984; Koyama in Dassanayake & Fosberg, Rev. Handb. Fl. Ceylon 5: 396. 1985; Karthik. *et al.*, Fl. Ind. Enum. Monocot. 38. 1989. **Fig. 5.**

Type : Sri Lanka, *Thwaites c.p.* 2630.

Illus. : Boott, Illus. Carex. 1: 9, t.26. 1858.

Tufted perennials with short, woody rhizome. Stems erect, obtusely trigonous, 1.2-2.3 mm thick below, smooth. Leaves mostly subradical, 1-few on lower part of the stem and spaced, shorter than to nearly equalling the stem, linear, acute at apex, 2-8 mm wide; sheaths ferrugineous or pale with dark red spots. Spikes 3- 5, erect, upper 3-4 subfastigiate, subsessile or short-peduncled; lowest slightly spaced, long-peduncled. Bracts leafy; lower 2 overtopping the inflorescence, sheathing at base; upper most bract setaceous, hardly sheathing at base. Terminal spike staminate, linear-cylindrical, 1.5-4.5 cm long, 1-2 mm thick, light brown when dry; male glumes oblong, acute at apex, *ca* 3.5 x 1.5 mm, membranous. Stamens 3; filaments up to 3.5 mm long. Lateral spikes pistillate, rarely with a short staminate part at apex, cylindrical, 1-4.5 cm long, 2.5 - 4 mm thick, densely many-flowered, dark brown when dry. Female glumes oblong-ovate, obtusish to acute at apex, *ca* 2 x 1 mm, not membranous, semitranslucent; midvein and sideveins not usually extending to apex. Utricle trigonous or compressed to planoconvex, broadly ellipsoid, beaked at apex, 2.5-3 x *ca* 1.5 mm, faintly to strongly nerved on faces, dark brown when dry. Nut prominently trigonous, ellipsoid or obovoid, subobtuse and with a minute beak at apex, minutely stipitate at base, 1.5-1.8 x *ca* 1 mm; style slightly thickened at base; stigmas 3.

Fls. & Frts. : May.

Habitat : Wet areas in high range forests.

Distrib. : Sri Lanka, China, Malesia, Japan, Formosa. INDIA: South India and North East India. KARNATAKA: Hassan (Hooper, *l.c.*).

Note : This species is included on the authority of Hooper, *l.c* .

6. **Carex speciosa** Kunth, Enum. Pl. 2: 504. 1837; Clarke in Hook.f., Fl. Brit. India 6: 729 1894; Cooke, Fl. Pres. Bombay 2: 906. 1908 (3: 421.

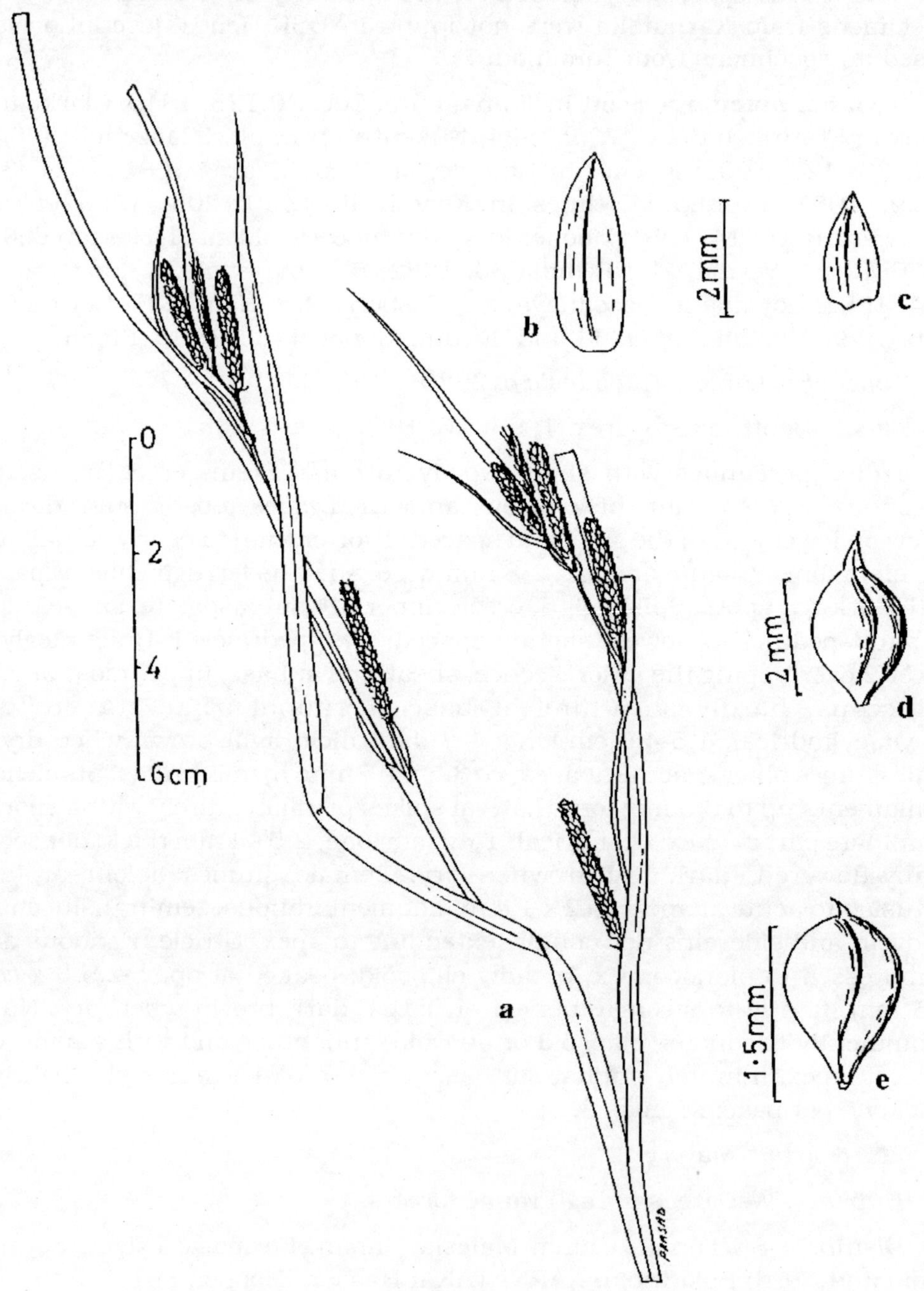

Fig. 5. *Carex maculata* Boott
a. Habit, **b.** Male glume, **c.** Female glume, **d.** Utricle, **e.** Nut

1958, repr.ed.); Fischer in Gamble, Fl. Pres. Madras 1686. 1931 (3: 1168. 1957, repr.ed.); Nelmes in Reinwardtia 1: 390. 1951; Rao & Verma, Cyp. NE India 79. 1982; Sharma *et al.*, Fl. Karnataka. 304. 1984; Karthik. *et al.*, Fl. Ind. Enum. Monocot. 42. 1989. **Fig. 6.**

Type : India, *Wallich* 3391.

Illus. : Boott, Illus. 1: 53. 1858; Matthew, Furth. Illus. Fl. Tamilnadu Carnatic Pl. 629. 1988.

Tufted perennials with short, woody rhizome, 18-50 cm high. Stems erect, slender, prominently trigonous with narrowly winged angles, 1-1.5 mm thick, more or less smooth or finely scaberulous on the angles. Leaves mostly subbasal, usually longer than stem, 3-7 mm wide, gradually narrowed to the long-attenuated trigonous apex; sheaths membranous on the ventral side, lower most sheaths splitting into fibres; ligule membranous. Inflorescence usually of one terminal and 1-2 axillary spikes (according to Nelmes 1-3 (-4) at nodes). Peduncles trigonous with acute angles, smooth. Bracts foliaceous, usually up to the terminal spike or overtopping, sheathing; bract of terminal spike glumiform. Spikes androgynous, subcylindric or slightly compressed, 1.5-3 cm long; female part usually longer than male part. Male part 6-11mm long, 1.5-2 mm thick, acute at apex; glumes densely packed, oblong or oblong-ovate, obtuse at apex, muticous, 2.5-3 x 1-1.6 mm, with a prominent midvein. Stamens 3; filaments united except at apex; anthers oblong, 0.7-0.8 mm long. Female part 10-18 mm long, 2-3 mm thick; glumes subdensely arranged, broadly oblong-ovate, obtuse or subacute at apex, muticous, *ca* 3 x 2 mm, whitish-stramineous, with a prominent midrib and many veined sides, with minutely erose-ciliolate margins. Style 1.2-1.8 mm long, trigonous, pyramidally thickened at base, persistent on the nut; stigmas 3, shorter than style. Utricle trigonous, ellipsoid, gradually narrowed to base, hardly stipitate, beaked at apex, 4.5-5 x 1.5-2 mm, coriaceous, slightly inflated, hispidulous especially on the margins, many nerved; beak *ca* 0.5 mm long, shortly bilobed at apex. Nut prominently triquetrous, obovoid- ellipsoid, *ca* 3.5 x 1.8 mm, stipitate at base, minutely beaked at apex.

Fls. & Frts. : Oct. & Apr.

Habitat : Humid soil in moist forest floors.

Distrib. : Indo-China and Malesia. INDIA: Peninsular India, Central and Eastern India, North-East and Eastern to Central Himalaya. KARNATAKA: Uttara Kannada.

Specimens examined : Uttara Kannada: Hugli, *Talbot* 2282, 10.10.1890 and 3142, 28.4.1894 (BSI).

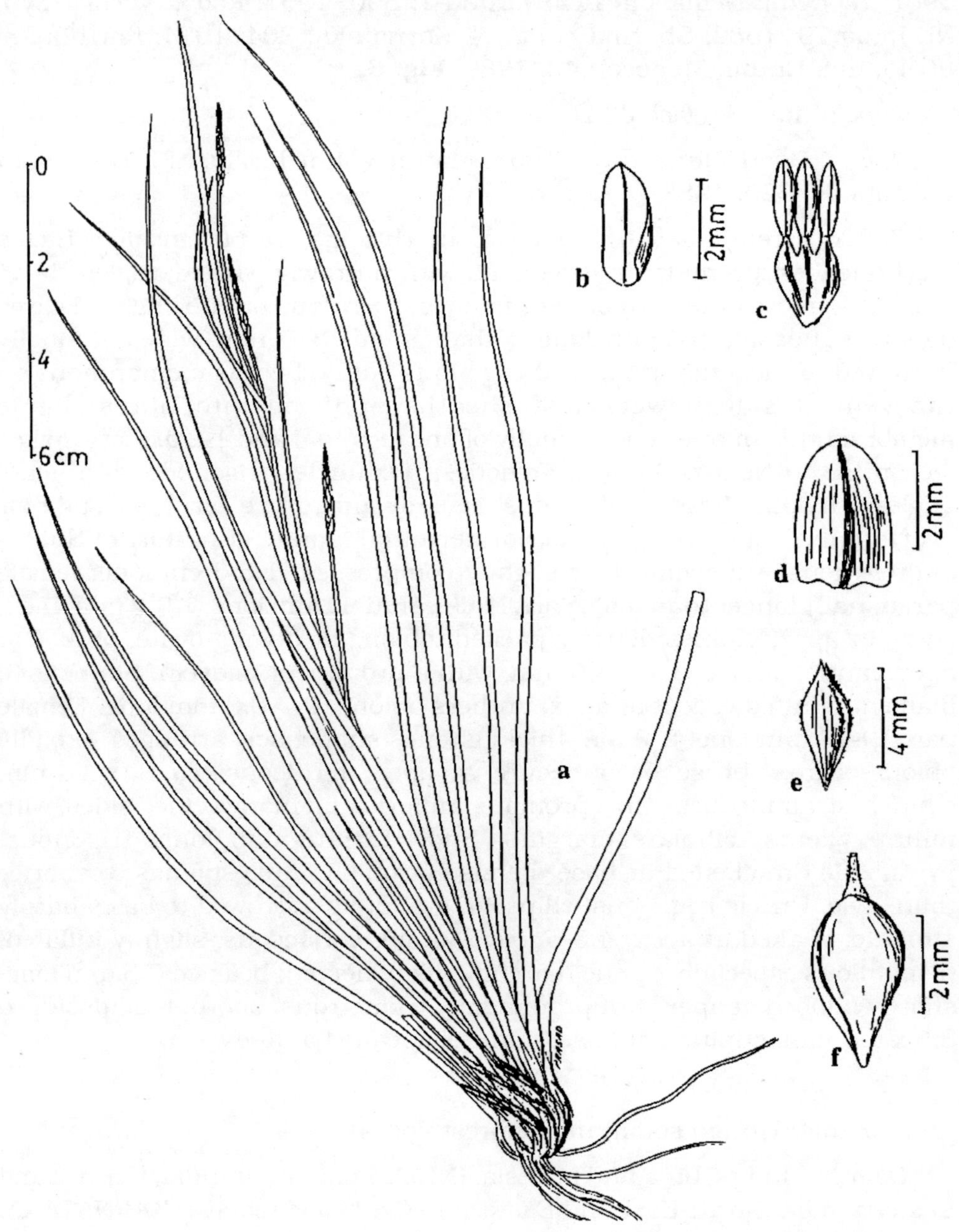

Fig. 6. *Carex speciosa* Kunth
a. Habit, **b.** Male glume, **c**. Stamens, **d.** Female glume, **e.** Utricle, **f.** Nut

5. COURTOISINA

J. Sojak in Cas. Nar. Muz. (Prague) 148: 193. (1979) 1980. *Courtoisia* Nees in Linnaea 9: 286. 1834; Bentham & Hooker, Gen. Pl. 3: 1046. 1880; Santapau & Henry, Dict. Fl. Pl. India 46. 1973, non *Courtoisia* Marchand, Bijdr. Nauark. Wetensch. 5: 191. 1830. *Indocourtoisia* Bennet & Raizada in Indian For.107:432. 1981.

Glabrous annuals with erect stems. Leaves basal, grass-like. Inflorescence an umbel of globose spikes. Spikelets compressed, few-flowered; rachilla scarcely winged, disarticulating at the base. Glumes distichous, persistent, lowest 2 empty; 1-3 above them bisexual, nut-bearing, winged on the keel; uppermost sterile. Perianth structures absent. Stamens 3, linear. Style shorter than stamens, persistent; stigmas 3, linear, longer than style. Fruit a trigonous nut, acute at apex.

2 species distributed in tropical Africa and Madagascar. One in India.

Courtoisina cyperoides (Roxb.) J. Sojak in Cas. Nar. Muz. (Prague) 148: 193. (1979) 1980. *Kyllinga cyperoides* Roxb., Fl. Ind. 1: 187. 1820. *Courtoisia cyperoides* (Roxb.) Nees in Linnaea 9: 286. 1834; Clarke in Hook.f., Fl. Brit. India 6: 625. 1893; Cooke, Fl. Pres. Bombay 2: 877. 1908 (3: 391. 1958, repr.ed.); Fischer in Gamble, Fl. Pres. Madras 1645. 1931 (3: 1143. 1957, repr.ed.); Hooper in Saldanha & Nicolson, Fl. Hassan 661. 1976; Sharma *et al.*, Fl. Karnataka 304. 1984; Singh, Fl. E. Karnataka 2: 628. 1988. *Cyperus pseudokyllingoides* Kuekenth. in Engl., Pflanzenr. 4 (20), Heft 101: 501. 1936; Ramaswamy & Razi, Fl. Bangalore 92. 1973; Rao & Verma, Cyp. NE India 21, f. 8-8a. 1982. *Indocourtoisia cyperoides* (Roxb.) Bennet & Raizada in Ind. For. 107 : 432. 1981; Karthik. *et al.*, Fl. Ind. Enum. Monocot. 57. 1989; Keshava Murthy & Yoganarasimhan, Fl. Coorg 510. 1990.*Bilsanna jambu hullu.* **Fig. 7.**

Illus. : Rao & Verma, *l.c.*

Tufted annuals with fibrous roots, 15-75 cm high. Stems slender, trigonous, 1.2 - 3 mm thick, glabrous. Leaves often as long as or exceeding the stem, linear, gradually narrowed to an acute apex, 2-6 mm broad; sheaths somewhat loosely enclosing the stem; lowest 1 or 2 leaves often reduced to bladeless sheaths. Inflorescence compound to decompound, up to 13 cm across. Involucral bracts 3-7, leafy, rather distantly placed, spreading, very long (except the uppermost 2-3); lowest up to 40 cm long. Primary rays 3-10, slender, suberect or reflexed; lowest 3-10 cm long. Secondary rays 2-5, up to 3 cm long. Spikes subglobose, 5-10 mm across, greenish-yellow, brownish when dry; spiklelets many, congested, strongly compressed, elliptic or oblong-elliptic, 3-4.5 x 1.2-1.5 mm, 1-2-flowered. Glumes elliptic, mucronate at apex, 3-4 mm long, membranous towards margins; keel strong, broadly winged, yellow. Stamens 3; filaments elongate

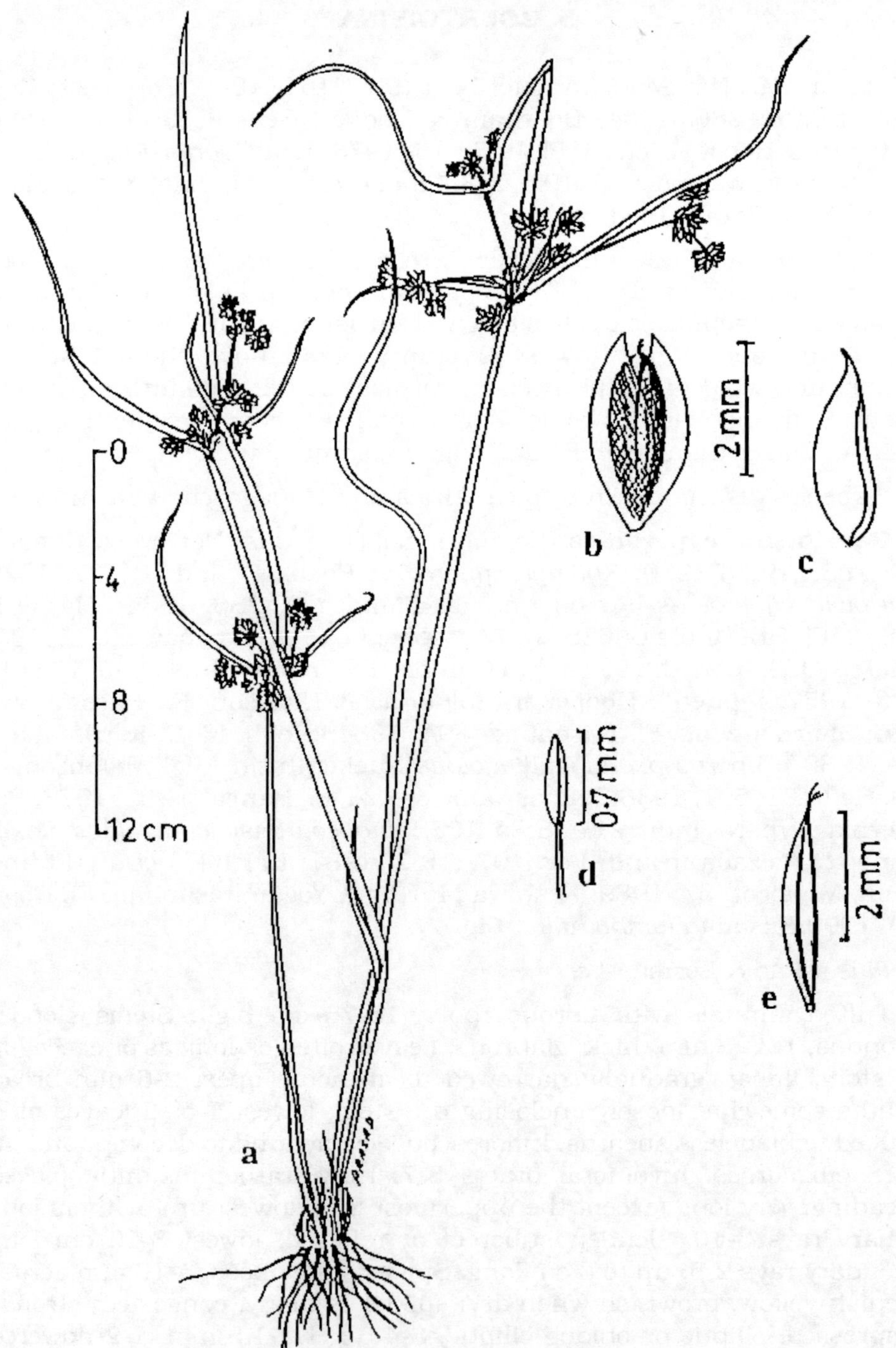

Fig. 7. *Courtoisina cyperoides* (Roxb.) J. Sojak
a. Habit, **b.** Spikelet, **c.** Glume, **d**. Stamen, **e**. Nut

up to 3 mm; anthers oblong, *ca* 0.7 mm long. Style *ca* 0.4 mm long; stigmas 3, *ca* 1 mm long. Nut trigonous, linear-oblong, narrowed to both ends, *ca* 3 x 0.7 mm, puncticulate, ultimately dark brown.

Fls. & Frts. : Oct. - May.

Habitat : A common weed in rice fields, along the bunds of rice fields and lakes, in swampy areas, wet sandy or muddy soil, in puddles along roadside and edges of ponds or canals or streams.

Distrib. : Tropical Africa, Madagascar. INDIA: Throughout (except North-West). KARNATAKA: Bangalore, Belgaum, Bellary, Chikmagalur, Dharwar, Hassan, Kodagu, Kolar, Mandya, Mysore (Fischer, *l.c.*), Shimoga, Uttara Kannada.

Specimens examined : Bangalore: Sringarpura, *Hooper & Saldanha* 18073, 18.11.1971 (JCB). Belgaum: Sutagatti, *Mahajan* 24971, without date (BSI); Manjapur Nallah, Khanapur, *Prasad* 172853, 14.12.1994 (BSI); Gottni Nallah, Jambotti R.F., *Prasad* 172880, 15.12.1994 (BSI); Dudwad village near Londa R.F., *Prasad* 172899, 17.12.1994 (BSI). Bellary: Kudligi-Sandur Road, *Sreenath & Ramesh* 4283, 14.11.1978 (JCB). Chikmagalur: Hilikeri village, along Sringeri Road, Koppa taluk, *Prasad* 173815, 7.11.1995 (BSI). Dharwar: Salikkinikuppa lake, Dharwar, *Prasad* 172921, 18.12.1994 (BSI). Hassan: Outskirts of Hassan town, *Saldanha* 16089, 22.1.1970 (JCB); Malvinakulu, Hanbal Road, Hassan, *Hooper & Gandhi* 2472, 13.11.1971 (JCB); Bourdalboore state forest, Hassan, *Saldanha et al.* 5462, 26.12.1978 (JCB). Kodagu: Nagarhole, *Arora* 46280, 5.1.1959 (BSI); Kottamudi, Napoklu, *Rao* 74952, 5.10.1961 (BSI); Kambile, Saunticoppa, *Rao* 86103, 3.3.1963 (BSI); Kadakare, Mercare, *Bhat* 729, 18.12.1980 (MGH); Kirugoor, *Bhat* 931, 26.1.1981 (MGH). Kolar: Sidlaghatta-Chintamani Road 14th km, *Singh* 142036, 3.1.1976 (BSI). Mandya: Narayana Surya, *Dinesh* 713, 4.12.1984 (MGH). Shimoga: Varahi falls, Hulical, *Raghavan* 80806, 20.5.1962 (BSI). Uttara Kannada: Yellapur, *Talbot* 1563, 10.10.1884 (BSI); Ekambi, *Saldanha* 8653, 24.10.1962 (JCB); Ulletikeri, Halyal, *Prasad* 172995, 29.10.1995 (BSI).

6. CYPERUS

L., Sp. Pl. ed. 1, 1: 44. 1753; Gen. Pl. ed. 5, 26. 1754. *Anosporum* Nees in Wight, Contr. Bot. India 70. 1834. *Sorostachys* Steud. in Flora 33: 229. 1850 *et* Syn. Fl. Glum. 2: 71. 1854. *Juncellus* (Griseb.) Clarke in Hook. f., Fl. Brit. India 6:594. 1893.

Type: Cyperus esculentus L.

Annual or perennial sedges, often rhizomatous or stoloniferous. Stems tufted or solitary, triquetrous, trigonous or sometimes subterete, usually leafy at base. Leaves tristichous, narrowly linear, grass-like, at times reduced to sub-phyllous sheaths; lower leaves often scale-like, covering base of the stem and the rhizome. Inflorescence terminal, usually anthelate, simple to decompound, sometimes capitate. Involucral bracts leaf- like, not sheathing. Spikelets sessile, prophyllate at base, usually compressed, bearing few to many glumes; rachilla persistent, often winged. Glumes distichous, usually 2 basal ones empty. Flowers bisexual; upper most often staminate or barren. Hypogynous scales or bristles absent. Stamens 3, 2 or 1, with connective of the anthers often produced into an apical appendage. Style continuous with the ovary, not jointed at base, 3-fid or 2-fid, rarely almost undivided. Nut trigonous, triquetrous or lenticular.

A cosmopolitan genus with *ca* 500 species, mostly distributed in tropics. *Ca* 70 species in India, 38 in Karnataka.

Literature : **KOYAMA, T.** (1961) The genus Cyperus in Eastern Asia, in *Quart. Journ. Taiwan Mus.* 14: 163. **KUKENTHAL, G.** (1935-1936) Scirpoideae - Cypereae, in: Engler, *Pflanzenr.* 101 (IV. 20): 1-671. **MERCADO, B. L.** (1979) A monograph on *Cyperus rotundus* L. in *Biotrop. Bull.* No. 15: 1-63.

Key to the species

1a. Stigmas 2 (rarely 3 in some flowers); nut planoconvex or biconvex (rarely trigonous when stigmas 3) 2

1b. Stigmas always 3 (usually with undivided style in *C. cephalotus* and *C. meeboldii*); nut always trigonous or triquetrous (if planoconvex, with an angular dorsal surface) 4

2a. Stems stout, 5-10 mm thick; inflorescence compound to decompound with well developed rays and cylindrical spikes of spikelets; rachilla winged . . . . **1. C. alopecuroides**

2b. Stems narrow or slender, 0.5-2 mm thick; inflorescence a contracted head of glomerules or a pseudolateral cluster of spikelets; rachilla wingless 3

3a. Annuals with tufted, trigonous stems; leaves well developed; inflorescence a terminal head of glomerules; stamens 1 or 2 . . . . **33. C. pygmaeus**

3b. Perennials with cylindrical stems in a row along the creeping rhizome; leaves reduced to papery sheaths; inflorescence a pseudolateral cluster of spikelets; stamens 3 . . . . **20. C. laevigatus**

4a. Style entire or obscurely 3-lobed; nut with the stipe developed into corky ridges which cover lower part of the nut . . . . **8. C. cephalotus**

4b. Style always divided into 3 stigmas; nut without corky ridges at base 5

5a. Spikelets spicately arranged (subdigitate in *C. compressus* and *C. macer*) . 6

5b. Spikelets digitately or stellately arranged 25

6a. Rachilla narrowly to broadly winged (at times caducous) 7

6b. Rachilla wingless or hardly winged 22

7a. Stems slender (0.5-) 1-2 (-2.5) mm thick 8

7b. Stems stout (2-) 3-10 mm thick 15

8a. Spikelets 3-5 mm broad (including the mucro) . . . . **9. C. compressus**

8b. Spikelets 1-2.5 mm broad 9

9a. Spikelets subquadrangular (in cross section); nut triquetrous, *ca* 2 mm long . . . . **37. C. tenuiculmis**

9b. Spikelets flattened or subcompressed; nut trigonous or planoconvex, 1-1.8 mm long 10

10a. Nut planoconvex, often with a concave ventral surface . . . . **36. C. stoloniferus**

10b. Nut trigonous, not with concave ventral surface. 11

11a. Stems tufted; leaves much shorter than stem; involucral bracts shorter, never overtopping the inflorescence; spikelets subdigitate . . . **22. C. macer**

11b. Stems usually solitary, arising from a tuber, at times few together; leaves as long as stem or longer, at times shorter; involucral bracts longer, at least the lowest usually overtopping the inflorescence; spikelets distinctly spicate . 12

12a. Rachilla straight; nut oblong to ovate- oblong, 1.2-1.3 mm long . . . . **25. C. mitis**

12b. Rachilla slightly or prominently flexuous; nut obovoid, ellipsoid or oblong-obovoid, 1.5-1.8 mm long 13

13a. Tubers bulb-like, clothed with a coriaceous coat which split into blackish segments (initially whitish); involucral bracts distinctly spaced from each other; glumes 9-11-nerved . . . . **6. C. bulbosus**

13b. Tubers not bulb-like, covered with membranous scales when young which disintegrate into fibres; involucral bracts not distinctly spaced (lower most often a little distant); glumes 5-7-nerved 14

14a. Stolons very slender, yellowish, disappearing after tuber formation; spikelets usually golden yellow to pale brown; rachilla slightly flexuous . . . . **15. C. esculentus**

14b. Stolons somewhat stout, dark brown, persistent; spikelets usually dark brown, rarely yellowish; rachilla prominently flexuous . . . . **34. C. rotundus**

15a. Stems terete at least below; leaves reduced to sheaths (upper 1 or 2 at times with a short blade in *C. corymbosus*) 16

15b. Stems trigonous to triquetrous; leaves usually well developed (lower ones often reduced to sheaths) 17

16a. Stems transversely septate; spikelets 1.5-1.8 mm broad . . **5. C. articulatus**

16b. Stems not septate; spikelets *ca* 1 mm broad . . . . **10. C. corymbosus**

17a. Stems deeply triquetrous, almost 3- winged; nut narrowly oblong .... **23. C. malaccensis**

17b. Stems not deeply triquetrous, never 3-winged; nut oblong to oblong obovoid 18

18a. Rhizome long-creeping; stems usually in a row along the creeping rhizome; leaves usually reduced to sheaths .... **28. C. pangorei**

18b. Rhizome short, not creeping; stems tufted or solitary; leaves usually well developed 19

19a. Spikelets suberect; nut triquetrous .... **27. C. nutans**

19b. Spikelets ultimately right angle to the rachis; nut trigonous 20

20a. Spikes broadly ovoid-pyramidal; rachilla flexuous, with *ca* 1 mm long internodes; nut oblong, 1.2-1.5 mm long, ultimately dark brown .... **14. C. distans**

20b. Spikes cylindrical; rachilla straight with 0.4 - 0.5 mm long internodes; nut ellipsoid or oblong-obovoid, 0.7-1 mm long, yellowish-brown 21

21a. Spikelets slightly compressed; glumes elliptic, acute at apex; nut *ca* 1 mm long .... **13. C. digitatus**

21b. Spikelets flattened; glumes ovate to broadly ovate, mucronate at apex; nut 0.7-0.8 mm long .... **16. C. exaltatus**

22a. Annuals, without stolons; stems tufted; stamens 2 (rarely 3 in *C. iria*); style minute, inconspicuous; nut 1.5-2 mm long 23

22b. Perennials, with long stolons; stems usually solitary; stamens 3; style 1.2 - 3 mm long; nut 1-1.5 mm long 24

23a. Rachis of spikes glabrous; spikelets *ca* 2 mm broad; rachilla internodes 0.5 - 0.8 mm long; glumes 1.3-1.5 x *ca* 1.5 mm, not spinulose-ciliate on the keel .... **19. C. iria**

23b. Rachis of spikes hispid on angles; spikelets 2.5 - 3 mm broad; rachilla internodes *ca* 1 mm long; glumes *ca* 2 x 2 mm, often with spinulose ciliate keel .... **2. C. alulatus**

24a. Inflorescence compound; rachis of the spikes usually densely hispidulous; spikelets 1.5-2 mm broad; anthers *ca* 0.5 mm long .... **29. C. pilosus**

24b. Inflorescence simple or subcompound; rachis of the spikes smooth; spikelets 2.5 - 3.5 mm broad; anthers 1.5 - 2 mm long ... **31. C. procerus**

25a. Inflorescence contracted to a head of few to numerous spikelets 26

25b. Inflorescence open, simple, compound or decompound (rarely reduced to a single head in *C. castaneus* and *C. cuspidatus*) 30

26a. Stamen 1 27

26b. Stamens 3 28

27a. Spikelets oblong; glumes with the keel excurrent into a conspicuous mucro; style undivided or with 3 stigmas .... **24. C. meeboldii**

27b. Spikelets elliptic or ovate; glumes without mucro; style always divided into 3 stigmas .... **32. C. pulchellus**

28a. Annuals with fibrous roots; stems tufted; glumes 2-3 mm long; anthers *ca* 0.5 mm long; nut *ca* 1 mm long .... **35. C. rubicundus**

28b. Perennials with long-erect or horizontally creeping rhizome; stems usually solitary or closely in a row; glumes 4-5 mm long; anthers 1.2 - 2 mm long; nut 1.7-2 mm long 29

29a. Rhizome long-creeping or erect, slender; stems not with swollen base; anthers *ca* 2 mm long; stigmas 2-3 mm long; nut planoconvex with an angular dorsal surface .... **4. C. arenarius**

29b. Rhizome short-creeping, thick, woody; stems with swollen base; anthers *ca* 1.2 mm long; stigmas 1-1.5 mm long; nut triquetrous ...... **26. C. niveus**

30a. Annuals with fibrous roots 31

30b. Perennials with short rhizome 36

31a. Spikelets 2-5 mm broad (including the mucro) 32

31b. Spikelets 1-1.5 mm broad 34

32a. Rachilla winged, with *ca* 0.7 mm long internodes; glumes 3.5 - 4 mm long; anthers *ca* 1 mm long; nut 1.4 - 1.7 mm long .... **9. C. compressus**

32b. Rachilla wingless, with *ca* 0.3 mm long internodes; glumes 1- 2 mm long; anthers *ca* 0.2 mm long; nut 0.5-1 mm long 33

33a. Stamen 1; nut oblong with parallel sides .... **7. C. castaneus**

33b. Stamens 2 or 3; nut obovoid to oblong-obovoid .... **11. C. cuspidatus**

34a. Nut obtuse-trigonous, broadly ovoid or subglobose, *ca* 0.3 mm long .... **38. C. tenuispica**

34b. Nut trigonous or triquetrous, oblong-obovoid or elliptic-obovoid, 0.5 - 0.7 mm long. 35

35a. Spikes globose, greenish-brown to blackish, bearing numerous spikelets; glumes orbicular or broadly obovate, 0.5-0.9 mm long; stamens 2(rarely 1?); style very short, *ca* 0.15 mm; nut triquetrous .... **12. C. difformis**

35b. Spikes hemispheric, orange-brown or ferrugineous, bearing 3-16 spikelets; glumes broadly oblong-ovate or elliptic-oblong, *ca* 1.1 mm long; stamen 1; style *ca* 0.6 mm; nut trigonous .... **3. C. amabilis**

36a. Rachilla winged 37

36b. Rachilla wingless 38

37a. Rhizome slender, creeping; inflorescence usually simple; involucral bracts shorter than inflorescence; glumes remote; nut trigonous, oblong-obovoid .... **22. C. macer**

37b. Rhizome short, woody; inflorescence decompound; involucral bracts atleast the lowest, overtopping the inflorescence; glumes densely imbricate; nut triquetrous, ellipsoid .... **21. C. laxus**

38a. Stems trigonous to subterete; involucral bracts many, distinctly spaced, somewhat equal in length .... **18. C. involucratus**

38b. Stems triquetrous; involucral bracts 2-8, not distinctly spaced, unequal in length 39

39a. Stems 1-2.5 mm thick; leaves not septate-nodulose; glumes 1-1.3 mm long; stamen 1; nut not with thickened corky angles .... **17. C. haspan**

39b. Stems 3-8 mm thick; leaves septate-nodulose; glumes 2-2.5 mm long; stamens 3; nut with thickened corky angles .... **30. C. platystylis**

1. **Cyperus alopecuroides** Rottb., Descr. Pl. Rar. Progr. 20. 1772 & Ic. Rar. Nov. Pl. 38, t. 8, f. 2. 1773; Cooke, Fl. Pres. Bombay 2: 860. 1908 (3. 373. 1958, repr.ed.); Kukenth. in Engl., Pflanzenr. 4 (20), Heft 101: 71. 1935; Ramaswamy & Razi, Fl. Bangalore 80. 1973; Kern in van Steenis, Fl. Males. 1, 7: 603, 1974; Hooper in Saldanha & Nicolson, Fl. Hassan 663. 1976; Rao & Razi, Fl. Mysore 557. 1981; Sharma *et al.*, Fl. Karnataka 304. 1984; Koyama in Dassanayake & Fosberg, Rev. Handb. Fl. Ceylon 5: 213. 1985; Singh, Fl. E. Karnataka 2: 629. 1988; Keshava Murthy & Yoganarasimhan, Fl. Coorg 501. 1990. *Juncellus alopecuroides* (Rottb.) Clarke in Hook.f., Fl. Brit. India 6: 595. 1893; Fischer in Gamble, Fl. Pres. Madras 1629. 1931 (3: 1133. 1957, repr.ed.); Karthik. *et al.*, Fl. Ind. Enum. Monocot. 57. 1989.

Type : Arabia, *Forskal.*

Illus. : Rottb., *l.c.;* Mahesh., Illus. Fl. Delhi f. 212. 1966.

Tufted perennials with short rhizhome, up to 150 cm high. Stems stout, trigonous, 5-10 mm thick, smooth. Leaves shorter than stem, broadly linear, gradually tapering to long-acuminate tip, 30-70 cm long, 5-17 mm wide, coriaceous, scabrous on margins and nerves, midrib prominent beneath, 2 lateral nerves strongly marked above; sheaths up to 25 cm long, cinnamomeous to fuscous. Inflorescence compound to decompound, up to 30 cm long, stramineous. Involucral bracts 4-7, lower 2-3 much overtopping the inflorescence, up to 65 cm long, and 10 mm wide. Primary rays 6-10, unequal, trigonous, up to 17 (-20) cm long, smooth; secondary rays short, up to 6 cm long. Spikes digitately arranged, cylindrical or oblong-cylindrical, 2-4 cm long, 6-15 mm broad, with densely arranged spikelets. Spikelets spicately arranged, ultimately spreading, ovate to oblong-lanceolate, 3-6 x 2-3 mm, 10-30-flowered; rachilla straight, flattened, very narrowly winged, persistent. Glumes membranous, broadly ovate to elliptic-ovate, apiculate at obtuse apex, 1.5-2 x 1-1.2 mm, not keeled, 5-7-nerved on the obtuse back; margins hyaline, yellowish or with purplish-brown stripes, ultimately inrolled. Stamens 2(-3); filaments up to 2 mm long; anthers linear-oblong, *ca* 0.5 mm long. Style 1-1.2 mm long; stigmas 2 (rarely 3). Nut dorsiventrally compressed, planoconvex (in trigynous flowers with a dorsal angle), elliptic to slightly obovate, shortly apiculate at apex, 0.8 - 1 x 0.5 - 0.6 mm, yellowish to light brown.

Fls. & Frts. : July - March.

Habitat : In and around shallow water of tanks, paddy fields, lakes etc; often partially submerged or on wet banks of water bodies.

Distrib. : Sri Lanka, Malesia, North Africa to tropical Africa, Madagascar, West Indies and Australia. INDIA: Throughout. KARNATAKA: Bangalore, Bidar, Bijapur, Chikmagalur (Sharma *et al., l.c.*), Dharwar, Gulbarga, Hassan, Kodagu, Mandya, Mysore (Rao & Razi, *l.c.*), Uttara Kannada.

Specimens examined : Bangalore: Bangalore, *Camaron* 593, 1.2.1891 (MH); Bangalore, *R.D.A.* B. 390, July 1918 (MH); Guttahalli, *Gandhi* 1, 20.9.1970 (JCB). Bidar: Malhar-Martant temple, Khanapur, *Singh* 117254, 8.2.1975 (BSI). Bijapur: Kendur Thalab, Badami, *Prasad* 172953, 21.12.1994 (BSI). Dharwar: Kalekeri village, near R.F., *Prasad* 172948, 19.12.1994 (BSI). Gulbarga: Gulbarga tank, *Singh* 138665, 24.3.1974 (BSI). Hassan: Road to Holenarsipur, *Saldanha* 16299, 21.1.1970 (JCB). Kodagu: Kushalnagar, *Bhat* 940, 16.2.1981 (MGH). Mandya: Ranganathittu, *Ahmed* 363, 4.3.1978 (JCB). Uttara Kannada: Vincholi, *Talbot s.n.* (Acc. No. 665), April 1884 (BSI); Devikope, *Talbot* 1064, 29.10.1884 (BSI); Yellapur, *Talbot* 1064A, 6.11.1884 (BSI).

2. **Cyperus alulatus** Kern in Reinwardtia 1: 463, f. 1. 1952; Sharma *et al.*, Fl. Karnataka 304.1984; Singh, Fl. E. Karnataka 2: 630. 1988; Karthik. *et al.*, Fl. Ind. Enum. Monocot. 43. 1989. *Cyperus iria* var. *rectangularis* Kuekenth. in Engl., Pflanzenr. 4 (20), Heft 101: 152. 1935. *Cyperus rectangularis* (Kukenth.) Bennet in Indian For. 95: 692. 1969. **Fig. 8.**

Type: Punjab, without exact locality, *Drummond* 24733 (Herb. Bogor.).

Illus. : Kern, *l.c.*

Annuals with fibrous roots, 20-35 cm high. Stems tufted, erect, triquetrous, 1-3 mm thick, smooth. Leaves few, shorter or longer than stem, long-acuminate towards apex, 1-5 mm wide, scabrous on upper margins, sheaths reddish-brown towards base. Inflorescence simple to subcompound, lax, 3-9 cm long. Involucral bracts 2-7, obliquely spreading, lower ones overtopping the inflorescence. Rays 2-9, unequal, up to 7 cm long; longer ones often paniculately branched to 2-5 spikes. Spikes ovate or oblong-ovate, 1-3 cm long, loosely or subdensely bearing 2-13 spikelets; rachis hispid on angles. Spikelets spicately arranged, compressed, spreading in right-angles, oblong to linear-oblong, obtuse at apex, 5-10 x 2.5-3 mm, 6-16-flowered; rachilla wingless, with *ca* 1 mm long internodes. Glumes somewhat loosely arranged, membranous, *ca* 1/3 imbricate, orbiculate, emarginate and mucronate at apex, *ca* 2 x 2 mm; keel prominent, 7-nerved, green, often spinulose-ciliate; sides yellowish-brown, single nerved. Stamens 2; filaments up to 2 mm long; anthers oblong to linear-oblong, 0.5 - 1 (-1.5) mm long. Style minute; stigmas 3, up to 8 mm

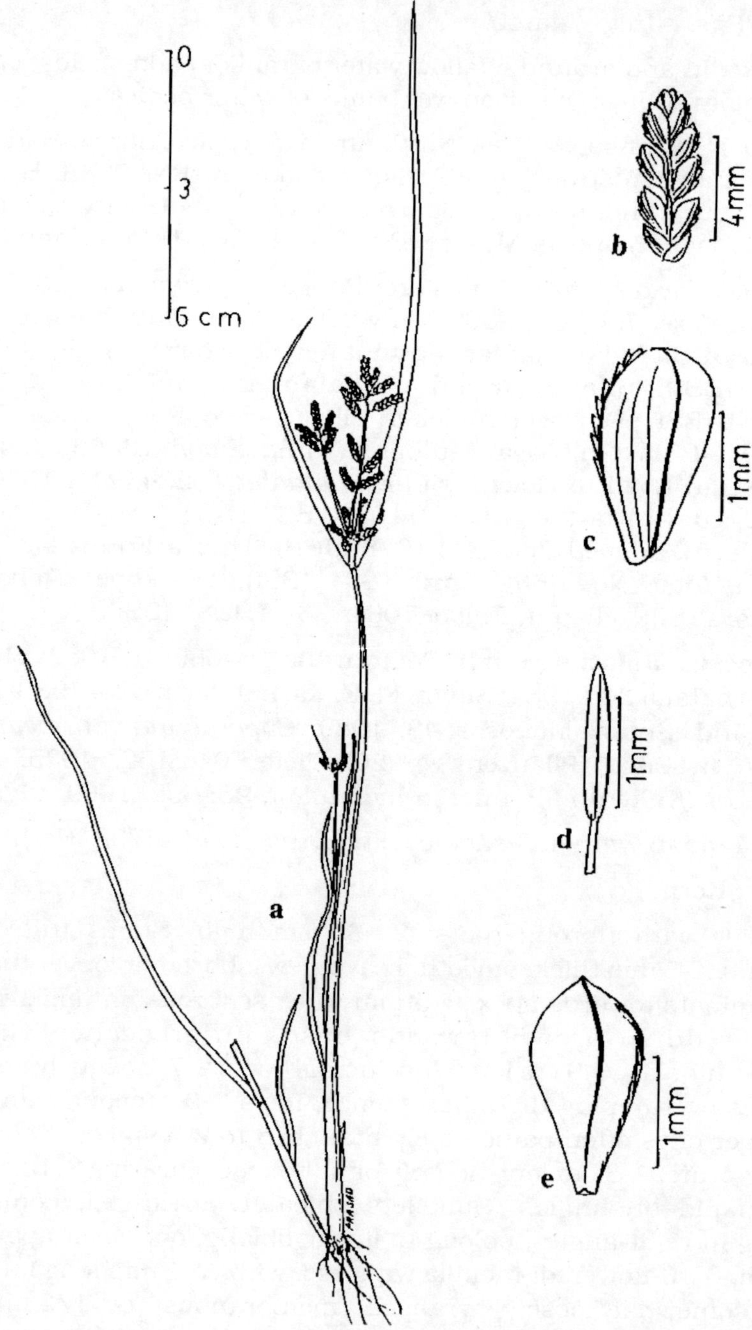

Fig. 8. *Cyperus alulatus* Kern
a. Habit, **b**. Spikelet, **c.** Glume, **d.** Stamen, **e.** Nut

long. Nut triquetrous, obovoid, with concave sides, mucronate at apex, 1.8-2 x *ca* 1 mm, dark brown when mature.

Fls. & Frts. : Sept. - Nov.

Habitat : Sandy soil near streams and canals.

Distrib.: Pakistan. INDIA: Probably throughout. KARNATAKA: Bijapur, Chikmagalur (Sharma *et al., l.c.*), Mysore, Shimoga.

Specimens examined : Bijapur: Badami, *Paranjpye s.n.* (Acc. No. 2464), 1.9.1912 (BSI). Mysore: Hirakulguda State Forest, *Wadhwa* 44587, 24.11.1957 (BSI); Kotayal, *Wadhwa* 44963, 1.10.1958 (BSI). Shimoga: Nalur, *Raghavan* 68039 A, 1.11.1960 (BSI).

Note: This species is very similar to *Cyperus iria* L. in appearance. But can be identified by their broader spikelets, larger glumes, anthers and nuts; also by the spinulose-ciliate keel of the glumes. Herbarium specimens of this species can be found along with that of *Cyperus iria* because of their close resemblance.

3. **Cyperus amabilis** Vahl, Enum. Pl. 2: 318. 1806; Clarke in Hook.f., Fl. Brit. India 6: 598. 1893; Kukenth. in Engl., Pflanzenr. 4 (20), Heft 101: 265, f. 29 A-D. 1936; Karthik. *et al.*, Fl. Ind. Enum. Monocot. 57. 1989. **Fig. 9.**

Illus. : Kukenth., *l.c.*

Annuals with fibrous roots, 15-20 cm high. Stems tufted, slender, compressed, obsoletely trigonous near apex, 0.5-1 mm thick, sulcate, smooth. Leaves 2-4 per stem, shorter than stem, canaliculate, linear, acuminate at apex, up to 8 cm long, 1-2 mm wide; sheaths reddish towards base, many-nerved. Inflorescence simple, rarely compound or reduced to a single spike, lax, 4-8 x 4-11 cm. Involucral bracts 3-6, shorter than inflorescence, lowest up to 7 cm long. Primary rays 1-9, slender, ultimately spreading, up to 7 cm long; secondary rays if present *ca* 1.5 cm long. Spikes usually hemispherical, 1.2-2.5 x 1.5-4.5 cm, bearing 3-16 spikelets. Spikelets digitately arranged, spreading, strongly compressed, linear, subacute at apex, 7-20 x *ca* 1 mm, orange-brown or ferrugineous, numerous (12-56)-flowered; rachilla slender, flexuous, wingless, with *ca* 0.35 mm long internodes. Glumes subdensely imbricate, patent in later stage, subchartaceous, broadly oblong-ovate or elliptic-oblong, truncate and often slightly sinuate at apex, *ca* 1.1 x 0.8 mm; keel green, 3-nerved, excurrent into a short suberect mucro; sides usually shining orange-brown, nerveless. Stamen 1; filament elongate up to 1 mm; anther very small, oblong, *ca* 0.3 mm long. Style *ca* 0.6 mm long; stigmas 3, shorter than style. Nut trigonous, oblong-obovoid, obtuse and minutely apiculate at apex, *ca* 0.6 x 0.3 mm, brownish, densely puncticulate.

Fls. & Frts. : Nov.

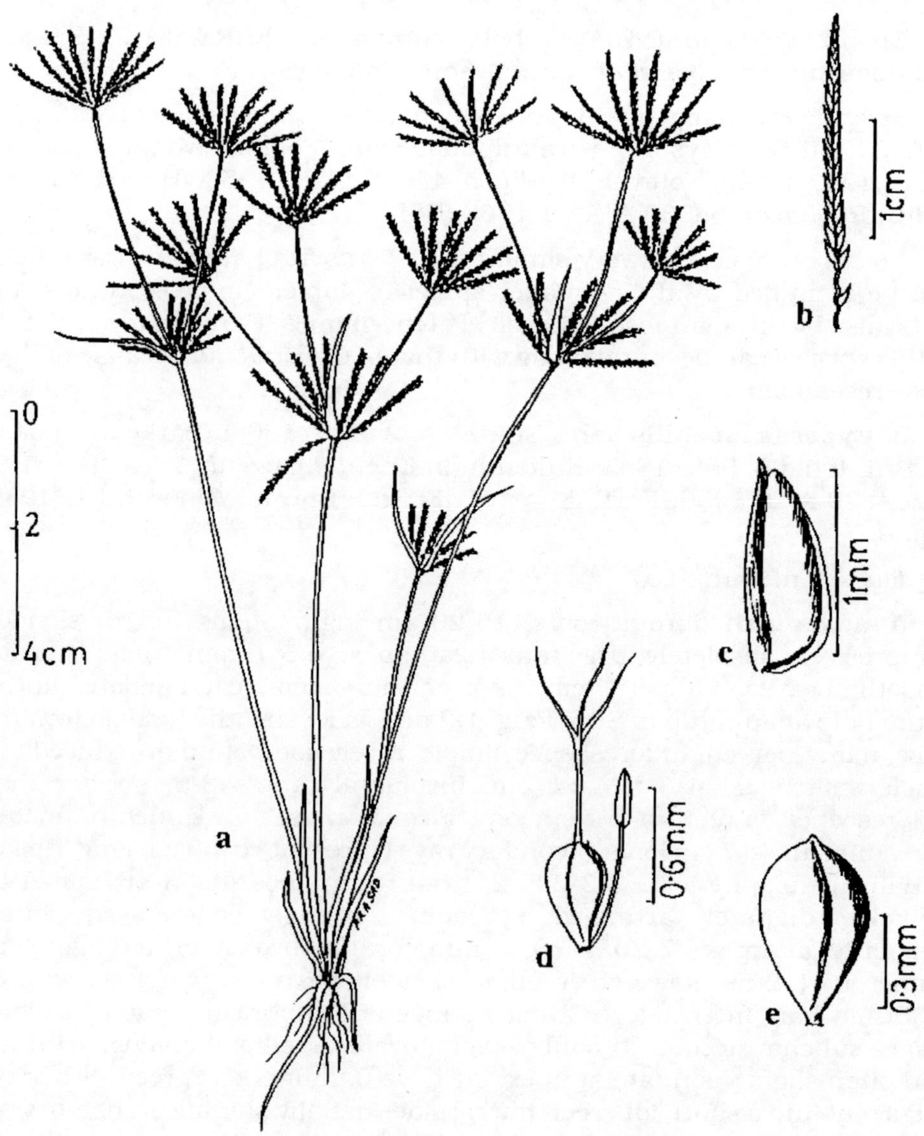

Fig. 9. *Cyperus amabilis* Vahl
a. Habit, **b.** Spikelet, **c**. Glume, **d.** Flower (later stage), **e.** Nut

Habitat : Cultivated fields.

Distrib. : Tropical regions of Asia, Africa and America. INDIA: South, Central, East and North India. KARNATAKA: Raichur.

Specimen examined : Raichur: Raichur-Hyderabad Road, 19th km, Raichur, *Singh* 141726, 15.11.1975 (BSI).

Note : This species was reported from Karnataka for the first time as a result of the present study (Prasad & Singh, 2001).

4. **Cyperus arenarius** Retz., Obs. Bot. 4: 9. 1786; Clarke in Hook.f., Fl. Brit. India 6: 602. 1893; Cooke, Fl. Pres. Bombay 2: 864. 1908 (3:378.1958, repr.ed.); Fischer in Gamble, Fl. Pres. Madras 1640. 1931 (3: 1140. 1957, repr. ed.); Kukenth. in Engl., Pflanzenr. 4 (20), Heft 101: 270. 1936; Sharma *et al.*, Fl. Karnataka 304. 1984; Koyama in Dassanayake & Fosberg, Rev. Handb. Fl. Ceylon 5: 208, f. 13. 1985; Singh, Fl. E. Karnataka 630. 1988; Karthik. *et al.*, Fl. Ind. Enum. Monocot. 43. 1989.

Type : India, *Koenig.*

Illus. : Koyama, *l.c.;* Matthew, Furth. Illus. Fl. Tamilnadu Carnatic Pl. 630. 1988.

Perennials, 3-35 cm high, often growing in patches. Rhizome long-erect or horizontally long- creeping, slender, 1-2 mm thick, branching, brownish, clothed with brown scales. Stems usually solitary, at times 2-3 together, obtusely trigonous, 1-3 mm thick above, smooth, densely covered with leaf sheaths towards the base. Leaves few to many per stem, shorter than to shortly overtopping the inflorescence, linear, recurved, thick, canaliculate or at times with fused margins, 2-30 cm long, 1-2 mm thick, often scabrous on the margins towards base; lower leaves reduced to bladeless sheaths, stramineous to reddish-brown, striate, spliting into fibres in later stage. Inflorescence a dense head of few to many spikelets, 1-2.5 cm across. Involucral bracts 1-3, suberect to reflexed; lowest 2-7.5 cm long, usually much overtopping the inflorescence. Spikelets somewhat compressed, ovate to ovate-lanceolate, subacute at apex, 7-15 x 4-5 mm, whitish-stramineous to light brown, 10-18-flowered; rachilla straight, wingless, excavated. Glumes densely imbricated, oblong-ovate to broadly ovate, obtuse or shallowly emarginate and mucronate at apex, 4-5 x 2-3 mm, obtuse at back, stramineous to reddish-brown, 9-11-veined, membranous and hyaline at margins in the upper half. Stamens 3; filaments elongate up to 5.5 mm; anthers oblong, *ca* 2 mm long. Style 1.5-5 mm long; stigmas 3, 2-3 mm long. Nut planoconvex with angular dorsal surface, obovoid to elliptic-obovoid, apiculate at obtuse apex, 1.7-2 x 1-1.2 mm, blackish-brown when mature.

Fls. & Frts. : July - Dec.

Habitat : Sandy, coastal areas and inland, saline, sandy areas.

Distrib : Sri Lanka, Pakistan, Southern Iran, Vietnam. INDIA: Peninsular India, East, North and North-West India. KARNATAKA: Bellary, Dakshina Kannada, Uttara Kannada.

Specimens examined : Bellary: Benkal Plantation, Moka, *Singh* 132903, 12.9.1974 (BSI); Hagari, *Sreenath & Ramesh* 4336, 15.11.1978 (JCB). Dakshina Kannada: Kudlur, without coll. name, 16755, 1.9.1920 (MH); Malpe, *Bhat* 298, 23.9.1976 (MGH); Hejmady, Mangalore - Mulki Road, *Saldanha & Sreenath* 4926, 9.12.1978 (JCB); Suratkal, *Saldanha* 10578, 27.12.1979 (JCB); Ucchil, Mangalore taluk, *Sheriff & Suresh* 579, 11.9.1985 (MH); Bengare beach, Mangalore, *Prasad* 173881, 10.11.1995 (BSI). Uttara Kannada: Karwar, *Talbot* 563, 30.7.1883 (BSI); Karwar, *Talbot* 1533, 20.9.1885 (BSI).

Note: An abnormal specimen was found in BSI, collected from Mumbai coast (not from Karnataka) in which 2-3 additional heads were found on 1-4 cm long rays. Normally inflorescence in this species is a single terminal head of spikelets. But Koyama, *l.c.* has reported rays up to 1 cm long.

A halophilous species growing in sandy, coastal areas (rarely in inland, saline, sandy areas also) and can be easily recognised by the well developed stoloniferous rhizomes and the whitish-ferrugineous head inflorescence.

5. **Cyperus articulatus** L., Sp. Pl. 44. 1753; Clarke in Hook. f., Fl. Brit. India 6: 611. 1893; Fischer in Gamble, Fl. Pres. Madras 1641. 1931 (3: 1140. 1957, repr.ed.); Kukenth. in Engl., Pflanzenr. 4 (20), Heft 101: 77. 1936; Ramaswamy & Razi, Fl. Bangalore 81. 1973; Hooper in Saldanha & Nicolson, Fl. Hassan 663. 1976; Rao & Razi, Fl. Mysore 558. 1981; Sharma *et al.*, Fl. Karnataka 305. 1984; Koyama in Dassanayake & Fosberg, Rev. Handb. Fl. Ceylon 5: 176. 1985; Singh, Fl. E. Karnataka 2: 630. 1988; Karthik. *et al.*, Fl. Ind. Enum. Monocot. 43. 1989. *Yeletollu jambu hullu.*

Type : Jamaica.

Illus. : Subramanyam, Aqua. Angio. 105, f. 59. 1962; Matthew, Illus. Fl. Tamilnadu Carnatic Pl. 780. 1982.

Perennials with creeping rhizome, 90-160 cm or more high. Rhizome stout, woody, long-creeping, thickened at base of the stems, clothed with brownish to blackish sheaths or their fibrous remains, producing new plants. Stems often in a row along the creeping rhizome or 2-3 together, spongy, terete, stout, narrowed towards apex, 6-10 mm thick in the middle portion, transversely septate. Leaves reduced to 2-3 papery sheaths, up to 30 cm long, brownish to purplish-brown towards base, at times with a short foliaceous limb. Inflorescence usually compound, loose, 5-13 (-18) cm long, 3-13 cm broad. Involucral bracts 2-3, very short, lanceolate, narrowed to a sharp-acute tip; lowest 8-14 mm long. Primary rays 6-10, suberect or often curved, very unequal, 1-10 (-15) cm long, terminating in corymbs of 8-15 densely arranged spikelets. Spikelets spicately arranged, suberect to obliquely patent, subcompressed, linear, acute at apex, 10-30 x 1.5-1.8 mm, pale brown, 16-38-flowered; rachilla flexuous, promminently winged; internodes *ca* 1 mm long; wings lanceolate, whitish-hyaline. Glumes subdensely imbricate, membranous, oblong-ovate or elliptic-ovate, obtuse or subacute at apex, keeled, 2.5-3 x 1.5-2 mm, 5-7-nerved; keel narrow, green; sides hyaline towards margins. Stamens 3; filaments up to 3 mm long; anthers linear, *ca* 1 mm long, with a smooth connective appendage. Style 0.5 - 0.7 mm long; stigmas 3, 2-2.5 mm long. Nut trigonous, oblong to narrowly elliptic-oblong, apiculate at apex, 1.2-1.5 x *ca* 0.5 mm, blackish when mature.

Fls. & Frts. : Throughout the year.

Habitat : Common in shallow standing water or wet banks of canals, ponds, lakes and irrigated areas with muddy bottom, often forming large patches or along with other marshy species like *Commelina* sp. and *Colocasia* sp.

Distrib : Sri Lanka, Indo-China, Mediterranean region, tropical Africa and South-Eastern U.S.A. to Central & South America. INDIA: Peninsular India and Eastern India. KARNATAKA: Bangalore, Bellary, Bijapur, Chitradurga, Dharwar, Hassan, Mysore, Raichur, Uttara Kannada.

Uses : Dry stems can be used for roofing and mat making.

Specimens examined : Bangalore: Bangalore, *Camaron* 500, 24.3.1891 (MH); Bidadi, *Ramesh* 14249, 17.2.1982 (JCB). Bellary: Hospet-Koppal Road, 26th km, *Singh* 141510, 9.11.1975 (BSI). Bijapur: Badami-Anantapur, *Jain* 20867, 28.7.1957 (BSI); Varathikavlu, Kendur, *Prasad* 172967, 21.12.1994 (BSI). Chitradurga: Siregere- Sasalu Road, *Ramesh & Sreenath* 7296, 21.4.1979 (JCB). Dharwar: Kelgeri, without coll. name, 2612, 15.9.1891 (BSI); Salikkinikuppa lake, Dharwar, *Prasad* 172916, 18.12.1994 (BSI). Hassan: Bourdalboore State Forest, *Saldanha et al.*, 5467, 26.12.1978 (JCB); Bourdalboore S.F., *Saldanha & Sreenath* 9103, 17.9.1979 (JCB). Mysore: Gundlapet area, *Naithani* 23264, 19.4.1965 (MH); Kukkanahalli tank, *Dasappa s.n.*, 20.1.1967 (MGH); Chamundi, *Bhat* 1, 6.7.1970 (JCB). Raichur: Gangavati on Kanakgiri Road, 6th km, *Singh* 147021, 8.5.1977 (BSI). Uttara Kannada: Mundgod, *Talbot s.n.* (Acc. No. 638), June 1884 (BSI); Dastikope, *Talbot* 1062, 29.10.1884 (BSI); Devikope, *Talbot* 1062, 30.10.1884 (BSI); Dastikope, *Talbot s.n.* (Acc. No. 53297), 1.1.1885 (BSI).

Note : This tall sedge can be easily distinguished by the septate stems, absence of well developed leaf-lamina and the inflorescence with very short bracts.

6. **Cyperus bulbosus** Vahl, Enum. Pl. 2: 342. 1806; Clarke in Hook.f., Fl. Brit. India 6: 611. 1893; Cooke, Fl. Pres. Bombay 2: 871. 1908 (3: 384.

1958, repr.ed.); Fischer in Gamble, Fl. Pres. Madras 1641. 1931 (3: 1140. 1957, repr.ed.); Kukenth. in Engl., Pflanzenr. 4 (20), Heft 101: 125, f.15 C-E. 1935; Kern in van Steenis, Fl. Males. 1,7: 605. 1974; Rao & Razi, Fl. Mysore 558. 1981; Sharma *et al.*, Fl. Karnataka 305. 1984; Koyama in Dassanayake & Fosberg, Rev. Handb. Fl. Ceylon 5: 179. 1985; Karthik. *et al.*, Fl. Ind. Enum. Monocot. 44. 1989.

Illus. : Kukenth., *l.c.*

Perennials with filiform stolons which disappear after producing bulb-like tubers, 15-30 cm high. Tubers ovoid to fusiform, 8-15 x 3-8 mm, clothed with coriaceous striate coat which split into caducous blackish segments, whitish initially, ultimately shining black. Stems solitary, arising from a tuber, slender, triquetrous, 0.5-1 mm thick, smooth. Leaves few to several, as long as or longer than the stem, often recurved, narrowly linear, gradually acuminate at apex, 1-2 mm wide, often scabrid on upper margins. Inflorescence simple, often reduced to a single spike, 1.5 - 3 (-4.5) x 1-3 (-7.5) cm. Involucral bracts 2-3(-5), 1-4 mm spaced, patent to reflexed, lowest 1 or 2 overtopping the inflorescence, up to 12 cm long. Rays if present 2-5, 0.5-3 (-5) cm long. Spikelets spicately arranged on a short rachis, patent, compressed, linear, acute at apex, 8-20 x *ca* 2 mm, 8-20-flowered, reddish-brown, rachilla flexuous, winged, with *ca* 1 mm long internodes. Glumes imbricate, chartaceous, suberect, ovate or oblong-lanceolate, acute and usually mucronulate at apex, 3-4 x *ca* 2 mm; keel green; sides reddish-brown to castaneous, 9-11-nerved. Stamens 3; filaments up to 4 mm long; anthers linear, 1.5 - 2 mm long. Style 1.8 - 2 mm long; stigmas 3, up to 4 mm long. Nut trigonous, obovoid to ellipsoid, obtuse and apiculate at apex, 1.5-1.8 x *ca* 0.5 mm, black when mature.

Fls. & Frts : June; Sept. - Dec.

Habitat : Along sandy sea shores and other sandy areas in plains.

Distrib. : South Asia, Malesia, tropical Africa and tropical Australia. INDIA: Throughout the coastal areas and plains. KARNATAKA: Bangalore (Sharma *et al., l.c.*), Mysore (Rao & Razi, *l.c.*), Uttara Kannada.

Uses : The bulb like tubers are edible.

Specimen examined : Uttara Kannada: Karwar, *Talbot* 1262, 25.6.1885 (BSI).

7. **Cyperus castaneus** Willd., Sp. Pl. 1: 278. 1797; Clarke in Hook. f., Fl. Brit. India 6: 598. 1893; Cooke, Fl. Pres. Bombay 2: 861. 1909 (3: 374. 1958, repr.ed.); Fischer in Gamble, Fl. Pres. Madras 1639. 1931 (3: 1139. 1957, repr.ed.); Kukenth. in Engl., Pflanzenr. 4 (20), Heft 101: 264. 1936; Kern in van Steenis, Fl. Males. 1, 7: 630. 1974; Sharma *et al.*, Fl. Karnataka 305. 1984; Koyama in Dassanayake & Fosberg, Rev. Handb. Fl. Ceylon 5: 207. 1985; Karthik. *et al.*, Fl. Ind. Enum. Monocot. 44. 1989.

Type : India.

Illus : Matthew, Furth. Illus. Fl. Tamilnadu Carnatic Pl. 631. 1988.

Tufted annuals with fibrous roots, up to 17 cm high. Stems slender, trigonous or triquetrous, 0.3-0.9 mm thick, smooth. Leaves often longer than stem, setaceous, canaliculate, linear, gradually narrowed to apex, 0.5 - 1 mm wide, scaberulous in the upper part; sheaths membranous, striate, reddish-brown. Inflorescence simple, at times reduced to a single head. Involucral bracts 3-5, patent, longer ones much overtopping the inflorescence. Rays 2-6, slender, up to 3 cm long. Spikelets digitately clustered, up to 20 in a cluster, strongly compressed, linear, 5-20 mm long, *ca* 2 mm wide (including the mucros), 10-40-flowered; rachilla persistent, wingless, with *ca* 0.3 mm long internodes. Glumes subchartaceous, ultimately patent, oblong to oblong-lanceolate, emarginate or subobtuse at apex, strongly mucronate, 1.5-2 mm long (including the 0.3 - 0.5 mm mucro), 0.7 - 1 mm wide; keel acute, greenish, usually purple-lineolate, 3-nerved, excurrent into somewhat recurved mucro; sides nerveless, castaneous. Stamen 1 (rarely 3 in the upper glumes); filament up to 1 mm long; anther elliptic- oblong, *ca* 0.2 mm long. Style 0.7-0.9 mm long; stigmas 3 (mostly 2 in the specimens cited below), shorter than style. Nut trigonous, oblong with parallel sides or oblong-obovoid, shortly apiculate, 0.7-1 x *ca* 0.3 mm, brownish.

Fls. & Frts. : Sept. - Jan.

Habitat : Open sandy areas and cultivated fields.

Distrib : Sri Lanka, Nepal, Indo-China, Malesia and tropical Australia. INDIA: Peninsular India, Central and East Himalayas. KARNATAKA: Bangalore, Dakshina Kannada, Mandya, Mysore (Sharma *et al., l.c.*), Raichur.

Specimens examined : Bangalore: Bangalore, *Camaron* 540, 24.11.1890 (MH). Dakshina Kannada: Mangalore, *Barber* 4863 A, January 1902 (MH); Nileshwar, without coll. name, 15285, 7.11.1917 (MH); Malpe, *Bhat* 295, 23.9.1976 (MGH); Kapu, *Bhat* 448, 15.1.1977 (MGH). Mandya: Ranganathittu, *Bhat* 93, 20.12.1970 (JCB). Raichur: Raichur-Hyderabad Road, 19th km, Raichur, *Singh* 141726 A, 15.11.1975 (BSI).

Note : It is closely allied to *C. cuspidatus* Kunth, but can be distinguished by the narrow (*ca* 1.5 mm wide) spikelets, chestnut brown glumes and single stamen.

Sharma *et al., l.c.* reported this species from Shimoga district based on wrongly identified specimens.

8. **Cyperus cephalotes** Vahl, Enum. Pl. 2: 311. 1806; Clarke in Hook. f., Fl. Brit. India 6: 597. 1893; Kukenth. in Engl., Pflanzenr. 4 (20), Heft 101: 291. 1936; Fischer in Gamble, Fl. Pres. Madras 1639. 1931 (3: 1139.

1957, repr. ed.); Kern in van Steenis, Fl. Males. 1, 7: 633, f. 61 & 62, 1974; Rao & Verma, Cyp. NE India 11. 1982; Sharma *et al.*, Fl. Karnataka 305. 1984; Koyama in Dassanayake & Fosberg, Rev. Handb. Fl. Ceylon 5: 212. 1985. *Anosporum cephalotes* (Vahl) Kurz, J. Asiat. Soc. Bengal 45 (2): 159. 1876; Karthik. *et al.*, Fl. Ind. Enum. Monocot. 32. 1989.

Type : India, Nicobar Islands.

Illus : Clarke, Illus. Cyp. t. 6, f. 1-7. 1909; Kern, *l.c.*

Perennials with short rhizome and slender stolons, up to 40 cm high; stolons rooting at nodes. Stems solitary, rigidulous, trigonous, 1.5 - 2 mm thick, smooth. Leaves few, shorter than to overtopping the inflorescence, canaliculate towards base, flat above, linear, gradually narrowed to a triquetrous and scabrid apex, 2-4 mm wide; lower sheaths cinnamomeous to reddish-brown. Inflorescence an ovoid, conical or globose, lobed head bearing 1-4 glomerules of 3-8 spikelets, 5-15 mm across. Involucral bracts 3-5, patent to reflexed, longest up to 30 cm long, dilated and purplish-striate at base. Spikelets ovate to ovate-lanceolate, subacute at apex, at times slightly recurved, 4-7 x 2-4 mm, brownish to dark brown, 10-16-flowered; rachilla wingless. Glumes subcoriaceous, densely imbricate, ovate, acute and mucronate at apex, prominently keeled, 3-4 x 1.5-2 mm, pale brown to chestnut brown; keel acute to narrowly winged, 3-nerved, often scabrid, produced into the mucro; sides obscurely 3-5-nerved. Stamens 3; filaments up to 4 mm long; anthers linear, 1-1.5 mm long. Style 3-4 mm long, undivided (or obscurely 3-lobed - not found in Karnataka specimen). Nut trigonous but dorsally compressed, oblong-ovoid, apiculate at apex, long-stipitate, 1-1.5 x 0.8-1 mm, persistent style base *ca* 1 mm long; stipe *ca* 1 mm long, initially 3-winged, wings ultimately developing into corky ridges which cover lower part of the nut also.

Fls. & Frts : Aug.

Habitat : Growing in large patches as floating vegetation.

Distrib.: Sri Lanka, South China, Malesia and tropical Australia. INDIA: South, Central, East and North-East India, also in Andaman & Nicobar Islands. KARNATAKA: Uttara Kannada. Rare; only one, more than 100 years old specimen was found.

Specimen examined : Uttara Kannada: without exact locality, *Talbot* 1125, 5.8.1885 (BSI).

Note : Corky ridges of the stipe which cover the basal part of the nut also in later stage is characteristic to this species which enables dispersal of the nut through water.

9. **Cyperus compressus** L., Sp. Pl. 46. 1753; Clarke in Hook.f., Fl. Brit. India 6:605. 1893; Cooke, Fl. Pres. Bombay 2: 866. 1908 (3: 379. 1958, repr.ed.); Fischer in Gamble, Fl. Pres. Madras 1640. 1931 (3: 1140. 1957,

repr.ed.); Kukenth. in Engl., Pflanzenr. 4 (20), Heft 101: 156. 1935; Ramaswamy & Razi, Fl. Bangalore 82. 1973; Kern in van Steenis, Fl. Males. 1, 7: 617. 1974; Hooper in Saldanha & Nicolson, Fl. Hassan 663. 1976; Rao & Razi, Fl. Mysore 558. 1981; Rao & Verma, Cyp. NE India 15. 1982; Sharma *et al.*, Fl. Karnataka 305. 1984; Koyama in Dassanayake & Fosberg, Rev. Handb. Fl. Ceylon 5: 194. 1985; Singh, Fl. E. Karnataka 630. 1988; Karthik. *et al.*, Fl. Ind. Enum. Monocot. 44. 1989.*Usumani hullu.*

Illus. : Matthew, Furth. Illus. Fl. Tamilnadu Carnatic Pl. 632.1988.

Tufted annuals with fibrous roots, 5-40 cm high. Stems triquetrous, 1-2 mm thick, smooth, striate. Leaves usually shorter than stem, at times somewhat longer in dwarf plants, flat or folded, linear, gradually narrowed to an acute apex, 2-4 mm wide, usually scabrous on the margins towards apex; sheaths striate, pale brown to reddish-brown towards base. Inflorescence simple, open, up to 22 cm broad, at times reduced to a single cluster of spikelets. Involucral bracts 3-5, lower ones overtopping the inflorescence, up to 20 cm long. Rays 0-5, patent, 1.5-15 cm long. Spikes broadly ovoid, subdigitately bearing 3-12 spikelets on a short axis, 2-5 cm across. Spikelets patent or spreading, strongly compressed, oblong-lanceolate to linear, subacute at apex, 10-30 (-40) x 3-5 mm, usually yellowish-green, up to 44-flowered; rachilla flexuous, with caducous wings, with *ca* 0.7 mm long internodes. Glumes somewhat rigid, ovate, acute and mucronate at apex, keeled, 3.5-4.2 x 2-2.5 mm, 4-nerved on both sides, with hyaline margins; keel 3-nerved, green; mucro up to 1 mm long, often slightly excurved. Stamens 3; filaments up to 3.5 mm long; anthers linear-oblong, up to 1 mm long, with slightly produced connective appendage. Style 1-1.5 mm long; stigmas 3, *ca* 2 mm long. Nut trigonous, broadly obovoid, often slightly oblique, obtuse at apex, minutely apiculate, 1.4-1.7 x *ca* 1 mm, shining dark brown.

Fls. & Frts. : Aug - March.

Chrom. No. : 2n = 48 (*Taxon* 21: 341. 1972).

Habitat : Common as a weed in rice fields, sunflower fields, waste lands, open grasslands, muddy banks of canals, edges of rice fields, along roadsides, damp sandy soils, gravelly soils, dry areas as well as marshy places, sea shores, lowlands and hilly areas.

Distrib. : Widely distributed in tropical, subtropical and temperate regions of the whole world. INDIA: Throughout. KARNATAKA: Bangalore, Belgaum, Bellary, Bijapur, Chikmagalur (Rao *et al.*, 2000) Dakshina Kannada, Hassan, Kolar, Mandya, Mysore, Raichur, Shimoga, Tumkur, Uttara Kannada.

Specimens examined : Bangalore: Bangalore, *Camaron* 525, 2.11.1890 (MH); Bangalore, *Anstead* 83370, Nov. 1916 (MH); Sangam-Makedat, *Ramesh & Bhasker* 419, 10.3.1978 (JCB). Belgaum: Sutagatti, *Mahajan* 24972, 4.11.1957 (BSI); Sutagatti, *Mahajan* 28680, without date (BSI);

Dinman Hassur, Belgaum-Gokak Road, *Prasad* 172847, 13.12.1994 (BSI). Bellary: Ramandurg, without coll. name, 15926, 9.10.1919 (MH). Bijapur: Badami hills, *Paranjpye s.n.,* 1.9.1912 (BSI); Guledagudda, *Ramesh & Ramesh* 9269, 14.9.1979 (JCB); Varathi Kavlu, Kendur, *Prasad* 172964, 21.12.1994 (BSI). Dakshina Kannada: Kudlu, without coll. name, 16739, 28.8.1920 (MH); Udyavara, *Bhat* 267, 8.8.1976 (MGH); Suratkal, *Saldanha, Ramesh & Sreenath* 8851, 4.8.1979 (JCB); Near Netravati river, Kadakar, *Prasad* 173872, 10.11.1995 (BSI). Hassan: Bagi, *Nicolson et al.* 2268, 23.10.1971 (JCB). Kolar: Tumkur-Chickballapur Road, 72nd km, *Singh* 142017, 3.1.1976 (BSI). Mandya: Sreerangapatana, *Doddappa* 120, 10.12.1964 (MGH); Paschimavahini, *Padma Rani* 21, 8.8.1970 (MGH). Mysore: Kupgal, *Wadhwa* 44897, 29.9.1959 (BSI); Ketedevargudi vicinity, *Kammatty* 73550, 4.9.1961 (BSI); Bandipur, *Naithani* 21163, 24.8.1964 (MH); Kukkanahalli, *Bhat* 61, 10.8.1970 (MGH); Ilwal, *Krishna Murthy* 15, 10.10.1976 (MGH). Raichur: Koppal-Kushtagi Road, 38th km, *Singh* 141525, 9.11.1975 (BSI); Deodurga, *Ramesh & Sreenath* 4419, 16.11.1978 (JCB). Shimoga: Hosur, Bhadravathi-Shimoga Road, *Sreenath & Murthy* 2880, 26.9.1978 (JCB). Tumkur: Kamplapura Road, Huliyurdurga, Kunigal, *Singh* 140848, 22.10.1975 (BSI). Uttara Kannada: Kumta, *Chibber s.n.,* 1.10.1810 (BSI); Without locality, *Talbot* 1254, Aug. 1885 (BSI); Karwar, *Talbot* 1305, 10.9.1885 (BSI); Karka bridge, Dandeli, *Prasad* 173730 & 173731, 30.10.1995 (BSI); Karwar Beach, *Prasad* 173745, 31.10.1995 (BSI); Kinnare, a small island in Kalinadi, Karwar, *Prasad* 173761, 1.11.1995 (BSI).

Note : It shows lots of variation in the size of the plant, inflorescence and spikelets, also in the number of glumes per spikelet.

10. **Cyperus corymbosus** Rottb., Descr. Ic. Rar. Nov. Pl. 42, t. 7, f. 4. 1773; Clarke in Hook.f., Fl. Brit. India 6: 612. 1893; Cooke, Fl. Pres. Bombay 2: 870. 1908 (3: 383. 1958, repr.ed.); Fischer in Gamble, Fl. Pres. Madras 1641. 1931 (3: 1140. 1957, repr.ed.); Kukenth. in Engl., Pflanzenr. 4 (20), Heft 101: 80, f.10. 1935; Ramaswamy & Razi, Fl. Bangalore 82. 1873; Rao & Razi, Fl. Mysore 558. 1981; Rao & Verma, Cyp. NE India 18. 1982; Sharma *et al.*, Fl. Karnataka 305. 1984; Koyama in Dassanayake & Fosberg, Rev. Handb. Fl. Ceylon 5: 175. 1985; Karthik. *et al.*, Fl. Ind. Enum. Monocot. 44. 1989; Keshava Murthy & Yoganarasimhan, Fl. Coorg 501. 1990. *Bhandre hullu, Jake.*

Illus. : Kukenth., *l.c ;* Matthew, Furth. Illus. Fl. Tamilnadu Carnatic Pl. 633. 1988.

Perennials with creeping rhizome, 70-140 cm high. Rhizome stout, woody at base of the stem, clothed with dark brown scales. Stems in a series along the creeping rhizome, erect, spongy, terete below, subtrigonous above, gradually narrowed to apex, 4-7 mm thick near the middle portion, smooth, not or scarsely septate when dry, clothed at base with 2-3 papery or membranous sheaths. Leaves reduced to sheaths or the upper 1 or 2 with a short, up to 12 cm long blade which terminate in an acute apex; sheaths loosely covering the stem, up to 25 cm long; purplish-brown towards base. Inflorescence compound, 5-10 (-17) x 2-6 (-9) cm. Involucral bracts 2-3; lowest shorter than to slightly overtopping the inflorescence, up to 9 cm long. Primary rays 4-10, rather unequal; secondary rays short, terminating in spikes of 4-10 spikelets. Spikelets spicately arranged, suberect to spreading, subterete, linear, subacute at apex, 10-25 x *ca* 1 mm, rusty brown, 12-24-flowered; rachilla somewhat flexuous, broadly winged; internodes 0.8-1 mm long; wings white-hyaline. Glumes imbricate, membranous, oblong-ovate, obtuse at apex, 2.2 - 2.5 x *ca* 1.5 mm, 3-nerved on the obtuse back; sides nerveless, scarious towards margins. Stamens 3; filaments elongate up to 3 mm; anthers linear, *ca* 1 mm long. Style 0.7-1 mm long; stigmas 3, up to 4 mm long. Nut trigonous, narrowly elliptic-oblong, apiculate at apex, *ca* 1.3 x 0.4 mm, greyish-brown when mature.

Fls. & Frts. : Aug. - March.

Chrom. No. : n = 54 (*Taxon* 20: 612. 1971).

Habitat : On the muddy banks of rivers, lakes and canals; shallow water near rivers and lakes, rocky river beds.

Uses : Stems used for making mats and screens, also for thatching. In Sri Lanka this species is often cultivated for the stems which are used for making mats (Koyama, *l.c.*).

Distrib. : Sri Lanka, Nepal, Myanmar, Vietnam, Madagascar, tropical West Africa and tropical South America. INDIA: Throughout. KARNATAKA: Bangalore (Ramaswamy & Razi, *l.c.*), Belgaum, Chikmagalur (Sharma *et al.*, *l.c.*), Dharwar, Mandya, Mysore, Shimoga (Kukenth. *l.c.*), Uttara Kannada.

Specimens examined : Belgaum: Ghataprabha dam side, *Prasad* 172801, 10.12.1994 (BSI). Dharwar: Salikkinikuppa lake, Dharwar, *Prasad* 172916 A, 18.12.1994 (BSI). Mandya: Ranganathittu, *Rao* 870, 28.8.1969 (MGH); Ranganathittu, *Bhat* 68, 28.8.1970 (MGH); Sreerangapatana, *Dinesh* 783, 26.1.1984 (MGH); Sreerangapatana, *Ushakumari* 81, without date (MGH). Mysore: Nageena halli, *Suma* 4, 17.8.1968 (MGH). Uttara Kannada: Ulletikeri, Halyal, *Prasad* 172996, 29.10.1995 (BSI).

11. **Cyperus cuspidatus** Kunth in H.B.K., Nov. Gen. & Sp. Pl. 1: 204. 1815; Clarke in Hook.f., Fl. Brit. India 6: 598. 1893; Kukenth. in Engl., Pflanzenr. 4 (20)), Heft 101: 261, f. 29 E. 1936; Kern in van Steenis, Fl. Males. 1, 7: 629, f. 59. 1974; Hooper in Saldanha & Nicolson, Fl. Hassan 664. 1976; Rao & Verma, Cyp. NE India 12. 1982; Sharma *et al.*, Fl. Karnataka 305. 1984; Koyama in Dassanayake & Fosberg, Rev. Handb. Fl. Ceylon 5: 208. 1985; Karthik. *et al.*, Fl. Ind. Enum. Monocot. 44. 1989.*C. angustifolius* Nees in Wight, Contr. Bot. India 79. 1834. *C. uncinatus* auct.

non Poir. 1806; Cooke, Fl. Pres. Bombay 2: 862. 1909 (3: 375. 1958, repr. ed.); Fischer in Gamble, Fl. Pres. Madras 1639. 1931 (3: 1139. 1957, repr. ed.); Ramaswamy & Razi, Fl. Bangalore 96. 1973.

Illus. : Kukenth, *l.c.*; Kern, *l.c.*.

Annuals with fibrous roots. Stems tufted, trigonous, 2-15 cm long, up to 1 mm thick. Leaves rigid, setaceous, gradually narrowed to apex, 0.5-1 mm wide, scaberulous in the upper part. Inflorescence simple, often reduced to a single head. Involucral bracts 3-4, patent, longer ones overtopping the inflorescene. Rays 0-4, up to 3 cm long, smooth. Spikelets digitately arranged, up to 20 in each cluster, strongly compressed, linear, 0.5-1.5 cm long, 2-3 mm wide (including the mucro), 12-40-flowered; rachilla persistent, wingless, with *ca* 0.3 mm long internodes. Glumes subchartaceous, ultimately obliquely patent, oblong-spathulate, emarginate at apex, strongly mucronate, 1-1.5 x 0.5-1 mm (including the *ca* 0.8 mm long mucro); keel acute, greenish, usually purple-lineolate, 3-nerved, excurrent into a strong, spreading and recurved mucro; sides nerveless, ferrugineous to castaneous. Stamens (2-) 3; filaments up to 1 mm long; anthers elliptic or broadly oblong, *ca* 0.2 mm long. Style *ca* 0.3 mm long; stigmas 3, longer than style. Nut trigonous, obovoid to oblong-obovoid, minutely apiculate, 0.5-0.75 x 0.3-0.4 mm, brown to castaneous.

Fls. & Frts. : July - Nov.

Chrom. No. : n=56 (*Taxon* 20: 612. 1971).

Habitat : Open, moist, sandy soils, in cultivated fields, damp sandy places, in dry deciduous forests.

Distrib. : Pantropical. INDIA: Throughout. KARNATAKA: Bangalore (Ramaswamy & Razi, *l.c.*), Dakshina Kannada (Fischer, *l.c.*), Hassan, Kodagu, Mandya.

Specimens examined : Hassan: Bourdalboore, *Nicolson et al.* 2347, 26.10.1921 (JCB). Kodagu: Kottamudi, along cauvery river, Napoklu, *Rao* 74955, 5.10.1961 (BSI). Mandya: Melukotte, *Dinesh* 363, 7.11.1983 (MGH).

12. **Cyperus difformis** L. in Torner, Cent. Pl. 2: 6. 1756; Clarke in Hook.f., Fl. Brit. India 6: 599. 1893; Cooke, Fl. Pres. Bombay 2: 862. 1908 (3: 376. 1958, repr.ed.); Fischer in Gamble, Fl. Pres. Madras 1640. 1931 (3: 1139. 1957, repr.ed.); Kukenth. in Engl., Pflanzenr. 4 (20), Heft 101: 237, f. 27 F-H. 1936; Ramaswamy & Razi, Fl. Bangalore 84. 1973; Kern in van Steenis, Fl. Males. 1,7: 629. 1974; Hooper in Saldanha & Nicolson, Fl. Hassan 664. 1976; Rao & Razi, Fl. Mysore 558. 1981; Rao & Verma, Cyp. NE India 12. 1982; Sharma *et al.*, Fl. Karnataka 305. 1984; Koyama in Dassanayake & Fosberg, Rev. Handb. Fl. Ceylon 5: 206. 1985; Singh, Fl. E. Karnataka 631. 1988; Karthik. *et al.*, Fl. Ind. Enum. Monocot. 44. 1989; Keshava Murthy & Yoganarasimhan, Fl. Coorg 501. 1990. *Karisanna jambuhullu.*

Type : India.

Illus. : Kukenth., *l.c.*; Matthew, Illus. Fl. Tamilnadu Carnatic. Pl. 781. 1982.

Annuals with reddish, fibrous roots, 8-60 cm high. Stems tufted, erect, triquetrous, somewhat weak, 1.5 - 4 mm thick, smooth. Leaves shorter or longer than stem, gradually tapering to an acute apex, 2-6 mm wide, smooth; sheaths stramineous to pale brown. Inflorescence simple or compound, usually lax, 2-6 (-9) cm wide, at times congested in a globose cluster of spikelets. Involucral bracts 2 or 3; lower 2 much overtopping the inflorescence; lowest up to 27 cm long, at times (in early stage) suberect and then the inflorescence looks like lateral. Primary rays 3-9, 1-5 cm long, smooth. Spikes dense, globose, 5-20 mm across, bearing numerous spikelets, greenish, brownish or blackish. Spikelets stellately arranged, compressed, linear, linear-oblong or oblong-ovate, obtuse at apex, 1.5 - 5 x 1-1.2 mm, 10-30-flowered; rachilla persistent, wingless, with *ca* 2 mm long internodes. Glumes very small, membranous, somewhat loosely arranged, orbicular or broadly obovate, obtuse or slightly emarginate at apex, slightly keeled, 0.5 - 0.9 x 0.7- 1 mm; keel greenish, faintly 3-nerved; sides nerveless, often purplish-brown, with whitish hyaline margins. Stamens 2 (rarely 1); filaments elongate up to 0.7 mm; anthers very small, oblong, *ca* 0.1 mm long. Style short, *ca* 0.15 mm long; stigmas 3, slightly longer than style. Nut triquetrous, elliptic-obovoid, minutely apiculate, shortly stipitate, nearly equalling the glume, 0.5-0.7 x *ca* 0.3 mm, yellowish.

Fls. & Frts.: Throughout the year.

Chrom. No. : n = 18 (*Taxon* 20: 612. 1971).

Habitat : Wet sandy soil in paddy fields, along canal banks, river beds and streams, muddy bottom of puddles, canals and lakes; shallow running water of narrow irrigation canals; also found as a weed in rice fields. Very common.

Distrib. : Widely distributed in the tropical and subtropical regions of the old world. INDIA: Throughout. KARNATAKA: Bangalore, Belgaum, Bijapur, Chikmagalur, Chitradurga, Dakshina Kannada, Dharwar, Gulbarga, Hassan, Kodagu, Kolar, Mysore, Raichur, Tumkur, Uttara Kannada.

Specimens examined : Bangalore: Bangalore, *Camaron* 563 & 580, March 1891 (MH); Kankapura-Sangam Road, *Saldanha et al.* 24, 28.1.1978 (JCB). Belgaum: Sutagatti, *Mahajan* 24975, without date (BSI); Near Ghataprabha dam, *Prasad* 172809, 10.12.1994 (BSI); Gokak falls, Gokak, *Prasad* 172816, 11.12.1994 (BSI); Manjapur nallah, Khanapur, *Prasad* 172855, 14.12.1994 (BSI); Malaprabha river, near the bridge, *Prasad* 172866 & 172867, 14.12.1994 (BSI). Bijapur: Varathi Kavlu, Kendur, *Prasad* 172960, 21.12.1994 (BSI); Khanapur, Hubbli-Bijapur Road, *Prasad*

172979, 22.12.1994 (BSI); Hossur Forest Research Nursery, Badami taluk, *Prasad* 172990, 23.12. 1994 (BSI). Chikmagalur: Kadur, *Murthy & Ramesh* 6291, 10.3.1979 (JCB). Chitradurga: Chitradurga fort, *Sreenath & Ramesh* 7255, 21.4.1979 (JCB). Dakshina Kannada: Kannadekatte, *Bhat* 181, 6.5.1975 (MGH); Kulur, *Saldanha & Sreenath* 4885, 9.12.1978 (JCB). Dharwar: Hallikeri, near Reserve Forest, *Prasad* 172930, 18.12.1994 (BSI); Sadankeri, on Alnavar Road, *Prasad* 172943, 19.12.1994 (BSI). Gulbarga: Bidar-Chincholi Road, 33rd mile, *Singh* 128614, 9.2.1975 (BSI); Hassan: Harnalli village, *Puri* 19891, 8.6.1957 (BSI); Channarayapatana, *Saldanha et al.* 13975, 28.11.1981 (JCB). Kodagu: Kottamudi, on the side of Cauvery river at Napoklu, *Rao* 74953, 5.10.1961 (BSI); Mercara, *Bhat* 785, 18.12.1980 (MGH); Kushalnagar, *Bhat* 941, 16.2.1981 (MGH). Kolar: Sidlaghatta-Chintamani Road, 14th km, *Singh* 142041, 3.1.1976 (BSI); Uddappanahalli, *Ramesh & Ravindra* 1475, 7.7.1978 (JCB). Mandya: Ranganathittu, *Dinesh* 532, 19.9.1983 (MGH). Mysore: Sivasamudram, without coll. name 10407, 12.5.1914 (MH); Kottathuhalla, Biligirirangan hill ranges, *Rao* 80176, 21.4.1962 (BSI); Kukenchalli tank, *Suma* 6, 1.8.1968 (MGH); Mandakalli, *Bhat* 21, 12.7.1970 (JCB). Raichur: Bankaldoddi R.F. (Deodurg), *Singh* 141706, 14.11.1975 (BSI); Sanapur R.F. (Gangavati), *Singh* 147011, 7.5.1977 (BSI). Tumkur: Devarayanadurga, *Ravindra & Ramesh* 1546, 14.7.1978 (JCB). Uttara Kannada: Vincholi, *Talbot* 1000, April 1884 (BSI); Karwar, *Talbot s.n.*, Sept. 1885 (BSI); without locality, *Talbot* 2271, without date (BSI); Kilgerry, *Talbot* 2619, 15.8.1891 (BSI); on the way to Dandeli, *Ahmed* 1093, 25.5.1978 (JCB); Jappinamogeru village, *Prasad* 173861 & 173863, 10.11.1995 (BSI); Near Netravati river, Kadakar village, *Prasad* 173875, 10.11.1995 (BSI). Without exact locality (Mysore & Karnatic), *Thomson s.n.* (Acc. No. 73473), without date (MH).

13. **Cyperus digitatus** Roxb., (Hort. Beng. 81. 1814, *nom. nud.*) Fl. Ind. 1, 1: 209. 1820; Clarke in Hook.f., Fl. Brit. India 6: 618. 1893; Cooke, Fl. Pres. Bombay 2: 873. 1908 (3: 387. 1958, repr.ed.); Fischer in Gamble, Fl. Pres. Madras 1642. 1931 (3: 1141. 1957, repr.ed.); Kukenth. in Engl., Pflanzenr. 4 (20), Heft 101: 55. 1935; Ramaswamy & Razi, Fl. Bangalore 84. 1973; Kern in van Steenis, Fl. Males. 1, 7: 601. 1974; Rao & Verma, Cyp. NE India 20. 1982; Sharma *et al.*, Fl. Karnataka 305. 1984; Koyama in Dassanayake & Fosberg, Rev. Handb. Fl. Ceylon 5: 170. 1985; Karthik. *et al.*, Fl. Ind. Enum. Monocot. 44. 1989.

Type : Bengal.

Key to the varieties

1a. Spikelets comparatively loosely arranged; glumes yellowish to pale reddish-brown. . . . . var. **digitatus**

1b. Spikelets very densely arranged; glumes golden brown. . . . . . var. **khasianus**

var. **digitatus**

Tall perennials with short, woody rhizome, up to 150 cm high. Stems tufted or solitary, trigonous, often triquetrous above, 3-4 mm thick, smooth. Leaves shorter than to equalling the stem, coriaceous, flat or slightly plicate, 5-10 mm wide, scarious on the margins, with a very prominent midrib below and usually with 2 lateral nerves above; sheaths purplish-brown towards base. Inflorescence compound or decompound, 10-25 (-40) cm across. Involucral bracts 3-9 (-12), spreading, mostly overtopping the inflorescence, up to 60 cm long. Primary rays 4-10, unequal, longer ones up to 22 cm long, smooth; secondary rays up to 3 cm long; spikes 3-9, digitately arranged (a few solitary on long peduncles), cylindrical, 2-6 x 1-3 cm, subloosely bearing many spikelets; rachis narrowly winged, glabrous. Spikelets spicately arranged, ultimately at right angle to the rachis, slightly compressed, linear, 5-10 (-20) x 1-1.5 mm, 8-20 (-40)-flowered; rachilla straight, winged, with *ca* 0.5 mm long internodes; wings lanceolate, *ca* 1 x 0.2 - 0.3 mm, yellowish-hyaline, caducous. Glumes membranous, elliptic, acute and apiculate at apex, keeled, 1.8-2.2 x *ca* 1 mm; keel greenish or brownish, 3-5-nerved, sides yellowish to pale reddish-brown, nerveless. Stamens 3; filaments elongate up to 2 mm; anthers linear, *ca* 0.5 mm long, with a short connective appendage. Style 1-1.2 mm long; stigmas 3, up to 1.3 mm long. Nut trigonous, ellipsoid to oblong-obovoid, apiculate, *ca* 1 x 0.5 mm, yellowish-brown.

Fls. & Frts. : Aug. - Oct., Feb.

Habitat : Wet river banks and other marshy areas.

Distrib. : Pantropic. INDIA: Peninsular India, Eastern India and North-West India. KARNATAKA: Bangalore (Ramaswamy & Razi, *l.c.*), Belgaum (Londa, Cooke, *l.c.*), Chikmagalur.

Specimen examined : Chikmagalur: Bababudan hills, *Talbot* 2480, Oct. 1890 (BSI).

var. **khasianus** (Clarke) Kern in Blumea, Suppl. 4: 165. 1958; Karthik. *et al.*, Fl. Ind. Enum. Monocot. 44. 1989.*Cyperus auricomus* Sieb. var. *khasianus* Clarke in Journ. Linn. Soc. Bot. 21: 189. 1884. *Cyperus hookeri* Boeck. in Linnaea 36: 308. 1870. *C. digitatus* Roxb. var. *hookeri* (Boeck.) Clarke in Hook.f., Fl. Brit. India 6: 618. 1893; Kukenth. in Engl., Pflanzenr. 4 (20), Heft 101: 56. 1935; Hooper in Saldanha & Nisolson, Fl. Hassan 664. 1976; Sharma *et al.*, Fl. Karnataka 305. 1984.

In this variety spikelets are very densely arranged, rigid, somewhat broader, ultimately bright brown and nut little more acuminate at apex.

Fls. & Frts. : November (*Hooper, l.c.*).

Habitat : Wet places in Ghats (Hooper, *l.c.*)

Distrib. : Sri Lanka, Myanmar. INDIA: Southern Peninsular India and North East India. KARNATAKA: Hassan (Hooper, *l.c.*).

Note : This variety is included on the authority of Hooper (*l.c.*). There are no specimens in B.S.I.

14. **Cyperus distans** L.f., Suppl. Sp. Pl. 103. 1781; Clarke in Hook.f., Fl. Brit. India 6: 607. 1893; Fischer in Gamble, Fl. Pres. Madras 1640. 1931 (3: 1140. 1957, repr.ed.); Kukenth. in Engl., Pflanzenr. 4 (20), Heft 101: 137. 1935; Ramaswamy & Razi, Fl. Bangalore 85. 1973; Kern in van Steenis, Fl. Males. 1, 7: 610. 1974; Hooper in Saldanha & Nicolson, Fl. Hassan 665. 1976; Rao & Razi, Fl. Mysore 558. 1981; Rao & Verma Cyp. NE India 16. 1982; Sharma *et al.*, Fl. Karnataka 305. 1984; Koyama in Dassanayake & Fosberg, Rev. Handb. Fl. Ceylon 5: 189. 1985; Karthik. *et al.*, Fl. Ind. Enum. Monocot. 44. 1989; Keshava Murthy & Yoganarasimhan, Fl. Coorg. 501. 1990.

Key to the varieties

1a. Spikelets horizontally spreading . . . . var. **distans**

1b. Spikelets suberect even in fruits . . . . var. **pseudonutans**

var. **distans**

Perennials with short rhizome, 25-90 cm high. Stems solitary to subcaespitose, triquetrous in the upper half, trigonous below, 2-4 mm thick in the middle portion, smooth. Leaves 2-3 per stem, shorter than to as long as the stem, flat, linear, gradually acuminating to apex, 3-8 mm wide, scabrous on upper margins; sheaths up to 15 cm long, purplish at base. Inflorescence compound to decompound, usually open and large, up to 25 cm or more long. Involucral bracts 4-6, spreading; lower ones overtopping the inflorescence. Primary rays 6-12, patent, very unequal, up to 17 cm long, smooth; secondary rays up to 6 cm long. Spikes broadly ovoid-pyramidal, bearing up to 20 spikelets; rachis glabrous. Spikelets spicately arranged, spreading, subterete, narrowly linear, 10-20 x *ca* 1 mm, 6-10-flowered; rachilla flexuous, winged; internodes *ca* 1 mm long. Glumes membranous, distantly arranged on both sides of rachilla, oblong-ovate or ellliptic, obtuse at apex, 1.8-2 x *ca* 1 mm; keel not prominent, greenish, 3-5-nerved; sides usually reddish-brown, white hyaline on upper margins. Stamens 3; filaments up to 2 mm long; anthers oblong, *ca* 0.5 mm long. Style *ca* 0.5 mm long; stigmas 3, up to 1.5 mm long. Nut trigonous, oblong, apiculate at apex, 1.2-1.5 x 0.3-0.4 mm, minutely puncticulate, ultimately dark brown.

Fls. & Frts. : Aug. - March.

Chrom. No. : n = 29 (*Taxon* 20: 612. 1971).

Habitat : Wet soil along the bank of streams and rivers, wet rocky areas, small ditches in grasslands, wet rice fields in ghats.

Distrib. : Widely distributed in the warmer regions of the whole world. INDIA: Throughout. KARNATAKA: Bangalore, Belgaum, Chikmagalur (Raizada *et al.*, 1979), Dakshina Kannada, Hassan, Kodagu, Kolar, Mysore, Shimoga, Uttara Kannada.

Specimens examined : Bangalore: Bannargatta, *Hooper* & *Saldanha* 18043, 7.11.1971 (JCB). Belgaum: Sutagatti, *Mahajan* 24978, 4.11.1957 (BSI); Gottni nallah, Jambotti R.F., *Prasad* 172878, 15.12.1994 (BSI). Dakshina Kannada: Udyavara, *Bhat* 265, 8.8.1976 (MGH); Panamboor, Mangalore, *Bhat* 485, 5.11.1977 (MGH). Hassan: Outskirts of Hassan town, *Saldanha* 17929, 16.9.1969 (JCB). Kodagu: Katakere, Mercara, *Bhat* 731, 18.12.1980 (MGH); Abbe falls, *Bhat* 867, 23.12.1980 (MGH). Kolar: Kendatti hills, *Prakash* & *Sreenath* 2769, 21.9.1978 (JCB). Mysore: Himavad Gopalaswamy hills, *Padma Rani* 36, 1.11.1970 (MGH). Shimoga: Kavledurga, *Raghavan* 82925, 3.10.1962 (BSI); Agumbe, *Bhat* 701, 26.9.1980 (MGH); Jog falls, *Prasad* 173765, 3.11.1995 (BSI). Uttara Kannada: Castle Rock, *Gammie* 15573, 24.10.1902 (BSI); Alnawar, *Jain* 16015, 8.3.1957 (BSI); Dandeli, *Jain* 16108, 9.3.1957 (BSI); without exact locality and coll. name, 1514, without date (BSI). Without locality (Mysore & Carnatic), *Thomson s.n.* (Acc. No. 73555), without date (MH).

var. **pseudonutans** Kukenth. in Engl., Pflanzenr. 4 (20), Heft 101: 140. 1935; Kern in van Steenis, Fl. Males. 1, 7: 610. 1974; Hooper in Saldanha & Nicolson, Fl. Hassan 665. 1976; Rao & Verma, Cyp. NE India 16. 1982; Sharma *et al.*, Fl. Karnataka 305. 1984; Karthik. *et al.*, Fl. Ind. Enum. Monocot. 44. 1989; Keshava Murthy & Yoganarasimhan, Fl.Coorg 502. 1990. *Cyperus nutans* auct. non Vahl,1806; Clarke in Hook.f., Fl. Brit. India 6: 607. 1893, *p.p.*

It differs from the typical variety by the suberect spikelets even in fruit, hence the penicillate (brush-like) appearance of the spikes.

Fls. & Frts. : Sept. - Dec.

Habitat : Wet soil along the bank of streams, rivers, paddy fields, slopes near waterfalls and rocky river beds.

Distrib. : Sri Lanka, Taiwan, Malesia, Formosa, Africa. INDIA: South India, North-East and North-West India: KARNATAKA: Belgaum, Chikmagalur, Dakshina Kannada, Hassan, Kodagu, Mysore, Shimoga, Uttara Kannada.

Specimens examined : Belgaum: Gottni nallah, Jambotti, R.F., *Prasad* 172875, 15.12.1994 (BSI). Chikmagalur: Kemmangudi, *Raghavan* 103679, 20.10.1964 (BSI). Dakshina Kannada: Gundia, without coll. name, 18230, 28.11.1927 (MH). Hassan: Kenchankumri, *Hooper* & *Gandhi* 2444, 12.11.1971 (JCB). Kodagu: Karwala badiga, West of College, *Rao* 74527,

22.9.1961 (BSI); Kottamudi, *Rao* 74950, 5.10.1961 (BSI); Shanivarasanthe, *Yoganarasimhan* 4167, 9.8.1983 (RRCBI). Mysore: Ketedevergudi vicinity, *Rao* 73571, 4.9.1961 (BSI). Shimoga: Venakeabbi falls, Agumbe, *Raghavan* 68108, 3.11.1960 (BSI); Jog falls, *Prasad* 173791, 4.11.1995 (BSI). Uttara Kannada: Yellapur, *Talbot s.n.* (Acc. No. 684), 10.10.1884 (BSI); Castle Rock, *Gammie* 15764, 28.10.1902 (BSI); Karka river bed, Dandeli, *Prasad* 173718, 30.10.1995 (BSI).

Note: As the name indicates, this variety can be confused with *C. nutans* Vahl as they are very similar in habit. But can be separated by the distantly arranged glumes on the linear spikelets. In *C. nutans* spikelets are *ca* 2 mm broad and glumes *ca* 1/3 imbricate.

15. **Cyperus esculentus** L., Sp. Pl. 45. 1753; Clarke in Hook.f., Fl. Brit. India 6: 616. 1893; Fischer in Gamble, Fl. Pres. Madras 1641. 1931 (3: 1141. 1957, repr. ed.); Kukenth. in Engl., Pflanzenr. 4 (20), Heft 101: 116. 1935; Ramaswamy & Razi, Fl. Bangalore 86. 1973; Kern in van Steenis, Fl. Males. 1, 7: 605. 1974; Sharma *et al.*, Karnataka 306. 1984; Karthik. *et al.*, Fl. Ind. Enum. Monocot. 45. 1989. *Nelabadami hullu.*

Perennials with short rhizome and very slender stolons, *ca* 35 cm high. Stolons yellowish, clothed with sheaths, ending in ovoid to globose tubers, often disappearing after tuber formation; sheaths with striate, ovate-lanceolate scales towards apex, acute at apex; tubers (not seen) *ca* 1 cm thick, transversely zoned when young, covered with a grey tomentum when mature. Stems slender, erect, triquetrous, 1-2 mm thick, smooth. Leaves few to several, shorter or longer than stem, linear, gradually narrowed to an acuminate apex, 3-4 (-6) mm wide; sheaths stramineous to reddish-brown towards base, often purple dotted on the papery ventral side. Inflorescence simple or compound, loose or somewhat dense, 5-10 x 4-6 cm. Involucral bracts 3-6, obliquely spreading, lower 1 or 2 overtopping the inflorescence. Primary rays 3-6 (-8), unequal, up to 6 (-10) cm long. Spikes ovoid, with few to many spikelets; rachis glabrous or sparsely scabrous on the angles. Spikelets spicately arranged, compressed, linear-oblong, subobtuse at apex, 6-15 x *ca* 2 mm, 12-16-flowered; rachilla winged, with *ca* 0.8 mm long internodes. Glumes membranous, often somewhat loosely imbricate, obliquely erect, elliptic-ovate or oblong-ovate, obtuse at apex, at times minutely mucronulate, *ca* 2.5 x 1.5 mm, yellowish to pale brown, with whitish-hyaline margins, 7-nerved; keel not prominent. Stamens 3; filaments elongate up to 3 mm; anthers linear, *ca* 1.5 mm long, with a red connective appendage. Style short, *ca* 1 mm long; stigmas 3, much longer than style, *ca* 3 mm long. Nut trigonous, obovoid to oblong-obovoid, obtuse at apex, hardly apiculate, *ca* 1.5 x 1 mm.

Fls. & Frts.: June

Habitat : Sandy soil, probably near coastal areas.

Distrib. : Malesia, from Mediterranean region to South Africa, Australia and America. Also cultivated. INDIA: Peninsular India, Central, North and North-West India. KARNATAKA: Bangalore (Ramaswamy & Razi, *l.c.*), Uttara Kannada.

Specimen examined : Uttara Kannada: Karwar, *Talbot* 1255, 25.6.1885 (BSI).

Note: This species seems to be very rare in the state. Only the above mentioned specimen was found in BSI from Karnataka. Though very similar to *C. rotundus* L., it can be distinguished by the slender stolons which never become hard but usually disintegrate after tuber formation. In the above mentioned specimen in some of the spikelets a peculiar growth of the floral parts was found within the glumes. This malformation which enclose the ovary resembling the utricle of *Carex* sp. may be due to some disease caused by some micro-organisms.

16. **Cyperus exaltatus** Retz., Obs. Bot. 5: 11. 1789; Clarke in Hook.f., Fl. Brit. India: 6: 617. 1893; Cooke, Fl. Pres. Bombay 2: 872. 1908 (3: 386. 1958, repr.ed.); Fischer in Gamble, Fl. Pres. Madras 1642. 1931 (3: 1141. 1957, repr.ed.); Kukenth. in Engl., Pflanzenr. 4 (20), Heft 101: 64, f. 9 A - E. 1935; Ramaswamy & Razi, Fl. Bangalore 86. 1973; Kern in van Steenis, Fl. Males. 1, 7: 602. 1974; Hooper in Saldanha & Nicolson, Fl. Hassan 665. 1976; Rao & Razi, Fl. Mysore 559. 1981; Rao & Verma, Cyp. NE India 19. 1982; Sharma *et al.*, Fl. Karnataka 306. 1984; Koyama in Dassanayake & Fosberg, Rev. Handb. Fl. Ceylon 5: 173. 1985; Karthik. *et al.*, Fl. Ind. Enum. Monocot. 45. 1989; Keshava Murthy & Yoganarasimhan, Fl. Coorg 502. 1990. *Kempajambu hullu.*

Type : India, Tranquebar.

Illus. : Kukenth., *l.c.*

Key to the varieties

1a. Spikelets obliquely patent to almost horizontal to the rachis, brownish to reddish-brown .... var. **exaltatus**

1b. Spikelets horizontally spreading, golden yellow .... var. **dives**

var. **exaltatus**

Tall perennials with short, woody rhizome, 80-180 cm high. Stems tufted, trigonous, 4-10 mm thick, smooth. Leaves often equalling the stem, at times longer, coriaceous, flat or plicate, 6-15 mm wide, with a prominent midrib beneath and 2 lateral nerves above, scabrous on the margins; sheaths spongy, purplish-brown to blackish towards base. Inflorescence compound or decompound, 12-25 (-35) cm long. Involucral bracts 4-6, lower 2-4 overtopping the inflorescence, up to 70 cm long. Primary rays 6-9, slender, erect-patent, unequal, up to 23 cm long, smooth; secondary rays up to 7 cm long. Spikes digitately arranged but a few solitary on slender peduncles, cylindrical, 2-6 cm long, 7-15 mm wide, subdensely to loosely bearing up to 44 spikelets, rachis visible among the spikelets, glabrous. Spikelets spicately arranged, obliquely patent to almost right angle to the rachis, flattened, oblong to oblong-lanceolate, 4-10 x 1.5-2 mm, 8-26-flowered, usually reddish-brown or brownish; rachilla straight, with narrow caducous wings and *ca* 0.4 mm long internodes. Glumes membranous, ovate to broadly ovate, mucronate at apex, 1.5-2 x 1.2-1.5 mm; keel green, 3-5-nerved; sides nerveless, brownish to shining brown. Stamens 3; filaments up to 2 mm long; anthers oblong, 0.5 - 0.7 mm long, with a shortly produced connective appendage. Style 1-1.5 mm long; stigmas 3, *ca* 1 mm long. Nut trigonous, ellipsoid, minutely apiculate, 0.7-0.8 x 0.3-0.5 mm, yellowish-brown.

Fls. & Frts. : July-March.

Habitat : Wet localities, flowing water in narrow canals, marshy edges of stagnant water bodies, muddy banks of rivers and canals, muddy lake beds, shallow water in ponds, often along with other marshy plants like *Coix lacryma-jobi, Hygrophila auriculata, Limnophila* sp., etc.

Distrib. : Sri Lanka, Indo-China, East China, Malesia, Japan, tropical Africa and South Australia. INDIA: Peninsular India, North-East and North-West India. KARNATAKA: Bangalore, Belgaum, Chikmagalur (Sharma *et al., l.c.*), Dharwar, Hassan, Kodagu, Mysore, Uttara Kannada.

Specimens examined : Bangalore: On the way to Ramanagaram, *Ramesh* 236, 23.2.1978 (JCB). Belgaum: Amod, *Saldanha & Prakash* 3516, 25.10.1978 (JCB); On the way to Ghataprabha dam from Gokak, *Prasad* 172836, 12.12.1994 (BSI). Dharwar: Haveri, *Talbot* 2227, 2.1.1890 (BSI); without exact locaity, *Talbot* 2614, 15.9.1891 (BSI); Kiligerry, *Talbot* 2950, 15.9.1891 (BSI); Pond near Dharwar Railway gate, Dharwar, *Prasad* 172915, 18.12.1994 (BSI); Salikkinikuppa lake, Dharwar, *Prasad* 172917, 18.12.1994 (BSI); Hallikeri, adjacent to R.F., *Prasad* 172931, 18.12.1994 (BSI); Kalikeri village, near R.F., *Prasad* 172949, 19.12.1994 (BSI). Hassan: 3 km before Hassan from Dudda, *Hooper & Gandhi* 2383, 10.11.1971 (JCB). Kodagu: Kushalnagar, *Bhat* 1036, 21.9.1981 (MGH). Mandya: Sreerangapatana, *Dinesh* 1013, 6.7.1984 (MGH). Mysore: Chamundi, *Bhat* 8, 10.7.1970 (JCB). Uttara Kannada: Small tank near Halyal bus stand, *Prasad* 172994, 29.10.1995 (BSI). Without locality (Mysore & Carnatic), *Thomson s.n.* (Acc. No. 73616), without date (MH).

var. **dives** (Del.) Clarke in Journ. Linn. Soc. 21: 186. 1884 *et* in Hook. f., Fl. Brit. India 6: 617. 1893; Cooke, Fl. Pres. Bombay 2: 872. 1908 (3: 386. 1958, repr. ed.); Rao & Razi, Fl. Mysore 559. 1981; Sharma *et al.*, Fl. Karnataka 306. 1984; Karthik. *et al.*, Fl. Ind. Enum. Monocot. 45. 1989.

Cyperus dives Del., Fl. Egypte 149, t. 4, f. 3. 1812; Kukenth. in Engl., Pflanzenr. 4 (20), Heft 101: 68. 1935.

This variety differs from the typical variety by the horizantally spreading, turgid, golden-yellow spikelets.

Fls. & Frts : Oct. - March.

Habitat : Marshy areas and shallow water tanks.

Distrib. Pakistan, Africa. INDIA: Western Peninsular India and Eastern India. KARNATAKA: Bellary, Chikmagalur (Sharma *et al., l.c.*), Mysore, Shimoga (Talguppa, Kukenth., *l.c.*) Uttara Kannada.

Specimens examined : Bellary: Banderi, *Manohar & Ramesh* 5856, 18.1.1979 (JCB). Mysore: Mysore, *Bhat* 91, 25.9.1970 (JCB). Uttara Kannada: Javalli village, Halyal range, *Prasad* 173743, 30.10.1995 (BSI).

17. **Cyperus haspan** L., Sp. Pl. 45. 1753; Clarke in Hook.f., Fl. Brit. India 6: 600. 1893; Cooke, Fl. Pres. Bombay 2:863. 1908 (3:376. 1958, repr.ed.); Fischer in Gamble, Fl. Pres. Madras 1640. 1931 (3: 1139. 1957, repr.ed.); Kukenth. in Engl., Pflanzenr. 4 (20), Heft 101: 247, f. 28 E-G. 1936; Kern in van Steenis, Fl. Males 1, 7: 624, f. 56-57. 1974 (*C. halpan* L.); Hooper in Saldanha & Nicolson, Fl. Hassan 665. 1976; Rao & Razi, Fl. Mysore 559. 1981; Rao & Verma, Cyp. NE India 13. 1982 (*C. halpan*); Sharma *et al.*, Fl. Karnataka 306. 1984; Koyama in Dassanayake & Fosberg, Rev. Handb. Fl. Ceylon 5: 203. 1985; Singh, Fl. E. Karnataka 2: 631. 1988; Karthik. *et al.*, Fl. Ind. Enum. Monocot. 45. 1989; Keshava Murthy & Yoganarasimhan, Fl. Coorg 502. 1990. *Chikkatungu hullu.*

Type : India and Ethiopia.

Illus. : Kukenth., *l.c.*; Kern, *l.c.*

Key to the subspecies

1a. Plants comparatively short, up to 35 cm; rhizome not creeping; spikelets 5-10 mm long; glumes 1-1.3 mm long; stamen 1; anther *ca* 0.5 mm long; nut *ca* 0.5 mm long. .... subsp. **haspan**

1b. Plants comparatively tall, up to 70 cm ; rhizome usually long-creeping; spikelets up to 20 mm long; glumes 1.3 - 2 mm long; stamens 2-3 (rarely 1 in same spikelet); anthers 0.5-1 mm long; nut 0.6 - 0.8 mm long. .... subsp. **juncoides**

subsp. **haspan**

Perennials with short rhizome. Stems solitary or tufted, triquetrous, 10-35 cm long, 1-2.5 mm thick, smooth. Leaves usually shorter than stems, flat, gradually narrowed to an acute apex, 2-4 mm wide, smooth or scaberulous on the margins towards apex; lower sheaths membranous, reddish-brown or cinnamomeus; at times all the leaves reduced to a lanceolate appendage of the sheaths. Inflorescence compound to decompound, loose to subdense, up to 12 cm across. Involucral bracts 2 or 3, usually all smaller than the inflorescence; lowermost at times overtopping, up to 10 cm long. Primary rays many, slender, unequal, up to 10 cm long, smooth; secondary rays up to 2.5 cm long. Spikelets digitately arranged in clusters of 3-6, strongly compressed, linear-oblong to linear-lanceolate, acute at apex, 5-10 x 1-1.5 mm, 10-24-flowered; rachilla hidden by densely arranged glumes, persistent, wingless. Glumes membranous, suberect, oblong-ovate, somewhat obtuse and muticous at apex, 1-1.3 x 0.5-0.7 mm; keel 3-nerved, green; sides pale brown to purplish-brown, at times variegated, nerveless. Stamen 1; filament up to 1.2 mm long; anther oblong, *ca* 0.5 mm long, with bristly connective appendage at apex. Style *ca* 0.5 mm long, narrowed to base; stigmas 2, almost as long as or slightly longer than style. Nut trigonous, broadly obovoid, minutely apiculate at apex, prominently stipitate, *ca* 0.5 mm long and as wide, grannulate-verruculose, creamish-yellow.

Fls. & Frts. : Sept.- Feb.

Chrom. No. : n = 18 (*Taxon* 20: 612. 1971).

Habitat : Open wet places along canals and streams, usually found as a weed in flooded rice fields.

Distrib. : Tropical and subtropical regions of the whole world. INDIA: Throughout. KARNATAKA: Bangalore, Belgaum, Chikmagalur, Dakshina Kannada, Hassan, Kodagu, Mandya, Mysore, Uttara Kannada.

Specimens examined : Bangalore: Kanakapura - Sangam Road, *Saldanha et al.* 26, 28.1.1978 (JCB). Belgaum: Manjapur nallah, Khanapur, *Prasad* 172856, 14.12.1994 (BSI); Gottni nallah, Jambotti R.F., *Prasad* 172873, 15.12.1994 (BSI). Chikmagalur: Bababudan hills, *Saldanha* 9536, 27.9.1979 (JCB). Dakshina Kannada: Shiradi, *Hooper & Saldanha* 2572, 24.11.1971 (JCB); Thumbe, *Raghavan* 146477, 20.4.1977 (BSI); Panamboor, Mangalore, *Bhat* 493, 5.11.1977 (MGH); Netravati river side, Kadakar village, *Prasad* 173877, 10.11.1995 (BSI). Hassan: Kempuhole, *Nicolson et al.* 2334, 25.10.1971 (JCB). Kodagu: On the way to makut, *Ramesh & Prakash* 3160, 10.11.1978 (JCB); Mercara, *Bhat* 789, 18.12.1980 (MGH). Mandya: Sreerangapatana, *Raghavendra* 91, 15.8.1967 (MGH); Ranganathittu, *Bhat* 94, 20.11.1970 (JCB). Mysore: St. Philomina College, *Suma* 14, 3.8.1968 (MGH). Uttara Kannada: Along Kadenkette river, *Ahmed* 1160, 29.5.1978 (JCB); Virnoli range, Dandeli, *Manohar* 6035, 14.2.1979 (JCB).

Note: Sharma *et al.* (*l.c.*), reported subsp. *haspan* from Shimoga and Uttara Kannada districts most probably based on wrongly identified specimens. Specimens in BSI from these districts are of subsp. *juncoides*. The former is not common and usually found in disturbed habitats like paddy fields as a weed. Though reported by Singh (*l.c.*) from Kolar district, specimens cited are of *C. tenuispica* Steud.

subsp. **juncoides** (Lamk.) Kukenth. in Fedde. Repert. 23: 184. 1926 *et* in Engl., Pflanzenr. 4 (20), Heft 101. 249. 1936; Ramaswamy & Razi, Fl. Bangalore 88. 1973 (as *C. haspan*); Kern in van Steenis, Fl. Males. 1, 7: 625. 1974; Hooper in Saldanha & Nicolson, Fl. Hassan 666. 1976; Sharma *et al.*, Fl. Karnataka 306. 1984; Karthik. *et al.*, Fl. Ind. Enum. Monocot. 45. 1989. *C. juncoides* Lamk., Illus. Gen. 1: 147. 1791.

Rhizome usually long-creeping. Stems often single from each node of the long rhizome, up to 65 cm long. Floral parts longer; spikelets 10-18 x 1.5-1.8 mm. Glumes 1.3 - 2 mm long. Stamens 2-3 (rarely 1 in the same spikelet); filaments up to 1.8 mm long; anthers oblong to linear-oblong, 0.5-1 mm long. Style 0.5 - 0.7 mm long; stigmas up to 1 mm long. Nut 0.6 - 0.8 x 0.5 mm. Other characters as in the typical subspecies.

Fls. & Frts. : Oct. - July.

Habitat : Moist soil near streams, canals and lakes, muddy crevices of rocks in streams, damp open areas and muddy areas in forests. Common.

Distrib. : Pantropical. INDIA: Probably as of subsp. *haspan*. KARNATAKA: Bangalore, Belgaum, Chikmagalur, Dakshina Kannada, Hassan, Shimoga, Uttara Kannada.

Specimens examined : Bangalore: Bannarghata, *Saldanha & Nicolson* 4865, 2.12.1978 (JCB). Belgaum: Manjapur nallah, Khanapur, *Prasad* 172854 & 172856 A, 14.12.1994 (BSI); Gottni nallah, Jambotti R.F., *Prasad* 172870, 15.12.1994 (BSI); Dudhwawada R.F., Londa, *Prasad* 172890, 17.12.1994 (BSI). Chikmagalur: Bababudan hills, *Saldanha & Ramesh* 1732, 26.6.1978 (JCB). Dakshina Kannada: Kannadekatte, *Bhat* 168, 6.5.1975 (MGH); Shirvadi Ghat, *Saldanha & Prakash* 4029, 11.11.1978 (JCB). Hassan: Road to Devarunde, *Saldanha* 12429, 28.1.1969 (JCB). Shimoga: Agumbe-Nalur Road, *Raghavan* 62618, 17.5.1960 (BSI); Chyatramane, Agumbe, *Raghavan* 80575, 14.5.1962 (BSI); Ghatibagh, Agumbe, *Raghavan* 80592, 14.5.1962 (BSI); Agumbe, *Bhat s.n.*, 26.9.1980 (MGH); Sharavati river, above Jog falls, *Prasad* 173769, 3.11.1995 (BSI). Uttara Kannada: Yellapur, *Talbot s.n.* (Acc. No. 673), 28.2.1884 (BSI); Castle Rock, *Gammie* 15765, 27.10.1902 (BSI); Castle Rock, *Bhide s.n.* (Acc. Nos. 2457, 2458 & 2459), 10.11.1911 (BSI).

Note: It seems subsp. *juncoides* is more common and prefer more natural habitat. But subsp. *haspan* is usually found in disturbed habitats like paddy fields.

18. **Cyperus involucratus** Rottb., Descr. Pl. Rar. Progr. 22.1772; Baijnath in Kew Bull. 30 : 522. 1975. *C. flabelliformis* Rottb., Descr. Pl. Rar. Progr. 22. 1772 *et* Descr. & Ic. 42, t. 12, f. 2. 1773; Kern in van Steenis, Fl.

Males. 1, 7: 618. 1974; Rao & Razi, Fl. Mysore 559. 1981; Sharma *et al.*, Fl. Karnataka 306. 1984; Karthik. *et al.*, Fl. Ind. Enum. Monocot. 45. 1989. *C. alternifolius* L. subsp. *flabelliformis* Kukenth. in Engl., Pflanzenr. 4 (20), Heft 101: 193. 1936; Koyama in Dassanayake & Fosberg, Rev. Handb. Fl. Ceylon 5: 201. 1985. *C. alternifolius* auct. non L. 1771; Sharma *et al.*, Fl. Karnataka 304. 1984.

Type : Arabia, *Forsskal.*

Illus. : Rottb., *l.c.*

Perennials with stout horizontal rhizome, up to 170 cm high. Stems subtufted or arranged in a series, stout, obtusely trigonous to subterete below, sulcate, 4-8 mm thick, scabrous below the inflorescence. Leaves on the flowering stems reduced to long, pale sheaths, brownish towards base; uppermost sheath with a very short lamina. Inflorescence decompound, 6-10 (-30) cm, across. Involucral bracts many, distinctly spaced, all much overtopping the inflorescence, nearly equal in length, obliquely patent to reflexed, flat, abruptly acuminate at apex, 16-39 (-50) cm long, 5-12 (-15) mm wide, scabrous on the margins. Primary rays numerous, slender, subequal in length, 2-4 (-10) cm long, smooth, bearing 4-10 secondary rays. Spikelets in clusters of 3-10, flattened, ovate, elliptic or oblong-lanceolate, 4-9 x 2-2.5 mm, 8-18 (-40)-flowered; rachillla straight, wingless, with *ca* 0.25 mm long internodes. Glumes membranous, ovate, acute at apex, *ca* 2 x 1.5 mm, 3-5-nerved; keel prominent, green; sides ferrugineous to red-tinged, with broadly hyaline margins. Stamens 3; filaments elongate up to 2.5 mm; anthers linear, *ca* 1 mm long, with an acute connective appendage. Style *ca* 1 mm long; stigmas 3, *ca* 1 mm long. Nut trigonous, ellipsoid, apiculate at apex, *ca* 1 x 0.5 mm, yellowish-brown.

Fls. & Frts. : Oct., Jan.

Habitat : Wet river banks, waste places, often cultivated in gardens and ponds as an ornamental.

Distrib. : Native of Arabia and tropical Africa; often cultivated in many countries. INDIA: Usually cultivated in different parts of the country as an ornamental, also found as escape. KARNATAKA: Belgaum, Dharwar, Mysore (*Rao & Razi, l.c.*); can be found in other districts also as a cultivated ornamental plant.

Specimens examined: Belgaum: Belgaum, *Talbot* 3819, Oct. 1896 (BSI); Near Gokak falls, *Prasad* 172821, 11.12.1994 (BSI). Dharwar: Dharwar, *Naik* 28, 13.1.1984 (MGH).

Note : Besides the morphological characters, a comparative anatomical study of vegetative parts and scanning electron microscope (S.E.M.) study of the nut surface of *Cyperus alternifolius* and *C. involucratus* revealed that these can be treated as seperate species (Baijnath, 1975).

19. **Cyperus iria** L., Sp. Pl. 45. 1753; Clarke in Hook.f., Fl. Brit. India 6: 606. 1893; Cooke, Fl. Pres. Bombay 2: 867. 1908 (3:380. 1958, repr.ed.);

Fischer in Gamble, Fl. Pres. Madras 1640. 1931 (3: 1140. 1957, repr.ed.); Kukenth. in Engl., Pflanzenr. 4 (20): Heft 101: 150. 1935; Ramaswamy & Razi, Fl. Bangalore 89. 1973; Kern in van Steenis, Fl. Males. 1, 7: 616. 1974; Hooper in Saldanha & Nicolson, Fl. Hassan 666. 1976; Rao & Razi, Fl. Mysore 559. 1981; Rao & Verma, Cyp. NE India 15. 1982; Sharma *et al.*, Fl. Karnataka 306. 1984; Koyama in Dassanayake & Fosberg, Rev. Handb. Fl. Ceylon 196. 1985; Singh, Fl. E. Karnataka 2: 631. 1988; Karthik. *et al.*, Fl. Ind. Enum. Monocot. 45. 1989; Keshava Murthy & Yoganarasimhan, Fl. Coorg 502. 1990. *C. paniciformis* Franch. & Sav., Pl. Jap. 2: 103. 537. 1879. *C. iria* var. *paniciformis* (Franch. & Sav.) Clarke in Hook.f., Fl. Brit. India 6: 607. 1893. *Dabbojambu hullu.*

Type: India, *Osbeck.*

Illus. Clarke, Illus. Cyp. t. 14, f. 1-2. 1909; Matthew, Furth. Illus. Fl. Tamilnadu Carnatic. Pl. 640. 1988.

Annuals with yellowish fibrous roots, up to 65 (-90) cm high. Stems usually tufted, at times solitary, slender or somewhat stout, triquetrous, (1-) 2-4 mm thick, smooth. Leaves usually shorter than stems, at times somewhat longer, flat or canaliculate, linear, gradually narrowed to an acute apex, (1-) 3-6 mm wide in the middle portion, scabrous on the margins towards apex; sheaths reddish-brown. Inflorescence compound, usually loose, variable in size and shape, 3-15 cm long. Involucral bracts 3-5, lower ones overtopping the inflorescence, up to 40 cm long and 8 mm wide. Rays 3-8, unequal, patent, slender, up to 11 cm long; longer ones paniculately branched into 3-5 spikes. Spikes usually oblong-ovoid, densely or loosely bearing 5-26 spikelets; rachis glabrous. Spikelets spicately arranged, strongly compressed, erect-patent, oblong to linear-oblong, obtuse at apex, 4-8 x *ca* 2 mm, 6-16-flowered; rachilla straight, wingless, with 0.5 - 0.8 mm long internodes. Glumes somewhat loosely arranged, membranous, broadly obovate or orbiculate, obtuse to emarginate and very shortly mucronate at apex, sharply keeled, 1.3-1.5 x *ca* 1.5 mm; keel 3-5-nerved, green; sides nerveless, yellowish-stramineous, with hyaline margins towards apex. Stamens 2 (-3); filaments up to 1 mm long; anthers oblong, *ca* 0.3 mm long. Style minute; stigmas 3, *ca* 0.5 mm long. Nut triquetrous, almost as long as the glume, elliptic-obovoid, minutely apiculate at apex, broadly stipitate, *ca* 1.5 x 0.7 mm, ultimately blackish-brown.

Fls. & Frts. : July - Feb.

Chrom. No. : n = 64 (*Taxon* 20: 612. 1971).

Habitat : Moist soil near rivers, streams, dams, lakes and tanks; along roadsides, bunds of paddy fields, usually as a weed in paddy fields and other cultivated soils, at times along sea shores.

Distrib. : Widely distributed in tropical Asia, extending to East Africa and Australia. Introduced and naturalised in South East United States, West Indies and South America. INDIA: Throughout. KARNATAKA:

Bangalore, Belgaum, Bellary, Bijapur, Chikmagalur, Chitradurga, Dakshina Kannada, Dharwar, Hassan, Kodagu, Mandya, Mysore, Raichur, Shimoga, Tumkur, Uttara Kannada.

Specimens examined : Bangalore: Bangalore, *Camaron* 442, Jan. 1890 (MH) & 525, 2.11.1890 (MH); Bannarghatta, *John* 9, 6.12.1979 (JCB). Belgaum: Sutagatti *Mahajan* 28681, 3.11.1957 (BSI); Ghataprabha dam, *Prasad* 172807, 10.12.1994 (BSI); Dinman Hassur, Belgaum-Gokak Road, *Prasad* 172848, 13.12.1994 (BSI); Malaprabha river, near the bridge, *Prasad* 172867, 14.12.1994 (BSI); Gottni nallah, Jambotti R.F., *Prasad* 172877A, 15.12.1994 (BSI); Dudwawada village, near Londa R.F., *Prasad* 172900, 17.12.1994 (BSI); Malaprabha dam, Saundati, *Prasad* 172950, 20.12.1994 (BSI). Bellary: Ramadurg, Without coll. name, 15936, 9.10,.1919 (MH). Bijapur: Mangalore R.F., Badami, *Singh* 141629, 12.11.1975 (BSI); Varathi kavlu, Kendur, *Prasad* 172965, 21.12.1994 (BSI). Chikmagalur: Near Forest Guest House, Koppa, *Prasad* 173814, 7.11.1995 (BSI); Hilikeri village, along Sringeri Road, *Prasad* 173817 & 173824, 7.11.1995 (BSI). Chitradurga: Haripar-Harpanhalli Road, 14th km, Davangre, *Singh* 141878, 30.12.1975 (BSI); Jogimath S.F., *Singh* 142270, 10.1.1976 (BSI). Dakshina Kannada: Udyavara, *Bhat* 264, 8.8.1976 (MGH); Jappinamogeru village, *Prasad* 173854, 10.11.1995 (BSI); Near Netravati river, Kadakar village, *Prasad* 173873, 10.11.1995 (BSI). Dharwar: Dharwar, *Patil* 4750, 25.9.1956 (BSI); Salikkinikuppa lake, Dharwar, *Prasad* 172923 & 172926, 18.12.1994 (BSI); Hallikeri, adjacent to R.F., *Prasad* 172933, 18.12.1994 (BSI). Hassan: Bourdalboore forest, *Agarwal* 2, 18.9.1979 (JCB). Kodagu: Kottamudi, along Cauveri river, near Napoklu, *Rao* 74954, 5.10.1961 (BSI); Somvarpet-Kushalnagar, *Ramesh* 2377, 24.8.1978 (JCB); Kirugoor, *Bhat* 924, 26.1.1981 (MGH); Kushalnagar, *Bhat* 959, 16.2.1981 (MGH). Mandya: Ranganathittu, *Padma Rani* 17, 2.8.1970 (MGH); Sreerangapatana, *Ramaswamy* 5, 6.12.1979 (MGH); T.M. Hosur, Ranganathittu, *Dinesh* 737, 8.12.1983 (MGH); Ranganathittu, *Dinesh* 799, 1.2.1984 (MGH). Mysore: Katedevargudi vicinity, *Rao* 73603, 4.9.1961 (BSI); Kuntugudi, Biligirirangan temple vicinity, *Rao* 73676, 5.9.1961 (BSI); Hirakulgudd S.F., Arsikeri, *Rao* 73867, 11.9.1961 (BSI); Bandipur, *Naithani* 21161, 24.8.1964 (MH); Mandakalli, *Bhat* 14, 12.7.1970 (JCB); Biligirirangan hills, *Ramesh* & *Manohar* 8648, 1.8.1979 (JCB). Raichur: Bankaldoddi R.F., *Singh* 141706, 14.11.1975 (BSI). Shimoga: Hosur, *Sreenath* & *Murthy* 2881, 26.7.1978 (JCB); Gaurikeri, Thalaguppa, Sagar Taluk, *Prasad* 173803, 5.11.1995 (BSI). Tumkur: Ippadi S.F., Haliyurdurga, Kunigal, *Singh* 140891, 23.10.1975 (BSI). Uttara Kannada: Yellapur, *Talbot s.n.*, 20.9.1884 (BSI); Darhala-Balmame, *Udayakumar* & *Gurudev Singh* 14641, 9.7.1982 (JCB); Halyal town, *Prasad* 172993, 29.10.1995 (BSI); Forest Nursery, Khurigadha, Halyal, *Prasad* 173704, 29.10.1995 (BSI); Karka river bed, Dandeli R.F., *Prasad* 173719, 30.10.1995 (BSI); Javalli village, Halyal range, *Prasad* 173739, 30.10.1995 (BSI); Karwar beach, *Prasad* 173749, 31.10.1995 (BSI). Without exact locality (Mysore-Carnatic), *Thomson s.n.*, without date (MH).

20. **Cyperus laevigatus** L., Mant. Pl. 2: 179. 1771; Cooke, Fl. Pres. Bombay 2: 860. 1908 (3: 373. 1958, repr.ed.); Kukenth. in Engl., Pflanzenr. 4 (20), Heft 101: 321. 1936; Hooper in Saldanha & Nicolson, Fl. Hassan 666. 1976; Rao & Razi, Fl. Mysore 560. 1981; Sharma *et al.*, Fl. Karnataka 306. 1984; Singh, Fl. E. Karnataka 2: 632. 1988. *Juncellus laevigatus* (L.) Clarke in Hook. f., Fl. Brit. India 6: 596. 1893; Fischer in Gamble, Fl. Pres. Madras 1629. 1931 (3: 1133. 1957, repr.ed.); Karthik. *et al.*, Fl. Ind. Enum. Monocot. 57. 1989.

Illus.: Matthew, Illus. Fl. Tamilnadu Carnatic Pl. 783. 1982.

Perennials with long creeping rhizome, 15-45 cm high. Rhizome terete, 1.5-3.5 mm thick, covered with brown scales. Stems usually in a row from the nodes of the long rhizome, at times crowded, rigid, erect or reflexed in the upper portion, cylindrical, 1-2 mm thick. Leaves reduced to papery sheaths except the upper most; upper most leaf with a short blade, terete, obtuse to acute at apex, 0.5 - 3.5 cm long; sheaths cylindrical, 1-6 cm long. Inflorescence a pseudolateral cluster of 2-18 sessile spikelets, 5-18 mm across, straw-coloured. Involucral bracts 2; longest erect, looks like continuation of the stem, acute at apex, 1.5-5.5 cm long; second bract short, spreading. Spikelets compressed, oblong-lanceolate, acute at apex, 5-10 x *ca* 2 mm, 10-22-flowered; rachilla quadrangular, wingless. Glumes distichous, closely imbricate, membranous, broadly ovate, acute and minutely mucronulate at apex, *ca* 2 x 2 mm, usually with reddish-brown spots; midrib prominent towards apex. Stamens 3; filaments up to 2.2 mm long; anthers linear-oblong, acute at apex, *ca* 1 mm long, yellowish with reddish-brown tip. Style 1-1.3 mm long; stigmas 2, shorter than style. Nut planoconvex, obovate or elliptic, 1-1.3 x 0.8-1 mm, pale greyish-brown.

Fls. & Frts. : June - Feb.

Habitat : Saline marshy areas, dried up muddy areas near marshes, moist sandy soil along rivers, nallahs, ponds etc., wet fields along roadsides and waste places, also in cultivated fields.

Distrib. : West Asia, Mediterranean Europe, South America, Africa and Australia. INDIA: South, West, Central, East and North India. KARNATAKA: Belgaum, Bellary, Chitradurga, Hassan, Mysore, Raichur, Tumkur.

Specimens examined : Belgaum: on the way to Ghataprabha from Gokak, *Prasad* 172834 & 172838, 12.12.94 (BSI). Bellary: Harihar-Harpanhalli Road, 21st km, *Singh* 141286, 2.11.1975 (BSI); Hospet-Koppal Road, 26th km, *Singh* 141512, 9.11.1975 (BSI). Chitradurga: Chitradurga-Hiriyur Road, 12th km, *Singh* 132607, 24.2.1975 (BSI). Hassan: Tank near Dandiganahalli, *Hooper & Gandhi* 2406, 11.11.1971 (JCB). Mysore: T. Narsipur, *Bhat* 87, 18.10.1970 (JCB); Bandipur, *Teylor* 14196, 19.8.1981 (JCB). Raichur: Shorapur-Lingsugur Road, 13th mile, *Singh* 129554, 13.2.1975 (BSI); Murgmaddi, Gunthgola

R.F., *Singh* 129578, 14.2.1975 (BSI); Ron-Kustagi Road, 25th mile, *Singh* 129788, 18.2.1975 (BSI); Gunthgola Block, near Amreshwar, *Singh* 132776, 8.9.1974 (BSI); Lingsugur, 4th km on Humnabad Road, *Singh* 141665, 13.11.1975 (BSI); Koppal-Kushtagi Road, 21st km, *Singh* 142510, 16.1.1976 (BSI); Lingsugur-Sindhnoor Road, 40th km, *Singh* 143046, 18.8.1976 (BSI); Near Deodurga, *Ramesh & Sreenath* 4405, 16.11.1978 (JCB). Tumkur: Shabiharaha- Jogihalli, on the way to Tumkur, *Rao* 73224, 27.8.1961 (BSI).

21. **Cyperus laxus** Lamk., Illus. 1: 146. 1791, *emend* Poir. in Lamk., Encycl. 7: 260. 1806; Raynal in Adansonia 2, 17: 277. 1978; Karthik. *et al.*, Fl. Ind. Enum. Monocot. 45. 1989. *C. diffusus* Vahl, Enum. Pl. 2: 321. 1805; Clarke in Hook.f., Fl. Brit. India 6: 603. 1893; Fischer in Gamble, Fl. Pres. Madras 1639. 1931 (3: 1139. 1957, repr.ed.); Kukenth. in Engl., Pflanzenr. 4 (20), Heft 101: 208. 1936; Kern in van Steenis, Fl. Males. 1, 7: 619. 1974; Arora *et al.*, Bot. S. Kanara 61. 1981; Rao & Verma, Cyp. NE India 14. 1982; Koyama in Dassanayake & Fosberg, Rev. Handb. Fl. Ceylon 5: 197. 1985.

Key to the subspecies

1a. Spikelets oblong, up to 7 mm long, 6-10-flowered; rachilla visible in later stages; glumes not closely imbricate .... subsp. **laxus**

1b. Spikelets linear, up to 14 mm long, 12-30-flowered; rachilla not visible even in later stages; glumes closely imbricate. .... subsp. **macrostachyus**

subsp. **laxus**

Perennials with short, woody rhizome, 30-50 cm high. Stems solitary or tufted, somewhat stout, triquetrous towards apex, 2-3 mm thick, smooth. Leaves many, shorter or longer than stem, flat, linear, acuminate at apex, 6-10 mm wide, scabrous on the margins, with 3 prominent nerves; sheaths purplish-brown towards the base, splitting into fibres in later stage. Inflorescence decompound, diffuse, 6-12 cm across. Involucral bracts 4-10, spreading, longer ones much overtopping the inflorescence, up to 14 mm wide. Primary rays many, spreading, unequal, up to 6 cm long, smooth; secondary rays up 2 cm long. Spikelets digitately arranged, 2-5 together, at times solitary, sessile when in cluster, somewhat compressed, oblong, subacute at apex, 4-7 x *ca* 2 mm, 6-10- flowered. Rachilla straight, visible when the nuts mature, narrowly winged, persistent; internodes 0.7 - 0.8 mm long. Glumes membranous, broadly ovate to suborbicular, obtuse and mucronate at apex, keeled, 1.5 -2 x *ca* 1.5 mm (including *ca* 0.5 mm long, slightly excurved mucro); keel strongly 3-nerved; sides faintly nerved, whitish-hyaline at margins. Stamens 3; filaments elongate up to 1.5 mm; anthers linear-oblong, *ca* 0.7 mm long, with a produced connective appendage at apex. Style minute; stigmas 3, *ca* 1 mm long, exserted from the glume. Nut surpassing the glume and hence clearly visible in the spikelet, triquetrous, ellipsoid, acute and minutely apiculate at apex, *ca* 1.5 x 0.7 mm, ultimately blackish-brown.

Fls. & Frts. : Nov.

Habitat : River banks.

Distrib. : Sri Lanka, South China, Malesia, Formosa. INDIA: Throughout. KARNATAKA: Dakshina Kannada (Arora *et al.*, *l.c.*).

Note : Presence of this subspecies in Karnataka has to be verified, as all the specimens found in BSI are of the subspecies *macrostachyus*. It is included in this flora as *Arora et al.* (*l.c.*) reported the species without ascribing the infraspecific taxon. The description provided above is based on specimens from Kerala.

subsp. **macrostachyus** (Boeck.) V.P. Prasad & N.P. Singh in Phytotaxonomy 1 : 64. 2001. *Cyperus diffusus* Vahl var. *macrostachyus* Boeck. in Linnaea 35: 534. 1868; Kukenth. in Engl., Pflanzenr. 4 (20), Heft 101: 209. 1936; Kern in van Steenis, Fl. Males. 1,7: 619, f. 54. 1974; Rao & Verma, Cyp. NE India 14. 1982. *C. pubesquama* Steud. [in Zoll. Syst. Verz. 1: 62. 1854, *nom. nud.*] Syn. Pl. Glum. 2: 20. 1855; Clarke in Hook.f., Fl. Brit. India 6: 604. 1893; Fischer in Gamble, Fl. Pres. Madras 1639. 1931 (3: 1139. 1957, repr.ed.); Hooper in Saldanha & Nicolson, Fl. Hassan 668. 1976; Sharma *et al.*, Fl. Karnataka 307. 1984. *C. diffusus* Vahl subsp. *macrostachyus* (Boeck.) Koyama in Gard. Bull. Singapore 30: 139. 1977 *et* in Dassanayake & Fosberg, Rev. Handb. Fl. Ceylon 5: 198.1985. *C. laxus* Lamk. var. *macrostachyus* (Boeck.) Karthik. in Karthik. *et al.*, Fl. Ind. Enum. Monocot. 46. 1989.

Type : Java.

Illus. : Kern, *l.c.*.

Comparatively larger habit, up to 75 cm high. Stems up to 5 mm thick. Inflorescence decompound to supradecompound, 10-30 cm across. Primary rays up to 20 cm long; secondary rays up to 4 cm and tertiary up to 2.5 cm long. Spikelets up to 9 together, less compressed, linear, 4-14 x *ca* 2 mm, 12-30-flowered. Rachilla completely covered by glumes even while fruiting; internodes *ca* 0.5 mm long. Glumes closely imbricate even at maturity, hence nuts not visible. Nut usually *ca* 1.2 mm long. Other characters are same as the typical subspecies.

Fls. & Frts. : July - Feb.

Habitat : Common in evergreen forests near streams, rocky areas near waterfalls, swampy areas, (partly submerged) along streams.

Distrib. : Sri Lanka, Myanmar, South China, Malesia, Vietnam,

Philippines, Formosa, New Guinea. INDIA: Western Peninsular India and North-East India. KARNATAKA: Chikmagalur, Dakshina Kannada, Hassan, Mysore (Fischer, *l.c.*), Shimoga.

Specimens examined : Chikmagalur: Kemmangudi, *Radhakrishna* 85, without date (MGH). Dakshina Kannada: Charmadi R.F., without coll. name, 018166, 23.11.1927 (MH); without exact locality, coll. name and number (Acc. No. 73467), without date (MH). Hassan: Kenchankumri State forest, *Saldanha* 14218, 18.7.1969 (JCB); Shiradi ghat, *Saldanha* 15857, 17.12.1969 (JCB). Shimoga: Pond near Silviculturist Office, Agumbe, *Raghavan* 68111, 4.11.1960 (BSI); Emkalkare, Agumbe, *Raghavan* 69348, 5.2.1961 (BSI); Varahi falls, Hulical, *Raghavan* 83046, 8.10.1962 (BSI); Hulical ghat area, *Raghavan* 86273, 15.12.1963 (BSI); Varahi forests, Hulical, *Raghavan* 90235, 26.8.1963 (BSI); Agumbe, *Bhat* 716, 7.12.1980 (MGH).

Note : It seems subspecies *macrostachyus* is common in the evergreen forests of Western Ghats. Only a single specimen of typical *C. laxus* was found in BSI, that also from the Kallada valley of Kerala.

22. **Cyperus macer** Clarke in Journ. Linn. Soc. 21: 160. 1884 *et* in Hook.f., Fl. Brit. India 6: 613. 1893; Kukenth. in Engl., Pflanzenr. 4 (20), Heft 101: 83. 1935; Rao & Verma, Cyp. NE India 18. 1982; Sharma *et al.*, Fl. Karnataka 306. 1984; Karthik. *et al.*, Fl. Ind. Enum. Monocot. 46. 1989.

Perennials with short, creeping, slender rhizome, 50-80 cm high. Stems tufted, slender, trigonous or almost triquetrous towards apex, thickened at base, 2-2.5 mm thick. Leaves much shorter than stem, 5-12 cm long, *ca* 2 mm wide. Inflorescence usually simple some times subcompound, loose, narrow. Involucral bracts 2-3, shorter than inflorescence. Rays 3-6, unequal, 4-8 cm long. Spikes bearing 2-5 spikelets. Spikelets subdigitate, suberect, subcompressed, linear-oblong, 6-20 x 1.5-2 mm, 8-24-flowered; rachilla with linear-oblong, hyaline wings. Glumes remote, not imbricate in fruit, obliquely spreading, oblong-ovate, obtuse at apex, *ca* 3 mm long, 3-5-nerved on the green back; sides brown, nerveless, with inrolled margins. Stamens 3; anthers linear-oblong, *ca* 1 mm long. Style shorter than nut; stigmas 3, long-exserted. Nut trigonous, oblong-obovoid, apiculate, about half as long as the glumes, ashy-black.

Fls. & Frts. : Not known from Karnataka.

Habitat : Rocky river beds.

Distrib. : Bangladesh, Myanmar. INDIA: Peninsular India, Central India and North- East India. KARNATAKA: Uttara Kannada (Sharma *et al., l.c.*).

Note : This species is included here as reported by Sharma *et al., l.c.* As no specimens are available in BSI, description provided above is compiled from Clarke, *l.c.*, Kukenth., *l.c.* and Rao & Verma, *l.c.*

23. **Cyperus malaccensis** Lamk., Illus. 1: 146. 1791; Clarke in Hook.f., Fl. Brit. India 6: 608. 1893; Cooke, Fl. Pres. Bombay 2: 869. 1908 (3:382. 1958, repr.ed.); Fischer in Gamble, Fl. Pres. Madras 1641. 1931 (3: 1140. 1957, repr.ed.); Kukenth. in Engl., Pflanzenr. 4 (21), Heft 101: 86. 1935; Kern in van Steenis, Fl. Males. 1, 7: 613. 1974; Karthik. *et al.*, Fl. Ind. Enum. Monocot. 46. 1989. **Fig. 10**

Perennials with stout stolons, up to 1.5 m or more high. Stolons clothed with scarious, ovate, dark brown scales. Stems robust, spongious, triquetrous, almost 3-winged above, 6-10 mm thick in the middle portion. Lower leaves reduced to purplish to blackish sheaths; upper leaves usually short-laminate, with long sheaths; lamina abruptly acuminate, 5-10 mm wide. Inflorescence usually compound, broader than long, at times congested, 3.5 - 8 x 4 - 13 cm. Involucral bracts 3-4, much overtopping the inflorescence, abruptly acuminate at apex; lowest often erect, up to 32 cm long, 7-13 mm wide. Primary rays 3-9, unequal, spreading, slender, 1.5 - 10 cm long, smooth; secondary rays 1.5-2 cm long. Spikes broadly ovoid, bearing 6-12 spikelets; rachis glabrous. Spikelets spicately arranged, subterete, linear, at times slightly curved, 8-20 (-30) x 1-1.5 mm, 14-24-flowered; rachilla very narrowly winged, with *ca* 0.7 mm long internodes. Glumes chartaceous, ovate to elliptic, obtuse and muticous at apex, without keel, 2-2.3 x 1.2-1.5 mm, purplish-lineolate on both sides of the midrib, yellowish-hyaline towards margins; veins not prominent. Stamens 3; filaments up to 2.2 mm long; anthers linear, *ca* 1 mm long, with a prominent connective appendage. Style *ca* 1 mm long; stigmas 3, *ca* 1.8 mm long. Nut trigonous, narrowly oblong, minutely apiculate at apex, 1.8 - 2 x *ca* 0.5 mm, dark brown to blackish.

Fls. & Frts. : Nov. - Jan.

Habitat : Muddy shores of brackish water, along with mangrove vegetation, shallow canals with muddy bottom near back waters, usually forming large patches.

Distrib.: South China, Iran, Malesia, Japan and Northern Australia. INDIA: Western Peninsula and Eastern India. KARNATAKA: Dskahina Kannada, Uttara Kannada.

Specimens examined : Dakshina Kannada: Mangalore, *Foulker* 4866, Jan. 1902 (MH); Ullal bridge, Mangalore, *Hooper & Saldanha* 2534, 25.11.1971 (JCB); Near Netravati river, Kadakar, Mangalore, *Prasad* 173870, 10.11.1995 (BSI). Uttara Kannada: Kaliguj, Karwar, *Prasad* 173753, 1.11.1995 (BSI).

Note: This tall sedge can be easily distinguished in muddy areas of saline waters and in muddy river-mouths, where it forms large patches.

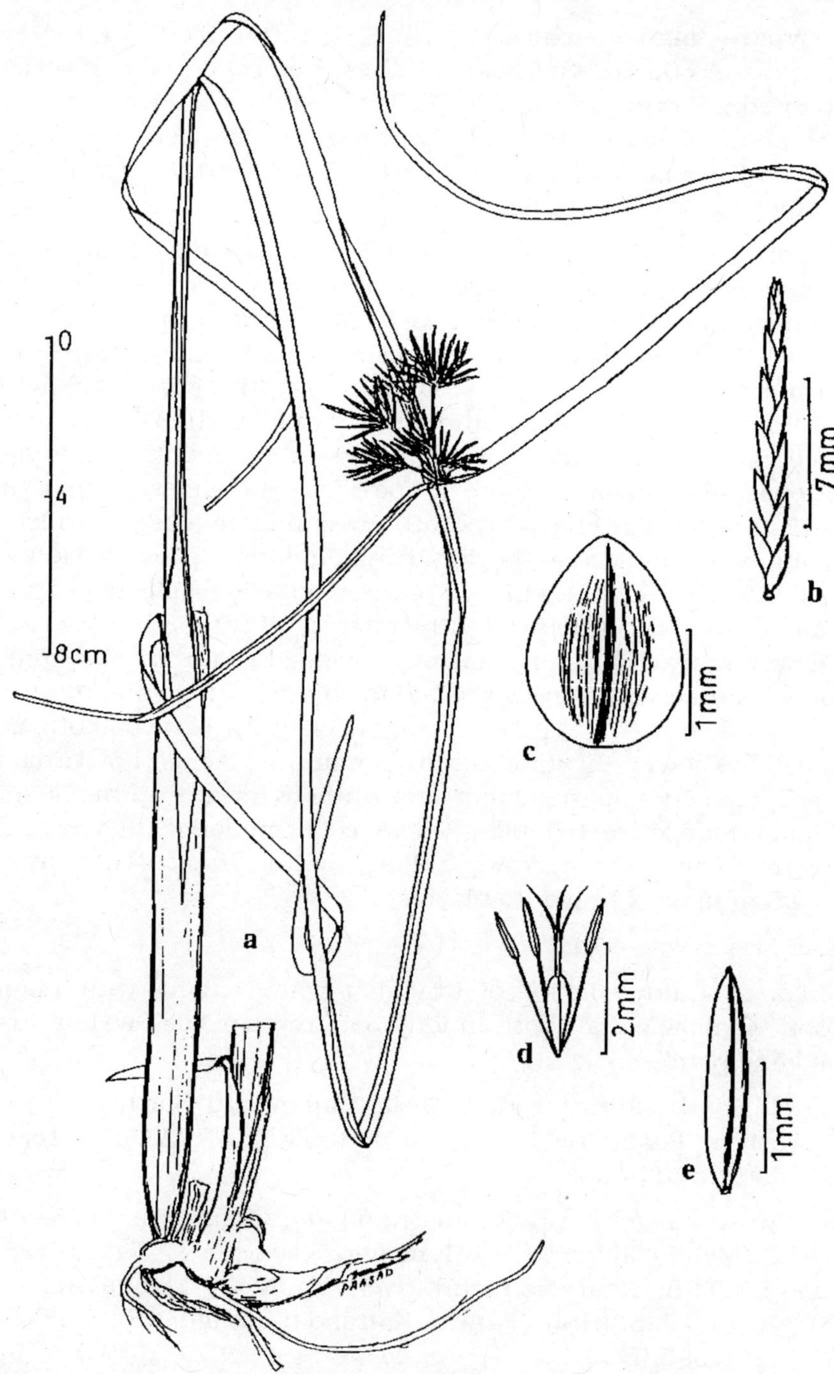

Fig. 10. *Cyperus malaccensis* Lamk.
a. Habit, **b.** Spikelet, **c.** Glume, **d.** Flower, **e**. Nut

24. **Cyperus meeboldii** Kukenth. in Fedde. Repert. 18: 345. 1922 *et* in Engl., Pflanzenr. 4 (20), Heft 101: 309. 1936; Sharma *et al.*, Fl. Karnataka 306. 1984; Karthik. *et al.*, Fl. Ind. Enum. Monocot. 46. 1989.

Type: India, Karnataka State, Bijapur Dist. Badami, *Meebold* 11257.

Rhizome short. Stems several, tufted, slender, slightly curved, triangular, 6-10 cm high, smooth; base bulbous-thickened by brown sheaths, later covered with fibrous remains of sheaths. Leaves few, equalling the stem, canaliculate. Inflorescence solitary dense globose head, 8-12 mm across. Involucral bracts 2-3, overtopping the inflorescence, spreading, often as if continuation of the stem, with dilated base. Spikelets compressed, oblong, *ca* 6 x 1.5 - 2 mm; rachilla straight, wingless. Glumes densely imbricate, membranous, ovate; keel 3-nerved, green, slightly excurrent. Stamen 1. Style long, undivided or with 3 stigmas. Nut very small, trigonous, broadly obovate suddenly cuneate towards base, apiculate, punctate, dark brown.

Fls. & Frts. : Not known from Karnataka.

Habitat : Not known.

Distrib. : East Africa. INDIA: South India and North-West India. KARNATAKA: Bijapur (Badami, Kukenth., *l.c.*), Uttara Kannada (Sharma *et al., l.c.*).

Note : This species is included on the authority of Kukenth., *l.c.* and Sharma *et al., l.c.* As no specimens are available in BSI, description provided above is translated from Kukenth., *l.c.* (Pflanzenr.).

25. **Cyperus mitis** Steud., Syn. Pl. Glum. 2: 316. 1855; Kukenth. in Engl., Pflanzenr. 4 (20), Heft 101: 103. 1935; Hooper in Saldanha & Nicolson, Fl. Hassan 669. 1976 (in note); Sharma *et al.*, Fl. Karnataka 307. 1984; Koyama in Dassanayake & Fosberg, Rev. Handb. Fl. Ceylon 184. 1985; Karthik. *et al.*, Fl. Ind. Enum. Monocot. 46. 1989. *C. stenostachyus* Benth. var. *indica* Clarke in Hook.f., Fl. Brit. India 6: 614. 1893.

Type: Bengal.

Perennials with long, slender stolons which terminate in tubers; tubers globose, 3-12 mm across. Stems usually solitary, trigonous to triquetrous, 50-70 cm high, 1.2 - 2 mm thick, smooth, forming a globose corm-like enlargement at base. Leaves few, shorter than to as long as the stem, subrigid, flattish-plicate, narrowly linear, gradually narrowed to an acuminate apex, up to 30 cm long, 1-3 mm wide; sheaths rather long, up to 15 cm long, reddish-brown stained or sanguineous, basal ones subaphyllus, dusky or purple-brown, disintegrating into dark brown fibres. Inflorescence terminal or often quasilateral, simple or subcompound in parts, rather lax, 2-8 cm long, 2-6 cm wide. Involucral bracts 2 to 4; lower most much overtopping the inflorescence, often erect. Primary rays up to

6, erect-patent, filiform, unequal, up to 6 cm long. Spikes broadly obovate, subloosely bearing 3-8 spikelets; rachis 2-5 mm long, smooth. Spikelets obliquely patent, flattened, linear, acute at apex, 7-30 x 1-1.5 mm, yellowish-brown or more or less ferrugineous, 10-35-flowered; rachilla straight, winged, with 0.8-1 mm long internodes. Glumes erect-patent, membranous, narrowly ovate to ovate-elliptic or ovate-oblong, subobtuse at slightly incurved apex, 2.5-3 x 1.2-1.4 mm, light yellowish-brown and stained with brown or reddish-brown on lower half, with 2(or 3) parallel nerves on both sides, with narrowly pale-brown hyaline margins. Stamens 3; anthers 1-1.25 mm long. Style long; stigmas 3. Nut trigonous, oblong to ovate-oblong, 1.2-1.3 x 0.5-0.6 mm.

Fls. & Frts. : Not known from Karnataka.

Habitat : May be as of *Cyperus rotundus* L.

Distrib. : Sri Lanka, Indo-China, Malesia, Madagaskar, Australia. INDIA: Peninsular India. KARNATAKA: Without exact locality, outside Hassan district (Hooper, *l.c.*).

Note: This species is included on the authority of Hooper who says it occurs just outside the Hassan District. As no specimens are available in BSI, description provided above is as given by Koyama (*l.c.*).

26. **Cyperus niveus** Retz., Obs. Bot. 5: 12. 1789; Clarke in Hook.f., Fl. Brit. India 6: 601. 1893; Cooke, Fl. Pres. Bombay 2: 864. 1908 (3: 377. 1958, repr.ed.); Fischer in Gamble, Fl. Pres. Madras 1640. 1931 (3: 1139. 1957, repr.ed.); Kukenth. in Engl., Pflanzenr. 4 (20), Heft 101: 288. 1936; Kern in Blumea 10: 644. 1960 *et* in Reinwardtia 6: 60. 1961; Ramaswamy & Razi, Fl. Bangalore 90. 1973; Rao & Verma, Cyp. NE India 14. 1982; Sharma *et al.*, Fl. Karnataka 307. 1984; Karthik. *et al.*, Fl. Ind. Enum. Monocot. 46. 1989.

Illus. : Clarke, Illus. Cyp. t. 7, f. 4-7.1909.

Perennials with short- creeping, thick, woody rhizome, 20-40 cm high. Stems usually closely in a row, slender, with swollen base, triangular, 1-2 mm thick, smooth. Leaves shorter than to equalling the stem, linear, gradually narrowed to an acute apex, *ca* 2 mm wide, very minutely scabrous towards apex. Inflorescence contracted into a single head of 3-20 spikelets, 1-2.5 cm across. Involucral bracts 2-3, widely spreading or reflexed, lower 1 or 2 much overtopping the inflorescence; lowest up to 8 cm long, not or hardly dilated at base. Spikelets stellately spreading, flattened, oblong-lanceolate, subobtuse at apex, 7-15 x 3-5 mm, pale stramineous to whitish, 10-30-flowered; rachilla wingless. Glumes closely imbricate, subchartaceous, ovate-lanceolate, acute and usually shortly sinuate at apex, 4-4.5 x *ca* 3 mm, acutely keeled towards apex; keel produced into a minute mucro; sides usually whitish-stramineous, rarely purplish-lineolate, many-nerved, with hyaline margins. Stamens 3;

filaments up to 4.5 mm long; anthers linear-oblong, *ca* 1.2 mm long, exserted from the glume. Style variable, 1-1.5 mm long; stigmas 3, 1-1.5 mm long. Nut triquetrous, obovoid, obtuse and apiculate at apex, *ca* 2 x 1 mm, dark brown to blackish.

Fls. & Frts. : Aug.

Chrom. No. : n = 36 (*Taxon* 20: 612. 1971).

Habitat : Dry grasslands and open forest areas; forming thick patches.

Distrib. : Myanmar, Pakistan, Afganistan, China, North Thailand. INDIA: Throughout. KARNATAKA: Bangalore (Ramaswamy & Razi, *l.c.*).

Note: This species is included on the authority of Ramaswamy & Razi, *l.c.* Description given above is based on 4 specimens available from Gujarat and Uttar Pradesh.

27. **Cyperus nutans** Vahl, Enum. Pl. 2: 363. 1806; Clarke in Hook. f., Fl. Brit. India 6: 607. 1893; Cooke, Fl. Pres. Bombay 2: 868. 1908 (3: 381. 1958, repr.ed.) ; Fischer in Gamble, Fl. Pres. Madras 1640. 1931 (3: 1140. 1957, repr.ed.); Kukenth. in Engl., Pflanzenr. 4 (20), Heft 101: 144. 1935; Kern in van Steenis, Fl. Males. 1, 7: 609. 1974; Rao & Razi, Fl. Mysore 560. 1981; Rao & Verma, Cyp. NE India 16. 1982; Sharma *et al.*, Fl. Karnataka 307. 1984; Koyama in Dassanayake & Fosberg, Rev. Handb. Fl. Ceylon 5: 190. 1985; Karthik. *et al.*, Fl. Ind. Enum. Monocot. 46. 1989.

Type: India (?), *Koenig*.

Key to the subspecies

1a. Primary rays erect to spreading; spikes loosely arranged; spikelets not up to base of the spike; rachilla internodes *ca* 0.8 mm long . . . . . .subsp. **nutans**

1b. Primary rays suberect; spikes densely arranged; spikelets almost up to base of the spike; rachilla internodes *ca* 0.5 mm long . . . . . subsp. **eleusinoides**

subsp. **nutans**

Perennials with short, thick rhizome, up to 100 cm or more high. Stems tufted, trigonous, triquetrous below the inflorescence, 2.5 - 5 (-10) mm thick in the middle portion, smooth. Leaves few, shorter than stems, subcoriaceous, linear, narrowed to an acute apex, 3-6 (-15) mm wide, scabrous on upper margins, with very prominent midrib; sheaths reddish-brown towards base, basal ones with very short lamina. Inflorescence compound to decompound, lax, up to 20 cm or more long. Involucral bracts 4-6; lower ones much overtopping the inflorescence. Primary rays 6-10, very unequal, erect to spreading, up to 20 (-30) cm long, smooth; secondary rays 3-7, up to 6 cm long. Spikes narrow, more or less penicellate, up to 4 cm long, 5-10 mm wide, often spreading, bearing 10-25 spikelets; rachis glabrous. Spikelets spicately arranged, suberect, compressed, linear or linear-oblong, 6-15 x *ca* 2 mm, 8-14-flowered; rachilla slightly flexuous, winged, persistent; internodes *ca* 0.8 mm long; wings whitish-hyaline, persistent. Glumes membranous, obliquely erect to patulose, 1/3 imbricate, oblong, obtuse at apex, keeled, 2-2.5 x *ca* 1 mm, dark brown to yellowish-brown, 3-nerved on both sides, with broad whitish-hyaline margins in the upper half; midnerve extending to an excurrent mucro from the incised tip of the glume. Stamens 3; filaments up to 2.5 mm long; anthers linear-oblong, *ca* 1 mm long, with a distinct and smooth connective appendage. Style short; stigmas 3. Nut triquetrous, oblong to oblong-ovoid, apiculate at apex, 1.2-1.5 x *ca* 0.5 mm, brown.

Fls. & Frts. : Sept. - Jan., May.

Habitat : Along streams, wet rice fields and other moist places.

Distrib. : Sri Lanka, S. China, Malesia. INDIA: South, North-West and North-East India. KARNATAKA: Chikmagalur (Sharma *et al., l.c.*), Dakshina Kannada, Hassan, Mandya (Dinesh *et al.*, 1990); Mysore, Shimoga, Uttara Kannada.

Specimens examined : Dakshina Kannada: Mangalore, *Foulker* 4861, Jan. 1902 (MH). Hassan: Bannuhalla, *Saldanha* 15500, 3.10.1969 (JCB). Mysore: Himavad Gopapalaswami hill, *Padma Rani* 35, 1.11.1970 (MGH). Shimoga: Tunga river, Tirthahalli, *Raghavan* 82806, 29.9.1962 (BSI). Uttara Kannada: Gersoppa falls, *Chibber s.n.* (Accn. No. 2300), 17.5.1911 (BSI).

subsp. **eleusinoides** (Kunth) Koyama in Gard. Bull. Singapore 30: 136. 1977 *et* in Dassanayake & Fosberg, Rev. Handb. Fl. Ceylon 5: 191. 1985. *Cyperus eleusinoides* Kunth, Enum. Pl. 2: 39. 1937; Clarke in Hook.f., Fl. Brit. India 6: 608. 1893; Cooke, Fl. Pres. Bombay 2: 868. 1908 (3: 382. 1958, repr.ed.); Fischer in Gamble, Fl. Pres. Madras 1640. 1931 (3: 1140. 1957, repr.ed.); Kukenth. in Engl., Pflanzenr. 4 (20), Heft 101: 144. 1935; Ramaswamy & Razi, Fl. Bangalore 85. 1973. *C. nutans* Vahl var. *eleusinoides* (Kunth) Haines, Bot. Bihar Orissa 5: 898. 1924; Kern in van Steenis, Fl. Males. 1, 7: 610. 1974; Hooper in Saldanha & Nicolson, Fl. Hassan 667. 1976; Rao & Razi, Fl. Mysore 560. 1981; Rao & Verma, Cyp. NE India 16. 1982; Sharma *et al.*, Fl. Karnataka 307.1984.

Type : Nepal, *Wallich* 3346 B.

Illus. : Matthew, Furth. Illus. Fl. Tamilnadu Carnatic 641. 1988.

Vegetative characters as in typical subspecies, but appear different in inflorescence. Inflorescence less compound; primary rays suberect. Spikes densely arranged, shorter, 2-3 cm long, 5- 10 mm wide, with spikelets almost up to the base. Spikelets with fewer (6-10) flowers; rachilla internodes *ca* 0.5 mm long. Glumes more patent, more prominently mucronate. Nut obovoid- oblong, 1-1.2 x 0.5 - 0.7 mm.

Fls. & Frts. : May - Dec.

Habitat : Along streams and canals, near puddles along roadside, paddy fields; in shallow canals with rocky or muddy bottom.

Distrib. : South and South-East Asia, Formosa, tropical Africa and tropical Australia. INDIA: Throughout. KARNATAKA: Bangalore, Belgaum, Bidar, Bijapur, Chikmagalur, Chitradurga, Hassan, Mysore, Tumkur.

Specimens examined : Bangalore: Tippagondanahalli, *Manohar & Murthy* 2823, 23.9.1978 (JCB). Belgaum: Gokak Road, *Saldanha* 7874, 23.5.1979 (JCB); Dinman Hassur, Belgaum-Gokak Road, *Prasad* 172840, 13.12.1994 (BSI). Bidar: Humnabad-Gulbarga Road, 10th km, *Singh* 142858, 11.8.1976 (BSI). Bijapur: Zalki, on Sholapur-Bijapur Road, *Singh* 138741, 4.9.1974 (BSI); Sindgi-Bijapur Road, 50th km, *Singh* 142918, 13.8.1976 (BSI); Varathi kavlu, Kendur, *Prasad* 172966, 21.12.1994 (BSI); Hossur Forest Nursery, Badami Taluk, *Prasad* 172988 & 172992, 23.12.1994 (BSI); Chikmagalur: Sakrepatana, *Ahuja* 59354, 18.5.1959 (BSI). Chitradurga: Sira-Hiriyur Highway, 31st km, *Singh* 143253, 24.8.1976 (BSI); Donnehalli-Chikka Ujjani Road, *Murthy & Manohar* 7767, 16.6.1979 (JCB). Hassan: Mokali, on Road to Konnanur, *Hooper & Gandhi* 2420, 11.11.1971 (JCB). Mysore: Nagenhalli-Mysore Road, *Suma* 15, 17.8.1968 (MGH); Mandakalli, *Bhat* 25, 12.7.1970 (JCB); Biligiri Rangan hills- Hassanur Road, *Saldanha & Ramesh* 2570, 6.9.1978 (JCB): Tumkur: Haliyurdurga State Forest, *Prakash et al.* 3224. 29.9.1978 (JCB).

28. **Cyperus pangorei** Rottb., Descr. IC. Rar. Nov. Pl. 31, t. 7, f. 3. 1773; Fischer in Gamble, Fl. Pres. Madras 1641. 1931 (3: 1140. 1957, repr.ed.); Kukenth. in Engl., Pflanzenr. 4 (20), Heft 101: 76. 1935; Ramaswamy & Razi, Fl. Bangalore 91. 1973; Hooper in Saldanha & Nicolson, Fl. Hassan 667, 1976; Rao & Verma, Cyp. NE India 18. 1982; Sharma *et al.*, Fl. Karnataka 307. 1984; Koyama in Dassanayake & Fosberg, Rev. Handb. Fl. Ceylon 5: 176, f. 9. 1985; Karthik. *et al.*, Fl. Ind. Enum. Monocot. 46. 1989; Keshava Murthy & Yoganarasimhan, Fl. Coorg 503. 1990. *C. tagetum* Roxb., Fl. Ind. 1: 208. 1832; Clarke in Hook.f., Fl. Brit. India 6: 613. 1893; Cooke, Fl. Pres. Bombay 2: 870. 1908 (3: 384. 1958, repr.ed.). *Murumule doddajumbu hullu.*

Type : India, Tranquebar.

Illus. : Clarke, Illus. Cyp. t. 17, f. 1. 1909; Koyama, *l.c.*; Matthew, Furth. Illus. Fl. Tamilnadu Carnatic Pl. 642. & 643. 1988.

Perennials with creeping rhizome, 50-150 cm high. Rhizome stout, woody, clothed with brownish scales; roots somewhat thick, usually branched into blackish fibrous hairs. Stems usually in a row along the creeping rhizome or subtufted, erect, trigonous to triquetrous towards apex, narrowed to apex, (2-) 3-7 mm thick in the middle portion, smooth. Leaves usually reduced to 2-4 subphyllus sheaths surrounding the basal portion of the stem, at times developed into up to 12 cm long blades; sheaths long, brownish to purplish-brown towards base, uppermost up to 30 cm long. Inflorescence compound, lax, 5-22 x 3-18 cm. Involucral bracts 3-5, patent, usually much overtopping the inflorescence, up to 35 cm long. Primary rays short, up to 4 cm long, ending in corymb-like spikes of 4-10 spikelets, with smooth rachis. Spikelets spicately arranged, compressed, linear, subacute at apex, 10-20 x 1-1.8 mm, usually reddish-brown, at times stramineous, 12- 30-flowered; rachilla straight, winged, with *ca* 1 mm long internodes; wings lanceolate, reddish-brown. Glumes membranous, somewhat loosely to closely imbricate, oblong to somewhat oblong- lanceolate, 2.5-3 x 1-1.5 mm, hardly keeled, 5-nerved on the obtuse back; sides nerveless, usually reddish-brown. Stamens 3; filaments up to 3 mm long; anthers linear, *ca* 1 mm long. Style 1-1.5 mm long; stigmas 3, longer than style. Nut trigonous, oblong or oblong-obovoid, minutely apiculate at apex, 1.2-1.5 x 0.3-0.5 mm, yellowish-brown to blackish-brown.

Fls. & Frts. : Aug. - Jan.

Habitat : Common in Marshy areas along canals, streams and rivers; rocky river beds, and muddy river banks.

Distrib. : Sri Lanka, Pakistan, Nepal, Myanmar; cultivated in Mauritius. INDIA: Throughout. KARNATAKA: Bangalore (Ramaswamy & Razi, *l.c.*), Belgaum, Bijapur, Chikmagalur (Rao *et al.*, 2000), Dakshina Kannada, Dharwar, Hassan, Kodagu, Mysore, Uttara Kannada.

Uses : Stems can be used for mat making. In Mauritius it is cultivated for this purpose.

Specimens examined : Belgaum: Manjapur nallah, Khanapur, *Prasad* 172862, 14.12.1994 (BSI). Bijapur: Malaprabha river bed, Pattadakkal, *Prasad* 172968, 21.12.1994 (BSI); Malaprabha river, Cholachgudd, Badami taluk, *Prasad* 172983, 23.12.1994 (BSI). Dakshina Kannada: Jamalabad, Belthangady, *Bhat* 356, 9.11.1976 (MGH); Kapu, *Bhat* 444, 15.1.1977 (MGH). Dharwar: Kiligeri, *Talbot* 2613, Sept. 15, without year (BSI). Hassan: Devarunde Road, *Saldanha* 15473, 29.10.1969 (JCB). Kodagu: Kushalnagar, *Bhat* 1043, 21.9.1981 (MGH). Mysore: Without locality, Coll. Name and number (Acc. No. 73575), without date (MH). Uttara Kannada: Halyal, *Talbot* 2219, 5.1.1890 (BSI); Kaltikau forest, *Fernandez s.n.* (Acc. No. 79749), 27.11.1950 (BLAT).

Note: It seems to be a variable species, sometimes very similar to *C. corymbosus* and difficult to distinguish from each other, as intermidiate forms (in the length of bracts and shape of stems) are found in the same population.

29. **Cyperus pilosus** Vahl, Enum. Pl. 2: 354. 1806; Clarke in Hook.f., Fl. Brit. India 6: 609. 1893, incl. var. *obliqua* (Nees) Clarke; Fischer in Gamble, Fl. Pres. Madras 1641. 1931 (3: 1140. 1957, repr.ed.); Kukenth.

in Engl., Pflanzenr. 4 (20), Heft 101: 92. 1935; Kern in van Steenis, Fl. Males. 1, 7: 611, f. 52. 1974; Hooper in Saldanha & Nicolson, Fl. Hassan 668. 1976; Rao & Razi, Fl. Mysore 560. 1981; Rao & Verma, Cyp. NE India 17. 1982; Sharma *et al.*, Fl. Karnataka 307. 1984; Koyama in Dassanayake & Fosberg, Rev. Handb. Fl. Ceylon 5: 193. 1985; Karthik. *et al.*, Fl. Ind. Enum. Monocot. 47. 1989. *Bimbul.* **Fig. 11.**

Type : India.

Illus. : Kern, *l.c.*; Matthew, Furth. Illus. Fl. Tamilnadu Carnatic Pl. 644. 1988.

Perennials with short rhizome emitting long, slender stolons clothed with scales, 30-70 cm high. Stems usually solitary, at times few together, erect, acutely triquetrous, 3-5 mm thick, smooth. Leaves shorter than to as long as the stem, canaliculate, gradually tapering to an acute apex, 4-10 mm wide, scabrous on the margins towards apex; sheaths purplish-brown towards base. Inflorescence compound, loose to subdense, variable in size. Involucral bracts 3-5, spreading, lower ones much overtopping the inflorescence. Primary rays 6-7, unequal, spreading, up to 5 (- 15) cm long, with 2-4 secondary rays. Spikes ovoid, 1-2.5 cm long, 1-2 cm wide, bearing 6-20 spikelets; rachis subdensely to densely hirsute. Spikelets spicately arranged, horizontal to reflexed, compressed, linear-oblong to oblong-lanceolate, subacute at apex, 5-8 x 1.5 - 2 mm, 7-13 (-20)-flowered; rachilla straight, wingless or almost wingless; internodes *ca* 0.8 mm long. Glumes subdensely imbricate, ovate, obtuse and usually mucronulate at apex, slightly keeled, 2-2.2 x 1-1.2 mm, 5-7-nerved; midrib green; sides stramineous to reddish-brown, with broad whitish-hyaline margins. Stamens 3; filaments up to 2 mm long; anthers oblong, *ca* 0.5 mm long. Style short, *ca* 0.3 mm long; stigmas 3, *ca* 1.5 mm long. Nut triquetrous, ellipsoid to obovoid, minutely apiculate at apex, 1-1.2 x *ca* 0.7 mm, yellowish-brown to dark brown.

Fls. & Frts. : May - Jan.

Habitat : Common in marshy areas, wet rice fields, river banks, open wet areas, wet slippery rocks near water bodies, wet banks of streams and canals; often along with thick spreading grasses.

Distrib. : Indian subcontinent, Central Asia, Malesia, Japan, tropical West Africa and tropical Australia. INDIA: Throughout. KARNATAKA: Belgaum, Chikmagalur, Dakshina Kannada, Hassan, Kodagu, Mandya, Mysore, Shimoga, Uttara Kannada.

Specimens examined : Belgaum: Anmod, *Saldanha & Prakash* 3515, 25.10.,1978 (JCB); Manjapur nallah, Khanapur, *Prasad* 172861, 14.12.1994 (BSI); Chappoli nallah, Jambotti R.F., *Prasad* 172883, 15.12.1994 (BSI); Dudwawada R.F., Londa, *Prasad* 172892, 17.12.1994 (BSI). Chikmagalur : Shiradi, just above Shiradi village, *Hooper & Saldanha*

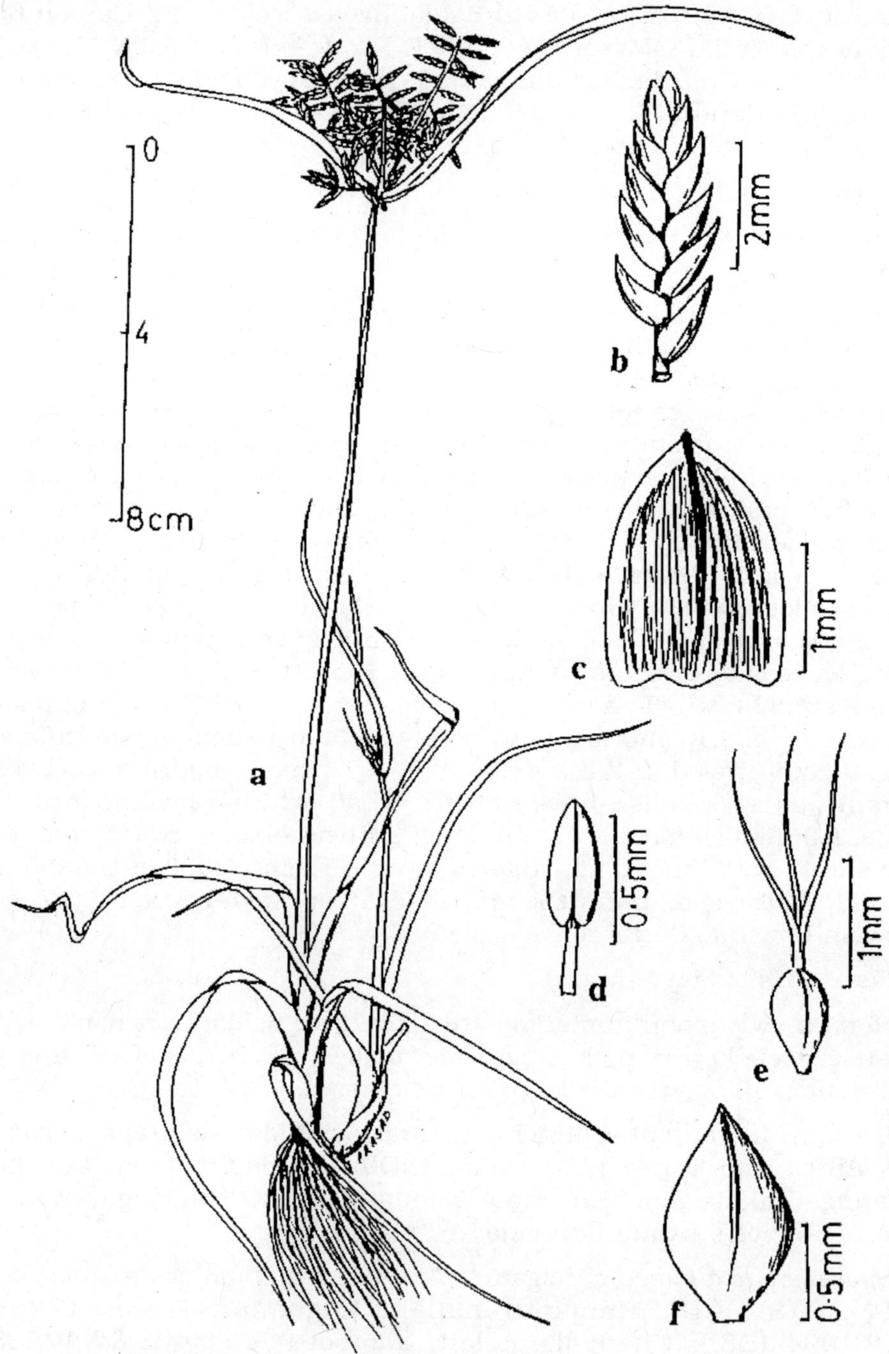

Fig. 11. *Cyperus pilosus* Vahl
a. Habit, **b**. Spikelet, **c**. Glume, **d**. Stamen, **e**. Nut with style, **f**. Nut

2571, 24.11.1971 (JCB). Dakshina Kannada: Sullia, *Saldanha* 6324, 28.10.1960 (JCB); Kannadekatte, *Bhat* 182, 6.5.1975 (MGH); Thumbe, *Raghavan* 146476, 20.4.1977 (BSI); Jappinamogeru village, *Prasad* 173867, 10.11.1995 (BSI). Hassan: Banaganahalla bridge, *Jarret & Ramamoorthy* 1033, 26.10.1970 (JCB). Kodagu: Bhagamandala, *Bhat* 804, 19.12.1980 (MGH); Abbe falls, Mercara, *Bhat* 879, 23.12.1980 (MGH); Ponnampet, *Bhat* 913, 21.2.1981 (MGH); Paschimavahini, *Padma Rani* 23, 8.8.1970 (MGH). Mysore: Ketedevargudi vicinity, *Rao* 73575, 4.9.1961 (BSI). Shimoga: Agumbe, *Raghavan* 80605, 14.5.1962 (BSI); Jog falls, *ca* 500 ft. below, *Prasad* 173786, 4.11.1995 (BSI); Gaurikeri, Thalaguppa, Sagar taluk, *Prasad* 173804, 5.11.1995 (BSI). Uttara Kannada: Kodasalli dam site, *Singh et al.* 14526, 3.6.1982 (JCB); Gersoppa, *Srirama* 97, without date (MGH).

Note: It is a variable species, but the hispid nature of the rachis is a good character for its correct identification.

30. **Cyperus platystylis** R. Br., Prodr. Fl. Nov. Holl. 214. 1810; Clarke, in Hook.f., Fl. Brit. India 6: 598 . 1893; Fischer in Gamble, Fl. Pres. Madras 1639. 1931 (3: 1139. 1957, repr.ed.); Kern in van Steenis, Fl. Males. 1, 7: 618. 1974; Rao & Verma, Cyp. NE India 12. 1982; Sharma *et al.*, Fl. Karnataka 307. 1984; Koyama in Dassanayake & Fosberg, Rev. Handb. Fl. Ceylon 5: 198, f. 12. 1985; Karthik. *et al.*, Fl. Ind. Enum. Monocot. 47. 1989. *Cyperus pallidus* Nees (Linnaea 9: 284. 1835, *nom. nud.*) *et* in Wight, Contr. Bot. India 79. 1834. *Anosporum pallidum* (Nees) Boeck. in Linnaea 36: 412. 1870.

Type; Australia.

Illus. : Koyama, *l.c.*

Perennials with short rhizome and thick roots; stolons absent. Stems usually solitary, at times 2-3 together, stout, triquetrous, 30-80 cm high, 3-8 mm thick, smooth, scabrous on the angles above. Leaves basal, few, as long as or longer than stem, coriaceous, flat or channelled, gradually narrowed to apex, 8-12 mm wide, septate-nodulose, scabrous on the margins and midrib; sheaths light brown, usually purplish shaded, lower ones usually bladeless. Inflorescence compound or decompound, very dense to somewhat loose, depressed corymbose or semiglobose, up to 20 cm across, with numerous spikelets. Involucral bracts 5-8, slightly spaced, spreading, lower ones much overtopping the inflorescence; longest up to 80 cm long. Primary rays 10-15, spreading, up to 10 cm long, smooth; secondary rays up to 4 cm long; bracteoles setaceous, short. Spikelets digitately arranged in clusters of 3-8, spreading, compressed, oblong-lanceolate or ovate-lanceolate, acute at apex, 5-15 x 2.5-3 mm, stramineous and brownish shaded, up to 60-flowered; rachilla persistent, wingless. Glumes thick- membranous, ovate to broadly ovate, subobtuse and mucronulate at apex, 2-2.5 x 1.5-2 mm, cellular-reticulate, with brownish sides and narrow hyaline margins; keel broad, strongly 3-nerved. Stamens 3; filaments up to 2.5 mm long; anthers linear, 0.5 - 1 mm long. Style flattened, narrowed to base, minutely papillose, *ca* 1 mm long (shorter than nut); stigmas 3, short. Nut trigonous, dorsally compressed, with concave ventral side, with corky thickened angles, ellipsoid to elliptic-ovoid; apiculate at apex, *ca* 2 x 1 mm, glaucous-brownish and with stramineous or yellow angles.

Fls. & Frts. : May, Nov.

Habitat : Wet swampy areas; also as floating vegetation in ponds and tanks.

Distrib. : Sri Lanka, Taiwan, Formosa, Malesia and Australia. INDIA: Throughout (except in North-West). KARNATAKA : Shimoga, Uttara Kannada. Not common.

Specimens examined : Shimoga: Kavaledurga, *Raghavan* 81111, 29.5.1962 (BSI). Uttara Kannada: Deviuram ghat, *Kulkarni s.n.* (Acc. No. 2496 & 2497), 20.11.1908 (BSI).

31. **Cyperus procerus** Rottb., Descr. IC. 29. 1773; Clarke in Hook.f., Fl. Brit. India 6: 610. 1893; Cooke, Fl. Pres. Bombay 2: 867. 1908 (3: 381. 1958, repr.ed.); Fischer in Gamble, Fl. Pres. Madras 1641. 1931 (3: 1140. 1957, repr.ed.); Kukenth. in Engl., Pflanzenr. 4 (21), Heft 101: 91. 1935; Ramaswamy & Razi, Fl. Bangalore 92. 1973; Kern in van Steenis, Fl. Males. 1, 7: 611. 1974; Rao & Razi, Fl. Mysore 560. 1981; Arora *et al.*, Bot. S. Kanara 61. 1981; Sharma *et al.*, Fl. Karnataka 307. 1984; Koyama in Dassanayake & Fosberg, Rev. Handb. Fl. Ceylon 5: 192. 1985; Karthik. *et al.*, Fl. Ind. Enum. Monocot. 47. 1989.

Illus. : Matthew, Furth. Illus. Fl. Tamilnadu Carnatic Pl. 645. 1988.

Perennials with short rhizome and long stolons, 50-80 (-150) cm high. Stolons slender, with short scale-like sheaths from the nodes. Stems usually solitary, erect, deeply triquetrous, 3.5 - 7 mm, thick, smooth. Leaves few, shorter than to surpassing the stem, canaliculate, spongy or subcoriaceous, linear, gradually narrowed to an acute apex, up to 8 mm wide, with scabrid margins near apex; basal sheaths brownish to purplish. Inflorescence simple or subcompound, dense or loose, 4-10 (-15) cm long. Involucral bracts 3-4, obliquely erect to spreading, lowest one or two overtopping the inflorescence. Primary rays 3-7, very unequal, suberect to spreading, smooth; longest 4-10 (-15) cm long; secondary rays if present very short. Spikes broadly ovoid, 2-4 cm long, loosely bearing 4-10 spikelets; rachis smooth. Spikelets spicately arranged, patent, compressed, often slightly curved, oblong to linear-oblong, subacute at apex, 10-30 x 2.5-3.5 (-4) mm, 16-40-flowered; rachilla straight, hardly winged, with 0.7- 0.8 mm long internodes. Glumes membranous, *ca* 1/2 imbricate, ovate, obtuse at apex, hardly keeled, 2.5-3 x *ca* 2 mm, 5 or 7-nerved, brownish to reddish-brown, with broad, white-hyaline margins. Stamens 3; filaments elongate up to 3.5 mm; anthers linear, with a smooth connective appendage, 1.5-2 mm long. Style 1.2-1.5 mm long; stigmas 3, much longer than style, up to 5 mm. Nut triquetrous, obovoid to ellipsoid, minutely apiculate at apex, 1.2-1.5 x 0.7-0.8 mm, brownish.

Fls. & Frts. : June - Dec., March.

Chrom. No. : n = 52 (*Taxon* 20: 612. 1971).

Habitat : Open moist areas, rice fields and other marshy areas, often near sea.

Distrib. : Sri Lanka, East China, Formosa, Malesia, Vietnam and Queensland. INDIA: Peninsular India & Eastern India. KARNATAKA: Bangalore, Dakshina Kannada (Arora *et al., l.c.*), Mysore (Rao & Razi, *l.c.*).

Specimens examined : Bangalore: Bangalore, *Camaron* 571, March 1891 (MH). Without actual locality (Mysore & Carnatic), Thomson *s.n.* (Acc. No. 73445), without date (MH).

Note : The specimens from Karnataka have the inflorescence with smooth rachis which is the character of typical *C. procerus*. According to Kern, *l.c.* and Koyama, *l.c.* the Malesian and Sri Lankan specimens respectively are with weakly hispid-scabrous rachis which may be of var. *lasiorrachis* Clarke.

32. **Cyperus pulchellus** R. Br., Prodr. Fl. Nov. Holland. 213. 1810; Kern in van Steenis, Fl. Males. 1, 7: 632. 1974; Hooper in Saldanha & Nicolson, Fl. Hassan 669. 1976; Sharma *et al.*, Fl. Karnataka 307. 1984. *C. leucocephalus* auct. non Retz. 1789; Clarke in Hook.f., Fl. Brit. India 6: 602. 1893; Cooke, Fl. Pres. Bombay 2: 864. 1908 (3: 377. 1958, repr.ed.); Fischer in Gamble, Fl. Pres. Madras 1640. 1931 (3: 1140. 1957, repr.ed.). *Sorostachys pulchellus* (R.Br.) Lye in Nor. Journ. Bot. 1: 189, f. 4. 1981; Karthik. *et al.*, Fl. Ind. Enum. Monocot. 73. 1989. **Fig. 12.**

Type: *R. Brown* 5915, Nova Hollandia Tropica (BM Holotype).

Illus. : Lye, *l.c.*

Perennials with short rhizome, 12-32 cm high; basal portion thickened, bulbous, covered with old blackish sheaths; stolons absent. Stems tufted, very slender, trigonous, 0.5 - 1 mm thick, smooth. Leaves basal, shorter than to as long as the stem, linear, narrowed to apex, 0.5 - 1 mm wide, smooth or scabrid on the margins above; sheaths brownish, withering to blackish fibres in later stage. Inflorescence a dense, globose head of numerous spikelets, 6-13 mm across, whitish. Involucral bracts 3 (-4), reflexed, much overtopping the inflorescence, somewhat dilated at base, linear; longest up to 12 cm long. Spikelets compressed, elliptic or ovate, 3-5 x 1.5-2 mm, 10-18-flowered; rachilla straight, slender, wingless, with *ca* 0.2 mm long internodes. Glumes membranous, subdense, oblong to

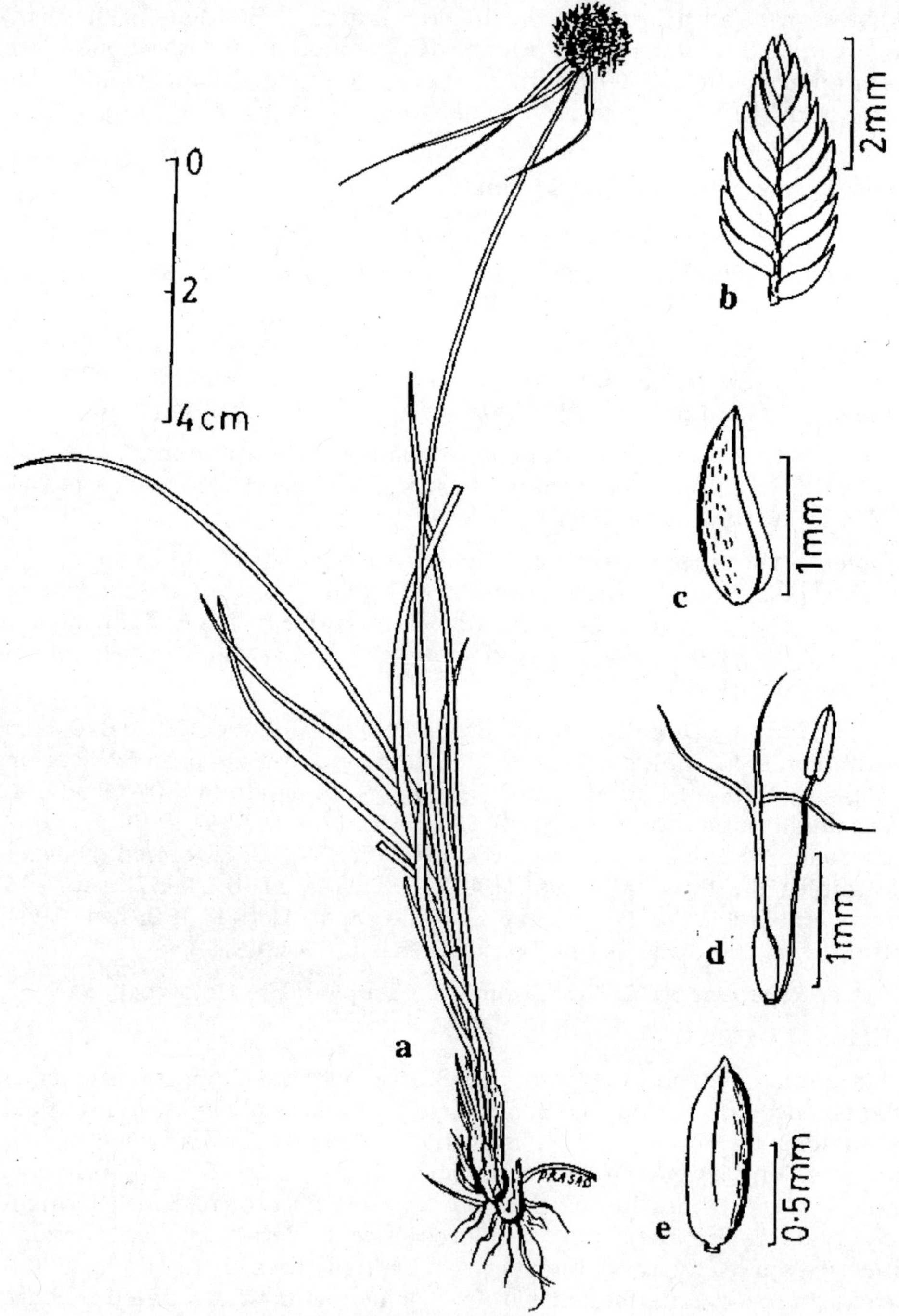

Fig. 12. *Cyperus pulchellus* R. Br.
a. Habit, **b**. Spikelet, **c**. Glume, **d**. Flower, **e**. Nut

oblong-lanceolate, obtuse at apex, slightly keeled, 1.5-2 x *ca* 0.7 mm, stramineous, faintly nerved, pale brown-lineolate, hyaline at margins. Stamen 1; filament elongate up to 2 mm; anther oblong, *ca* 0.6 mm long. Style 0.6- 1 mm long; stigmas 3, as long as the style. Nut trigonous, oblong-ellipsoid or oblong-obovoid, shortly apiculate at apex, *ca* 1 x 0.3 mm, yellowish-brown to dark brown, minutely puncticulate.

Fls. & Frts. : June - Nov.

Habitat : Open areas near water, swampy areas.

Distrib. : Tropical regions of Asia, Africa and Australia. INDIA: South, Central and Eastern India. KARNATAKA: Bellary, Chikmagalur, Hassan, Mysore, Tumkur, Uttara Kannada.

Specimens examined : Bellary: Kudligi - Sandur Road, *Sreenath & Ramesh* 4290, 14.11.1978 (JCB). Chikmagalur: Lokavalli, *Saldanha & Ramesh* 1774, 27.6.1978 (JCB). Hassan: 6 miles before Arsikeri on Tiptur-Arsikeri Road, *Nicolson et al.* 2223, 22.10.1971 (JCB). Mysore: Nagpure Rest House, near temple, Arsikeri, *Rao* 73896, 11.9.1961 (BSI). Tumkur: Devarayanadurga, *Saldanha* 2200, 15.8.1978 (JCB). Uttara Kannada: Karwar, *Talbot s.n.* (Acc. No. 679), 10.8.1885 (BSI); Balemane, *Udaya Kumar & Gurudev Singh* 14644, 9.7.1982 (JCB).

Note : It seems most of the specimens labelled as *Cyperus leucocephalus* Retz. in some Indian herbaria are actually of *C. pulchellus* R. Br. Hence a thorough study of the specimens of both the species is necessary to find out their actual distribution in India.

33. **Cyperus pygmaeus** Rottb., Descr. & Ic. Rar. Nov. Pl. 20, t. 14, f. 4-5. 1773; Cooke, Fl. Pres. Bombay 2: 859. 1908 (3: 372. 1958, repr.ed.); Kern in van Steenis, Fl. Males. 1, 7: 634. 1974; Koyama in Dassanayake & Fosberg, Rev. Handb. Fl. Ceylon 5: 214. 1985. *Juncellus pygmaeus* (Rottb.) Clarke in Hook. f., Fl. Brit. India 6: 596. 1893; Fischer in Gamble, Fl. Pres. Madras 1629. 1931 (3: 1133. 1957, repr.ed.); Karthik. *et al.*, Fl. Ind. Enum. Monocot. 58. 1989. *Cyperus michelianus* (L.) Delile subsp. *pygmaeus* (Rottb.) Aschers. & Graebn., Synops. Mitteleur. Fl. 2 (2): 273. 1903; Kukenth. in Engl., Pflanzenr. 4 (20), Heft 101: 312, f. 35 F-G. 1936; Ramaswamy & Razi, Fl. Bangalore 90. 1973; Hooper in Saldanha & Nicolson, Fl. Hassan 667. 1976; Rao & Razi, Fl. Mysore 560. 1981; Sharma *et al.*, Fl. Karnataka 307. 1984; Singh, Fl. E. Karnataka 2: 632. 1988.

Type : India, Tranquebar, *Koenig.*

Illus. : Rottb., *l.c.;* Kukenth., *l.c.;* Matthew, Furth. Illus. Fl. Tamilnadu Carnatic 646. 1988.

Annuals with fibrous roots, 5-25 cm high. Stems densely tufted, erect or patent, often pulvinate, trigonous, 0.5-1.5 mm thick, smooth. Leaves shorter than to slightly overtopping the stem, flat or canaliculate, linear, gradually narrowed to apex, 1-2 mm wide, scabrid on the margins in the upper part; basal sheaths often light brown or reddish. Inflorescence a contracted lobed head of several dense glomerules of spikelets, triangular-ovoid or subglobose, up to 15 mm across. Involucral bracts 2-7, patent, all overtopping the inflorescence, up to 15 cm long, dilated at base. Spikelets strongly compressed, often twisted, ovate to ovate-lanceolate, 3-5 x 1.5-2 mm, whitish-green to stramineous, 10-20-flowered; rachilla slightly flexuous, wingless. Glumes distichous, often looks irregular due to the twisting of rachilla, membranous, lanceolate, acute or short-mucronate at apex, keeled, 1.5-2 x *ca* 0.5 mm, pale whitish, subtransluscent; keel greenish, often spinulose. Stamens 1 or 2, filaments elongate up to 2 mm; anthers linear, 0.3-0.5 mm long. Style 0.5-0.7 mm long; stigmas 2 or 3, up to 1 mm long. Nut planoconvex or trigonous, oblong or elliptic-oblong, *ca* 1 x 0.3 mm, pale brown, minutely puncticulate.

Fls. & Frts. : Almost throughout the year.

Habitat: Dry river beds, dry muddy areas near ponds and other water bodies, near puddles along roadsides, wet soil in between rocks and in seasonal marshy areas.

Distrib. : Widely distributed in South and South-East Asia, Mediterranean region, Africa and Australia. INDIA: Throughout. KARNATAKA: Bangalore, Belgaum, Bijapur, Dakshina Kannada, Hassan, Mysore, Raichur, Tumkur, Uttara Kannada.

Specimens examined : Bangalore: Sangam to Makedat, *Bhaskar & Ramesh* 417, 10.3.1978 (JCB). Belgaum: Ghataprabha dam side, Belgaum, *Prasad* 172804, 10.12.1994 (BSI); Near Gokak falls, Gokak, *Prasad* 172822, 11.12.1994 (BSI); Western side of Malaprabha dam, near Urunkolla village, Saundati, *Prasad* 172952, 20.12.1994 (BSI). Bijapur: Bagalkot, without coll. name, *s.n.* (Acc. Nos. 2589 & 2590), 17.4.1906 (BSI); Badami-Ron Road, 8th mile, *Singh* 129774, 18.2.1975 (BSI). Dakshina Kannada: Near Mulki, *Saldanha & Prakash* 4090, 12.11.1978 (JCB). Hassan: Nagpuri, *Jaret, Saldanha & Ramamoorthy* 597, 3.9.1970 (JCB). Mysore: Bilikeri, *Bhat* 28, 20.7.1970 (JCB). Raichur: Shorapur - Lingsugur Road, 13th mile, *Singh* 129552, 13.2.1975 (BSI). Tumkur: Hiriyur-Sira Road, 19th mile, *Singh* 132622, 24.2.1975 (BSI). Uttara Kannada: without exact locality, *Talbot* 946, 4.4.1884 (BSI); Kampli, *Ramesh & Shivaprakash* 12949, 3.6.1981 (JCB). Without locality (Mysore & Carnatic), Thomson s.n. (Acc. No. 73450), without date (MH).

34. **Cyperus rotundus** L., Sp. Pl. 45. 1753; Clarke in Hook. f., Fl. Brit. India 6: 614. 1893; Cooke, Fl. Pres. Bombay 2: 871. 1908 (3: 385. 1958, repr.ed.); Fischer in Gamble Fl. Pres. Madras 1641. 1931 (3: 1140. 1957, repr.ed.); Kukenth. in Engl., Pflanzenr. 4 (20), Heft 101: 107, f. 13. 1935; Satyanarayan & Shankarnarayan in Ann. Arid Zone 2: 146. 1964; Ramaswamy & Razi, Fl. Bangalore 93. 1973; Kern in van Steenis, Fl. Males.

1, 7: 604, f. 49. 1974; Hooper in Saldanha & Nicolson, Fl. Hassan 669. 1976; Rao & Razi, Fl. Mysore 561. 1981; Rao & Verma, Cyp. NE India 19. 1982; Sharma *et al.*, Fl. Karnataka 307. 1984; Koyama in Dassanayake & Fosberg, Rev. Handb. Fl. Ceylon 5: 181, f. 10.1985; Singh, Fl. E. Karnataka 2: 633. 1988; Karthik. *et al.*, Fl. Ind. Enum. Monocot. 47. 1989; Keshava Murthy & Yoganarasimhan, Fl.Coorg 503.1990. *Kannari hullu, Tunge hullu, Bhadra hullu, Jake.*

Type : India.

Illus. : Kukenth., *l.c.;* Kern, *l.c.;* Koyama, *l.c.*

Key to the subspecies

1a. Stems 10-30 cm long; glumes subobtuse at apex, usually reddish-brown . . . . subsp. **rotundus**

1b. Stems 30-60 cm long; glumes acute at apex, ferrugineous. . . . . subsp. **tuberosus**

subsp. **rotundus**

Perennials with long slender stolons ending in tubers, 10-30 cm high. Tubers subglobose to ellipsoid, often zonate, ultimately blackish. Stems usually solitary, at times few together, slender, triquetrous, 1-2 mm thick, smooth, tuberous at base. Leaves few to several, usually shorter than stem, flat, linear, gradually narrowed to an acuminate apex, 2-5 mm wide, usually scabrid on the margins towards apex; sheaths usually pale brown, soon disintegrating into fibres. Inflorescence simple to compound, loose, 2.5 - 9 (-15) x 3-9 cm. Involucral bracts 2-5, shorter than inflorescence or the lowest often overtopping, scabrous on upper margins. Primary rays 3-9, slender, patent, unequal, lowest up to 6 (-10) cm long. Spikes densely to subloosely bearing 3-12 spikelets; rachis glabrous. Spikelets spicately arranged, suberect to spreading, strongly compressed, linear, acute at apex, 10-35 x 1.5-2.5 mm, 10-32 (-40)-flowered, usually reddish-brown; rachilla broadly winged, with *ca* 0.8 mm long internodes. Glumes membranous, obliquely erect, ovate, subobtuse at apex, keeled, 3-4 x *ca* 2.5 mm, 5-7-nerved, usually sanguineous, with narrow hyaline margins. Stamens 3; filaments elongate up to 4 mm; anthers linear, 1-2 mm long. Style short, up to 1 mm long; stigmas 3, much longer than style. Nut trigonous, oblong-obovoid, apiculate at apex, *ca* 1.5 x 0.7 mm, brownish.

Fls. & Frts. : Almost throughout the year.

Chrom. No. : n = 48 (*Taxon* 20: 612. 1971).

Habitat : Moist places in cultivated fields, along ponds and canals, muddy banks of rivers, moist humous soil; a troublesome weed in cultivated fields.

Distrib. : Widely distributed in the warmer parts of the whole world. INDIA: Throughout. KARNATAKA: Bangalore, Bellary (Satyanarayan & Shankaranarayan, *l.c.*), Belgaum, Bijapur, Chikmagalur (Swamy *et al.*, 1992), Dakshina Kannada, Dharwar, Gulbarga, Hassan, Kodagu, Kolar (Tirumalachar *et al.*, 1949), Mandya, Mysore, Shimoga (Sharma *et al., l.c.*), Tumkur (Gowda *et al.*, 1986), Uttara Kannada.

Uses : Used in local medicines mainly for worms, dysentry, fever etc.

Specimens examined : Bangalore: Bangalore, *Camaron* 432, Jan. 1890 (MH); Bangalore, *Camaron* 592, March 1891 (MH); Bangalore, without coll. name, B. 393, without date (MH); Hulimavu, *Hooper & Saldanha* 18032, 7.11.1971 (JCB). Belgaum: On way to Ghataprabha, near the bridge over Ghataprabha river, *Prasad* 172825, 12.12.1994 (BSI). Bijapur: Sindgi-Bijapur Road, 50th km, *Singh* 142917, 13.8.1976 (BSI); Bijapur-Bagewadi Road 35th km, *Singh* 142982, 16.8.1976 (BSI). Dakshina Kannada: Udyavara, *Bhat* 274, 8.8.1976 (MH); Panamboor, *Bhat* 481, 5.11.1977 (MGH); Suratkal, *Saldanha et al.* 8850, 4.8.1979 (JCB); Near the bridge over Netravati river, Kadakar village, *Prasad* 173878, 10.11.1995 (BSI). Dharwar: Agri. College, *Puri* 19955, 9.6.1959 (BSI); Dharwar, without Coll. name, 100450, 8.9.1964 (BSI). Gulbarga: Kadaganchi F.P., *Singh* 142884, 12.8.1976 (BSI). Hassan: Dudda, near Railway bridge, *Hooper & Gandhi* 2390, 10.11.1971 (JCB). Kodagu: Mercara, *Ahuja* 47532, 10.1.1959 (BSI); Kushalnagar, *Bhat* 1012, 21.9.1981 (MGH); Abbe falls, Mercara, *Yoganarasimhan & Murthy* 4649, 10.1.1984 (RRCBI). Mandya: Paschimavahini, *Bhat* 92, 2.11.1970 (MGH). Mysore: Hassan Road, *Puri* 19926, 6.6.1957 (BSI); Biligirirangan betta, *Rao* 80424, 28.4.1962 (BSI); Mysore University Bot. Garden, *Suma* 3, 3.9.1968 (MGH); Kukkanahalli tank, *Rao* 630, 14.5.1970 (JCB); Manasagangotri, *Bhat* 20, 6.7.1970 (MGH). Uttara Kannada: Without locality, *Puri s.n .* (Acc. No. 18119), without date (BSI); Forest nursery, Halyal, *Prasad* 173705, 29.10.1995 (BSI). Without locality (Mysore & Carnatic), *Thomson s.n.* (Acc. No. 78599), without date (MH).

subsp. **tuberosus** (Rottb.) Kukenth. in Engl., Pflanzenr. 4 (20), Heft 101: 113. 1935. *Cyperus tuberosus* Rottb., Descr. & Ic. Rar. Nov. Pl. 28, t. 7, f. 1. 1773; Clarke in Hook.f., Fl. Brit. India 6: 616. 1893, in part; Cooke, Fl. Pres. Bombay 2: 872. 1908 (3: 385.1958, repr.ed.); Karthik. *et al.*, Fl. Ind. Enum. Monocot. 47. 1989.

Illus. : Rottb., *l.c.*

Very similar to the typical subspecies. Can be distinguished by the following characters. Stems 30-60 cm long. Spikelets linear-oblong or linear-lanceolate, 2-2.5 mm broad. Glumes ovate-lanceolate, acute at apex, ferrugineous or dirty yellow. Nut broadly obovoid.

Fls. & Frts. : Not known.

Habitat : May be same as of typical subspecies.

Distrib. : Sri Lanka, Mauritius, Africa and Australia. INDIA: Peninsular India, Central and East India. KARNATAKA: Mysore & Carnatic (without exact locality).

Specimen examined: Without exact locality (Mysore & Carnatic), *Thomson s.n.* (Acc. No. 78599 A), without date (MH).

Note: This subspecies was reported from Karnataka for the first time as a result of the present study (Prasad & Singh, 2001).

35. **Cyperus rubicundus** Vahl, Enum. Pl. 2: 308. 1805; Hooper in Saldanha & Nicolson, Fl. Hassan 669. 1976; Sharma *et al.*, Fl. Karnataka 308. 1984; Karthik. *et al.*, Fl. Ind. Enum. Monocot. 47. 1989; Keshava Murthy & Yoganarasimhan, Fl. Coorg 503. 1990. *C. teneriffae* Poir. in Lamk., Encycl. 7: 245. 1806; Clarke in Hook.f., Fl. Brit. India 6: 601. 1893; Cooke, Fl. Pres. Bombay 2: 861. 1908 (3: 374. 1958, repr.ed.); Fischer in Gamble, Fl. Pres. Madras 1639. 1931 (3: 1139. 1957, repr.ed.); Kukenth. in Engl., Pflanzenr. 4 (20), Heft 101: 306, f. 34 H-K. 1936; Ramaswamy & Razi, Fl. Bangalore 95. 1973; Kern in van Steenis, Fl. Males. 1, 7: 631, f. 60. 1974; Rao & Razi, Fl. Mysore 561. 1981. **Fig. 13.**

Illus. : Kukenth., *l.c.*; Kern, *l.c.*; Matthew, Furth. Illus. Fl. Tamilnadu Carnatic Pl. 648. 1988.

Annuals with fibrous roots, 5-16 cm high. Stems tufted, slender, triquetrous, 0.5 - 1 mm thick, smooth or scabrid above; base surrounded by broad, striate, purplish sheaths. Leaves gradually acuminate to apex, 1-2 mm wide, smooth. Inflorescence a head of 4-12 spikelets, 1.5-3.5 cm across. Involucral bracts 2-3, spreading; lowest usually slightly overtopping the inflorescence, 1.5 - 4 cm long. Spikelets digitate, spreading, strongly compressed, oblong, obtusish at apex, 8-15 x 3-5 mm; rachilla wingless, with *ca* 0.5 mm long internodes. Glumes membranous, broadly ovate, acuminate or with a spreading or recurved mucro at apex, acutely keeled, 2-3 x 1.5-2 mm, with reddish-brown sides, 9-13-nerved. Stamens 3; filaments elongate up to 2.5 mm; anthers oblong, *ca* 0.5 mm long, with a short connective appendage. Style *ca* 1.5 mm long; stigmas 3, shorter than style. Nut triquetrous, obovoid, attenuate towards the base, minutely apiculate at apex, *ca* 1 x 0.7 mm, greyish-brown.

Fls. & Frts. : June-Nov.

Habitat : Rocky hills and dry areas.

Distrib. : Malesia, South Arabia, Africa, Madagascar and North Queensland. INDIA: Peninsular India. KARNATAKA: Bangalore, Chikmagalur (Sharma *et al.*, *l.c.*), Hassan, Mandya, Mysore.

Specimens examined : Bangalore: Kankanhalli, *Camaron* 635, Nov. 1891 (MH); Bangalore, *Saldanha* 8889, 15.8.1964 (JCB). Hassan: Arkavati river, Tumkur Road, *Nicolson et al.* 2221, 22.10.1971 (JCB). Mandya:

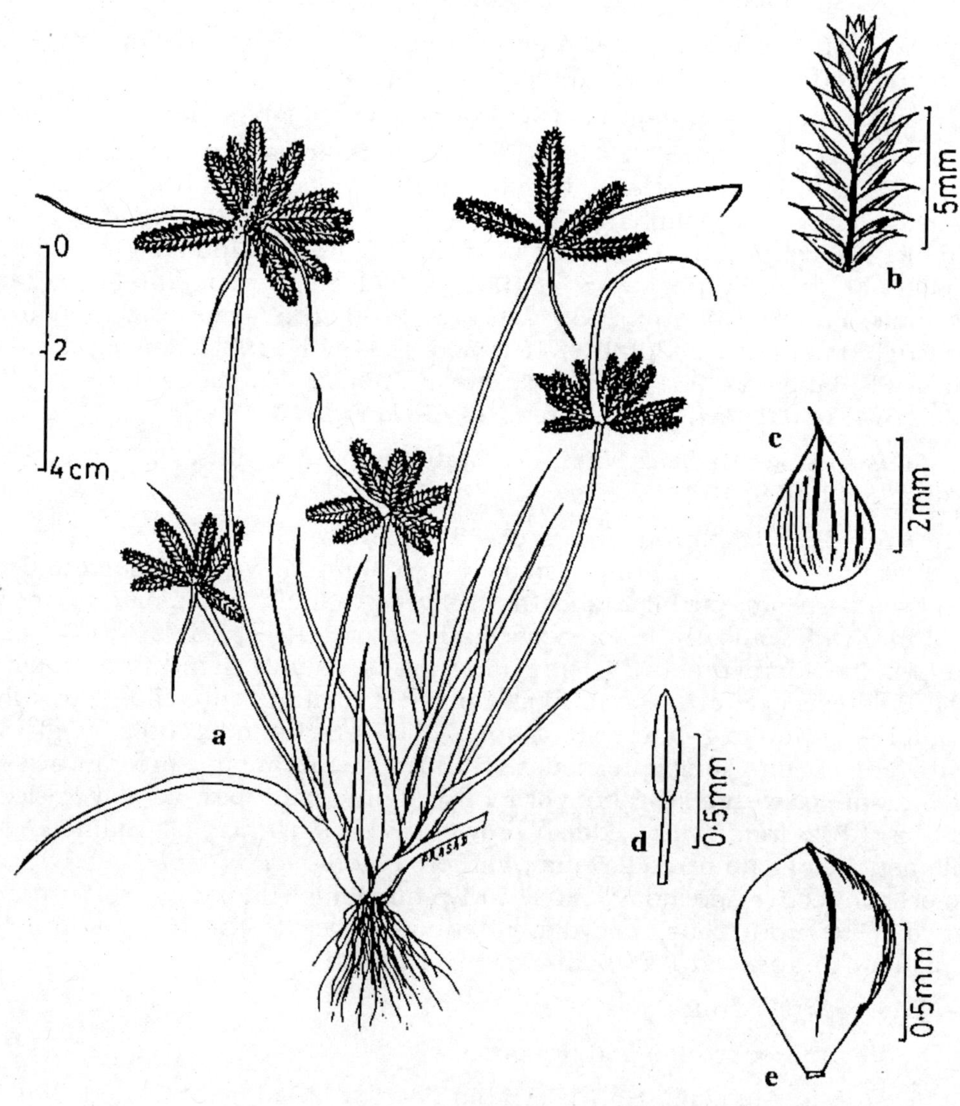

Fig. 13. *Cyperus rubicundus* Vahl
a. Habit, **b**. Spikelet, **c**. Glume, **d**. Stamen, **e**. Nut

Karighatta, *Dinesh* 631, 2.11.1983 (MGH); Melukotte, *Dinesh* 658, 7.11.1983 (MGH). Mysore: Bandipur, without coll. name, *s.n.* (Acc. No. 73464), Nov. 1889 (MH); Mandakalli, *Bhat* 13, 12.7.1970 (JCB).

36. **Cyperus stoloniferus** Retz., Obs. Bot. 4: 10. 1786; Clarke in Hook. f., Fl. Brit. India 6: 615. 1893; Fischer in Gamble, Fl. Pres. Madras 1641. 1931 (3: 1141. 1957, repr.ed.); Kukenth. in Engl., Pflanzenr. 4 (20), Heft 101: 106, 1935; Kern in van Steenis, Fl. Males. 1, 7: 606. 1974; Rao & Razi, Fl. Mysore 561. 1981; Sharma *et al.*, Fl. Karnataka 308. 1984; Koyama in Dassanayake & Fosberg, Rev. Handb. Fl. Ceylon 5: 180. 1985; Karthik. *et al.*, Fl. Ind. Enum. Monocot. 48. 1989. **Fig.14.**

Type : India, Tranquebar.

Illus. : Clarke, Illus. Cyp. t. 19, f. 1-3. 1909; Matthew, Furth. Illus. Fl. Tamilnadu Carnatic Pl. 649. 1988.

Perennials with thick woody rhizome and long-creeping stolons, 10-40 cm high. Rhizome tuberous, globose to ellipsoid, 8-13 mm thick, covered with brownish to blackish fibres; stolons clothed with scaly sheaths, becomes hard in later stages, with disintegrated scales. Stems usually solitary from the tuberous rhizome, slender, rigid, trigonous, 1-2 mm thick, smooth. Leaves usually shorter than stem, linear, narrowed to the acuminate apex, 2-4 mm wide, scabrid on the margins towards apex; sheaths pale, disintegrating in later stages leaving fibrous remains at base of the stem. Inflorescence usually simple, at times subcompound, dense or rather loose, rarely congested, 2-6 x 2-5 cm. Bracts 2-3, suberect; lowest usually much overtopping the inflorescence, at times erect giving the inflorescence pseudolateral appearance. Primary rays 2-5, obliquely erect, unequal, 0.5 - 5 cm long, smooth. Spikes ovoid to broadly ovoid, 1-2 cm long, bearing 3-8 spikelets on a short rachis. Spikelets spicately arranged, obliquely erect to spreading, somewhat thickened with blunt edges, linear-oblong or oblong-lanceolate, subacute or obtuse at apex, at times weakly curved, 5-17 x 2-2.5 mm, yellowish-brown to reddish-brown, 8-22-flowered; rachilla slightly flexous, broadly winged, with 0.5-0.7 mm long internodes. Glumes closely imbricate, obliquely erect, broadly ovate, obtuse at apex, 2.5-3 x 2-2.5 mm, 5-7-nerved, ferrugineous or rubigineous, often reddish variegated, hyaline at margins; keel absent. Stamens 3; filaments elongate up to 3 mm; anthers linear, up to 2 mm long, with a smooth connective appendage. Style very short, 0.2-0.5 mm long; stigmas 3, much longer than style, up to 2.5 mm long. Nut planoconvex, often with a concave ventral surface and a slightly raised angle on the dorsal convex surface, broadly obovate, obtuse at apex, 1-1.5 x *ca* 1 mm, dark brown to blackish.

Fls. & Frts. : May - Jan.

Habitat : Coastal sandy areas, margins and bunds of brackish water, wet sandy or muddy river beds.

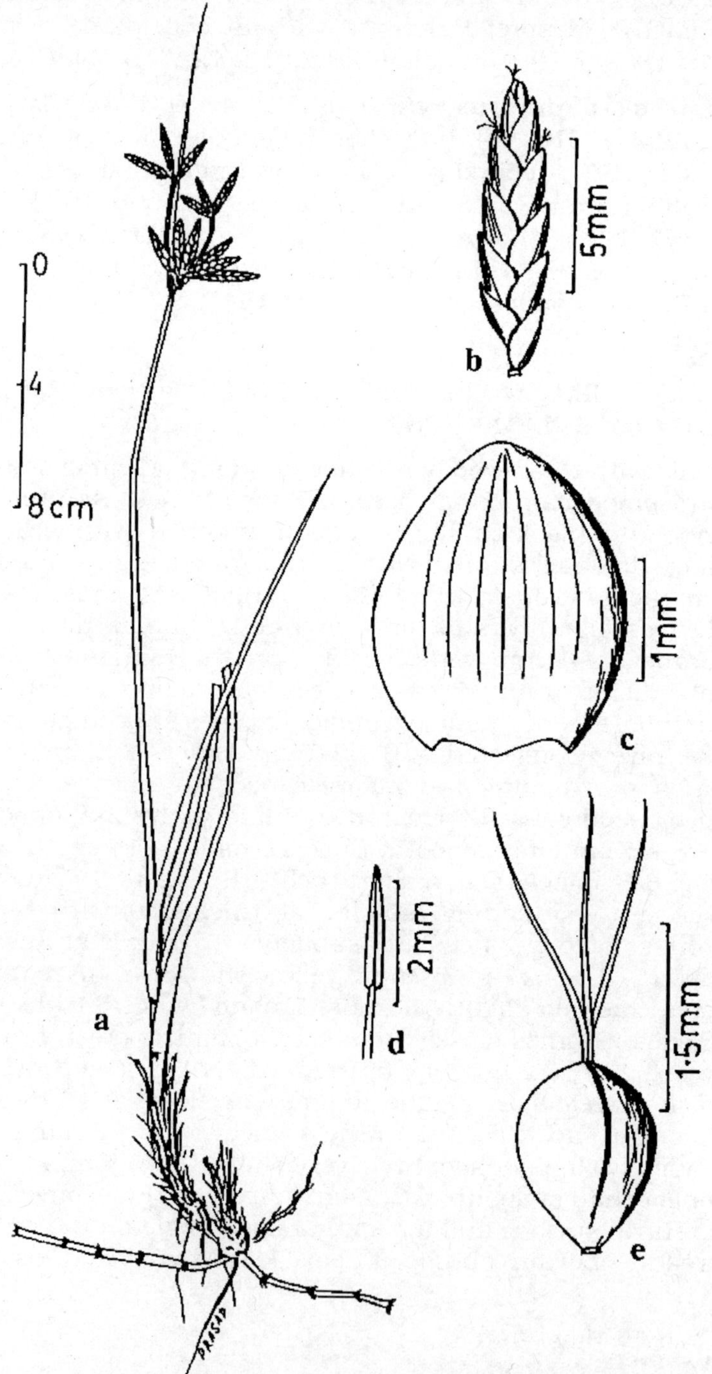

Fig. 14. *Cyperus stoloniferus* Retz.
a. Habit, **b.** Spikelet, **c.** Glume, **d.** Stamen, **e.** Nut with style.

Distrib. : Sri Lanka, Indo-China, Malesia, Mauritius, Madagascar and Australia. INDIA: Peninsular India, Andaman & Nicobar Islands. KARNATAKA: Bellary, Bijapur, Dakshina Kannada, Mandya, Mysore, Uttara Kannada.

Specimens examined : Bellary: Hagari, *Manohar & Ramesh* 5954, 20.1.1979 (JCB). Bijapur: Malaprabha river bed, Pattadakal, *Prasad* 172969, 21.12.1994 (BSI); Malaprabha river, Cholachgudd, Badami taluk, *Prasad* 172985, 23.12.1994 (BSI). Dakshina Kannada: Kapu, *Bhat* 450, 15.1.1977 (MGH); Panamboor, Mangalore, *Bhat*, 482, 5.11.1977 (MGH); Kulur, Mangalore-Suratkal Road, *Saldanha & Sreenath* 9145, 20.9.1979 (JCB); Suratkal, *Bhat* 611, 2.10.1979 (MGH); Gopady, Coondapur taluk, *Sheriff & Suresh* 341, 16.2.1985 (MH). Mandya: Shivasamudram, *Ramesh & Manohar* 8614, 31.7.1979 (JCB). Mysore: Without exact locality, *Bhat* 72, without date, 1970 (JCB). Uttara Kannada: Karwar beach, *Prasad* 173744, 31.10.1995 (BSI); Kaliguj, Karwar, *Prasad* 173758, 1.11.1995 (BSI). Without locality (Mysore & Carnatic), *Thomson s.n.* (Acc. No. 73612), without date (MH).

37. **Cyperus tenuiculmis** Boeck. in Linnaea 36: 286. 1870; Kern in Reinwardtia 3: 30. 1954 *et* in van Steenis, Fl. Males. 1, 7: 608. 1974; Hooper in Saldanha & Nicolson, Fl. Hassan 669. 1976; Rao & Verma, Cyp. NE India 18. 1982; Sharma *et al.*, Fl. Karnataka 308. 1984; Koyama in Dassanayake & Fosberg, Rev. Handb. Fl. Ceylon 5: 187, f. 11 A-F. 1985; Karthik. *et al.*, Fl. Ind. Enum. Monocot. 48. 1989; Keshava Murthy & Yoganarasimhan, Fl. Coorg 504. 1990. *C. zollingeri* auct. non Steud. 1855; Clarke in Hook.f., Fl. Brit. India 6: 613. 1893; Fischer in Gamble, Fl. Pres. Madras 1641. 1931 (3: 1141. 1957, repr.ed.); Kukenth. in Engl., Pflanzenr. 4 (20), Heft 101: 133. 1935.

Type : Nepal, *Wallich* 3321.

Illus. : Koyama, *l.c.*; Matthew, Furth. Illus. Fl. Tamilnadu Carnatic Pl. 650. 1988.

Perennials with short, woody, corm-like rhizome and short stolons, up to 85 cm high. Stems solitary or few together, slender but stiff, erect, triquetrous, 1-2 mm thick, smooth. Leaves few, shorter than stem, canaliculate, rigid, gradually narrowed to an acute apex, 2-3 (-6) mm wide, scabrid in the upper part; sheaths light brown. Inflorescence simple or subcompound, lax, at times reduced to a single cluster, up to 17 cm long and as broad. Involucral bracts 3-7, erect, patent; lowest equalling to overtopping the inflorescence, up to 21 cm long. Primary rays 4-6, erect-patent, slender, very unequal, longest up to 15 cm long. Spikes broadly ovoid, up to 4 x 5 cm, bearing 3-11 spikelets; rachis glabrous or slightly scabrid. Spikelets spicately arranged, patent to horizontally spreading, slightly compressed, subquadrangular, linear to linear-lanceolate, acute at apex, 10-25 x *ca* 2 mm, subloosely 4-14 (-20) -

flowered, stramineous; rachilla flexuous, broadly winged, with *ca* 1.5 mm long internodes; wings oblong-lanceolate, hyaline, caducous. Glumes chartaceous, appressed, elliptic-ovate, obtuse or acutish and muticous or mucronulate at apex, keeled, *ca* 3.5 x 2 mm, 7-9- nerved, stramineous, with hyaline margins; keel broad, green. Stamens 3; filaments elongate up to 3 mm; anthers linear, 1-1.5 mm long, with slightly produced connective appendage. Style 1-1.5 mm long; stigmas 3, up to 1.5 mm long. Nut triquetrous, ellipsoid or elliptic-obovoid, minutely apiculate, *ca* 2 x 1 mm, castaneous to blackish-brown.

Fls. & Frts. : Aug. - Sept.

Habitat: Open, moist grasslands and open fields along with grasses and other sedges; not common.

Distrib. : South and South-East Asia, tropical Africa and Australia. INDIA: Throughout the moist regions. KARNATAKA: Hassan, Kodagu.

Specimens examined : Hassan: Tagáre, *Saldanha* 14699, 23.6.1969 (JCB): Kodagu: Keelar, *Rao* 74822, 30.9.1961 (BSI); Kushalnagar, *Bhat* 10412, 21.9.1981 (MGH).

38. **Cyperus tenuispica** Steud., Syn. 2: 11. 1855; Fischer in Gamble, Fl. Pres. Madras 1640. 1931 (3:1139. 1957, repr.ed.); Kukenth. in Engl., Pflanzenr. 4 (20), Heft 101: 245, t. 28 A-D. 1936; Ramaswamy & Razi, Fl. Bangalore 95. 1973; Kern in van Steenis, Fl Males. 1, 7: 625, f. 58. 1974; Hooper in Saldanha & Nicolson, Fl. Hassan 670. 1976; Rao & Razi, Fl. Mysore 561. 1981; Rao & Verma, Cyp. NE India 13. 1982; Sharma *et al.*, Fl. Karnataka 308. 1984; Koyama in Dassanayake & Fosberg, Rev. Handb. Fl. Ceylon 5: 205. 1985; Singh, Fl. E. Karnataka 2: 633. 1988; Karthik. *et al.*, Fl. Ind. Enum. Monocot. 48. 1989; Keshava Murthy & Yoganarasimhan, Fl. Coorg 504. 1990. *C. flavidus* auct. non Retz. 1789; Clarke in Hook.f., Fl. Brit. India 6: 600. 1893; Cooke, Fl. Pres. Bombay 2: 862. 1908 (3: 375. 1958, repr.ed.). *Honnebu hullu.*

Type: India, Mangalore, *Hohanacker* 1607.

Illus. : Kukenth., *l.c.*; Kern, *l.c.*; Matthew, Furth. Illus. Fl. Tamilnadu Carnatic Pl. 651. 1988.

Annuals with reddish, fibrous roots, 8-35 cm high. Stems tufted, weak, triquetrous, 1-2 mm thick, smooth. Leaves usually smaller or equalling the stem, at times longer, linear, gradually narrowed to an acute apex, up to 30 cm long, 2-5 mm wide, smooth or scabrid on the upper margins; sheaths scarious, stramineous to reddish-brown, lower ones bladeless. Inflorescence compound or decompound, relatively large, lax, 2-17 cm across. Involucral bracts 2-3; longest usually overtopping the inflorescence, up to 15 cm long. Primary rays 5-17, suberect to patent, unequal, slender, up to 10 cm long; secondary rays up to 2 cm long. Spikelets digitate, in clusters of 3-9, flattened, linear-oblong, subacute at apex, 3-8 x 1-1.5 mm,

10-30-flowered; rachilla wingless, persistent, with *ca* 0.5 mm long internodes. Glumes subloosely arranged exposing part of rachilla and mature nuts, membranous, ovate to oblong-ovate, obtuse and slightly recurved at apex, 0.8-1 x 0.5-0.7 mm, slightly keeled towards apex; keel greenish, 3-nerved, midnerve excurved as a minute mucro; sides nerveless, hyaline or purplish-lineolate. Stamen 1(-2); filament up to 1 mm long; anther oblong, *ca* 0.3 mm long, with smooth connective appendage. Style *ca* 0.5 mm long; stigmas 3, as long as the style. Nut obtusely trigonous, broadly obovoid to subglobose, distinctly stipitate, very minutely apiculate, *ca* 0.3 x 0.25 mm, densely and minutely tuberculate.

Fls. & Frts. : Aug. - April.

Habitat : Common weed in wet paddy fields; also found along the banks of canals, streams, tanks, puddles, paddy fields, etc. and in swampy areas in open forests.

Distrib. : Widely distributed in Sri Lanka, Nepal, China, Malesia, South Japan, tropical Africa and tropical Australia. INDIA: Throughout. KARNATAKA: Bangalore, Belgaum, Chikmagalur, Dakshina Kannada, Dharwar, Hassan, Kodagu, Kolar, Mandya, Mysore, Shimoga, Tumkur, Uttara Kannada.

Specimens examined : Bangalore: Bangalore, *Camaron* 562, March 1891 (MH); Sringarpura, *Hooper & Saldanha* 18070, 18.11.1971 (JCB). Belgaum: Sutagatti, *Mahajan* 24977, without date (BSI); Gottni nallah, Jambotti R.F., *Prasad* 172877, 15.12.1994 (BSI). Chikmagalur: Hilikeri village, along Sringeri Road, *Prasad* 173822 & 173828, 7.11.1995 (BSI); Basavanikeri, Koppa taluk, *Prasad* 173847, 8.11.1995 (BSI). Dakshina Kannada: Sampagi, *Barbar* 2411, 15.11.1900 (MH); Mangalore, *Barbar* 4863, Jan. 1902 (MH); Charmadi ghat, *Saldanha* 6243, 27.10.1960 (JCB); Kannadekatte, *Bhat* 173, 6.5.1975 (MGH); Udyavara, *Bhat* 323, 24.10.1976 (MGH); Kapu, *Bhat* 345, 1.11.1976 (MGH); Jappinamogeru village, *Prasad* 173855, 10.11.1995 (BSI). Dharwar: Salikkinikuppa lake, Dharwar, *Prasad* 172922, 18.12.1994 (BSI). Hassan: Tank near Railway station, Hassan, *Hooper & Gandhi* 2379, 10.11.1971 (JCB). Kodagu: Kalhalla, *Arora* 46138, 5.1.1959 (BSI); Abbe falls, Mercara, *Bhat* 872, 23.12.1980 (MGH); Kirugoor, *Bhat* 932, 26.1.1981 (MGH); Abbe falls, Mercara, *Yoganarasimhan & Murthy* 4650, 10.1.1984 (RRCBI). Kolar: Sidlghatta-Chintamani Road, 14th km, *Singh* 142035, 3.1.1976 (BSI); Kendatti hills, *Prakash & Sreenath* 2772, 21.9.1978 (JCB). Mandya: Karigatta, *Taradevi* 18, 5.9.1965 (MGH); Karigatta, *Santosh* 3, 30.12.1965 (MGH); Ranganathittu, *Padma Rani* 14, 2.8.1970 (MGH); Ranganathittu, *Bhat* 45, 3.8.1970 (JCB); Srirangapattana, *Dinesh* 781, 26.1.1984 (MGH). Mysore: Kottathuhalla, Biligirirangan hill ranges, *Rao* 80176 A, 21.4.1962 (BSI). Shimoga: Gaurikeri, Thalaguppa, Sagar taluk, *Prasad* 173797, 5.11.1995 (BSI). Tumkur: Ippadi S.F., Huliyurdurga, Kunigal, *Singh* 140889, 23.10.1975 (BSI). Uttara Kannada: Yellapur, *Talbot* 925, 28.2.1884 & *s.n.* (Acc. No 671), Sept. 1884 (BSI);

Karwar, *Talbot s.n.* (Acc. No. 669), 10.8.1885 (BSI); Ulletikeri, Halyal, *Prasad* 173703, 29.10.1995 (BSI); Javalli village, Halyal range, *Prasad* 173740, 30.10.1995 (BSI). Without locality (Mysore & Carnatic) Thomson *s.n.* (Acc. No. 73491), without date (MH).

SPECIES EXCLUDED

1.**Cyperus alternifolius** L., Mant. 2: 28. 1771.

This species actually found in Madagascar and Mauritius is reported by Sharma *et al.* (1984) from Belgaum district. But a detailed study of the specimens from the district revealed that these are **C. involucratus** Rottb. which is very similar to *C. alternifolius* L. and the former is often treated as a subspecies of the latter. But in the present study plants found in Karnataka is treated as **C. involucratus** Rottb. which is also reported by *Sharma et al.* in the same Flora as *C. flabelliformis*. Hence, the name *C. alternifolius* L. is excluded from the state flora (Prasad & Singh, 2001).

Specimens examined: Belgaum: Belgaum, *Talbot* 3819, Oct. 1896 (BSI); Near Gokak falls, *Prasad* 172821, 11.12.1994 (BSI).

2. **Cyperus fuscus** L., Sp. Pl. 67. 1753.

Singh (1983 & 1988) reported this species from Raichur district as a new record for Karnataka State and subsequently included in the state flora by Sharma *et al., l.c.* But detailed study of the specimen cited revealed that it is *Cyperus amabilis* Vahl.

Hence, *Cyperus fuscus* L. which is actually found in Western Himalaya is excluded from the state flora of Karnataka (Prasad & Singh, 2001).

Specimen examined : Raichur: Raichur-Hyderabad Road, 19th km, *Singh* 141726, 15.11.1975 (BSI).

7. DIPLACRUM

R. Br., Prodr. Fl. Nov. Holl. 1, 240. 1810. *Scleria* sect. *Diplacrum* (R.Br.) Kern in Blumea 11: 208. 1961.

Type: Diplacrum caricinum R. Br.

Small, slender annuals. Leaves basal and cauline, linear, contraligule absent. Inflorescence head-like clusters of spikelets from the axil of leaf-like bracts, usually almost throughout the length of the stem. Spikelets as a rule bisexual, terminated by a female flower; nut-bearing spikelets with 2 glumes which closely envelope the nut and falling off along with the nut. Stigmas 3. Nut globular, bony, irregularly rugose or smooth. Disk (hypogynum) obsolete, completely adnate to base of the nut.

A small genus of 6 species distributed in the tropics of both the hemispheres. 3 species in India; 2 in Karnataka.

Key to the species

1a. Plants very slender with *ca* 0.5 mm thick stems; female glumes oblong-lanceolate, with 2 small lateral lobes near the middle portion, with 2 prominent nerves; nut broadly ellipsoid or oblong-ellipsoid, with many longitudinal ribs which are not interveined .... **1. D. africanum**

1b. Plants comparatively stout with 1 mm or more thick stems; female glumes broadly oblong, distinctly 3-lobed at apex, many-nerved; nut globular, irregularly ribbed .... **2. D. caricinum**

1. **Diplacrum africanum** (Benth.) Clarke in Durana & Schinz, Consp. Fl. Afr. 5: 668. 1895. *Scleria africana* Benth. in Benth. & Hook.f., Gen. Pl. 3: 1071. 1883; Jain & Raghavan in Bull. Bot. Surv. India 9: 301, f. 1-4. 1967; Jain in Ind. For. 95: 130. 1969; Napper in Kew Bull. 25: 445. 1971; Sharma *et al.*, Fl. Karnataka 316. 1984; Karthik. *et al.*, Fl. Ind. Enum. Monocot. 71. 1989. **Fig. 15.**

Type : Nigeria, Nupe, *Barter 1041 (K, Lectotype).*

Illus. : Jain & Raghavan, *l.c.*

Small, tufted annuals. Stems slender, trigonous, 2-12 cm long, *ca* 0.5 mm thick, glabrous. Leaves basal as well as cauline, short-linear, abruptly narrowed to an acute or acuminate apex, 1-3 cm long, 1.5-2.5 mm wide, very minutely scabrous on the margins towards apex; sheaths trigonous, not winged, slightly dilated upward, truncate at mouth. Inflorescence consisting of axillary and terminal clusters of spikelets, almost throughout the length of the stem (when the stem is well developed); clusters scarcely exserted from the subtending leaf-sheaths. Bracts leaf-like. Male spikelets always basal, linear-oblong, *ca* 1.5 mm long. Glumes usually 3, linear, oblong-lanceolate, *ca* 1.5 mm long, hyaline. Stamen 1; filament *ca* 1 mm long; anther oblong, *ca* 0.3 mm long. Female spikelets with glumes spreading in the upper half, 2-2.5 mm long, single flowered. Glumes paired, fused at base, oblong-lanceolate, acuminate at apex, 2-3 mm long, with 2 lateral lobes near the middle portion, with 3 prominent nerves. Nut broadly ellipsoid or oblong-ellipsoid, shortly beaked at apex, *ca* 0.8 mm long, with 3 prominent, longitudinal ribs and 5-7 faint ribs in between the prominent ribs; ribs not interveined. Disk obsolete, adnate to the nut base, obtusely triangular in outline.

Fls. & Frts. : Oct.

Habitat : Rocky slopes at 600 - 700 m altitude.

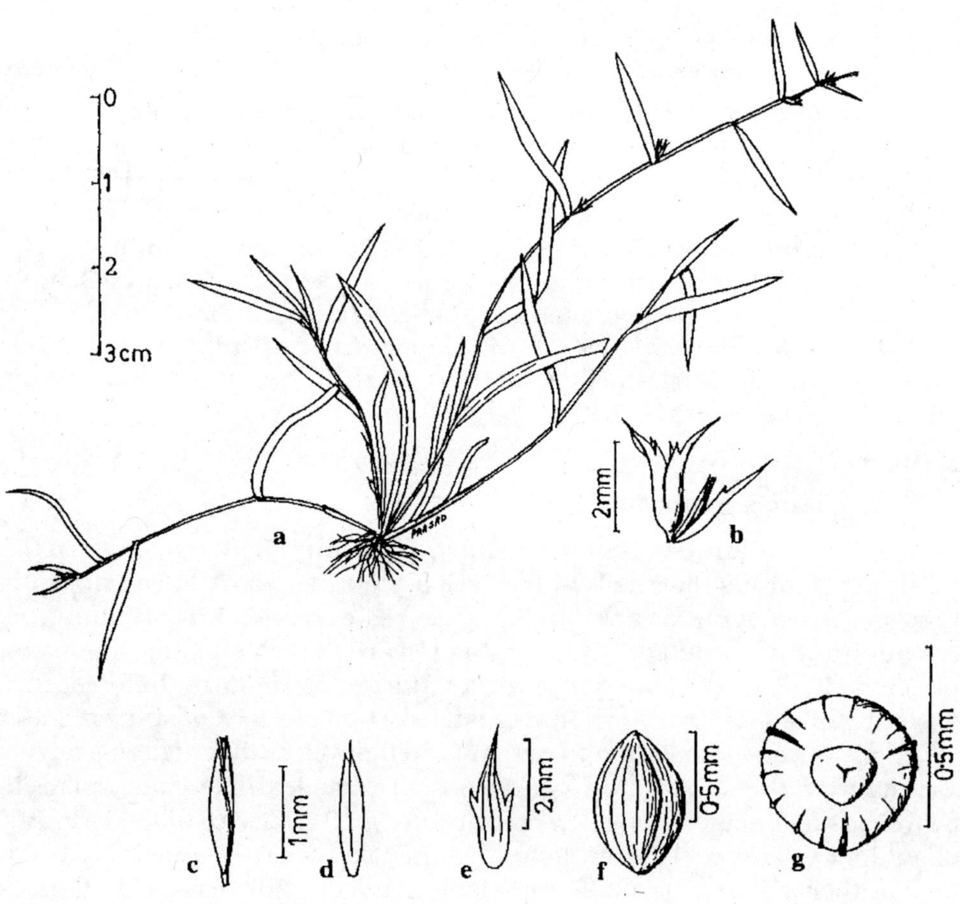

Fig. 15. *Diplacrum africanum* (Benth.) Clarke
a. Habit, **b.** Spikelets (male and female), **c.** Male spikelet, **d.** Male glume, **e.** Female glume, **f.** Nut, **g.** Basal view of nut showing the disk

Distrib. : Western tropical Africa. INDIA: East India (Orissa, see Jain, *l.c.*), Western peninsular India. KARNATAKA: Shimoga, Uttara Kannada.

Specimens examined : Shimoga: Agumbe, *Raghavan* 83330, 18.10.1962 (BSI); Thirthahalli, Raghavan 97483 A, 19.10.1964 (BSI). Uttara Kannada: Yellapur, Talbot *s.n.* (Acc. No. 822), 25.10.1883 (BSI) in part; (Acc. No. 820), 20.9.1884 (BSI); (Acc. No. 821), Sept. 1884 (BSI).

2. **Diplacrum caricinum** R. Br., Prodr. Fl. Nov. Holl. 1, 241. 1810; Fischer in Gamble, Fl. Pres. Madras 1678. 1931. (3: 1164. 1957, repr.ed.); Hooper in Saldanha & Nicolson, Fl. Hassan 670. 1976; Arora *et al.*, Bot. S. Kanara 61. 1981; Koyama in Dassanayake & Fosberg, Rev. Handb. Fl. Ceylon 5: 367. 1985; Karthik. *et al.*, Fl. Ind. Enum. Monocot. 48. 1989. *D. zeylanicum* Nees in Wight, Contr. Bot. India 119. 1834. *Scleria caricina* (R. Br.) Benth., Fl. Austral. 7: 426. 1878; Clarke in Hook.f., Fl. Brit. India 6: 688. 1894; Kern in Blumea 11: 208. f. 8e 1961 *et* in van Steenis, Fl. Males. 1,7: 749, f. 106 (33). 1974; Rao & Verma, Cyp. NE India 58. 1982; Sharma *et al.*, Fl. Karnataka 316. 1984. **Fig. 16.**

Type : Australia, Endeavour River, *Banks and Solander.*

Illus. : Kern, *l.c.*

Small, diffuse annuals. Stems slender, sharply 3-angled, *ca* 17 cm long, *ca* 1 mm thick, smooth, glabrous. Leaves basal as well as cauline, short-linear, abruptly narrowed to an acute tip, 1-4 cm long, *ca* 2 mm wide, minutely scabrous on the margins towards apex; sheaths 3-angled, not winged, slightly dilated upwards. Inflorescence consisting of axillary clusters of spikelets almost throughout the length of the stem; peduncles short, scarcely exserted from the leaf sheath. Primary bracts leaf-like. Male spikelets *ca* 1. 5 mm long, few-flowered. Stamen 1; filament up to 1.5 mm long; anther oblong, *ca* 0.4 mm long. Female spikelet *ca* 2 mm long. Glumes paired, broadly oblong, 3-lobed at apex, *ca* 2 mm long. Stigmas 3. Nut tightly enveloped by 2 subtending glumes and falling off with them, globular, umbonulate at apex, *ca* 1 mm across, with 3 prominent, longitudinal ribs, irregularly ribbed, slightly hispidulous at apex, whitish. Disk obsolete, adnate to base of the nut, obtusely triangular in outline.

Fls. & Frts. : Sept. - Dec.

Habitat : Marshy areas in deciduous forests; sea level to 700 m.

Distrib.: Sri Lanka, South China, Malesia, Japan, Micronesia and Australia. INDIA: Peninsular India, North-East India and Andaman & Nicobar Islands. KARNATAKA: Dakshina Kannada, Hassan, Mysore (Fischer, *l.c.*), Uttara Kannada.

Specimens examined: Dakshina Kannada: Karkala, *Bhat* 414, 20-11-1976 (MGH). Hassan: Byra, *Hooper & Gandhi* 2436, 12.11.1971 (JCB). Uttara Kannada : Yellapur, *Talbot s.n.* (Acc. No. 822), 25.10.1883

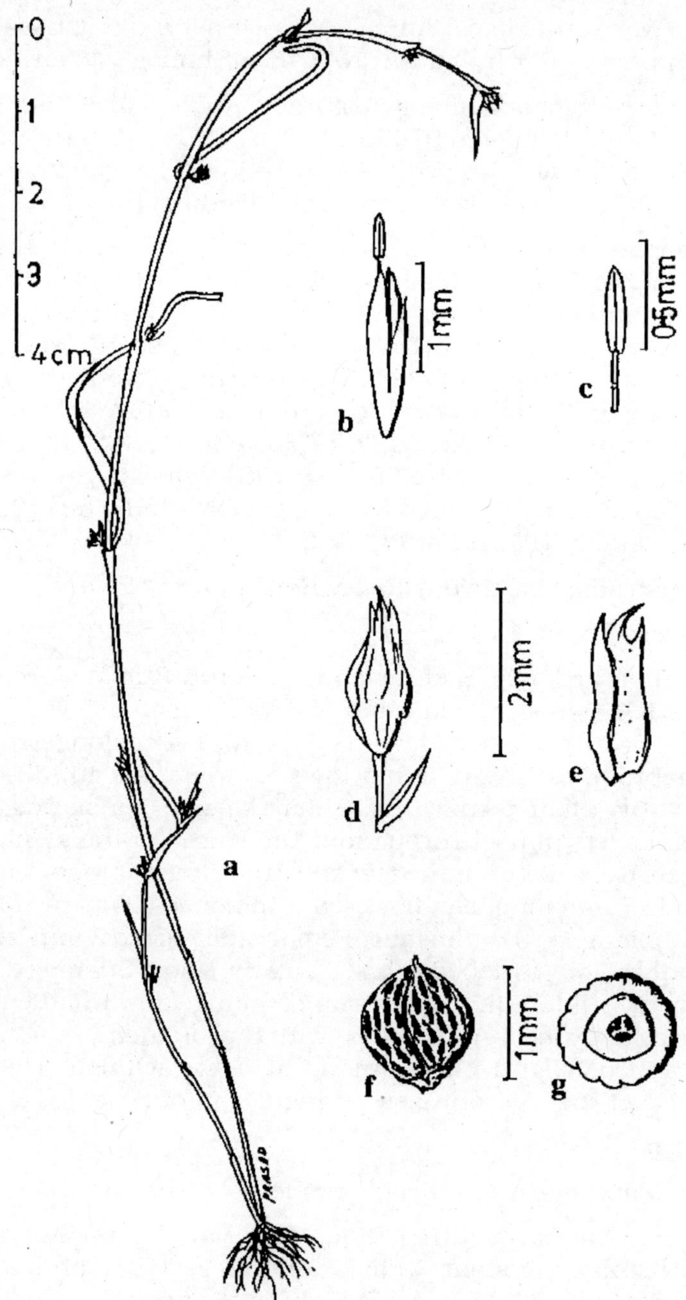

Fig. 16. *Diplacrum caricinum* R. Br.
a. Habit, **b**. Male spikelet, **c.** Stamen, **d.** Female spikelet, **e.** Female glume, **f.** Nut, **g.** Basal view of nut showing the disk.

(BSI) (in part); Yellapur *Talbot s.n.* (Acc No. 820), 20.9.1884 (BSI); Yellapur *Talbot s.n.* (Acc. No. 821), Sept. 1884 (BSI); Bilki, *Ramesh & Udayakumar* 13608, 18.9.1981 (JCB).

SPECIES EXCLUDED

Diplacrum reticulatum Holtt. in Gard. Bull. Singapore 11: 295. 1947. *Scleria reticulata* (Holtt.) Kern in Reinwardtia 6: 71. 1961.

Sharma *et al.* (1984) included this species in the Flora of Karnataka Analysis (from Uttara Kannada dt.), most probably based on two very old specimens collected by W.A. Talbot from Yellapur, (Acc. No. 820, Sept. 20. 1884 and Acc. No. 821, Sept. 1884). These specimens were initially identified as *Scleria caricina* (R.Br.) Benth. and later on corrected as *Scleria reticulata* by Veena Chandra. But a careful study of these specimens revealed that they are nothing but *Diplacrum africanum* (Benth.) Clarke. Hence, *D. reticulatum* Holtt. is excluded from the Flora of Karnataka (Prasad & Singh, 1996a). Thus the Indian distribution of this species might be restricted to Eastern India only.

8. ELEOCHARIS

R. Br., Prodr. 224. 1810. *Heleocharis* Lestib., Ess. Fam. Cyper. 41. 1819. *Chaetocyperus* Nees in Linnaea 9: 289. 1835. *Chlorocharis* Rikli, Jahrb. Wiss. Bot. 27: 564. 1895.

Type : *Eleocharis capitata* R. Br.

Annual or perennial herbs, with rhizome or stolons when perennial. Stems usually tufted, erect or arcuate, terete or triangular, often transversely septate. Leaves reduced to bladeless tubular sheaths, often with a short erect mucro at apex. Inflorescence a single, terminal, ebracteate spikelet, rarely proliferous. Spikelets cylindrical, ovoid, angular, rarely compressed, few to many-flowered. Rachilla persistent. Glumes spirally imbricate, sometimes distichous, membranous to subcoriaceous, lowest 1-2 usually empty. Flowers hermaphrodite. Perianth consists usually of 4-10 hypogynous bristles. Stamens 1-3; connective produced into a small subulate appendage. Style base persistent on the nut; stigmas 2-3. Nut obovoid, trigonous or biconvex; epidermal cells hexagonal, transversely or vertically oblong.

A widely distributed genus, from the tropical to temperate regions of both the hemispheres with *ca* 150 species; *ca* 21 in India; 7 in Karnataka.

Literature : **GOVINDARAJALU, E**. (1975) The systematic anatomy of South Indian Cyperaceae. *Eleocharis* R. Br., *Rhynchospora* Vahl and *Scleria* Berg. in *Adansonia* 14: 581-632. **SVENSON, H.K**. (1929-1939)

Monographic studies in the genus *Eleocharis*, in *Rhodora* 31: 121-135, 152-163, 167-191, 199-219, 224-242. 1929; *ibid.* 34: 193-203, 215-227. 1932; *ibid.* 36: 377-389. 1934; *ibid.* 39: 210-231, 236-273. 1937; *ibid.* 41: 1-19, 43-77, 90-110. 1939.

Note: Eleocharis is a natural genus characterised by the bladeless leaf sheaths and dilated style base articulated with the ovary which persist on the nut.

Key to the species

1a. Perennial herbs with stolons; stems somewhat (2-8 mm) thick; spikelets cylindrical, *ca* as broad as the stem 2

1b. Annual (rarely perennial?) herbs, without stolons; stems capillary to slender, 0.25- 1 mm thick; spikelets ovoid, oblong-ovoid or hemispherical (often slightly flat), much broader than the slender stem 4

2a. Stems terete, septate .... **4. E.dulcis**

2b. Stems triquetrous, not septate 3

3a. Stems slightly and gradually narrowed towards the apex; glumes tightly imbricate, broadly obovate, truncate at apex .... **7. E. spiralis**

3b. Stems not narrowed towards the apex; glumes comparatively loosely imbricate, broadly oblong or oblong-ovate, obtuse-acute at apex .... **1. E. acutangula**

4a. Spikelets never proliferate into rays from the base; stigmas 2; nut biconvex 5

4b. Spikelets often proliferate into rays from the base; stigmas 3; nut trigonous. 6

5a. Spikelets 3-5 x 3-4 mm; glumes tightly imbricate even during fruiting, without keel; stamens (2-) 3; nut *ca* 1 x 0.75 mm, shining blackish-brown; perianth bristles brownish while fruiting .... **5. E. geniculata**

5b. Spikelets 2-4 x *ca* 2 mm; glumes loosely imbricate during fruiting, with a green keel; stamens 1 or 2; nut *ca* 0.7 x 0.5 mm, shining black; perianth bristles whitish while fruiting .... **2. E. atropurpurea**

6a. Stems slender, 0.5-1 mm thick; nut neither longitudinally ribbed, nor with acute angles projecting from the apex; style base not lobed ..**3. E. congesta**

6b. Stems capillary to slender, *ca* 0.3 mm thick; nut longitudinally ribbed, with 3 acute angles projecting from the apex; style base 3-lobed .... **6. E. retroflexa** subsp. **chaetaria**

1. **Eleocharis acutangula** (Roxb.) Schult. in Roem. & Schult., Mant. 2: 91. 1824; Ramaswamy & Razi, Fl. Bangalore 101. 1973. Kern in van Steenis, Fl. Males. 1,7: 525. 1974; Hooper in Saldanha & Nicolson, Fl.

Hassan 671. 1976; Rao & Razi, Fl. Mysore 561. 1981; Rao & Verma, Cyp. NE India 23, f. 9-9a. 1982; Sharma *et al.*, Fl. Karnataka 308. 1984; Koyama in Dassanayake & Fosberg, Rev. Handb. Fl. Ceylon 5: 256. 1985; Singh, Fl. E. Karnataka 634. 1988; Karthik. *et al.*, Fl. Ind. Enum. Monocot. 48. 1989. *Scirpus acutangulus* Roxb., Fl. Ind. 1: 216. 1820. *Eleocharis fistulosa* Schult. in Roem. & Schult., Mant. 2: 89. 1824; Clarke in Hook.f., Fl. Brit. India 6: 626. 1893; Cooke, Fl. Pres. Bombay 2: 888. 1908 (3: 402. 1958, repr.ed.); Fischer in Gamble, Fl. Pres. Madras 1648. 1931 (3: 1145. 1957, repr.ed.).

Illus. : Clarke, Illus. Cyp. t. 35, f. 1-4, 1909; Rao & Verma, *l.c.*

Perennial herbs with short rhizome; stolons long, 2-3 mm thick, rooting at nodes. Stems triquetrous, acute at apex, 30-75 cm long, 3-5 mm thick, not transversely septate. Sheaths membranous, 5-17 cm long, oblique at mouth, purplish towards base. Spikelets cylindrical, acute at apex, 1.5 - 3.5 cm long, as broad as the stem, many-flowered. Glumes all fertile, comparatively loosely imbricate, broadly oblong or oblong-ovate, obtuse-acute at apex, 4.5 x 3 mm, not keeled, scarious towards margins; midvein prominent, with many faint veins on both sides. Perianth bristles 6, linear, slightly broader towards apex, *ca* 2.5 mm long, subequal, as long or slightly exceeding the nut, retrorsely scabrid. Stamens 3; filaments very long, *ca* 4.5 mm long, hyaline; anthers linear-oblong, *ca* 2 mm long, connective appendage very minute, dark brown. Ovary obovate, *ca* 1 mm long; style *ca* 4 mm long, divided into 3 stigmas up to 1.5 mm from the apex which are scabrid. Nut obovoid, slightly compressed, constricted to a conspecuous neck below the apex, *ca* 2 x 1.5 mm, yellowish-brown, with persistent style base and perianth bristles; style base conical, *ca* 0.5 mm long, dark brown; epidermal cells conspicuous, in *ca* 15 vertical rows on either face, transversely oblong.

Fls. & Frts. : July - May.

Chrom. No. : 2n = 54 (*Taxon* 21: 683. 1973).

Habitat : Shallow stagnant water in ponds, tanks, rice fields, ditches, marshes and margins of ponds and tanks. Often forming large patches or in association with other sedges and wild rice.

Distrib. : Nepal, Myanmar, Malesia, Japan, Formosa, tropical Africa, Madagascar, West Indies, North Australia and South America. INDIA: Western Peninsula and North-East India. KARNATAKA: Bangalore, Bijapur, Dakshina Kannada, Hassan, Mysore, Shimoga, Uttara Kannada.

Specimens examined: Bangalore: without locality, *Camaron,* 531, 16.11.1890 (BSI); Dodda bellapur, *Ramesh* 674, 30.3.1978 (JCB). Bijapur: Badami, *Bhide s.n.* (Acc. No. 2639), 8.9.1911 (BSI). Dakshnina Kannada: Kudlu, without coll. name, 16751, 1.9.1920 (MH). Hassan: Tank after Byra, *Saldanha* 13695, 29.5.1969 (JCB). Mysore: Chamundi hills, *Bhat* 9,

10.7.1970 (JCB) & (MGH). Shimoga; Tirthamuthru-Gersu, *Raghavan* 81030, 27.5.1962 (BSI); Gaurikeri, Thalaguppa, Sagar taluk, *Prasad* 173801, 5.11.1995 (BSI). Uttara Kannada : Yellapur, *Talbot*, 1040, 1.10.1884 (BSI); Castle Rock, *Bhide s.n.* (Acc. Nos. 2640, 2641, 2642), Oct. 1909 (BSI); Karlukatta tank, Halyal, *Prasad* 173709, 29.10.1995 (BSI).

2. **Eleocharis atropurpurea** (Retz.) Presl, Rel. Haenk. 1: 196. 1828; Clarke in Hook.f., Fl. Brit. India 6: 627. 1893; Cooke, Fl. Pres. Bombay 2: 889. 1908 (3: 403. 1958, repr.ed.); Fischer in Gamble, Fl. Pres. Madras 1648. 1931 (3: 1145. 1957, repr.ed.); Ramaswamy & Razi, Fl. Bangalore 101. 1973; Kern in van Steenis, Fl. Males. 1, 7: 536. 1974; Hooper in Saldanha & Nicolson, Fl. Hassan 671. 1976; Rao & Razi, Fl. Mysore 561. 1981; Rao & Verma, Cyp. NE India 25. 1982; Sharma *et al.*, Fl. Karnataka 308. 1984; Karthik. *et al.*, Fl. Ind. Enum. Monocot. 48. 1989; Keshava Murthy & Yoganarasimhan, Fl. Coorg 504. 1990. *Scirpus atropurpureus* Retz., Obs. 5: 14. 1789.

Illus. : Clarke, Illus. Cyp. t. 36, f. 6-9. 1909; Mahesh., Illus. Fl. Delhi f. 219. 1966.

Annual herbs with short, densely tufted stems. Stems capillary to slender, minutely angular-sulcate, 2-15 cm long; up to 0.25 mm thick. Sheaths membranous 1- 2.5 cm long, oblique at mouth, usually attenuate at apex, often purplish towards base. Spikelets ovoid to oblong-ovoid, 2-4 mm long, *ca* 2 mm broad. Glumes membranous, rather loosely arranged during fruiting, elliptic-oblong, obtuse at apex, *ca* 1.5 x 0.75 mm, with purplish bands on both sides of the green keel. Perianth bristles (3-) 4, persistent, *ca* 1 mm long, hyaline, whitish, minutely scabrid, shorter than to slightly exceeding the nut. Stamen 1 or 2; filament elongate, 0.5 to 1.25 mm, hyaline; anther linear-oblong, *ca* 0.5 mm long; connective appendage very minute, apiculate. Ovary obovate, *ca* 0.5 mm long; style *ca* 1 mm long, bifid about half the length. Nut biconvex, obovoid, *ca* 0.7 x 0.5 mm, shining black when mature, with persistent minute style base; persistent perianth bristles whitish, shorter than to slightly exceeding the nut.

Fls. & Frts. : Oct. - Feb.

Chrom. No. : 2 n = 20 (*Taxon* 21: 683. 1972).

Habitat : Wet open places, harvested rice fields, grassy areas on the edges of ponds and in water logged soil.

Distrib. : Pakistan, China, Malesia, Formosa, Japan, South Europe, Australia and America. INDIA: Peninsular India, Western India, Western Himalaya, North & North-East India. KARNATAKA: Bangalore, Dakshina Kannada, Hassan, Mysore, Uttara Kannada.

Specimens examined : Bangalore: Ramanagaram, *Ramesh* 247, 23.2.1978 (JCB). Dakshina Kannada: Sampaji, *Barber* 2407, 15.11.1900 (MH); Udyavara, *Bhat* 325, 24.10.1976 (MGH); Jappinamogeru village,

Mangalore, *Prasad* 173859, 10.11.1995 (BSI). Hassan: 10 km before Arsikere town on Tiptur-Arsikere Road, *Saldanha* & *Ramamoorthy s.n.*, 2.12.1970 (JCB). Kodagu: Kirugoor, *Bhat* 928, 26.1.1981 (MGH). Mysore: Yelwala, *Bhat* 98, 2.12.1970 (JCB). Uttara Kannada: Kumta, *Talbot s.n.* (Acc. No. 728 A), 10.12.1883 (BSI), in part.

3. **Eleocharis congesta** Don, Prodr. Fl. Nepal 41. 1825; Clarke in Hook.f., Fl. Brit. India 6: 630. 1893; Fischer in Gamble, Fl. Pres. Madras 1648. 1931 (3: 1145. 1957, repr.ed.); Kern in van Steenis, Fl. Males. 1, 7: 532. 1974; Hooper in Saldanha & Nicolson, Fl. Hassan 672. 1976; Rao & Verma, Cyp. NE India 25. 1982; Sharma *et al.*, Fl. Karnataka 308. 1984; Koyama in Dassanayake & Fosberg, Rev. Handb. Fl. Ceylon 5: 263. 1985; Karthik. *et al.*, Fl. Ind. Enum. Monocot. 48. 1989; Keshava Murthy & Yoganarasimhan, Fl. Coorg 505. 1990. *E. pellucida* Presl, Rel. Haenk. 1: 196. 1828; Miq., Fl. Ind. Bat. 3: 301. 1856. *E. afflata* Steud. in Zoll. Syst. Verz. 1: 62. 1854 (*nom.nud.*); Clarke in Hook.f., Fl. Brit. India 6: 629. 1893. *E. subvivipara* Boeck. in Linnaea 36: 424. 1870; Clarke in Hook.f., Fl. Brit. India 6: 629. 1893.

Illus. : Clarke, Illus. Cyp. t. 37, f. 13-16. 1909 (As *E. subvivipara* Boeck.).

Annual herbs with short, highly tufted stems. Stems slender, sulcate, 5-30 cm long, 0.5 - 1 mm thick. Sheaths membranous, 1.5 - 6 cm long, truncate-oblique at mouth, attenuate at apex, purplish towards base. Spikelets ovoid to oblong, acute at apex, 5-9 x 2-4 mm, many-flowered, often proliferate from the base into 1-3 slender rays which terminate in a small spikelet. Glumes membranous, oblong-ovate, obtuse at apex, *ca* 2.2 x 1 mm, 1-nerved and green in the middle portion, hyaline on both sides, usually purplish towards margins. Perianth bristles 6, subequal, longest *ca* 1.2 mm long, whitish, retrorsely scabrid. Stamens 2; filaments elongate while fruiting; anthers oblong, *ca* 0.75 mm long; connective appendage minute, apiculate. Ovary oblong, *ca* 0.5 mm long; style *ca* 1.5 mm long, divided up to 2/3 from the apex into 3 stigmas, conical at base; stigmas scabrid. Nut trigonous, obovoid, *ca* 1 mm long, smooth, yellowish; persistent style base conical, brownish; persistent perianth bristles slightly exceeding the nut, light brownish.

Fls. & Frts.: Sept.- Jan., Apr. - May.

Chrom. No. : 2n = 20 (*Taxon* 21: 683. 1972).

Habitat : Shallow water, swampy areas, wet rice fields, edges of pools, canals and ponds in hilly areas or ghats.

Distrib. : Sri Lanka, Nepal, Myanmar, China, Malesia, Japan, Madagascar. INDIA: Throughout. KARNATAKA: Belgaum, Hassan, Kodagu, Mysore, Uttara Kannada (Sharma *et al.*, *l.c.*)

Specimens examined : Belgaum: Londa, *Gammie* 2629 (A.C.), May 1900; Dudwa village near Londa R.F., *Prasad* 172914, 17.12.1994 (BSI);

Manjapur Nallah, Khanapur, *Prasad* 172851, 13.12.1994 (BSI). Hassan: Aglatta coffee estate, Hanbal Road, *Hooper & Gandhi* 2476, 13.11.1971 (JCB). Kodagu: Mercara, *Bhat* 1049, 25.9.1981 (MGH). Mysore: Tondaikerekan, Biligirirangan hill range, *Rao* 80387, 27-4. 1962. (BSI).

Note : *Eleocharis congesta* is very similar to *E. geniculata* in appearance. But can be differentiated from the later by its proliferous rays of the spikelet (if present), 3-fid style and trigonous, yellowish nuts. Often new plants are produced from tip of the flowering stem when it touches wet sandy soil. This way the plant propogates vegetatively [see specimen No. 172851 (BSI)].

4. **Eleocharis dulcis** (Burm.f.) Trin. ex Hensch., Vita Rumph. 186. 1833; Kern in van Steenis, Fl. Males. 1, 7: 529. 1974; Hooper in Saldanha & Nicolson, Fl. Hassan 672. 1976; Rao & Verma, Cyp. NE India 23. 1982; Sharma *et al.*, Fl. Karnataka 308. 1984; Koyama in Dassanayake & Fosberg, Rev. Handb. Fl. Ceylon 5: 259. 1985; Karthik. *et al.*, Fl. Ind. Enum. Monocot. 49. 1989. *Andropogon dulce* Burm. f., Fl. Ind. 219. 1768. *Eleocharis plantaginea* (Retz.) Roem. & Schult., Syst. 2: 150. 1817; Miq., Fl. Ind. Bot. 3: 302. 1856; Clarke in Hook. f., Fl. Brit. India 6: 625. 1893; Cooke, Fl.Pres. Bombay 2: 888. 1908 (3: 402. 1958, repr.ed.); Fischer in Gamble, Fl. Pres. Madras 1647. 1931 (3: 1145. 1957, repr.ed.). *Scirpus tuberosus* Roxb., Fl. Ind. 1: 213. 1820. *S. tumidus* Roxb., Fl. Ind. 1: 215. 1820. *Eleocharis equisetina* Presl, Rel. Haenk. 1: 195. 1828; Miq., Fl. Ind. Bat. 3: 302. 1856; Clarke in Hook. f., Fl. Brit. India 6: 626. 1893.

Illus.: Clarke, Illus. Cyp. t. 33, f 1-5, 1909.

Perennials with woody (when dry) rhizome, usually with brownish roots; stolons 2-4 mm thick. Stems terete, acute at apex, 50-150 cm long, (1-) 2-5 mm thick, septate, striate. Sheaths 7-26 cm long, light purplish-brown towards base. Spikelets cylindrical, obtuse-acute at apex, (1-) 2-3 cm long, as broad or slightly broader than stem, many-flowered. Glumes oblong-obovate, obtuse or acute-obtuse at apex, *ca* 6 x 3 mm, purplish-brown spotted inside, many-veined, scarious at margins. Perianth bristles 6-8, subequal up to 3 mm long, slightly broader towards base, retrorsely scabrid. Stamens 3; filaments elongate up to 4 mm; anthers 3-4 mm long, often purplish-brown spotted; connective appendage very prominent, acuminate, slightly recurved, 0.25-0.5 mm long. Ovary obovate, *ca* 1 mm long, whitish; style *ca* 4 mm long, flat and conical at base; stigmas 3, *ca* 2 mm long, minutely scabrid. Nut prominently convex on the anticous face, obovoid, *ca* 2 x 1.5 mm, with longitudinal furrows on the convex face, brownish; epidermal cells very small, hexagonal, in many vertical rows; style base triangular, dark-brown, usually persistent.

Fls. & Frts. : Apr., Aug.- Nov.

Habitat : Fresh water as well as brackish water swamps, pools, ponds, rice fields, other open marshy areas and muddy banks of lakes. Often forming large pure patches in shallow open water.

Distrib. : Sri Lanka, Pakistan, China, Malesia, Formosa, Japan, tropical West Africa, Madagascar and Australia. INDIA: Western Peninsula, North-East & North-West India. KARNATAKA: Chikmagalur, Dakshina Kannada, Dharwar, Hassan, Mysore (Fischer, *l.c.*), Shimoga, Uttara Kannada.

Uses: Tubers are edible. In Sri Lanka culms of this sedge are used for making a rough kind of mat known as 'ratan' mats (Koyama, *l.c.*). Boiled tubers are eaten in Malesia and also made into chips. In China this species is cultivated for its edible tubers. In Sumatra and N. Celebes also it is cultivated. Here also the stems are used for making sleeping mats. Some Papuan tribes use the stem for making women's skirts (Kern, *l.c.*).

Specimens examined : Chikmagalur: Hilikeri tank, along Sringeri road, *Prasad* 173826, 7.11.1995 (BSI). Dakshina Kannada: Kinimoolky, Udupi, *Bhat* 301, 10.10.1976 (MGH); Thumbe, *Raghavan* 146471, 20.4.1977 (BSI); Jappinamogeru village, Mangalore, *Prasad* 173866, 10.11.1995 (BSI). Dharwar: Salikkinikuppa lake, Dharwar, *Prasad* 172925, 18.12.1994 (BSI). Hassan: Hassan, *Saldanha* 9526, 27.4.1965 (JCB). Shimoga: Tunga forests, Thirthahalli, *Raghavan* 82865, 1.10.1962 (BSI). Uttara Kannada: Karwar, *Talbot s.n.* (Acc. No. 726), Aug. 1890 (BSI); Karlukatta tank, Halyal, *Prasad* 173716, 29.10.1995 (BSI).

Note: In this variable species, at times stems are very narrow (1-1.5 mm thick), more prominently striate, somewhat rigid and not flattened in the specimens. In such specimens spikelets are also short and narrow (10-15 x 2-2.5 mm). Such specimens might be from comparatively dry habitats like dried up marshy areas or dried up seasonal wetlands. Purplish-brown spots are usually found on the anther and on the inner surface of the glumes. But often specimens are found without purplish-brown spots on these parts.

5. **Eleocharis geniculata** (L.) Roem. & Schult., Syst. veg. 2: 150. 1817; Ramaswamy & Razi, Fl. Bangalore 102. 1973; Kern in van Steenis, Fl. Males. 1, 7: 536. 1974; Hooper in Saldanha & Nicolson, Fl. Hassan 672. 1976; Rao & Razi, Fl. Mysore 562. 1981; Sharma *et al.*, Fl. Karnataka 308. 1984; Koyama in Dassanayake & Fosberg, Rev. Handb. Fl. Ceylon 5: 264. 1985; Singh, Fl. E. Karnataka 2: 634. 1988; Karthik. *et al.*, Fl. Ind. Enum. Monocot. 49. 1989; Keshava Murthy & Yoganarasimhan, Fl. Coorg 505. 1990. *Scirpus geniculatus* L., Sp. Pl. 1: 48. 1753. *Eleocharis capitata* R. Br., Prodr. 225. 1810; Clarke in Hook.f., Fl. Brit. India 6: 627. 1893; Cooke, Fl. Pres. Bombay 2: 889. 1908 (3: 404. 1958, repr.ed.); Fischer in Gamble, Fl. Pres. Madras 1648. 1931 (3: 1145. 1957, repr.ed.). **Fig. 17.**

Annual herbs with short, highly tufted stems. Stems slender but comparatively rigid, erect or reflexed, angular-sulcate, 2-35 cm long, 0.6-1 mm thick. Sheaths membranous, 1-4 cm long, truncate- oblique at mouth, acute to attenuate at apex, usually purplish- brown towards base. Spikelets

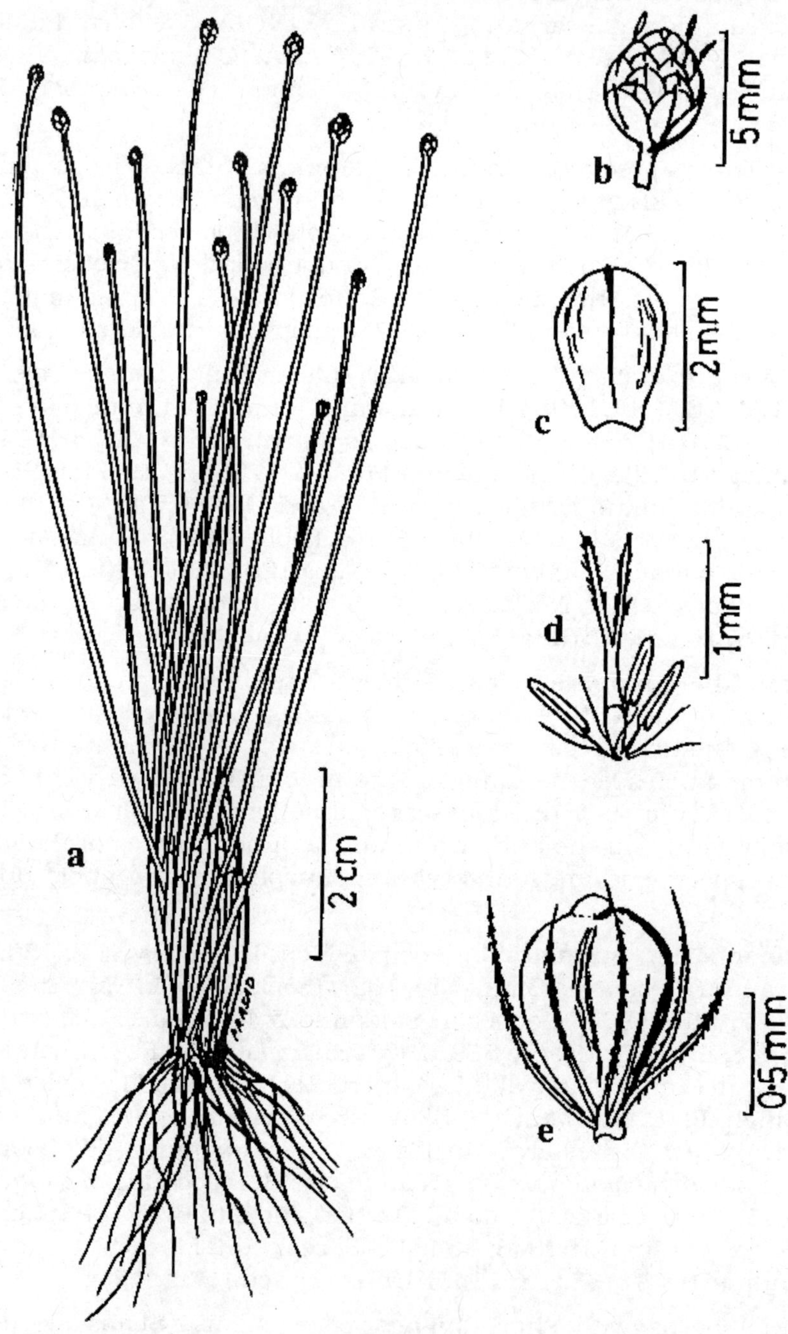

Fig. 17. *Eleocharis geniculata* (L.) Roem. & Schult.
a. Habit, **b.** Spikelet, **c.** Glume, **d.** Flower, **e.** Nut

broadly ovoid, oblong-ovoid or hemispherical, obtuse at apex, 3-5 x 3-4 mm, brownish, many-flowered. Glumes membranous, tightly imbricate even during fruiting, elliptic-oblong, obtuse at apex, *ca* 2 x 1 mm, without keel, often with few brown spots. Perianth bristles 4-7, *ca* 1.25 mm long, retrorsely, scabrid. Stamens (2-) 3; filaments elongate, 0.5-2 mm long; anthers oblong, *ca* 0.5 mm long, connective appendage very minute, apiculate. Ovary obovate, *ca* 0.5 mm long; style 1.5-2 mm long, divided half to three-fourth of the length into 2 stigmas. Nut obovoid, biconvex, *ca* 1 x 0.75 mm, shining blackish-brown; persistent style base minute, conical; persistent perianth bristles brownish, exceeding the nut.

Fls. & Frts. : Almost throughout the year.

Chrom. No. : 2n = 10 (*Taxon* 21: 683. 1972).

Habitat : Open wet sandy areas, swampy grasslands, wet rice fields, banks of ponds and forest streams and other water logged soil; often submerged and forming small patches; often near sea shore in saline soil.

Distrib. : Most widely distributed species of *Eleocharis*. Throughout the warmer parts of old and new world. INDIA: Peninsular India, East and North-East India. KARNATAKA: Bangalore, Belgaum, Bijapur, Chitradurga (Singh, *l.c.*), Dakshina Kannada, Dharwar, Gulbarga, Hassan, Kodagu, Mandya, Mysore, Raichur, Shimoga, Tumkur, Uttara Kannada.

Specimens examined : Bangalore : Bannerghatta, *Saldanha* 12264, 17.9.1980 (JCB). Belgaum: Gokak falls, Gokak, *Prasad* 172815, 11.12.1994 (BSI); Canal on the way to Ghataprabha from Gokak, *Prasad* 172830, 12.12.1994 (BSI); Dinman Hassur, Belgaum-Gokak Road, *Prasad* 172841 & 172841 A, 13.12.1994 (BSI). Bijapur: Bagalkot, without coll. name, *s.n.* (Acc No. 2628), 16.4.1906 (BSI); Kalas Canal, Badami, *Prasad* 172979, 22.12.1994 (BSI); Hossur Forest Nursery, Badami, *Prasad* 172991, 23.12.1994 (BSI). Dakshina Kannada: Malpe, *Bhat* 690, 25.8.1980 (MGH). Dharwar: Sadankeri, Alnavar Road, Dharwar, *Prasad* 172935 & 172943. 19.12.1994 (BSI). Gulbarga: Bidar-Chincholi Road 33rd mile, *Singh* 128611. 9.2.1975 (BSI). Hassan: Boudalboore, *Saldanha* 13566, 26.5.1969 (JCB). Kodagu: Kushalnagar, *Bhat* 952, 16.2.1981 (MGH). Mandya: Abalwadi, *Ahmad* 468, 29.3.1978 (JCB); Sreerangapatana, *Dinesh* 175, 26.1.1983 (MGH). Mysore: Sholapur, near Hipparga, *Jain* 19982, 20.7.1957 (BSI); Mandakalli, *Bhat* 15, 12.7.1970 (JCB) & (MGH). Raichur: Gurguntha R.F., *Singh* 129599, 14.2.1975 (BSI) ; Gunthgola block, *Singh* 132773, 8.9.1974 (BSI). Shimoga: Bara Kane, Agumbe, *Raghavan* 68120 A, 3.11.1960 (BSI); Jog falls, *ca* 500 ft. below the falls, *Prasad* 173785, 4.11.1995 (BSI). Tumkur: Sira-Hiriyur Highway 31st km, *Singh* 143256, 24.8.1976 (BSI). Uttara Kannada: Kumta, *Talbot s.n.* (Acc. No. 728 B), 10.12.1883 (BSI); Kerehoshalli, *Ramesh* & *Shivaprakash* 12977, 4.6.1982 (JCB); Karwar, *Vartak s.n.*, without date (MACS).

6. **Eleocharis retroflexa** (Poir.) Urb. subsp. **chaetaria** (Roem. & Schult.) Koyama in Bull. Nat. Sci. Mus. Tokyo 17: 68. 1974; Hooper in Saldanha &

Nicolson, Fl. Hassan 673. 1976; Sharma *et al.*, Fl. Karnataka 308, 1984; Koyama in Dassanayake & Fosberg, Rev. Handb. Fl. Ceylon 5: 265. 1985; Keshava Murthy & Yoganarasimhan, Fl. Coorg 505. 1990. *Eleocharis chaetaria* Roem. & Schult., Syst. Veg. 2: 154. 1817; Clarke in Hook. f., Fl. Brit. India 6: 629. 1893; Cooke, Fl. Pres. Bombay 2: 890. 1908 (3: 404. 1958, repr.ed.); Fischer in Gamble, Fl. Pres. Madras 1648. 1931 (3: 1145. 1957, repr.ed.). *E. retroflexa* sensu Kern in van steenis, Fl. Males. 1, 7: 534. 1974; Arora *et al.*, Bot. S. Kanara 61. 1981; Rao & Verma, Cyp. NE India 25. 1982; Karthik. *et al.*, Fl. Ind. Enum. Monocot. 49. 1989. **Fig.18.**

Annual herbs with very slender stems. Stems minutely angular-sulcate, 3-25 cm long, *ca* 0.3 mm thick, often curved. Sheaths membranous, 0.5 - 2 cm long, purplish towards the base, whitish- scarious towards apex; upper most usually with a well developed orifice. Spikelets ovoid, slightly flattened, 3-4 x 2-3 mm, few-flowered, often proliferating into rays from the base. Glumes membranous, loosely imbricate during fruiting, oblong-ovate, obtuse at apex, *ca* 2.75 x 1.75 mm, keeled, hyaline and thinly membranous towards margin. Perianth bristles 6, unequal; longest *ca* 0.5 mm long (in flower), whitish. Stamens 2(-3); filaments *ca* 0.7 mm long; connective appendage apiculate. Ovary obovoid, *ca* 0.7 mm long; style *ca* 3 mm long, conical at base, divided up to 2/3 into 3 stigmas, brownish; stigmas scabrous. Nut usually up to 4 in each spikelet, obovoid, trigonous, *ca* 1.1 x 0.7 mm, with 3 acute angles projecting from the apex, yellowish; epidermal cells very prominent, pitted. Persistent style base pyramidal, extending to the three acute projections of the nut (shoulders), brownish; persistent perianth bristles smaller to longer than nut.

Fls. & Frts. : Sept. - Dec., March.

Habitat : Wet grassy areas, drying swamps and moist rice fields, wet muddy soil and sandy edges of ponds, tanks and streams, along wet roadsides, shallow pools in ghats up to 1000 m; often forming small patches. In flowing water, stems are like long fine capillaries.

Distrib. : Old world tropics. INDIA: More or less throughout. KARNATAKA: Bangalore, Belgaum, Chickmagalur, Dakshina Kannada, Hassan, Kodagu, Mysore (Fischer, *l.c.*), Shimoga, Uttara Kannada.

Specimens examined : Bangalore: Bannergatta, *Saldanha* 12264, 17.9.1980 (JCB). Belgaum: Dudwawada R.F., Londa, *Prasad* 172889, 17.12.1994 (BSI). Chikmagalur: Hilikeri, along Sringeri road, *Prasad* 173830 & 173832, 7.11.1995 (BSI); Gadical village tank, Koppa Taluk, *Prasad* 173839, 8.11.1995 (BSI). Dakshina Kannada: Karkal, without coll. name, 11954, 16.3.1915 (MH); Uppinangady, *Hooper & Saldanha* 2527, 25.11.1971 (JCB); Kannadekatte, *Bhat* 167, 6.5.1975 (MGH); Panamboor, *Bhat* 489, 5.11.1977 (MGH); Jappinamogeru village, Mangalore, *Prasad* 173858, 10.11.1995 (BSI). Hassan: Streams between Devalkere and Devarunde, *Hooper & Gandhi* 2484, 13.11.1971 (JCB). Kodagu: Karnangere, *Puri*, 311735, 7.3.1958 (BSI); Mercara, *Bhat* 786, 18.12.1980 (MGH); Edur, near Somvarpet, *Yoganarasimhan et al.* 4504, 28.11.1983 (RRCBI). Shimoga: Agumbe, *Raghavan* 83317, 18.10.1962 (BSI);

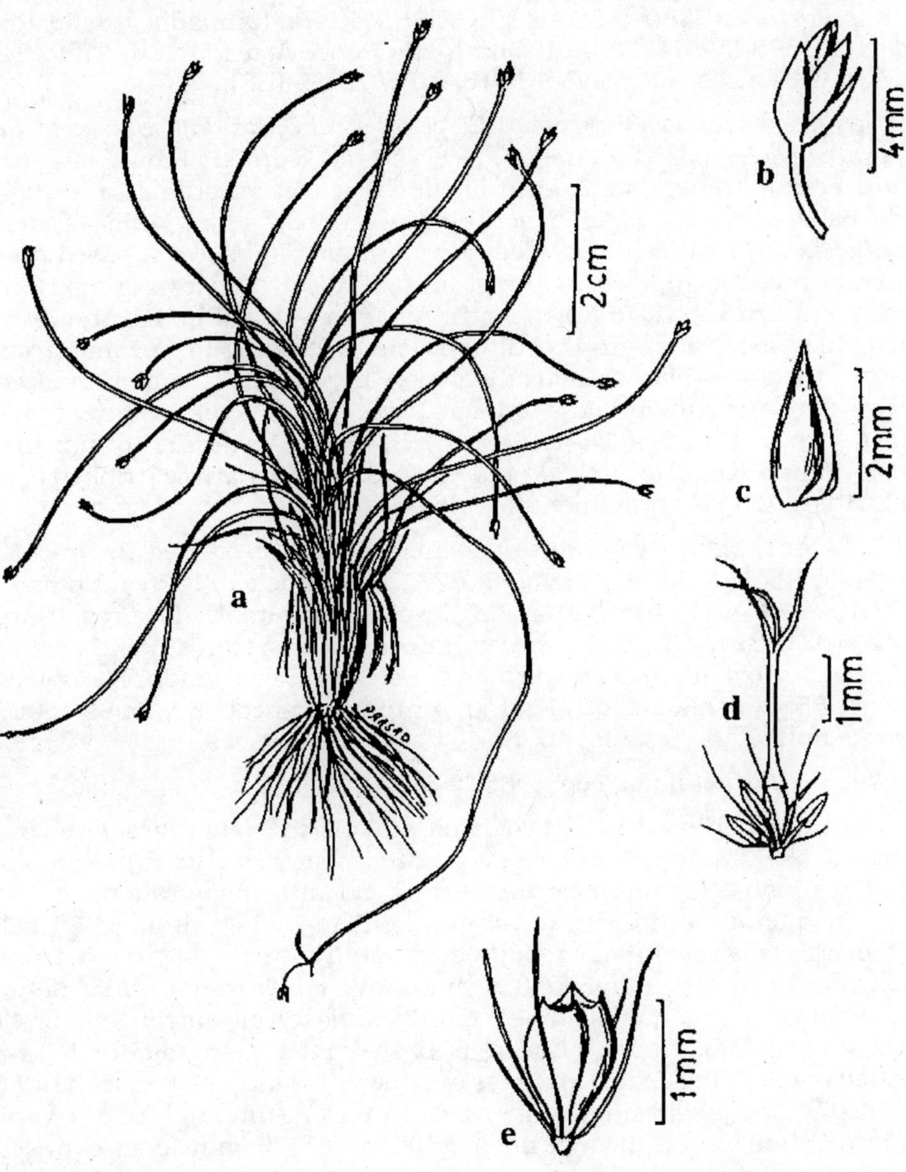

Fig. 18. *Eleocharis retroflexa* subsp. *chaetaria* (Roem. & Schult.) Koyama
a. Habit, **b**. Spikelet, **c.** Glume, **d**. Flower, **e.** Nut

Nishanigudd pond, Agumbe, *Raghavan* 69505, 8.2.1961 (BSI); Bara kana, Agumbe, *Raghavan* 68120, 3.11.1960 (BSI); Gaurikeri, Thalaguppa, Sagar Taluk, *Prasad* 173800, 5.11.1995 (BSI). Uttara Kannada: Castle Rock, *Gammie* 15738, 27.10.1902 (BSI); Castle Rock, *Almeida* 936, 27.12.1968 (BLAT); without locality *Talbot* 1316, 10.10.1889 (BSI).

Note : Svenson (1937) treated *Eleocharis retroflexa* and *E. chaetaria* as separate species based on the much lower and blunt style-base and larger and deeper markings on the nut of the later. But according to Kern (*l.c.*) style base is variable and hence he merged both the species under *E. retroflexa.* Koyama (*l.c.*) identified *chaetaria* as a subspecies based on the shape of the nut and size of the glumes. Typical *retroflexa* is distributed from Southern U.S.A. to Northern S. America only. Although the nuts in Karnataka specimens are broadly obovoid (according to Koyama broadly ovate), based on other characters like more broad (*ca* 0.7 mm), prominently ribbed nut with quite broad shoulders, here also this taxon is treated as a subspecies of *E. retroflexa* as done by Koyama. According to him in the typical subspecies nuts are 0.4 - 0.5 mm broad, with narrowly ribbed angles and with much less prominent shoulders.

7. **Eleocharis spiralis** (Rottb.) Roem. & Schult., Syst. Veg. 2: 155. 1817; Clarke in Hook. f., Fl. Brit. India 6: 627. 1893; Cooke, Fl. Pres. Bombay 2: 888. 1908 (3: 403. 1958, repr.ed.); Fischer in Gamble, Fl. Pres. Madras 1647. 1931 (3: 1145. 1957, repr.ed.); Kern in van Steenis, Fl. Males. 1, 7: 527. 1974; Koyama in Dassanayake & Fosberg, Rev. Handb. Fl. Ceylon 5: 257. 1985; Karthik. *et al.*, Fl. Ind. Enum. Monocot. 50. 1989. *Scirpus spiralis* Rottb., Descr. & Ic. 45, t. 15, f. 1. 1773. **Fig. 19.**

Illus. : Clarke, Illus. Cyp. t. 35, f. 5-7.1909.

Perennial herbs with short rhizome; stolons *ca* 3 mm thick, with scales. Stems densely tufted, triquetrous, gradually narrowed to an acute apex, 20-60 cm high, 2-4 mm thick, not septate. Sheaths membranous, 4-14 cm long, oblique at mouth with a 2-4 mm long appendage at apex. Spikelets cylindrical, slightly broader than apex of the stem, slightly narrowed towards the acute apex, 2-3.5 cm long, *ca* 3 mm broad, densely many-flowered, whitish. Glumes tightly and spirally arranged, broadly obovate, truncate at apex, cuneate at base, *ca* 3 x 3 mm, distinctly brown spotted towards the centre and scarious towards margins. Perianth bristles 4-6, linear, unequal in length, longest *ca* 1 mm long, minutely and retrorsely scabrid. Stamens 3; anthers linear-oblong, *ca* 1.5 mm long; connective appendage very minute. Ovary obovate, *ca* 0.5 mm long; style *ca* 2 mm long, usually divided into 3 stigmas up to slightly above the base, minutely scabrid; undivided portion *ca* 0.3 mm long, smooth, conical at base. Nuts (not seen) obovate to ovate, 1.5- 1.7 x 1.2-1.4 mm.

Fls. : July-Aug.

Habitat : Wet places in low altitudes, pools and swamps. Restricted to brackish and saline localities towards coastal areas.

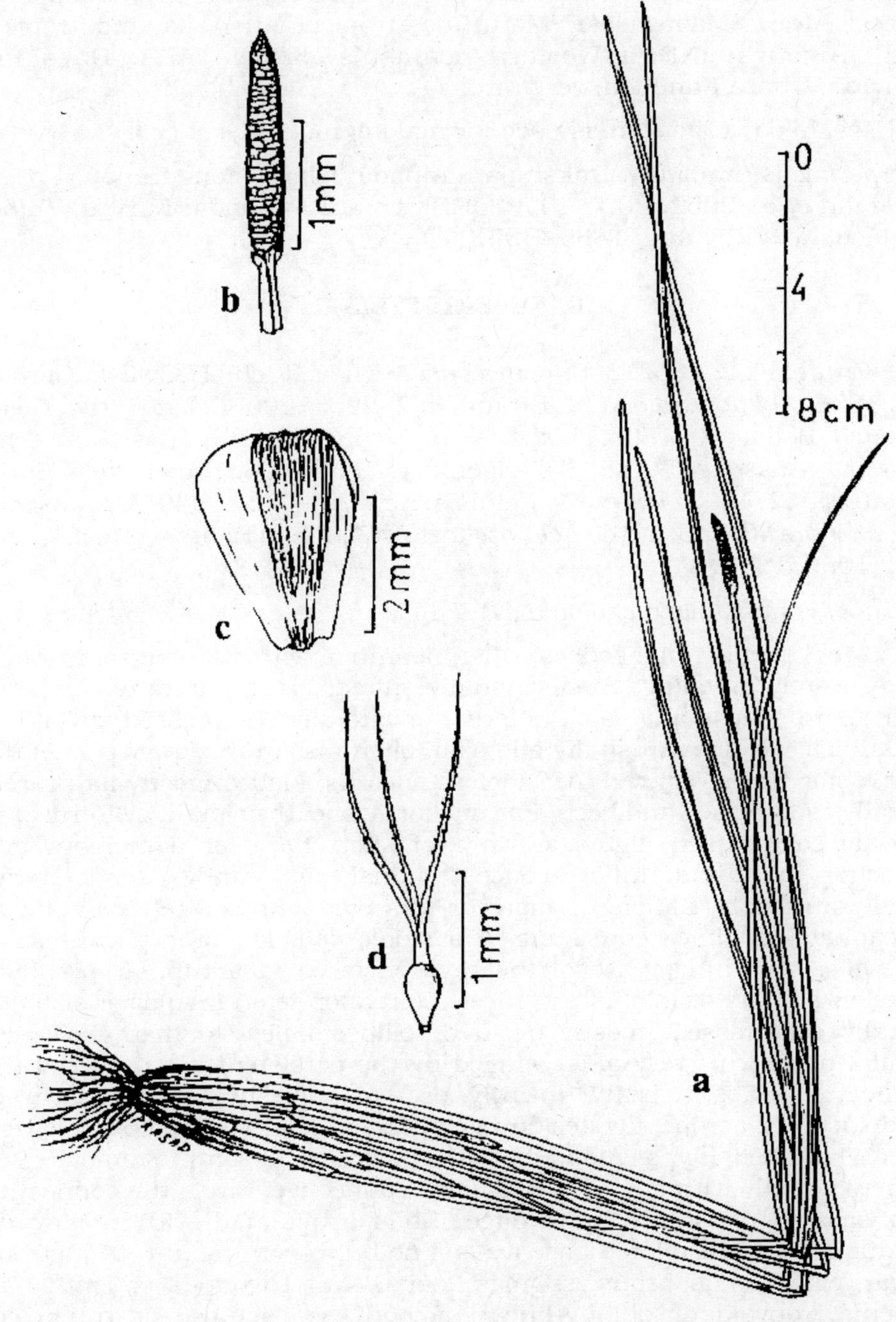

Fig. 19. *Eleocharis spiralis* (Rottb.) Roem. & Schult.
a. Habit, **b**. Spikelet, **c**. Glume, **d**. Nut with style (early stage)

Distrib.: Sri Lanka, Bangladesh, Myanmar, South China, Malesia, tropical Africa, Madagascar, Mauritius, tropical Australia and tropical South America. INDIA: Western peninsula. KARNATAKA: Dakshina Kannada, Uttara Kannada; very rare.

Uses: In Java the stem is used for making mats (Kern, *l.c.*).

Specimens examined: Dakshina Kannada: Island along Gangotri river, *Saldanha et al.* 2052, 30.7. 1978 (JCB). Uttara Kannada: Karwar, *Talbot s.n.* (Acc. No. 729), Aug. 1885 (BSI).

9. FIMBRISTYLIS

Vahl, Enum. Pl. 2: 285. 1806, *nom. cons. Iria* (L.C. Rich.) Hedwig f., Gen Pl. 360. 1806. *Abildgaardia* Vahl, Enum. Pl. 2: 296. 1806. *Echinolytrum* Desv. in Journ. Bot. 1: 20. 1808. *Trichelostylis* Beauv. in Lestib., Ess. Fam. Cyp. 40. 1819. *Gussonea* Presl., Rel. Haenk. 1: 183. 1828. *Pogonostylis* Bert., Fl. Ital. 1: 312. 1833. *Tylocarya* Nelms in Kew Bull. 139. 1949. *Mischospora* Boeck., Flora 43: 113. 1860. *Actinoschoenus* Benth. in Benth. & Hook., Gen. Pl. 3: 1058. 1883.

Type: Fimbristylis dichotoma (L.) Vahl

Annual or perennial sedges; when perennial with short rhizome, often woody, rarely creeping. Stems usually tufted, erect, rarely decumbent, solid, more or less angular, subterete or laterally compressed, striate or sulcate. Leaves linear, usually all basal, often also in the lower part of the stem, some times reduced to bladeless sheaths; blades mostly flat, rarely laterally compressed and becoming ensiform; sheaths closed, cylindrical or laterally compressed; ligule a fringe of short hairs or a membranous projection or absent. Inflorescence terminal, very rarely pseudolateral, usually anthelate, simple or more or less compound, rarely capitate or subcapitate, at times consisting of a single spikelet; bracts leaf-like or scale-like, sheathing; base of the rays enclosed in a tubular prophyll. Spikelets solitary or in clusters, rarely paired,terete or angular or strongly laterally compressed, mostly ovoid or ellipsoid, few to many-flowered; rachilla persistent, very often winged by the persistent basal part of the glumes. Glumes usually spirally or distichously arranged, rarely subdistichous, acropetally deciduous, at times basal 1-few empty. Flowers bisexual, naked (hypogynous bristles or scales absent). Stamens 1-3; anthers usually with a shortly produced connective; rarely the connective with smooth or bristly appendage. Style articulated with the ovary, deciduous with the dilated base leaving no button on the nut, triquetrous or flat, ciliate or glabrous; stigmas 2 or 3. Nut trigonous or lenticular, orbicular, obvoid or oblong-linear, smooth, verruculose or tubercled, reticulate by the hexagonal or roundish epidermal cells, or lineolate when the epidermal cells are linear, trabeculate with longitudinal ribs connected by cross bars or cancellate.

A large genus of *ca* 300 species distributed chiefly in tropics and subtropics of both hemispheres and some species extending to the warmer parts of the temperate regions; a great concentration of species is found in tropical Asia. In India *ca* 115 species; 39 in Karnataka.

Literature : **KOYAMA, T**. (1974) The genus *Fimbristylis* (Cyperaceae) in Ceylon, in *Bot. Mag. Tokyo* 87 (1008): 301- 331. **NIJALINGAPPA, B.H.M.** (1975) Cytological studies in *Fimbristylis* (Cyperaceae), in *Cytologia* 40 (1): 177-183 (Chrom. nos.). **TANAKA N.** (1939) Chromosome studies in Cyperaceae VII. Chromosome number and pollen development of *Fimbristylis, in Bot. Mag. Tokyo* 53: 480-487.

Notes: Vahl (1806) founded this genus by segregating species from *Scirpus* which have spirally arranged glumes and flat, ciliate, distigmatic, deciduous style with enlarged base. He created the genus *Abildgaardia* for the species having same flower structure but with distichous glumes. But there are some species which have spikelets with distichous glumes at the base and spiral glumes towards apex. The tristigmatic species were left in *Scirpus*. But there are species in which we find tristigmatic and distigmatic flowers in the same specimen. Moreover, Robert Brown (1810) recognised the deciduous style articulated with the nut as the essential character of the genus *Fimbristylis* and added several tristigmatic Australian species. Thereafter the genus was further expanded by the addition of more species to it.

Key to the species

1a. Inflorecence a single terminal spikelet (at times proliferate into 1-3, peduncled spikelets) 2

1b. Inflorescence simple to decompound, with few to numerous spikelets (rarely with a solitary spikelet in the early stages of flowering)..... 8

2a. Style usually flat; stigmas 2; nut biconvex 3

2b. Style usually triquetrous; stigmas 3; nut trigonous or oblong-cylindrical . . . 5

3a. Stems 0.3 - 0.8 mm thick; glumes *ca* 2 mm long; stamens 3 . . . . **30. F. schoenoides**

3b. Stems *ca* 1 mm thick; glumes 2.5 - 4.5 mm long; stamens 1 or 2 4

4a. Annuals or short-lived perennials; rhizome inconspicous; leaves reduced to sheaths; style *ca* 3 mm long; nut *ca* 1.7 x 1.5 mm, whitish to cream-coloured . . . . **1. F. acuminata**

4b. Perennials; rhizome short, clothed with brown scales; leaves smaller; style *ca* 1 mm long; nut *ca* 1 x 0.8 mm, greyish to black . . . . . **28. F. polytrichoides**

5a. Glumes distichous, *ca* 5 x 4-5 mm; style *ca* 3 mm long; nut *ca* 2.5 x 1.5 mm . . . . **27. F. ovata**

5b. Glumes spiral, closely imbricate, 1.75 - 4 x 1.5 - 3 mm; style 0.7 - 1.5 mm long; nut 0.8 - 2 x 0.5 - 1 mm 6

6a. Leaves usually reduced to 2-3 sheaths (uppermost often with a small blade); style *ca* 1.5 mm long; nut oblong-cylindrical, slightly curved, *ca* 2 mm long . . . . **36. F. tetragona**

6b. Leaves not reduced to sheaths, shorter than to as long as the stem; style 0.7-1 mm long; nut obovoid, never curved, 0.8 - 1.5 mm long 7

7a. Perennials; stems 0.5 - 1 mm thick; spikelets 7-12 x 4-6 mm; glumes *ca* 4 x 3 mm, keeled; style *ca* 1 mm long; stigmas *ca* 1.5 mm long; nut *ca* 1.5 x 1 mm . . . . **19. F. kingii**

7b. Annuals; stems 0.2 - 0.4 mm thick; spikelets 3-6 x 3-4 mm; glumes 1.75 - 2 x 1.5-1.8 mm, not distinctly keeled; style 0.7- 0.8 mm long; stigmas 0.7 - 0.8 mm long; nut 0.8-0.9 x 0.6-0.7 mm . . . . **25. F. monospicula**

8a. Style usually flat; stigmas 2; nut biconvex (terete in *F. dipsacea*) 9

8b. Style usually triquetrous; stigmas 3; nut trigonous 20

9a. Ligule a fringe of hairs 10

9b. Ligule absent 16

10a. Annuals; stems 0.5 - 1 mm thick; style 1-1.5 mm long 11

10b. Perennials with rhizome; stems 1-2 mm thick; style 1.5 - 3 mm long . . . 14

11a. Spikelets 3-4 x 1.5-2 mm; glumes *ca* 1.5 x 1.2 -1.5 mm; nut 0.7-1 x 0.4-0.8 mm, conspicously trabeculate 12

11b. Spikelets 4-7 x 2.5-4 mm; glumes 2-2.5 x 2-2.5 mm; nut 1- 1.7 x 0.75-1.1mm, sparsely tubercled or scaly verruculose 13

12a. Inflorescence usually with numerous spikelets; spikelets angled; glumes membranous, with a prominent keel; style ciliate; nut *ca* 0.7 x 0.4 mm . . . . **7. F. bisumbellata**

12b. Inflorescence with 3-few spikelets; spikelets terete; glumes chartaceous, without a prominent keel; style glabrous or with a few cilia at the tip; nut *ca* 1 x 0.8 mm . . . . **22. F. merrillii**

13a. Involucral bracts 1-2, lowest somewhat shorter than to twice as long as the inflorescence; spikelets 5-7 x *ca* 2.5 mm; stamen 1; style *ca* 1 mm long, ciliate in the upper part; nut 1-1.25 x 0.75-0.9 mm, scaly verruculose, stramineous, with a short stipe . . . . **3. F. alboviridis**

13b. Involucral bracts 3-5, lowest much longer than inflorescence; spikelets 4-6 x 3-4 mm; stamens 2; style *ca* 1.5 mm long, ciliate on the margins; nut 1.5 - 1.7 x 1-1.1 mm, sparsely tubercled, with a prominent stipe . . . . **37. F. tomentosa**

14a. Glumes *ca* 2 x 1.5 mm, glabrous; stamens 2; anthers *ca* 0.7 mm long; style *ca* 1.5 mm long; stigmas *ca* 1 mm long; nut verruculose, trabeculate . . . . **13. F. dichotoma**

14b. Glumes 3-5 x *ca* 3 mm, puberulous on the upper half; stamens 3; anthers 1-2 mm long; style 2-3 mm long; stigmas 1.5 - 1.8 mm long; nut smooth.. 15

15a. Cauline leaves much shorter than stem; basal sheaths coriaceous, shining brown; involucral bracts shorter than inflorescence; nut shortly stipitate . . . . **18. F. ferruginea**

15b. Cauline leaves comparatively longer (than *F. ferruginea*) but shorter than stem; basal sheaths membranous to papery, not shining brown; involucral bracts usually overtopping the inflorescence; nut distinctly stipitate . . . . **32. F. sieberiana**

16a. Nut subterete, linear-oblong, slightly curved, usually with few caducous clavate appendages on both margins . . . . **15. F. dipsacea**

16b. Nut biconvex, usually obovate, rarely ellipsoid, never curved, never with clavate appendages on the margins 17

17a. Perennials with short rhizome 18

17b. Annuals 19

18a. Stems sulcate, 1-2 mm thick; leaves much shorter than stem, 1-3 mm wide; inflorescence compound to decompound, loose to subcapitate; involucral bracts much shorter than inflorescence; rachilla prominently winged; glumes *ca* 1.5 x 1.2 mm; stamens 2; style glabrous; stigmas *ca* 1 mm long; nut *ca* 1 x 0.7 mm . . . . **12. F. cymosa**

18b. Stems trigonous, slightly compressed, 0.5- 1 mm thick; leaves shorter than to overtopping the inflorescence, 0.5-1 mm wide; inflorescence a head of few to many spikelets; involucral bracts much longer than inflorescence; rachilla narrowly winged; glumes *ca* 1 x 1 mm; stamen 1; style minutely ciliate towards apex; stigmas *ca* 0.6 mm; nut *ca* 0.6 x 0.5 mm . . . . . **5. F. argentea**

19a. Style base with a whorl of long pendant hairs . . . . **34. F. squarrosa** var. **esquarrosa**

19b. Style base not with a whorl of long pendant hairs . . . . **2. F. aestivalis**

20a. Ligule present 21

20b. Ligule absent 30

21a. Glumes distichous atleast in the lower half (spiral in the upper half in *F. pseudomicrocarya*) ; stigmas 0.2-0.3 mm long. 22

21b. Glumes spiral; stigmas more than 0.3 mm long 23

22a. Inflorescence with 1-3 spikelets; stamens 2; style *ca* 1 mm long; nut *ca* 1 mm long . . . . **8. F. breviculma**

22b. Inflorescence with 3-9 (-15) spikelets; stamen 1; style 0.4- 0.5 mm long; nut 0.5 - 0.6 mm long . . . . **29. F. pseudomicrocarya**

23a. Annuals; epidermal cells on the nut transversely elongated, in 4-6 (-8) vertical rows 24

23b. Perennials; epidermal cells on the nut transversely oblong, in many vertical rows 29

24a. Inflorescence simple to decompound; stamen 1 25

24b. Inflorescence always simple; stamens 2 or 3 26

25a. Inflorescence somewhat lax; spikelets solitary . . . . **23. F. microcarya**

25b. Inflorescence somewhat congested; spikelets usually in clusters of 3, at times solitary . . . . **39. F. woodrowii**

26a. Spikelets paired; stamens 2 . . . . **6. F. bispicula**

26b. Spikelets solitary; stamens 3 27

27a. Anthers 1.2-1.3 mm long; nuts dimorphic, basal 1-2 larger and asymmetric . . . . **14. F. dimorphonucifera**

27b. Anthers 0.2-0.6 mm long; nuts not dimorphic, equal in size and symmetric 28

28a. Inflorescence with 1-3 spikelets; nut trigonous, epidermal cells in *ca* 4 vertical rows on each face . . . . **33. F. simpsonii**

28b. Inflorescence with 6-9 (-12) spikelets; nut triquetrous, epidermal cells in 6-8 vertical rows on each face . . . . **38. F. virella**

29a. Stems flattened and ancipitous towards apex, 2-4 mm thick; leaves 2-4 mm wide; sheaths flattened, not rounded on the back (equitant in early stage); spikelets 1.5 - 2 mm broad; glumes 1.6 - 2 mm broad; nut smooth or verruculose . . . . **10. F. complanata**

29b. Stems slightly compressed, 0.5- 0.8 mm thick; leaves 1-1.5 mm wide; sheaths cylindrical, rounded on the back (not equitant in early stage); spikelets *ca* 1.2 mm broad; glumes 1-1.3 mm broad; nut sparsely tubercled . . . . **11. F. consanguinea**

30a. Spikelets in clusters, often forming a head, at times in pairs... 31

30b. Spikelets solitary 33

31a. Annuals; glumes distichous in lower half; style 1.2-1.3 mm long . . . . **31. F. semidisticha**

31b. Perennials; glumes spiral throughout; style 1.5-2 mm long 32

32a. Rhizome creeping; leaves 1-3 mm wide; spikelets in compound or subcompound clusters, often in pairs; glumes 2-3 mm broad; anther connective distinctly produced, with short white bristles. . . . . . . **17. F. falcata**

32b. Rhizome bulbous; leaves 0.5-0.7 mm wide; spikelets in a head; glumes *ca* 1.5 mm broad; anther connective neither distinctly produced, nor with bristles . . . . **20. F. lawiana**

33a. Glumes distichous, rarely subspiral 34

33b. Glumes spiral 36

34a. Annuals; inflorescence simple with 3-4 spikelets; anthers *ca* 0.5 mm long; epidermal cells on the nut broadly rounded-hexagonal (isodiametric) . . . . **26. F. narayanii**

34b. Perennials with woody rhizome; inflorescence compound or decompound, with few to numerous spikelets; anthers 1.3-2 mm long; epidermal cells on the nut linear-oblong or minutely isodiametric 35

35a. Leaves much shorter than stem, 1.5-3 mm wide; spikelets 5-8 x 2-3 mm; glumes distichous or subspiral, broadly ovate, 3.2 - 3.8 x 2.5 - 2.8 mm, shining glabrous; style 2.2 - 2.5 mm long; stigmas as long as the style . . . . **16. F. eragrostis**

35b. Leaves at least half as long as the stem, 0.3-0.5 mm wide; spikelets 4-5 x *ca* 1 mm; glumes distichous, ovate-lanceolate, 2.75 - 4 x 1.5-2 mm, with reddish gland-dots on the sides; style 3-3.5 mm long; stigmas much shorter than style . . . . **9. F .cinnamometorum**

36a. Perennials with short rhizome; stamens 3; anthers *ca* 0.8 mm long . . . . **4. F. aphylla**

36b. Annuals, without rhizome; stamens 1 or 2; anthers 0.3-0.7 mm long . . . 37

37a. Spikelets globose or subglobose, terete; rachilla not winged; glumes 1-1.2 x *ca* 0.8 mm . . . . **21. F. littoralis**

37b. Spikelets ovoid or ovoid-lanceolate, faintly to prominently angled; rachilla winged; glumes 1.7 - 2.5 x 1.7 - 2 mm 38

38a. Spikelets ovoid, 2-4 mm long; glumes *ca* 1.7 x 1.7 mm; style 0.5-0.7 mm long, minutely ciliate in the upper half; epidermal cells on the nut transversely linear-oblong, in *ca* 6 vertical rows on each face. **24. F. miliacea**

38b. Spikelets ovate-lanceolate, 4-6 mm long; glumes *ca* 2.5 x 2 mm; style *ca* 1.2 mm long, glabrous; epidermal cells on the nut transversely hexagonal, in 10 vertical rows on each face . . . . **35. F. tenera**

1. **Fimbristylis acuminata** Vahl, Enum. Pl. 2: 285. 1806; Clarke in Hook.f., Fl. Brit. India 6: 631. 1893; Fischer in Gamble, Fl. Pres. Madras 1658. 1931 (3: 1150. 1957, repr.ed.); Kern in van Steenis, Fl. Males. 1, 7: 588. 1973; Hooper in Saldanha & Nicolson, Fl. Hassan 675. 1976; Rao & Razi, Fl. Mysore 562. 1981; Rao & Verma, Cyp. NE India 29. 1982; Sharma *et al.*, Fl. Karnataka 308. 1984; Koyama in Dassanayake & Fosberg, Rev. Handb. Fl. Ceylon 5: 319. 1985; Karthik. *et al.*, Fl. Ind. Enum. Monocot. 50. 1989; Keshava Murthy & Yoganarasimhan, Fl. Coorg 506. 1990.

Type : Malacca, *Koenig.*

Illus. : Clarke, Illus. Cyp. t. 40, f. 5-7. 1909.

Annuals or short-lived perennials with an inconspicous rhizome and fibrous roots. Stems densely tufted, erect, slender, angled, sulcate, 6-30 cm long, *ca* 1 mm thick, glabrous. Leaves reduced to obliquely truncate sheaths, 1-4.5 cm long. Inflorescence a solitary, terminal spikelet. Bracts glume-like, triangular-ovate, *ca* 3 x 2 mm, keeled; keel slightly curved at apex towards inside. Spikelets ovoid to lanceolate, acute-acuminate at apex, 6-10 x 2-4 mm, few to many-flowered, greenish-white to pale-brown. Glumes broadly ovate, keeled, slightly mucronate at apex, 3-4.5 x 2-3 mm, usually brown spotted towards the middle portion, papery towards margins; lowest 1 or 2 empty. Stamens 2; filaments elongate up to 4 mm; anthers linear-oblong, 1-1.25 mm long, minutely apiculate. Ovary oblong, 0.75-1 mm long; style *ca* 3 mm long, minutely ciliate towards apex; stigmas 2, *ca* 1 mm long, minutely ciliate, narrowed towards apex. Nut biconvex, obovate, rounded towards apex, *ca* 1.7 x 1.5 mm, with 5-8 transverse ridges, whitish to cream-coloured; gynophore inconspicous.

Fls. & Frts. : Sept. - Oct.

Chrom. No. : 2n = 10 (*Taxon* 21:683.1972).

Habitat : Wet muddy areas, marshes, swamps, rice fields, bunds of rice fields and river banks.

Distrib.: South and South-East Asia (Sri Lanka, China, Malesia), tropical Australia. INDIA: Throughout. KARNATAKA: Dakshina Kannada, Hassan, Mandya, Mysore (Fischer *l.c.*), Shimoga, Uttara Kannada.

Specimens examined : Dakshina Kannada: Sampagi, *Barber* 2408, 15.11.1900 (MH); Uppinangady, *Hooper & Saldanha* 2929, 25.11.1971 (JCB); Kannadekatte, *Bhat* 389, 9.11.1977 (MGH); Panamboor, Mangalore, *Bhat* 491, 5.11.1977 (MGH). Hassan: Byra, *Nicolson et al.* 2254, 23.10.1971 (JCB). Mandya: Paschimavahini, *Bhat* 100, 25.12.1970 (JCB). Shimoga: Mattigu, near Yedur, *Raghavan* 83029, 6.10.1962 (BSI); Jog falls, *Vartak s.n.* (S. No. 9705), Nov. 1971 (MACS). Uttara Kannada: Yellapur, *Talbot s.n.* (Acc. No. 739), Sept. 1885 (BSI); Karwar, *Vartak s.n.* (S.No. 1334), without date (MACS).

2. **Fimbristylis aestivalis** (Retz.) Vahl, Enum. Pl. 2: 288. 1806; Clarke in Hook.f., Fl. Brit. India 6: 637. 1893; Cooke, Fl. Pres. Bombay 2: 881. 1908 (3:395. 1958, repr.ed.); Kern in van Steenis, Fl. Males. 1, 7: 584. 1974; Hooper in Saldanha & Nicolson, Fl. Hassan 675. 1976; Arora *et al.*, Bot. S. Kanara 61. 1981; Rao & Verma, Cyp. NE India 32. 1982; Sharma *et al.*, Fl. Karnataka 308. 1984; Koyama in Dassanayake & Fosberg, Rev. Handb. Fl. Ceylon 5: 313. 1985; Karthik. *et al.*, Fl. Ind. Enum. Monocot. 50. 1989; Keshava Murthy & Yoganarasimhan, Fl. Coorg 506. 1990. *Scirpus aestivalis* Retz., Obs.4: 12.1768.

Type : Sri Lanka, *Koenig.*

Illus. : Clarke, Illus. Cyp. t. 41, f. 14-15. 1909.

Small annuals with fibrous roots, 4-15 cm high. Stems slender, densely tufted, angular, 1-11 cm long, 0.2-0.7 mm thick. Leaves slender to filiform, shorter than to as long as the stem, up to 1 mm wide, flat or canaliculate, densely scabrous; ligule absent. Inflorescence a decompound corymb of few to numerous spikelets. Involucral bracts few, lowest 2-3 leafy, shorter or more often overtopping the inflorescence, with stout bristles at base. Primary rays short, up to 1.5 cm long, glabrous. Spikelets solitary, ovoid or oblong-lanceolate, angular, acute at apex, 2-5 x 1-1.5 mm, greenish- brown, many-flowered. Glumes spiral, membranous, ovate, acute at apex, 1.2-1.5 x *ca* 0.7 mm, keeled; keel green, extending to a sharp mucro of *ca* 0.3 mm. Stamen 1; filament elongate up to 1.2 mm; anther oblong, 0.3 - 0.5 mm long. Ovary oblong, *ca* 0.2 mm long, shortly stipitate; style flat, 0.5-0.8 mm long, ciliate towards the top, dilated at base; stigmas 2, 0.3-0.4 mm long. Nut biconvex, obovate to elliptic, 0.5-0.8 x 0.2 - 0.3 mm, brownish, often shining, usually with prominent margins.

Fls. & Frts. : Nov. - June.

Habitat : Wet areas along streams and ponds, near ditches, open damp places, swampy areas; common weed of paddy fields.

Distrib. : South to East Asia (Sri Lanka, China, Indo-China, Thailand, Malesia, Japan, Amurland), tropical Australia. INDIA: Almost throughout. KARNATAKA: Belgaum, Chikmagalur, Daksina Kannada, Dharwar (Sharma *et al., l.c.*), Hassan, Kodagu, Mandya, Mysore, Shimoga, Uttara Kannada.

Specimens examined : Belgaum: Chappoli nallah, Jambotti R.F., *Prasad* 172882, 15.12.1994 (BSI). Chikmagalur: Hasakere village forests, on way to Byrapura forests, *Yoganarasimhan* 1383, 16.11.1972 (RRCBI); Balihole-Magundi, *Ramesh & Prakash* 6568, 9.3.1979 (JCB). Dakshina Kannada: Kannadekatte, *Bhat* 179, 6.5.1975 (MGH), Padergudda, Malpe, *Bhat* 438, 18.12.1976 (MGH); Near Mulki, *Saldanha & Prakash* 4089, 12.11.1978 (JCB). Hassan: Bisle, *Saldanha* 13075, 14.3.1969 (JCB). Kodagu: Katakare, Mercara, *Bhat* 728, 18.12.1980 (MGH). Mandya: Sreerangapatana, *Dinesh* 779, 26.1.1984 (MGH). Mysore: Gathenahalla (Devagiribetta), Biligirirangan hill ranges, *Rao* 80033, 17.4.1962 (BSI); Tondaikerekan, base of Malkibetta, Biligirirangan hill ranges, *Rao* 80380, 27.4.1962 (BSI); B.R.Hills, *Fariheeddin* 1, 28.3.1985 (MGH). Shimoga: Agumbe-Someswar Road, Addaiguda, *Raghavan* 62741A, 20.5.1960 (BSI); Agumbe, *Raghavan* 62619, 18.5.1960 (BSI); Chyatsamane, Agumbe, *Raghavan* 80574, 14.5.1962 (BSI); Kavaldurga, *Raghavan* 80949A, 24.5.1960 (BSI). Uttara Kannada: Dandeli, *Talbot* 2270, Aug 11, without year (BSI); Vincholi, *Talbot* 943, 4.4.1884 (BSI); Yesle forest, *Mahajan* 1694, 2.5.1956 (BSI); Ekkambi, *Puri* 1125, 24.4.1956 (BSI); Siddapur, *Puri* 1982, 5.5.1956 and 1124B, 24.4.1956 (BSI).

3. **Fimbristylis alboviridis** Clarke in Hook.f., Fl. Brit. India 6: 638. 1893; Kern in Blumea 8: 140. 1955, *et* in van Steenis, Fl. Males. 1, 7:580, f. 46. 1974; Hooper in Saldanha & Nicolson, Fl. Hassan 675. 1976; Rao &

Verma, Cyp. NE India 35. 1982; Sharma *et al.*, Fl. Karnataka 309. 1984; Karthik. *et al.*, Fl. Ind. Enum. Monocot. 50. 1989.

Illus.: Kern, *l.c.*

Annuals with fibrous roots. Stems slender, tufted, compressed, obtusangular, 20-50 cm long, 0.5- 1 mm thick, glabrous and smooth. Leaves half as long as the stem, often slightly falcate, rather firm, flat, obtuse to rather acute at apex, 1-2 mm wide, glabrous, scabrid on the margins in the upper part, greyish-green; ligule a fringe of short, white hairs. Inflorescence simple or subcompound, loose, with few to several spikelets, 2-7 cm long. Involucral bracts 1-2, suberect, lowest somewhat shorter than to twice as long as the inflorescence, up to 12 cm. Primary rays few, suberect, compressed, up to 5 cm long, smooth. Spikelets solitary, terete, ovoid to oblong-ovoid, acute at apex, 5-7 x *ca* 2.5 mm, densely many-flowered, whitish or greyish-green, often brownish-variegated; rachilla narrowly winged. Glumes spiral, subchartaceous, broadly ovate, obtuse and muticous or scarcely apiculate at apex, *ca* 2 x 2 mm, with 3-nerved keel and nerveless sides, glabrous, brownish, with hyaline basal part and margins. Stamen 1; anther oblong, *ca* 0.5 mm long. Style slender, flat, slightly dilated at base, *ca* 1 mm long, ciliate in the upper portion; stigmas 2, shorter than style. Nut biconvex, obovate, shortly stipitate, umbonulate, 1-1.25 x 0.75-0.9 mm, scaly-verruculose and obtusely reticulate on each face, stramineous; epidermal cells transversely elliptic or oblong, not impressed, in 10-16 vertical rows on each face.

Fls. & Frts. : Oct.

Habitat : Dry grass-fields, edges of pools, wastelands, roadsides.

Distrib. : South and South-East Asia (India to Malesia) INDIA: Western Peninsula, Central and North-East India, Andaman & Nicobar Islands. KARNATAKA: Hassan (Hooper, *l.c.*).

Note : This species is included on the authority of Hooper (*l.c.*). As no specimens are available in BSI, description provided is as given by Kern, *l.c.*

4. **Fimbristylis aphylla** Steud., Syn. Pl. Glum. 2: 114. 1855; Kern in van Steenis, Fl. Males. 1,7: 552. 1974; Hooper in Saldanha & Nicolson, Fl. Hassan 675. 1976; Arora *et al.*, Bot. S. Kanara 61. 1981; Sharma *et al.*, Fl. Karnataka 309. 1984; Karthik. *et al.*, Fl. Ind. Enum. Monocot. 50. 1989. *F.quinquangularis* var. *crassa* Clarke in Hook.f., Fl. Brit. India 6: 644. 1893; Fischer in Gamble, Fl. Pres. Madras 1659. 1931 (3:1151. 1957, repr.ed.). **Fig.20.**

Perennials with short rhizome, up to 65 cm high. Stems tufted, smooth, glabrous, prominently 4-5-angled, gradually narrowed to apex, 1.5-3 mm thick in the middle portion. Leaves of sterile shoots well developed, flat, 2-3

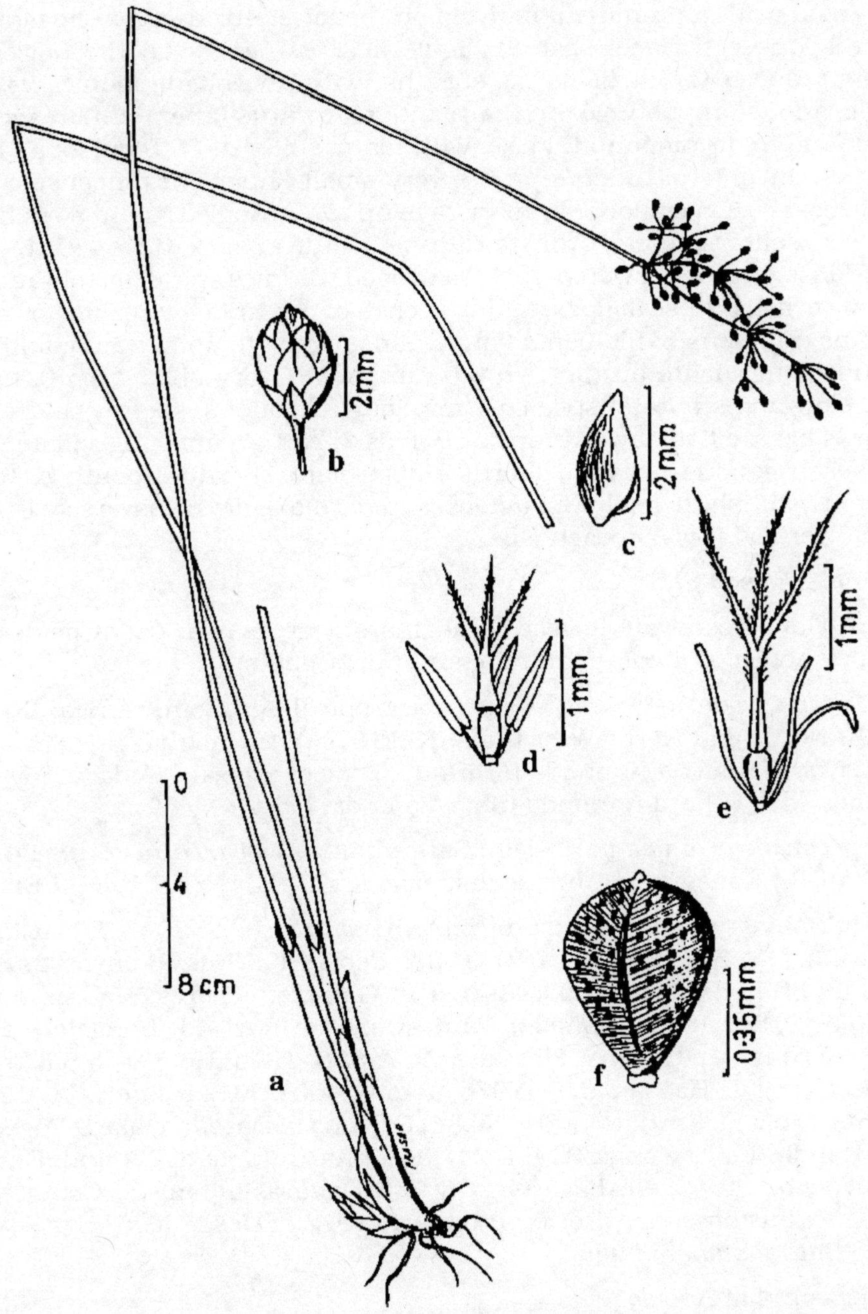

Fig. 20. *Fimbristylis aphylla* Steud.
a. Habit, **b.** Spikelet, **c.** Glume, **d.** Flower (early stage), **e.** Flower (later stage), **f.** Nut

mm broad, with a prominent midvein on the lower surface; margins folded towards upper surface, scabrid; ligule absent. Leaves on the flowering shoots reduced to 3-4 bladeless sheaths, with an oblique mouth, with a membranous appendage on both sides towards apex. Inflorescence compound to decompound, loose, with many spikelets, 5-8 cm long, 4-10 cm wide. Involucral bracts up to 8, very small, inner ones inconspicous, outer most *ca* 1.5 cm long. Primary rays up to 13, angled, up to 5 cm long. Spikelets solitary, terete, ovoid to ellipsoid, acute at apex, 2.5 - 4 x 1.5 - 2.5 mm, densely many- flowered; rachilla winged. Glumes spiral, membranous, broadly ovate, obtuse at apex, faintly keeled, *ca* 2 x 1.6 mm, hyaline towards margins. Stamens 3; filaments flat, hyaline, elongate up to 2 mm; anthers linear-oblong, acute at apex, *ca* 0.8 mm long. Ovary oblong, *ca* 0.3 mm long, elongate, stipitate; style *ca* 1 mm long, trigonous, slightly thickened towards base, ciliate towards apex; stigmas 3, *ca* 1.75 mm long, ciliate. Nut obtusely trigonous, obovoid, shortly umbonulate, shortly stipitate, *ca* 0.7 x 0.5 mm, whitish, densely verruculose; epidermal cells transversely linear, in *ca* 5 vertical rows on each face.

Fls. & Frts. : Sept. - Nov.

Habitat : Open water logged areas, marshy areas near paddy fields and swampy areas, at medium altitude and upper ghats.

Distrib. : South & South-East Asia, tropical and South Africa. INDIA: South, North and North-West India. KARNATAKA: Chikmagalur (Sharma *et al., l.c.*), Dakshina Kannada (Arora *et al., l.c.*) , Hassan, Kodagu, Mysore (Fischer, *l.c.*), Uthara Kannada (Sharma *et al., l.c.*).

Specimens examined : Hassan: Bannuhalla, *Nicolson et al.* 2298, 24.10. 1977 (JCB). Kodagu: Behind 'Roshanara', *Rao* 74711, 27.9.1961 (BSI).

5. **Fimbristylis argentea** (Rottb.) Vahl, Enum. Pl. 2: 294. 1806; Clarke in Hook.f., Fl. Brit. India 6: 640. 1893; Cooke, Fl. Pres. Bombay 2: 881. 1908 (3: 395. 1958, repr.ed.); Fischer in Gamble, Fl. Pres. Madras 1659. 1931 (3: 1151. 1957, repr.ed.); Ramaswamy & Razi, Fl. Bangalore 103. 1973; Kern in van Steenis, Fl. Males. 1, 7: 586. 1974; Hooper in Saldanha & Nicolson, Fl. Hassan 676. 1976; Rao & Razi, Fl. Mysore 562. 1981; Sharma *et al.*, Fl. Karnataka 309. 1984; Koyama in Dassanayake & Fosberg, Rev. Handb. Fl. Ceylon 5: 314, f. 24. 1985; Karthik. *et al.*, Fl. Ind. Enum. Monocot. 51. 1989; Keshava Murthy & Yoganarasimhan, Fl. Coorg 507. 1990. *Scirpus argenteus* Rottb., Progr. 27. 1972 *et* Descr. Ic. 51, t. 17, f. 6. 1773. *Bittdausumani hullu.*

Type: India, *Koenig.*

Illus.: Koyama, *l.c.*

Small, glabrous, tufted perennials with short rhizome, 4-20 cm high. Stems slightly compressed, trigonous, striate, 0.5-1 mm thick. Leaves usually shorter than or often overtopping the inflorescence, narrowly linear, subacute at apex, 0.5-1 mm wide, usually canaliculate, spinulose-scabrous towards apex; sheaths papery, much broader than leaf blade; ligule absent. Inflorescence a head of few to many spikelets, globose or subglobose, 6- 15 mm across. Involucral bracts 2-4, narrowly linear, usually canaliculate, lower 2-3 much longer than inflorescence, 0.8-4 cm long, slightly dilated at base. Spikelets sessile, oblong-ovoid to oblong, acute at apex, slightly angular, 4-6 x 1.5-2 mm, densely many-flowered; rachilla narrowly winged. Glumes membranous, spiral, broadly ovate, obtuse and minutely apiculate at apex, *ca* 1 x 1 mm, faintly keeled; nerves 3 in the middle portion. Stamen 1; filament flat, narrowed towards base, elongate up to 1.2 mm; anther linear-oblong, *ca* 0.5 mm long. Ovary broadly oblong, *ca* 0.4 mm long; style flat, dilated at base, *ca* 0.8 mm long, minutely ciliate towards apex; stigmas 2, *ca* 0.6 mm long, scabrous. Nut broadly biconvex, obovate, apiculate at apex, *ca* 0.6 x 0.5 mm, smooth or minutely verruculose, cream-coloured.

Fls. & Frts. : June - April.

Chrom. No. : 2n = 20 (*Taxon* 21: 683. 1972).

Habitat : Drying edges of open waters, paddy fields, wet sandy ground of open grasslands, wet or swampy areas, waste places, shady areas, clefts of rocks, along sea shores.

Distrib. : Sri Lanka, Malesia, Peninsular Thailand, Mauritius, Indo-China. INDIA: Western peninsula, Central, North and Eastern India (Bengal). KARNATAKA: Bangalore, Dakshina Kannada, Hassan, Kodagu, Mysore, Tumkur, Uttara Kannada.

Specimens examined : Bangalore: Sringarpura, *Hooper* & *Saldanha* 18067, 18.11.1971 (JCB). Dakshina Kannada: Udyavara, *Bhat* 327, 24.10.1976 (MGH); Malpe, *Bhat* 336, 31.10.1976 (MGH); Mavanthe, *Saldanha et al.* 6945, 9.4.1979 (JCB). Hassan: 7 miles before Arsikeri, *Saldanha* & *Ramesh* 1620, 24.6.1978 (JCB). Kodagu: Makuta, *Bhat* 891, 24.12.1980 (MGH). Mysore: Regional College of Education, Mysore, *Bhat* 6, 9.7.1970 (JCB); without locality, Coll. name, number (Acc. Nos. 73759, 73760, 73761) and date (BSI). Tumkur: Devarayanadurga, *Ravindra* & *Ramesh* 1548, 14.7.1978 (JCB). Uttara Kannada: Karwar, *Talbot* 1313, 20.1.1889 (BSI).

6. **Fimbristylis bispicula** Govind. in Rheedea 6 (2): 59, f.1. 1996.

Type: India, Karnataka state, Shimoga dt., Agumbe, Nishanigudda, *Govindarajalu* 8800A - Holotype (CAL).

Illus.: Govind., *l.c.*

Annuals. Stems rather slender, usually solitary, rarely tufted, 1-3 (-7), stiff, erect, tetragonous, smooth, leafy at base, 3-7 cm long, *ca* 0.5 mm thick. Leaves shorter than stem, flat, stiff, erect, subacute or obtuse at apex, 0.5-1 mm wide, ligulate, glabrous, with smooth margins; sheaths all laminiferous, brown. Inflorescence, simple, usually bearing a single pair of spikelets (rarely up to 4), 5-7 mm long. Involucral bracts usually 2, distichously paired, glumiform. Rays absent. Spikelets paired, sessile, ovoid-lanceolate, angular, acute at apex, 3-4 x 1-1.2 mm, dark brown, 8-10-flowered; rachilla raggedly winged. Glumes spiral, membranous, broadly ovate, acute and muticous at apex, *ca* 2 x 1.2 - 1.3 mm, actutely keeled; keel 3- nerved; sides nerveless, dark brown, translucent towards margins. Stamens 2; anthers very short, ovate, 0.2 - 0.3 mm long, brown. Style triquetrous with narrow pyramidal base, *ca* 1 mm long, glabrous; stigmas 3, slightly shorter than style. Nut trigonous, broadly obovoid or subgloboid, tricostulate with convex sides, obtuse at apex, *ca* 1 x 0.5 - 0.6 mm, somewhat verruculose, brownish; epidermal cells transversely elongated-hexagonal, in 4-6 regular rows on each face.

Fls. & Frts. : Nov.

Habitat : Growing alongwith grasses.

Distrib. : Endemic to Karnataka (Shimoga Dt.).

Note : This species is included on the authority of Govindarajalu, *l.c.*. As no specimen is available in BSI, description provided above is as given in the protologue.

7. **Fimbristylis bisumbellata** (Forsk.) Bub., Dodec. 30.1850; Fischer in Gamble, Fl. Pres. Madras 1898. 1936 (Corrig.) *et* 3: 1151. 1957 (repr.ed.); Ramaswamy & Razi, Fl. Bangalore 104. 1973; Kern in van Steenis, Fl. Males. 1, 7:579. 1974; Hooper in Saldanha & Nicolson, Fl. Hassan 676. 1976; Yoganarasimhan *et al.*, Fl. Chickmagalur 361. 1981; Rao & Razi, Fl. Mysore 562. 1981; Rao & Verma, Cyp. NE India 31. 1982; Sharma *et al.*, Fl. Karnataka 309. 1984; Koyama in Dassanayake & Fosberg, Rev. Handb. Fl. Ceylon 5: 312. 1985; Singh, Fl. E. Karntaka 635. 1988; Karthik. *et al.*, Fl. Ind. Enum. Monocot. 51. 1989; Keshava Murthy & Yoganarasimhan, Fl. Coorg 507. 1990. *Scirpus bisumbellata* Forsk., Fl. Aug. - Arab. 1: 15. 1775. *S. dichotomus* auct. non L. 1753; Rottb., Descr. & Ic. 57, t. 13, f. 1. 1786. *Fimbristylis dichotoma* auct. non Vahl, 1806; Kunth, Enum. Pl. 2: 225. 1837; Clarke in Hook.f., Fl. Brit. India 6: 635. 1893; Cooke, Fl. Pres. Bombay 2: 880. 1908 (3: 394. 1958, repr.ed.). *Scirpus palliscens* Roxb., Fl. Ind. 1: 229. 1820. *Kadu sabbasigai hullu, Niru sabbasigai hullu.* **Fig.21.**

Type: Egypt.

Annual herbs with fibrous roots (3-) 6-30 (-50) cm high (often very short due to stunted growth). Stems tufted, slender, triangular, striate, glabrous, (1-) 2.5 - 28 (-35) cm long, 0.5 - 1 mm thick. Leaves shorter to taller than stem, often falcate, linear, flat, obtuse or abruptly accuminate at apex, (1.5-) 3-8 (-30) cm long, 0.7-2 mm wide, usually glabrous above, scabrous below; sheaths chartaceous, 0.5 - 7 cm long; ligule a dense fringe of white short hairs. Inflorscence usually a decompound corymb of numerous loose

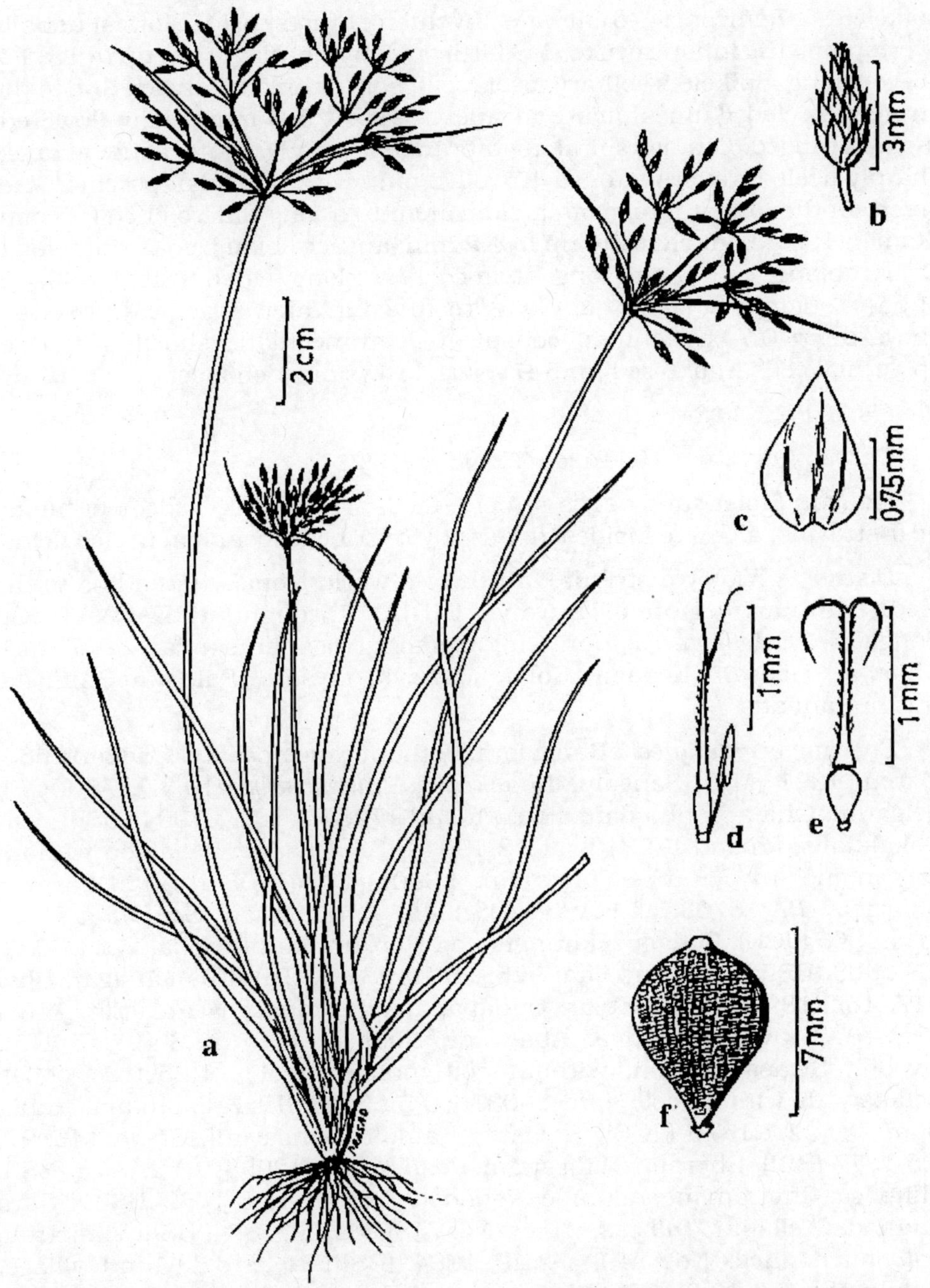

Fig. 21. *Fimbristylis bisumbellata* (Forsk.) Bub.
a. Habit, **b.** Spikelet, **c.** Glume, **d.** Flower (early stage),
e. Nut with style (early stage), **f.** Nut

spikelets, 1-7 cm long, 1-6 cm wide. Involucral bracts up to 5; lowest usually overtopping the inflorescence, 1-8 (-16) cm long. Primary rays up to 15, 1-2 (- 5) cm long. Spikelets solitary, ovoid, ellipsoid or oblong, angled due to the strongly keeled glumes, acute at apex, 3-4 x 1.5-2 mm, many-flowered; rachilla winged. Glumes spiral, membranous, broadly ovate, acute at apex, sharply keeled, mucronate, *ca* 1.5 x 1.2 mm, with brownish patches; keel green, in the lowest glume often elongate into a long mucro of *ca* 0.5 mm. Stamen 1; filament elongate up to 1.2 mm; anther oblong, *ca* 0.5 mm long. Ovary oblong, *ca* 0.3 mm long; style *ca* 1 mm long, ciliate, slightly dilated at base; stigmas 2, *ca* 0.6 mm long, recurved in later stage. Nut biconvex, obovate, *ca* 0.7 x 0.4 mm, conspicously trabeculate, shortly stipitate; epidermal cells impressed, in 5-9 vertical rows on each face.

Fls. & Frts. : Nov. - July

Chrom. No. : 2n = 10 (*Taxon* 21: 683. 1972).

Habitat: Moist sandy soils in open fields, river banks, edges of tanks, near streams, along roadsides, dry sandy river beds; common in rice fields.

Distrib. : Widely distributed in the old world tropics, extending to the Mediterannian region of Europe. INDIA: Throughout. KARNATAKA: Bangalore, Belgaum, Bijapur, Chikmagalur (Yoganarasimhan *et al., l.c.*), Dharwar, Hassan, Kodagu, Kolar, Mandya, Mysore, Raichur, Shimoga, Uttara Kannada.

Specimens examined : Bangalore: Without exact locality, *Camaron* 565, March, 1891 (MH); Sangam, *Bhaskar & Ramesh* 392, 10.3.1978 (JCB). Belgaum: Ghataprabha dam side, *Prasad* 172806, 10.12.1994 (BSI); near Gokak falls, *Prasad* 172813, 11.12.1994 (BSI). Bijapur: Bagalkot, without coll. name and number (Acc. No. 2690), 17.4.1906 (BSI). Dharwar: Bhartiola, *Puri* 37635, 2.6.1958 (BSI). Hassan: Hassan, *Saldanha* 9584, 27.4.1965 (JCB). Kodagu: Harangi river, Hardur, Saunticoppa, *Rao* 86659, 2.3.1963 (BSI); Kirugoor, *Bhat* 925, 25.1.1981 (MGH); Kushalnagar, *Bhat* 947, 16.2.1891 (MGH). Kolar: Uddapanapalli, *Ramesh & Ravindra* 1467, 29.3.1978 (JCB). Mandya: Abalwadi, *Ahamed* 469, 29.3.1978 (JCB). Mysore: Sivasamudram, without coll. name, 10411, 12.5.1914 (MH); Belakawadi, without coll. name, 10356, 9.5.1914 (MH); Kukkanahalli tank, *Bhat* 33, 22.7.1970 (JCB). Raichur: Kambli (Gangavati), *Singh* 146997, 6.5.1977 (BSI); Shorapur-Lingsugur Road, *Singh* 12953, 13.2.1975 (BSI). Shimoga: Chytramane, Agumbe, *Raghavan* 80573, 14.5.1962 (BSI). Uttara Kannada: Yellapur, *Talbot s.n.* (Acc. No. 748), March, 1884 (BSI); Vimictroli, *Talbot s.n.* (Acc. No. 749), April, 1884 (BSI); Near old Gund village, *Fernandez* J.F., 13.5.1950 (BLAT); 3 miles above Nagargali, *Fernandez* J.F. (Acc. No. 802178), 7.4.1951 (BLAT); Alnawar, *Jain* 16013, 8.3.1957 (BSI); Ekkambi, *Puri* 1124 A, 14.4.1956 (BSI); Sirsi-Hulikkal Road, *Puri* 1244, 25.4.1956 (BSI): Bedthi river bank, *Ramesh et al.*, 12836, 11.5.1961 (JCB).

8. **Fimbristylis breviculma** Govind. in Rheedea 6 (2): 61, f.2. 1996.

Type : India, Karnataka state, Shimoga dt., Agumbe, Nishanigudd, *Govindarajalu* 8870 A - Holotype (CAL).

Illus.: Govind., *l.c.*

Annuals. Stems tufted, erect, tetragonous, 4-ribbed, sulcate, 5- 7 cm long, *ca* 0.3 mm thick, glabrous, leafy at base. Leaves many, filiform, divergent, usually acuminate at apex, *ca* 1 mm wide, ligulate, glabrous; sheaths membranous, with transversely truncate or obliquely open mouth. Inflorescence simple, bearing 1-3 spikelets, 6-7 mm long; bracts stiff, erect, much shorter than inflorescence, 3.5 - 4 mm long. Rays absent. Spikelets solitary, flat, broadly ellipsoid, subacute at apex, 3.5 x 1.2-1.5 mm, brown, 6-8-flowered; rachilla excavated, winged. Glumes distichous, membranous, trullate or triangular- ovate, acute and mucronate at apex, 1.2 - 1.3 x 1-1.2 mm, with distinct scarious glabrous margins; keel distinct, 3- nerved. Stamens 2; anthers apiculate, spurred at base, *ca* 0.5 mm long. Style more or less triquetrous, slightly dilated at base, usually *ca* 1 mm long; stigmas 3, slender, much shorter than style, 0.2-0.3 mm long. Nut triquetrous, tricostulate, narrowly obovoid, obtuse at apex, stipitate, *ca* 1 x 0.6- 0.7 mm, indistinctly tubercled on the faces; epidermal cells distinct, transversely elongated-hexagonal, in 4-5 regular rows on each face.

Fls. & Frts. : Nov.

Habitat : Very common on rocky slopes near rivulets.

Distrib. : Endemic to Karnataka (Shimoga Dt.)

Note : This species is included on the authority of Govindarajalu, *l.c.* As no specimens are available in BSI, description provided above is as given in the protologue.

9. **Fimbristylis cinnamometorum** (Vahl) Kunth, Enum. Pl. 2: 229. 1837; Kern in Blumea 8: 123. 1955, 15: 433. 1967 *et* in van Steenis, Fl. Males. 1, 7: 565. 1974; Hooper in Saldanha & Nisolson, Fl. Hassan 676. 1976; Rao & Verma, Cyp. NE India 39. 1982; Sharma *et al.*, Fl. Karnataka 309. 1984; Koyama in Dassanayake & Fosberg, Rev. Handb. Fl. Ceylon 5: 278. 1985; Karthik. *et al.*, Fl. Ind. Enum. Monocot. 51. 1989. *Scirpus cinnamometorum* Vahl, Enum. Pl. 2: 278. 1806. *Fimbristylis cyperoides* R. Br., Prodr. Fl. Nov. Holl. 228. 1810; Clarke in Hook.f., Fl. Brit. India 6: 650. 1893. *F. cyperoides* var. *cinnamometorum* Clarke in Hook.f., Fl. Brit. India 6: 650. 1893; Fischer in Gamble, Fl. Pres. Madras 1659. 1931 (3: 1151. 1957, repr.ed.).

Type : Sri Lanka "habitat in cinnamometis Zeylonae", *Koenig.*

Illus. : Clarke, Illus. Cyp. t. 44, f. 1-4. 1909 (*F. cyperoides*).

Perennials with creeping, woody, 1-1.5 mm thick rhizome. Stems solitary, very slender, compressed-angular, smooth, 25-50 cm long, 0.5 - 1 mm thick. Leaves at least half as long as the stem, erect, setaceous, rigid, acute at apex, 0.3 - 0.5 mm wide, antrorsely scabrid at top; ligule absent; lower sheaths horny. Inflorescence compound or decompound, loose, with many to numerous spikelets, 3-6 cm long. Involucral bracts 1-2, erect, much shorter to slightly longer than inflorescence; lowest up to 4 cm long. Primary rays 5-8, filiform, erect or obliquely erect, smooth. Spikelets solitary, oblong or linear-oblong, strongly compressed, acute at apex, 4-5 x 1 mm, few-flowered; rachilla winged. Glumes distichous, thinly membranous, erect, ovate- lanceolate, acute and muticous or apiculate at apex, sharply keeled, 2.75 - 4 x 1.5 - 2 mm, with 3-nerved keel and nerveless, densely reddish gland-dotted sides. Stamens 3; anthers linear, (1-) 1.5 - 2 mm long. Style slender, 3-3.5 mm long, pyramidally thickened at base, glabrous above, with short hairs at base; stigmas 3, much shorter than style. Nut trigonous with somewhat convex sides, obovoid or oblong-obovoid, shortly stipitate, umbonulate, 0.75-0.9 x 0.4-0.7 mm, verruculose, finely transversely lineolate on the faces by the linear-oblong epidermal cells.

Fls. & Frts. : May-July; Oct.

Chrom. No. : 2 n = 10 (*Taxon* 21:683. 1972).

Habitat : In swamps, open marshy areas of dry deciduous forests, wet places at low altitude.

Distrib. : South and South-East Asia (Sri Lanka, Myanmar, South China, Indo-China, Thailand, Malesia) to tropical Australia. INDIA: Western Peninsula, Central and North-East India. KARNATAKA: Dakshina Kannada (Fischer *l.c.*), Hassan, Uttara Kannada.

Specimens examined : Hassan: Byra, *Nicolson et al.* 237, 6.5.1970 (JCB). Uttara Kannada: Balemane, *Udayakumar & Gurudev Singh* 14613, 7.7.1982 (JCB).

Note: Blake (Journ. Arn. Arb. 35:220. 1954) has clearly pointed out that the Australian *F. cyperoides* R. Br. and the Asian *F. cinnamometorum* are not different species, though Clarke (*l.c.*) reported these two based on the annual habit of the latter and the perennial habit of the former. It is observed that these plants become annual especially on unstable sandy ground, where the water level fluctuates remarkably (Koyama, *l.c.*).

10. **Fimbristylis complanata** (Retz.) Link, Hort. Berol. 1: 292. 1827; Clarke in Hook.f., Fl. Brit. India 6: 646. 1893; Cooke, Fl. Pres. Bombay 2: 885. 1908 (3: 400. 1958, repr.ed.); Fischer in Gamble, Fl. Pres. Madras 1659. 1931 (3: 1151. 1957, repr.ed.); Ramaswamy & Razi, Fl. Bangalore 105. 1973; Kern in van Steenis, Fl. Males. 1, 7: 548. 1974; Hooper in Saldanha & Nicolson, Fl. Hassan 677. 1976; Rao & Razi, Fl. Mysore 562. 1981; Rao & Verma, Cyp. NE India 37. 1982; Sharma *et al.*, Fl. Karnataka 309. 1984; Koyama in Dassanayake & Fosberg, Rev. Handb. Fl. Ceylon 5:

284. 1985; Karthik. *et al.*, Fl. Ind. Enum. Monocot. 51. 1989; Keshava Murthy & Yoganarasimhan, Fl. Coorg 507. 1990. *Scirpus complanatus* Retz., Obs. 5: 14. 1789. **Fig. 22.**

Type. : India.

Perennials with short rhizome, 30- 80 cm high. Stems strongly 4 angled towards base, flattened and ancipitous towards upper part, scabrous on the margin of the wing-like portion towards apex, 2-4 mm thick. Leaves shorter than stem, flat, abruptly acuminate at apex, 9-42 cm long, 2-4 mm wide, striate; margins thickened, densely scabrous towards apex; sheaths compressed, flattened and equitant in early stage, 3-16 cm long; ligule a dense fringe of short, white hairs; orifice membranous to papery, often ciliate on the margin, with few bristles near apex; lower leaves reduced to bladeless, sheaths. Inflorescence compound, 4-10 cm long, 2.5 - 6 cm wide, with many spikelets. Involucral bracts 3-4, shorter than inflorescence, acuminate at apex, densely scabrous on margins; lowest 2-6.5 cm long. Primary rays few, unequal, compressed, up to 6 cm long, sulcate, scabrous. Spikelets solitary, lanceolate or ovoid-lanceolate, angled, acute at apex, 4-6 x 1.5- 2 mm, few-flowered. Glumes spirally arranged, membranous, ovate or oblong-ovate, acute and mucronulate at apex, keeled, 2-2.8 x 1.6-2 mm, hyaline towards margins. Stamens 3; filaments flat, hyaline, elongate up to 2.5 mm; anthers linear-oblong, apiculate, *ca* 1.3 mm long. Ovary oblong, *ca* 0.4 mm long, shortly stipitate; style 1-2 mm long, triquetrous, pyramidally thickened at base, glabrous; stigmas 3, as long as the style, scabrous. Nut trigonous, broadly obovoid, shortly stipitate and minutely umbonulate, 0.8 - 1 x 0.4 - 0.5 mm, smooth or verruculose, creamish-white; epidermal cells very minute, in many rows.

Fls. & Frts. : Most part of the year.

Chrom. No. : 2n = 10 (*Taxon* 21: 683. 1972).

Habitat : Moist soils, muddy river banks, swampy grass fields, wet rice fields and on their margins; sea-level to 1000 m.

Distrib. : Pantropic; very common in South-East Asia. INDIA: Throughout. KARNATAKA: Bangalore, Bellary, Chikmagalur, Hassan, Kodagu, Kolar (Sharma *et al., l.c.*) , Mandya, Mysore.

Specimens examined : Bangalore: Without locality, *Camaron* 628, Oct. 1891 (MH); Ghati, *Ramesh* 627, 29.3.1978 (JCB); Pearl valley, *Ravindra* 1440, 27.6.1978 (JCB). Bellary: Kudligi-Bandri Road, *Manohar & Ramesh* 5853, 18.1.1979 (JCB). Chikmagalur: Sakrepatana, *Ahuja* 5936, 18.5.1959 (BSI); Bababuddan hills, *Saldanha & Ramesh* 1733, 26.6.1978 (JCB). Hassan: Just South of Channarayapatana, *Saldanha* 14273, 4.8.1969 (JCB). Kodagu: Kushalnagar, *Bhat* 942, 16.2.1981 (MGH). Mandya: Hemagiri, *Dinesh* 1038, 12.7.1984 (MGH). Mysore: Srirangapatanam,

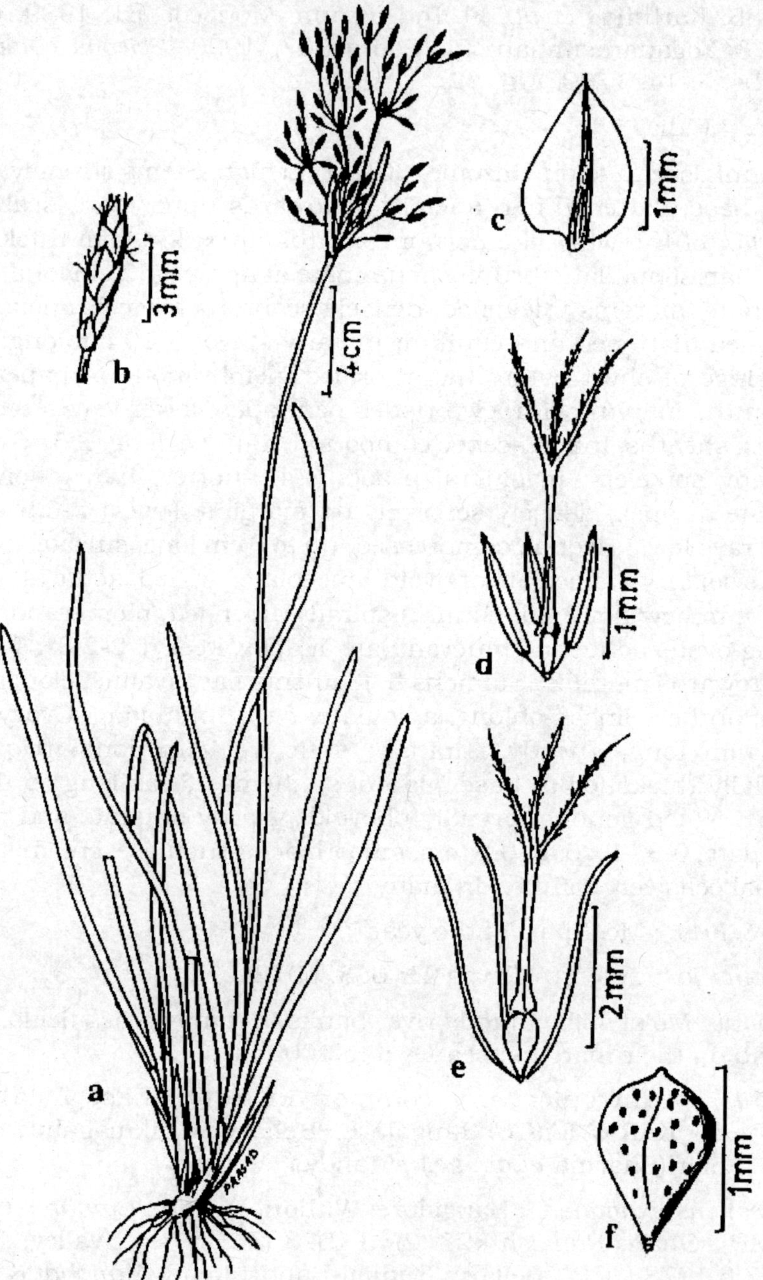

Fig. 22. *Fimbristylis complanata* (Retz.) Link
a. Habit, **b.** Spikelet, **c**. Glume, **d.** Flower (early stage),
e. Flower (later stage), **f.** Nut

Sebastine 18730, 11.3.1964 (MH); Mysore, *Raghavendra* 87, 20.5.1967 (MGH): Biligiri, *Bhat* 5, 8.7.1970 (MGH).

Note: Plants are polymorphic. In short plants (*ca* 30 cm high) nut is smaller (*ca* 0.8 mm long) and verruculose. In taller plants (*ca* 75 cm high) nut is *ca* 1 mm long and smooth.

11. **Fimbristylis consanguinea** Kunth, Enum. Pl. 2: 228. 1837; Kern in Blumea 8: 110. 1955, in Back & Bakh. f., Fl. Java 3: 464. 1968, *et* in van Steenis, Fl. Males. 1, 7: 549. 1974; Koyama in Dassanayake & Forsberg, Rev. Handb. Fl. Ceylon 5: 285. 1985; Karthik. *et al.*, Fl. Ind. Enum. Monocot. 51. 1989. *F. kraussiana* Hochst. ex Krauss, Flora 28: 757. 1845 (invalid name published as synonym); Trimen, Fl. Ceylon 5: 63. 1900. *F. connectens* Thw., Enum. Pl. Zeyl. 349 & 433. 1864. *F. complanata* Link var. *kraussiana* Clarke in Hook. f., Fl. Brit. India 6: 646. 1893. **Fig. 23.**

Type: S. Africa, Cape of Good Hope, *Drege s.n.* Isotype (L)

Perennials with short rhizome, 7-15 cm high. Stems tufted 3-4 angled, slightly compressed, 0.5-0.8 mm thick. Leaves shorter than stem, flat, linear, abruptly acuminate at apex, 1-1.5 mm wide, slightly thickened on the margins, smooth; cauline leaves usually with reduced blades; sheaths cylindrical, up to 2.3 cm long, not thickened on the back; ligule a fringe of short hairs; orifice membranous, oblique, ciliate. Lowest leaves usually reduced to sheaths. Inflorescence simple to compound, with 3-few spikelets, 1-2 cm long, 0.5 - 2 cm wide. Involucral bracts 3, shorter than inflorescence; lowest 6-8 mm long. Primary rays compressed, 3-angled, striate, longest up to 1.5 cm long. Spikelets solitary, ovoid or oblong-ovoid, acute or subacute at apex, 3-4 x *ca* 1.2 mm, faintly angled, few-flowered; rachilla winged. Glumes spiral, membranous, ovate, acute-mucronate at apex, 1.6 - 2 x 1-1.3 mm, keeled, hyaline towards margins. Stamens 3; filaments initially *ca* 0.3 mm long, elongatge up to 1.8 mm; anthers linear-oblong, apiculate at apex, *ca* 0.7 mm long. Style elongate up to 2.2 mm, trigonous, pyramidally thickened towards base, glabrous; stigmas 3, slightly shorter than style. Nut trigonous, broadly obovoid, umbonulate, shortly stipitate, 0.8-1 x 0.5 - 0.6 mm, with smooth or sparsely tubercled faces; epidermal cells not clear on the smooth faces; transversely oblong, in many vertical rows.

Fls. & Frts. : Nov.

Habitat : Swampy areas, wet grasslands.

Distrib. : Sri Lanka, China, Malesia (Java), South Africa and Madagascar. INDIA: Southern peninsula. KARNATAKA: Shimoga (Rare).

Specimen examined : Shimoga: Nalur, Agumbe, *Raghavan* 68041, 1.11.1960 (BSI).

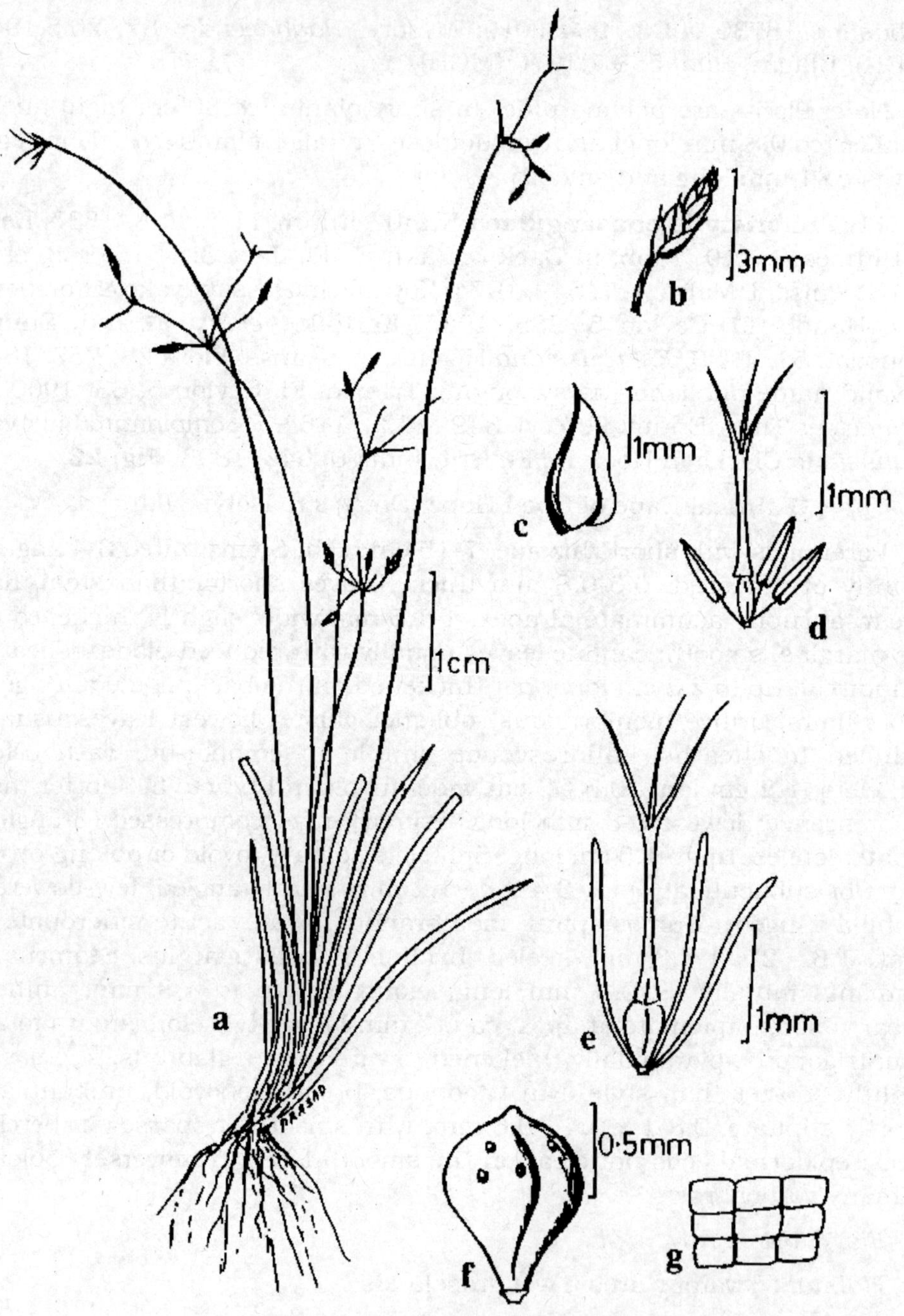

Fig. 23. *Fimbristylis consanguinea* Kunth
a. Habit, **b.** Spikelet, **c.** Glume, **d.** Flower (early stage), **e.** Flower (later stage), **f.** Nut, **g.** Epidermal cells

Note. : This species was reported for the first time from Karnataka during the present study (*Prasad & Singh*, 1996a). The description is based on a single specimen available in BSI. It seems to be closely related to *F.complanata* but can be recognised easily by the narrow stems, cylindrical leaf- sheaths with rounded back and less compound inflorescence with few spikelets.

According to Kern (*l.c.*) this species occurs at high altitude in Malesia, Sri Lanka and Madagascar.

12. **Fimbristylis cymosa** R. Br., Prodr. 228. 1810; Ramaswamy & Razi, Fl. Bangalore 106. 1973; Kern in van Steenis, Fl.Males. 1, 7: 557. 1974; Rao & Razi, Fl. Mysore 562. 1981; Arora *et al.*, Bot. S. Kanara 61. 1981; Sharma *et al.*, Fl. Karnataka 309. 1984; Koyama in Dassanayake & Forberg, Rev. Handb. Fl. Ceylon 5: 301. 1985 (as subsp. *spathacea* (Roth) Koyama); Singh, Fl. E. Karnataka 2: 635. 1988; Karthik. *et al.*, Fl. Ind. Enum. Monocot. 51. 1989. *F. spathacea* Roth, Nov. Pl. sp. 24. 1821; Clarke in Hook. f., Fl. Brit. India 6: 640. 1893; Cooke, Fl. Pres. Bombay 2: 882. 1908 (3:396.1958, repr.ed.); Fischer in Gamble, Fl. Pres. Madras 1659. 1931 (3: 1151. 1957, repr.ed.).

Type : India.

Illus. : Mahesh., Illus. Fl. Delhi f. 224. 1966.

Perennials with short rhizome, 13-65 cm high. Stems densely tufted, sulcate, rarely subterete, striate, 1-2.5 mm thick. Leaves densely tufted, coriaceous, much shorter than stem, usually canaliculate, at times flat, acute-acuminate at apex, 4-25 cm long, 1-3 mm wide; basal sheaths 1-3 cm long, 5-8 mm broad; ligule absent. Inflorescence decompound, 1.5-4 cm long, 2-3 cm wide, with numerous spikelets. Involucral bracts 2-3, much shorter than inflorescence, lowest 0.5 - 2.5 cm long, dilated at base; primary rays very unequal in size, 0.6 - 3 cm long. Spikelets usually solitary, often clustered, ovoid or ellipsoid, obtuse or obtuse-acute at apex, 2-5 x 1.2-2 mm, slightly angled, densely many-flowered; rachilla prominently winged. Glumes spiral, broadly ovate, obtuse, slightly notched or minutely apiculate at apex, *ca* 1.5 x 1.2 mm, faintly keeled, scarious towards margins. Stamens usually 2, maturing at different times; filaments flat, elongate up to 2 mm; anthers linear-oblong, *ca* 0.7 mm long, connective minutely apiculate. Ovary oblong, *ca* 0.5 mm long; style glabrous, *ca* 0.7 mm long, dilated at base; stigmas 2, *ca* 1 mm long, minutely ciliate. Nut biconvex, obovate, *ca* 1 x 0.7 mm, minutely verruculose, cream-coloured to blackish; stipe not conspicuous.

Fls. & Frts. : May - Dec.

Chrom. No. : 2n = 48 (reported as *F.spathacea* in *Taxon* 2: 683. 1972).

Habitat: Crevices of rocks near sea; clayish areas near sea, sandy soil in beaches and near back waters, muddy areas of mangroves, river banks, marshy areas, gravelly soil near water bodies; at times among grasses in wet muddy areas and as a weed in agricultural fields.

Distrib. : Pantropic (Sri Lanka, Pakistan, China, Singapore, Malesia, Arabian Countries, Ethiopia). INDIA: Western Peninsular India, East India, Anadman & Nicobar Islands. KARNATAKA: Bangalore, Belgaum, Bellary, Bijapur, Dakshina Kannada, Dharwar, Hassan, Mandya, Mysore, Raichur, Uttara Kannada.

Specimens examined : Bangalore: Hulimavu, *Hooper & Saldanha* 18025, 7.11.1971 (JCB). Belgaum: Near Gokak falls, *Prasad* 172814, 11.12.1994 (BSI); On the way to Ghataprabha from Gokak, *Prasad* 272832, 12.12.1994 (BSI); Dinman Hossur, on Belgaum-Gokak Road, *Prasad* 172842, 13.12.1994 (BSI). Bellary: Bellary-Kanekal Road 8th km, *Singh* 143105, 20.8.1976 (BSI). Bijapur: Hebballi seemi, Badami, *Prasad* 172975, 22.12.1994 (BSI); Khanapur, along Hubli-Bijapur Road, *Prasad* 182982, 22.12.1994 (BSI). Dakshina Kannada: Kudlu, without coll. name. 16745, 29.8.1920 (MH); Kapu, *Bhat* 452, 15.1.1977 (MGH); Surathkal, *Bhat* 607, 2.10.1979 (MGH); Coastal areas, *Sheriff & Suresh* 829, 13.12.1985 (MH); Near Kulur bridge, Mangalore, *Prasad* 173879, 10.11.1995 (BSI). Dharwar: Sadankeri, Alnavar Road, *Prasad* 172938, 19.12.1994 (BSI); Near Kalekari lake, *Prasad* 172946, 19.12.1994 (BSI). Hassan: Belluru, *Nicolson et al.* 2357, 26.10.1971 (JCB). Mandya: Sreerangapatana, *Bhat* 41, 26.7.1970 (MGH). Mysore: Kotagal, *Wadhwa* 44962, 1.10.1958 (BSI); Biligiri Rangan hills - Hassanur Road, *Saldanha & Ramesh* 2533, 6.9.1978 (JCB). Raichur: Guthgola block, near Amreshwar, *Singh* 132775, 8.9.1974 (BSI); Lingsugur-Sindnoor, *Singh* 143044, 18.8.1976 (BSI). Uttara Kannada: Karwar, *Talbot* 557. 20.7.1883 (BSI).

13. **Fimbristylis dichotoma** (L.) Vahl, Enum. Pl. 2: 287. 1806; Fischer in Gamble, Fl. Pres. Madras. 1898. 1936 (Corrig.) *et* 3: 1151. 1957 (repr.ed.); Ramaswamy & Razi, Fl. Bangalore 106. 1973; Kern in van Steenis, Fl. Males, 1, 7: 575. 1974; Hooper in Saldanha & Nicolson, Fl. Hassan 677. 1976; Rao & Razi, Fl. Mysore 562. 1981; Rao & Verma, Cyp. NE India 31. 1982; Sharma *et al.*, Fl. Karnataka 309. 1984; Koyama in Dassanayake & Fosberg, Rev. Handb. Fl. Ceylon 5: 306. 1985; Singh, Fl. E. Karnataka 2: 636. 1988; Karthik. *et al.*, Fl. Ind. Enum. Monocot. 51. 1989; Keshava Murthy & Yoganarasimhan, Fl. Coorg 507. 1990. *Scirpus dichotomus* L., Sp. Pl. 1: 50. 1753. *S.diphyllus* Retz., Obs. Bot. 5: 15. 1789. *Fimbristylis diphylla* (Retz.) Vahl, Enum. Pl. 2: 289. 1805; Clarke in Hook.f., Fl. Brit. India 6: 636. 1894; Cooke, Fl. Pres. Bombay 2: 882. 1908 (3: 396. 1958, repr.ed.). *F. annua* var. *diphylla* Kukenth., Act. Hort. Got. 5: 109. 1929. *F. annua* var. *paucispiculata* Blatt. & McC. in Journ. Bombay nat. Hist. Soc. 37: 544. 1934; Sharma *et al.*, Fl. Karnataka 309. 1984.

Illus. : Clarke, Illus. Cyp. t. 42, f.1-2. 1909 (*F. diphylla*)

Perennials (at times annuals ?), often with short woody rhizome, 15-70 cm high. Stems slender to stout, glabrous, angular, striate, 9-65 cm long, 1-2 mm thick. Leaves mostly from the base, flat, linear, acute to obtuse at apex, 3-40 cm long, 1.5 - 3 mm wide, glabrous to sparsely scabrous; ligule a dense fringe of short hairs. Sheaths 0.5 - 18 cm long, glabrous to densely puberulous. Inflorescence simple to decompound corymbs of few to many spikelets. Involucral bracts 2-5, much shorter to longer than inflorescence, 0.5-5 cm long. Primary rays 0.5-3 cm long, angled, striate. Spikelets ovoid to ellipsoid, terete, acute at apex, 4-7 x 2-4 mm, densely many-flowered; rachilla prominently winged. Glumes chartaceous, spiral, broadly ovate, 1.8-3 x 1.5-2 mm, keeled, glabrous, dark brown; keel protruding above the apex as a mucro. Stamens 2; filaments elongate up to 3 mm; anthers oblong-lanceolate, apiculate at apex, *ca* 0.7 mm long. Ovary oblong, *ca* 0.5 mm long, stipitate; style flat, 1-1.8 mm long, slightly dilated at base, ciliate at margin; stigmas 2, tapering towards apex, 0.5 - 1.5 mm long. Nut biconvex, obovate, 1-1.2 x 0.8-1 mm, shortly stipitate, verruculose, trabeculate, whitish or cream-coloured (one specimen was found with light brown nut).

Fls. & Frts. : April-Nov.

Chrom. No. : 2 n = 20 (*Taxon* 21: 683. 1972).

Habitat : Moist sandy areas, open waste places, grassy roadsides, along the banks of streams, edges of rice fields, moist ditches along roadsides.

Distrib. : Throughout the warmer parts of the world. One of the most widely distributed species. Very common in South and East Asia. INDIA: Throughout. KARNATAKA: Bangalore, Belgaum, Chikmagalur (Sharma *et al., l.c.*), Dakshina Kannada, Hassan, Kodagu, Kolar (Sharma *et al., l.c.*), Mandya, Mysore, Shimoga, Tumkur, Uttara Kannada.

Uses : Leaves can be used as forage.

Specimens examined : Bangalore: Near Solur, *Saldanha & Murthy* 3307, 15.10.1978 (JCB). Belgaum: Near Gokak falls, Gokak, *Prasad* 172813A, 11.12.1984 (BSI); Gottni nallah, Jambotti R.F., *Prasad* 172872, 15.12.1984 (BSI); Dudwawada R.F., Londa, *Prasad* 172896 & 172898, 17.12.1994 (BSI). Dakshina Kannada: Nileswar, without coll. name, 15324, 9.11.1917 (MH); Mangalore, *Foulker* 4868, Jan. 1902 (MH); Kannadekatte, *Bhat* 170, 6.5.1975 (MGH); Indrali, Udupi, *Bhat* 277, 22.8.1976 (MGH); Gundia, *Raghavan* 146338, 14.4.1977 (BSI). Hassan: Tagare, *Nicolson et al.* 2284, 23.10.1971 (JCB). Kodagu: Fraserpet, *Muddaiah s.n.* (Acc. No. 97089). 1955 (MH); Bridal path of Raja seat, *Rao* 744413, 20.9.1961 (BSI); Karawha Badiga, *Rao* 74516, 22.9.1961 (BSI); Mercara, *Bhat* 698, 28.9.1980 (MGH); Kushalnagar, *Bhat* 1040, 21.9.1981 (MGH); Talakaveri, *Bhat* 1074, 27.9.1981 (MGH). Mandya: Sreerangapatana, *Padma Rani* 5, 12.7.1970 (MGH); Malavalli-Kolligal Road, *Saldanha & Ramesh* 2405, 5.9.1978 (JCB). Mysore: Namadachilume, *Rao* 73257, 28.8.1961 (BSI); Eichalakolame betta, North of Ketadevargudi, *Rao* 80235, 22.4.1962 (BSI); Ketadevargudi vicinity, *Rao* 73604, 4.9.1962 (BSI); Bandipur, *Naithani* 21116, 24.8.1964 (MH); Chamundi, *Bhat* 11B, 10.7.1970 (JCB); without locality and coll. name 28245, without date (BSI). Shimoga: Batehalli, Agumbe, *Raghavan* 90388, 2.9.1963 (BSI); Shimoga, *Bhat* 698, 28.9.1980 (MGH). Tumkur: way to Devarayanadurga, *Saldanha* 2162, 15.8.1978 (JCB). Uttara Kannada: Karwar, *Talbot* 601, Aug. 1883 (BSI); Karwar, without coll. name and number, 10.8.1883 (BSI); Anatari forest, Katgal, *Jain* 15589 A, 17.11.1957 (BSI); Kargette, *Maqsood* 1168, 29.5.1978 (JCB).

Note : A highly polymorphic species with a lot of variations in the stems, leaves, inflorescence, flowers and nuts. Hence there are many synonyms and infraspecific taxa for this species.

14. **Fimbristylis dimorphonucifera** Govind. in Rheedea 7 (2): 122, f. 4. 1997.

Type : India, Karnataka state, S. Kanara, Kannadekatta, *Govindarajalu* 8860 G-Holotype (CAL).

Illus. : Govind, *l.c.*

Annuals with blackish-brown, fibrous roots. Stems tufted, stiff, erect, trigonous, 8-12 cm long, *ca* 1 mm thick, glabrous, usually up to 3-noded and bearing tubular sheaths; base covered with 2-3 ovate-lanceolate, distichously arranged scales; sheaths non keeled, with or without reduced blades. Cauline leaves when present 1.5-2.5 (-3.5) cm long. Leaves of sterile shoots shorter than stems, 1.5-2.5 mm wide, somewhat keeled; ligule a dense fringe of brown hairs. Inflorescence simple, sometimes 1-2 rays added, loose, with (3-) 5-8 (-10) spikelets, 1-2.5 cm long and broad. Involucral bracts 3-4, acute-acuminate at apex, usually shorter than inflorescence. Rays when present *ca* 1 cm long. Spikelets solitary, angular, ellipsoid or elliptic-ovoid, acute at apex, 4.5-5 x 1.5-2 mm, 10-20-flowered; rachilla excavated, winged. Glumes spiral, ovate, broadly rounded at apex, 1.5-2.2 x *ca* 0.75 mm, with 3-nerved keel; sides broad, translucent, eglandular. Stamens usually 3; anthers linear-oblong, 1.2-1.3 mm long. Style triquetrous, with narrow pyramidal base, up to 1.2 mm long; stigmas 3, 0.7-0.8 mm long. Nuts dimorphic, densely tubercled; 1-2 basal nuts larger, obscurely trigonous, asymmetric, sessile, minutely umbonulate or non umbonulate, 0.75-1mm long and broad, black, with obscure epidermal cells; the remainder small, triquetrous, symmetric, narrowly obovoid, shortly stipitate, non umbonulate, *ca* 0.75 x 0.5 mm, stramineous, with narrowly linear transversely elongated epidermal cells.

Fls. & Frts. : Nov.

Habitat : Marshy soil and wet habitats (Govind. *l.c.*).

Distrib. : Endemic to Karnataka (Dakshina Kannada Dt.).

Note : This species is included on the authority of Govindarajalu, *l.c.* As specimens are not available, description provided above is from the protalogue.

15. **Fimbristylis dipsacea** (Rottb.) Clarke in Hook.f., Fl. Brit. India 6: 635. 1893; Fischer in Gamble, Fl. Pres. Madras 1658. 1931 (3: 1151. 1957, repr.ed.); Kern in van Steenis, Fl. Males. 1, 7:590. 1974; Hooper in Saldanha & Nicolson, Fl. Hassan 678. 1976; Rao & Verma, Cyp. NE India 30. 1982; Sharma *et al.*, Fl. Karnataka 309. 1984; Koyama in Dassanayake & Fosberg, Rev. Handb. Fl. Ceylon 5: 321. 1985; Karthik. *et al.*, Fl. Ind. Enum. Monocot. 52. 1989. *Scirpus dipsacea* Rottb., Descr. & Ic. 56. t. 12, f. 1. 1773. *Echynolytrum dipsacum* Desv. in J. Bot. 1: 21, t. 1808. *Isolepis dipsacea* Roem. & Schult., Syst. 2: 119. 1811. **Fig. 24.**

Type: India, *Koenig.*

Illus. : Clarke, Illus. Cyp. t. 41, f. 4-7. 1909.

Annuals with fibrous roots, 4-10 cm high. Stems highly tufted, very slender, angular, sulcate, 0.5-7 cm long, *ca* 0.3 mm thick. Leaves short, filiform, tapering towards apex, 0.5 - 3 cm long, *ca* 0.3 mm wide; sheaths very small, lowest scale like, brownish; ligule absent. Basal leaves without blades. Inflorescence usually a simple umbel of few spikelets, solitary in early stage. Involucral bracts 3-5, filiform, 2.5-20mm long; rays up to 8 mm long, angled. Spikelets solitary, subglobose, oblong or ovoid, terete, obtuse at apex, 2-6 x 2-4 mm, densely many-flowered; rachilla densely and minutely winged. Glumes sprial, spreading, lanceolate, narrowed towards apex, 1.5 - 2 x 0.3 -0.5 mm (including the awn), keeled, thinly membranous towards margins; keel protruding beyond the apex forming a recurved awn. Stamen 1; filament elongate up to 1.5 mm; anthers oblong-lanceolate, apiculate at apex, *ca* 0.3 mm long. Ovary oblong, *ca* 0.2 mm long; style 0.6-0.8 mm long, slightly dilated at base; stigmas 2, as long as the style. Nut subterete, linear-oblong, obtuse at apex, slightly curved, 0.5 - 0.9 x *ca* 0.2 mm, brownish, with few caducuous, clavate appendages on both the margins; stipe minute, very narrow; epidermal cells minute, transversely oblong.

Fls. & Frts. : April-May; Oct.

Habitat : Muddy areas of lake shores, river banks, wet rice fields, open marshy areas and wet sandy areas at low altitude.

Distrib. : South and East Asia (Sri Lanka, Myanmar, Indo-China, Malesia), tropical Africa. INDIA: Western Peninsular India, Central India, North India, East and North-East India. KARNATAKA: Dakshina Kannada, Hassan, Mysore (Fischer, *l.c.*).

Specimens examined : Dakshina Kannada: Padubidri, *Saldanha & Prakash* 7041, 19.4.1979 (JCB); Barkur, *Bhat* 614, 4.10.1979 (MGH); without locality, *Raghavan* 155260 & 155263, without date (BSI). Hassan:

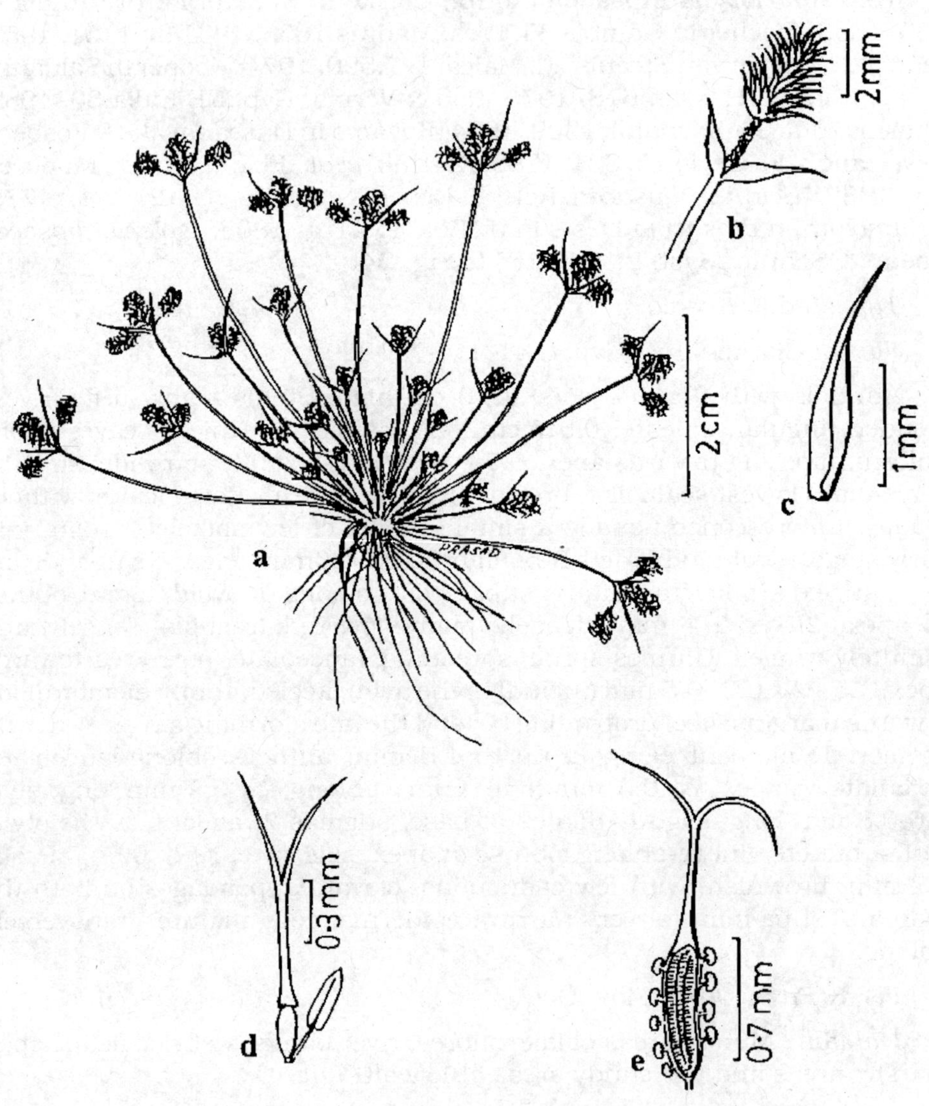

Fig. 24. *Fimbristylis dipsacea* (Rottb.) Carke
a. Habit, **b.** Spikelet, **c**. Glume, **d.** Flower, **e.** Nut with style

opposite to Vishnu Samudra, Belur, *Saldanha* 13530, 15.5.1969 (JCB). Without locality (Mysore & Carnatic), *Thomson s.n.* (Acc. No. 78707), without date (MH).

16. **Fimbristylis eragrostis** (Nees & Mey.) Hance in Journ. Linn. Soc. Bot. 13: 132. 1873; Kern in van Steenis, Fl. Males. 1, 7: 567. 1974; Rao & Verma, Cyp. NE India 38. 1982; Koyama in Dassanayake & Fosberg, Rev. Handb. Fl. Ceylon 5: 275. 1985; Karthik. *et al.*, Fl. Ind. Enum. Monocot. 52. 1989. *Abildgaardia eragrostis* Nees & May. ex Nees in Wight, Contr. Bot. India 95. 1834. *Fimbristylis nigrobrunnea* Thw., Enum. Pl. Zeyl. 434. 1864; Clarke in Hook.f., Fl. Brit. India 6: 648. 1893; Fischer in Gamble, Fl. Pres. Madras 1659. 1931 (3: 1151. 1957, repr.ed.). *F. eragrostis* var. *nigrobrunnea* (Thw.) Hooper in Saldanha & Nicolson, Fl. Hassan 678. 1976; Sharma *et al.*, Fl. Karnataka 309. 1984. **Fig. 25.**

Type : China, *Meyen*.

Perennials with woody rhizome, 6-42 cm high. Rhizome often clothed with old decayed leaf-sheaths. Stems usually solitary, at times tufted, 4-angled, sulcate, slightly compressed and often ancipitous below the inflorescence, smooth or scabrid on the flattened margins at top. Leaves many, much smaller than flowering stems, flat, stiff, usually subfalcate, linear, obtuse and apiculate at apex, 2-10 cm long, 1.5-3 mm wide, often spinulose-scabrous on the margins towards apex; sheaths dilated towards base, striate, 1.5-5 cm long; ligule absent; orifice chartaceous, not ciliate; upper leaves very small, often reduced to 2-3 tubular sheaths with very short lamina. Infloresence compound to decompound, 1.5 - 4 cm long, 1-5 cm wide, with few- many spikelets. Involucral bracts 2-4, very short, not spreading, acute at apex, dilated towards base; lowest 4-6 mm long. Primary rays 5-7, slightly compressed, striate, longest 2-3 cm long, smooth. Spikelets solitary, compressed or subterete, ovate, oblong-ovate or lanceolate, acute at apex, 5-8 x 2-3 mm, light to dark brown, few-flowered; rachilla winged. Glumes distichous or subspiral, chartaceous, broadly ovate, obtuse-acute and mucronulate at apex, keeled, 3.2 - 3.8 x 2.5 - 2.8 mm, shining glabrous. Stamens 3; filaments flat, elongate up to 4 mm; anthers linear, prominently apiculate, 1.3 - 2 mm long. Ovary oblong, *ca* 0.5 mm long; style triquetrous, pyramidally thickened towards base, 2.2-2.5 mm long, glabrous or sparsely ciliate at top; stigmas 3, as long as the style. Nut trigonous, obovoid, umbonulate, *ca* 1 x 0.7 mm, shortly stipitate, verruculose, greyish; epidermal cells minute, isodiametric, in *ca* 13 vertical rows.

Fls. & Frts. : May-Oct.

Habitat : Grassy hill sides on low mountains, Savannahs, Forest clearings, Hillocks and other rocky areas, also in stable sandy grounds. Wet deciduous forests and hilly areas in Peninsular India.

Distrib. : Sri Lanka, South China, Hainan, Formosa, Malesia and

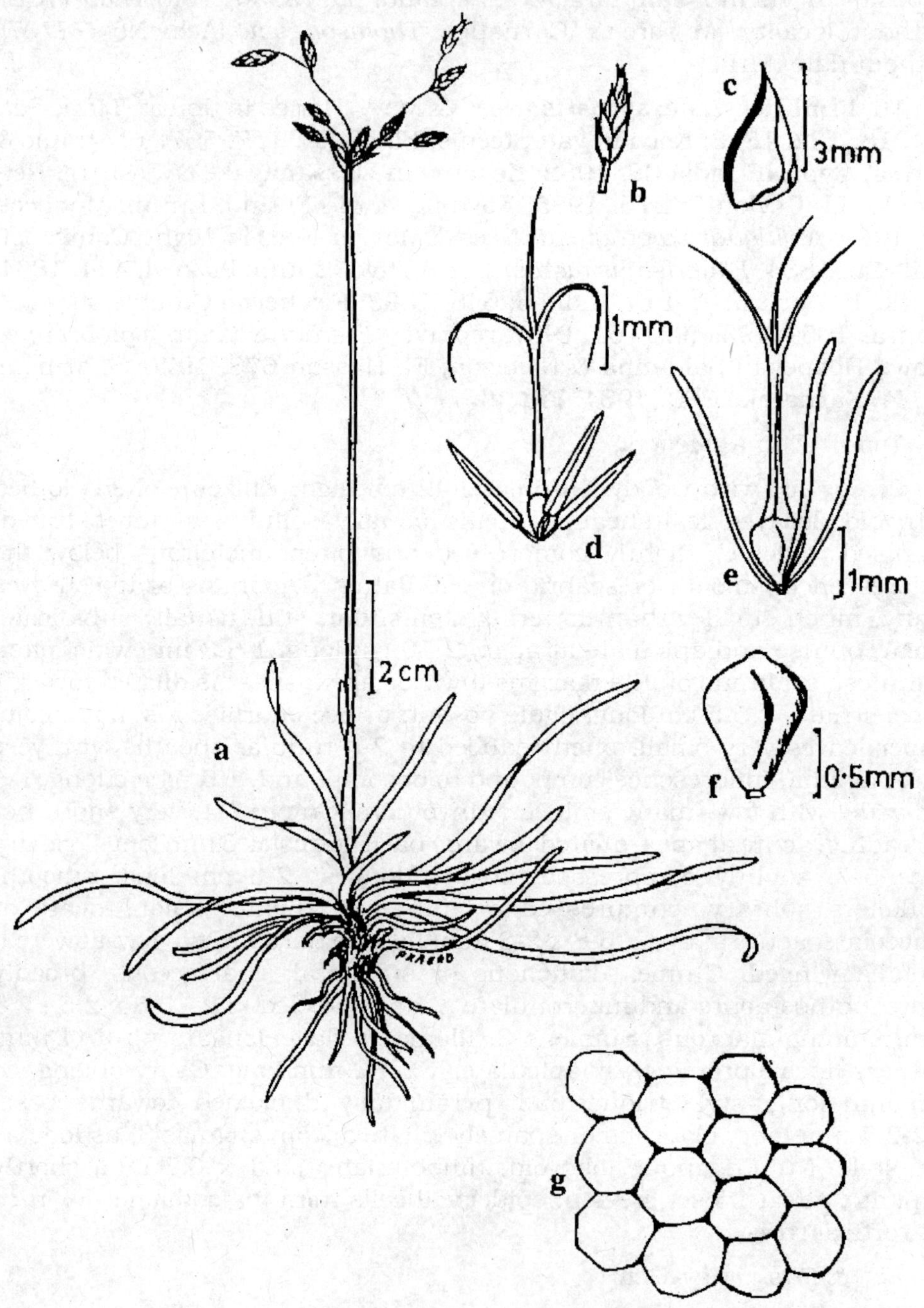

Fig. 25. *Fimbristylis eragrostis* (Nees & May.) Hance
a. Habit, **b.** Spikelet, **c.** Glume, **d.** Flower (early stage),
e. Flower (later stage), **f.** Nut, **g.** Epidermal cells.

tropical Australia. INDIA: Peninsular India, North-East India, Andaman & Nicobar Islands. KARNATAKA: Chikmagalur, Dakshina Kannada, Hassan, Kodagu, Shimoga, Uttara Kannada.

Specimens examined : Chikmagalur: Near Gangamula road, *Yoganarasimhan* 1267, 8.5.1972 (RRCBI). Dakshina Kannada: Without locality, *Raghavan* 155307, without date (BSI). Hassan: Hills above Panorama Point, Shiradi, *Saldanha* 14769, 3.9.1969 (JCB). Kodagu: Mercara-Bhogmandala, *Ramesh* 1237, 10.6.1978 (JCB). Shimoga: Hegganguda, Thirthahalli, *Raghavan* 90045A, 30.8.1963 (BSI). Uttara Kannada: Karwar, *Ahmad* 1138, 28.5.1978 (JCB).

17. **Fimbristylis falcata** (Vahl) Kunth, Enum. Pl. 2: 239. 1837; Kern in Blumea 8: 113. 1955; Ramaswamy & Razi, Fl. Bangalore. 106. 1973; Kern in van Steenis, Fl. Males. 1, 7: 557. 1974; Hooper in Saldanha & Nicolson, Fl. Hassan 678. 1976; Rao & Razi, Fl. Mysore 562. 1981; Rao & Verma, Cyp. NE India 38. 1982; Sharma *et al.*, Fl. Karnataka 309. 1984; Koyama in Dassanayake & Fosberg, Rev. Handb. Fl. Ceylon 5: 289. 1985; Karthik. *et al.*, Fl. Ind. Enum. Monocot. 52. 1989; Keshava Murthy & Yoganarasimhan, Fl. Coorg 508. 1990. *Scirpus falcatus* Vahl, Enum. Pl. 2: 275. 1806. *Fimbristylis junciformis* Kunth, Enum. Pl. 2: 239. 1837; Clarke in Hook.f., Fl. Brit. India 6: 647. 1894; Cooke, Fl. Pres. Bombay 2: 886. 1908 (3: 400. 1958, repr.ed.); Fischer in Gamble, Fl. Pres. Madras 1660. 1931 (3: 1152. 1957, repr.ed.). **Fig. 26.**

Type: India, *Koenig.*

Glabrous perennials, 10-40 cm high. Rhizome woody, 0.5 - 1 cm thick, usually creeping, clothed with remains of old leaf- sheaths and with rather stout fibrous roots. Stems more or less tufted, compressed, angular, sulcate, 0.75 - 1.5 mm thick, smooth. Leaves mostly basal, usually much shorter than stem, flat or with in-rolled margins, abruptly acuminate at apex, 2-15 cm long, 1-3 mm wide, often scabrous on the margins; ligule absent; orifice membranous, smooth, often with few bristles at top; lower sheaths subpersistent, covering the rhizome. Inflorescence compound, often subcompound, open, with few-many spikelets, 1-5 cm long, 1-3 cm wide. Involucral bracts 3-4, much shorter than inflorescence, acuminate at apex, often scabrous on the margins towards apex; lowest up to 1.8 cm long. Primary rays 4-7, slightly compressed, striate-sulcate, very unequal, up to 4 cm long. Spikelets usually in clusters, often in pairs, rarely solitary, ovoid, acute at apex, faintly angled, 3.5-6 x *ca* 1.5 mm, few to many-flowered, rachilla winged. Glumes spiral, broadly ovate, acute and minutely apiculate at apex, keeled, 2-3 mm long and as broad, scarious to hyaline towards margins; lowest glumes often long-mucronate. Stamens 3; filaments flat, hyaline, elongate up to 3.5 mm; anthers linear-oblong, connective distinctly produced, with short white bristles at tip. Ovary oblong, *ca* 0.5 mm long, minutely stipitate; style *ca* 2 mm long, triquetrous,

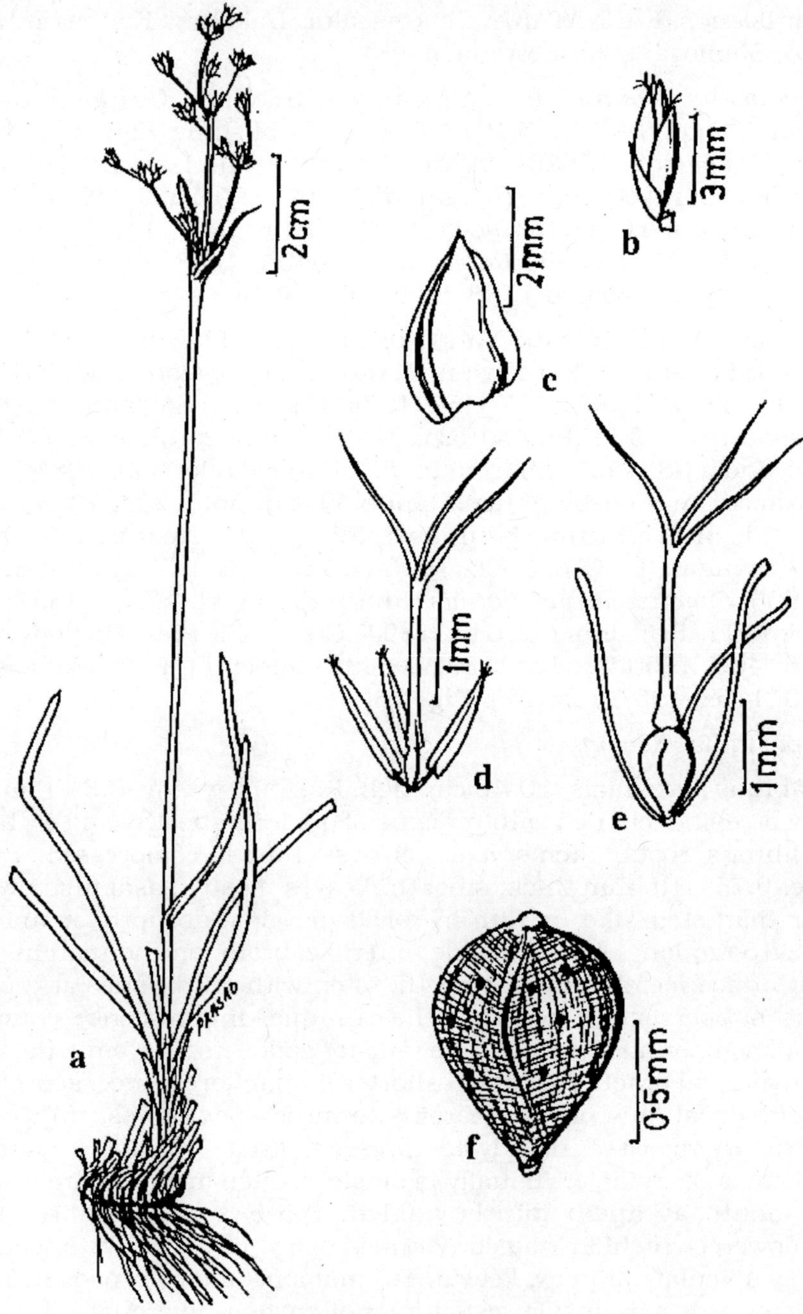

Fig. 26. *Fimbristylis falcata* (Vahl) Kunth
a. Habit, **b**. Spikelet, **c.** Glume, **d.** Flower (early stage), **e.** Flower (later stage), **f.** Nut

pyramidally thickened at base, glabrous or ciliate towards apex; stigmas 3, slightly shorter than to as long as the style. Nuts trigonous, obovoid, shortly stipitate, umbonulate, *ca* 1 x 0.7 mm, whitish, smooth or slightly verruculose; epidermal cells transversely linear-oblong.

Fls. & Frts. : April - Nov.

Habitat : Mountain peaks, rocky slopes, shady areas on hills, among grasses in moist forest clearings, moist humous soil, grasslands.

Distrib. : Sri Lanka, Pakistan, Nepal, Thailand, Indo-China, Malesia. INDIA: Throughout. KARNATAKA: Bangalore, Bellary, Chikmagalur, Dharwar, Hassan, Kodagu, Mysore, Shimoga, Uttara Kannada.

Specimens examined : Bangalore: Nandi hills, *Hooper & Saldanha* 18098, 4.12.1971 (JCB); Pearl valley-Muthyal Madugu, *Ravindra* 1430, 27.6.1978 (JCB). Bellary: Yeshvanthnagar, *Murthy & Manohar* 8148, 17.6.1979 (JCB). Chikmagalur: Lakavalli, *Saldanha & Ramesh* 1776, 27.6.1978 (JCB). Dharwar: Dharwar, *Talbot s.n.* (Acc. No. 758), 10.8.1890 (BSI). Hassan: Tank after Byra, *Saldanha* 13695, 29.5.1969 (JCB). Kodagu: Manipal, *Bhat* 469, 19.10.1977 (MGH); Kunjargiri, *Bhat* 624, 21.7.1980 (MGH); Talakaveri, *Bhat* 1000, 23.7.1981 (MGH). Mysore: Along stepped path to Yelandur, behind Biligirirangan temple, *Rao* 80326, 25.4.1962 (BSI); Uppanere, Biligirirangan hill ranges, *Rao* 80250, 23.4.1962 (BSI); Dasanahatti halla (Devagiribetta), Biligirirangan hill ranges, *Rao* 80045, 17.4.1962 (BSI); Bandipur, *Naithani*, 23863, 21.4.1965 (MH); Talkad, *Rao* 679, 23.5.1970 (MGH); Chamundi hills, *Rao* 1288, 22.5.1971 (JCB). Shimoga: Vanakeabbi falls, Agumbe, *Raghavan* 68109A, 3.11.1960 (BSI); Barakana, Agumbe, *Raghavan* 86122, 3.11.1960 (BSI); Sharavathi, *Puri* 19099, 23.5.1957 (BSI). Uttara Kannada: without exact locality, *Talbot* 867 (A), without date (BSI); without exact locality, *Talbot s.n.* (Acc. No.759), June 1883 (BSI); Kalsur, *Udaya Kumar & Ramesh* 13106, 25.6.1981 (JCB). Without locality and coll. name, 37642 and 37972, without date (BSI).

Note : Similar to *F. cymosa*, but this species can be distinguished by the thick creeping rhizome, scarious margins of the glume, bristly appendages on the anther and the pale nut.

18. **Fimbristylis ferruginea** (L.) Vahl, Enum. Pl. 2: 291. 1806; Cooke, Fl. Pres. Bombay 2: 881. 1908 (3: 396. 1958, repr.ed.); Fischer in Gamble, Fl. Pres. Madras 1658. 1931 (3: 1151. 1957, repr.ed.); Ramaswamy & Razi, Fl. Bangalore 107. 1973; Kern in van Steenis, Fl. Males. 1, 7: 572. 1974; Hooper in Saldanha & Nicolson, Fl. Hassan 679. 1976; Rao & Razi, Fl. Mysore 562. 1981; Sharma *et al.*, Fl. Karnataka 310. 1984; Koyama in Dassanayake & Fosberg, Rev. Handb. Fl. Ceylon 5: 305. 1985; Singh, Fl. E. Karnataka 2: 636. 1988; Karthik. *et al.*, Fl. Ind. Enum. Monocot. 52. 1989. *Scirpus ferrugineus* L., Sp. Pl. 50. 1753.

Type : Jamaica.

Illus. : Clarke, Illus. Cyp. t. 42, f. 9-10. 1909; Mahesh., Illus. Fl. Delhi f. 222. 1966.

Perennial herbs with short woody rhizome, 25-90 cm high. Stems tufted, rather stiff, glabrous, striate, 1.5-2 mm thick. Cauline leaves much shorter than stem, 7-35 cm long, 1-2 mm wide; basal sheaths coriaceous, glabrous, shining brown; ligule a dense fringe of short hairs. Inflorescence simple or often decompound, with few to many spikelets, 2-6 (-13) cm long. Involucral bracts usually 3, shorter than inflorescence; lowest stiff, dilated at base. Primary rays compressed, 1-2.5 (-8) cm long. Spikelets solitary, terete, ovoid, acute at apex, 4-14 x 3-5 mm, densely many- flowered, dull brown; rachilla winged. Glumes spiral, ovate or oblong-ovate, obtuse and apiculate at apex, *ca* 5 x 3 mm, slightly keeled, puberulous towards apex. Stamens 3; filaments *ca* 3 mm long; anthers linear-oblong, 1.5 - 2 mm long. Ovary oblong, *ca* 1 mm long, distinctly stipitate; style flat, *ca* 3 mm long, densely ciliate, faintly dilated at base; stigmas 2, *ca* 1.8 mm long. Nut biconvex, compressed, broadly obovate, obtuse-truncate at apex, *ca* 1.2 x 1 mm, smooth, shortly stipitate.

Fls. & Frts. : Throughout the year.

Chrom. No. : 2n = 10 (*Taxon* 21: 683. 1972).

Habitat : Saline areas near brackish waters and sea, in muddy fields, drying ponds, edges of tanks, dry areas with clayish soil; also near fresh water, among grasses and in running water.

Distrib. : Sri Lanka, China, Malesia, Japan, Polynesia; Pantropical. INDIA: Throughout (except in North-East). KARNATAKA: Bangalore (Ramaswamy & Razi *l.c.*), Belgaum, Chitradurga, Dakshina Kannada, Dharwar (*Sharma et al.,l.c.*) , Hassan, Mysore , Raichur, Uttara Kannada.

Specimens examined : Belgaum: Along Ghataprabha-Gokak Road, *Prasad* 172831 and 172837, 12.12.1994 (BSI). Chitradurga: Koradurga-Holalkeri Road, *Ramesh & Sreenath* 7244, 21.4.1979 (JCB). Dakshina Kannada: Mangalore, *Faulker* 4867, Jan. 1902 (MH); Ullal, Mangalore, *Bhat* 291, 29.8.1976 (MGH); Udyavara, *Bhat* 317, 24.10.1976 (MGH); Kulur-Mangalore, *Saldanha & Ravindra* 1327, 14.6.1978 (JCB). Babbukudra, *Raghavan* 156354, 6.6.1979 (BSI). Hassan: Near Addahalla, on Channarayapattana Road, *Hooper & Gandhi* 2399, 11.11.1971 (JCB). Mysore: Near Mandakalli tank, on way to Nanjangud, *Rao* 760, 12.7.1970 (JCB). Raichur: Gangavathi, *Manohar & Murthy* 8223, 19.6.1979 (JCB). Uttara Kannada: Kumta, *Chibber s.n.* (Acc. No. 2143), 1.11.1910 (BSI); Edges of Kalinadi, Karwar, *Bell s.n.* (Acc. No. 80301), Dec. 1920 (BLAT).

Note : This species seems to be adapted to halophytic conditions and normally found towards coastal areas and near brackish waters. In inland areas also it is found in saline soils. Similar observations were made also by Kern, *l.c.*

19. **Fimbristylis kingii** Clarke ex Boeck., Cyp. Nov. 2: 40. 1888 *et* in Hook.f., Fl. Brit. India 6: 633. 1893; Fischer in Gamble, Fl. Pres. Madras 1658. 1931 (3: 1150. 1957, repr.ed.); Ramaswamy & Razi, Fl. Bangalore 107. 1973; Sharma *et al.*, Fl. Karnataka 310. 1984; Karthik. *et al.*, Fl. Ind. Enum. Monocot. 53. 1989. **Fig. 27.**

Illus. : Clarke, Illus. Cyp. t. 40, f. 13-14.1909.

Perennial herbs with short rhizome. Stems densely tufted, slender, sulcate, 10-40 cm long, 0.5-1 mm thick. Leaves densely scabrous, much smaller than stem, 5-18 cm long, canaliculate; sheaths up to 4 cm long; ligule absent. Inflorescence a solitary, terminal spikelet. Bracts usually glume-like, oblong- ovate, *ca* 6 x 2 mm, strongly keeled, with *ca* 1 mm long mucro at apex; papery towards margins; often leafy, linear, 4-10 mm long. Spikelets ovoid to oblong-ovoid, obtuse to acute at apex, 6-12 x 4-6 mm, terete, purplish-brown, many-flowered. Glumes densely imbricate, broadly ovate, obtuse-acute at apex, *ca* 4 x 3 mm, slightly fimbriate towards apex, keeled. Stamens 3; filaments elongate up to 4 mm; anthers linear-oblong, minutely apiculate at apex, *ca* 2 mm long. Ovary oblong, *ca* 0.7 mm long; style *ca* 1 mm long, not dilated at base; stigmas 3, *ca* 1.5 mm long, densely ciliate. Nut trigonous, obovoid, obtuse at apex, cuneate towards base, *ca* 1.5 x 1 mm, greyish, sub-papillose, shortly-stipitate.

Fls. & Frts. : July-Dec.

Habitat : Hill stations, in exposed rocky areas in moist places.

Distrib. : A rare species endemic to South India. KARNATAKA: Bangalore, Shimoga.

Specimens examined : Bangalore: Nandi hills, *Hooper & Saldanha* 18095, 4.12.1971 (JCB). Shimoga: Malur-Kundadagurdda Road, *Narasimha Naik* 90431A, 3.9.1963 (BSI).

20. **Fimbristylis lawiana** (Boeck.) Kern in Reinwardtia 4: 96. 1956; Sharma *et al.*, Fl. Karnataka 310. 1984; Singh, Fl. E. Karnataka 2: 636. 1988; Karthik. *et al.*, Fl. Ind. Enum. Monocot. 53. 1989. *Scirpus lawiana* Boeck. in Linnaea 36: 479. 1869. *Fimbristylis digitata* Boeck. in Flora 61: 35. 1878; Clarke in Hook.f., Fl. Brit. India 6: 648. 1894; Cooke, Fl. Pres. Bombay 2: 884. 1908 (3: 398. 1958, repr.ed.). **Fig. 28.**

Small, erect perennials with bulbous rhizome and fibrous roots, 6-18 cm high. Rhizome clothed with remains of decayed leaf sheaths and fibrous roots. Stems very slender, angular-sulcate, usually triangular to quadrangular towards apex, striate, 0.3-1mm thick. Leaves radical, shorter than stem, flat, abruptly acuminate at apex, 2-13 cm long, 0.5 - 0.7 mm wide, usually scabrid on the margins towards base, smooth towards apex; sheaths 0.5 - 2.5 cm long, striate; ligule absent; leaves on the flowering stems reduced to bladeless sheaths. Inflorescence a head of 3-5 sessile spikelets, 5-10 mm long, 5-11 mm wide. Involucral bracts 3, glume-like,

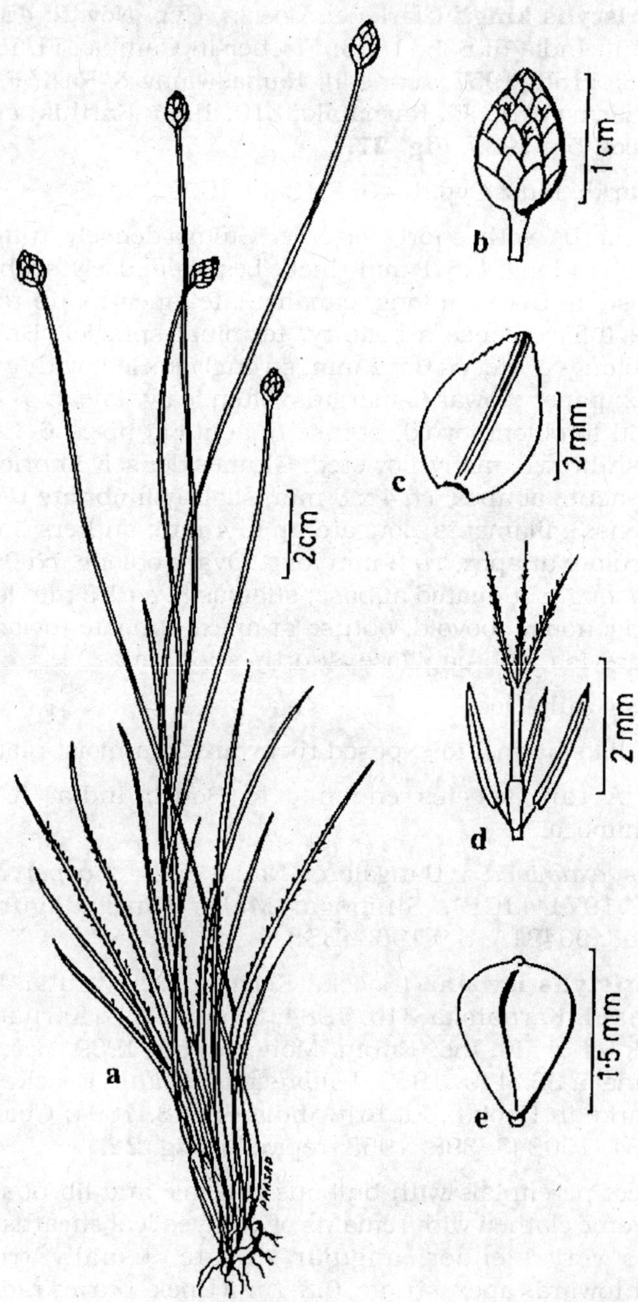

Fig. 27. *Fimbristylis kingii* Clarke ex Boeck.
a. Habit, **b.** Spikelet, **c.** Glume, **d.** Flower, **e.** Nut

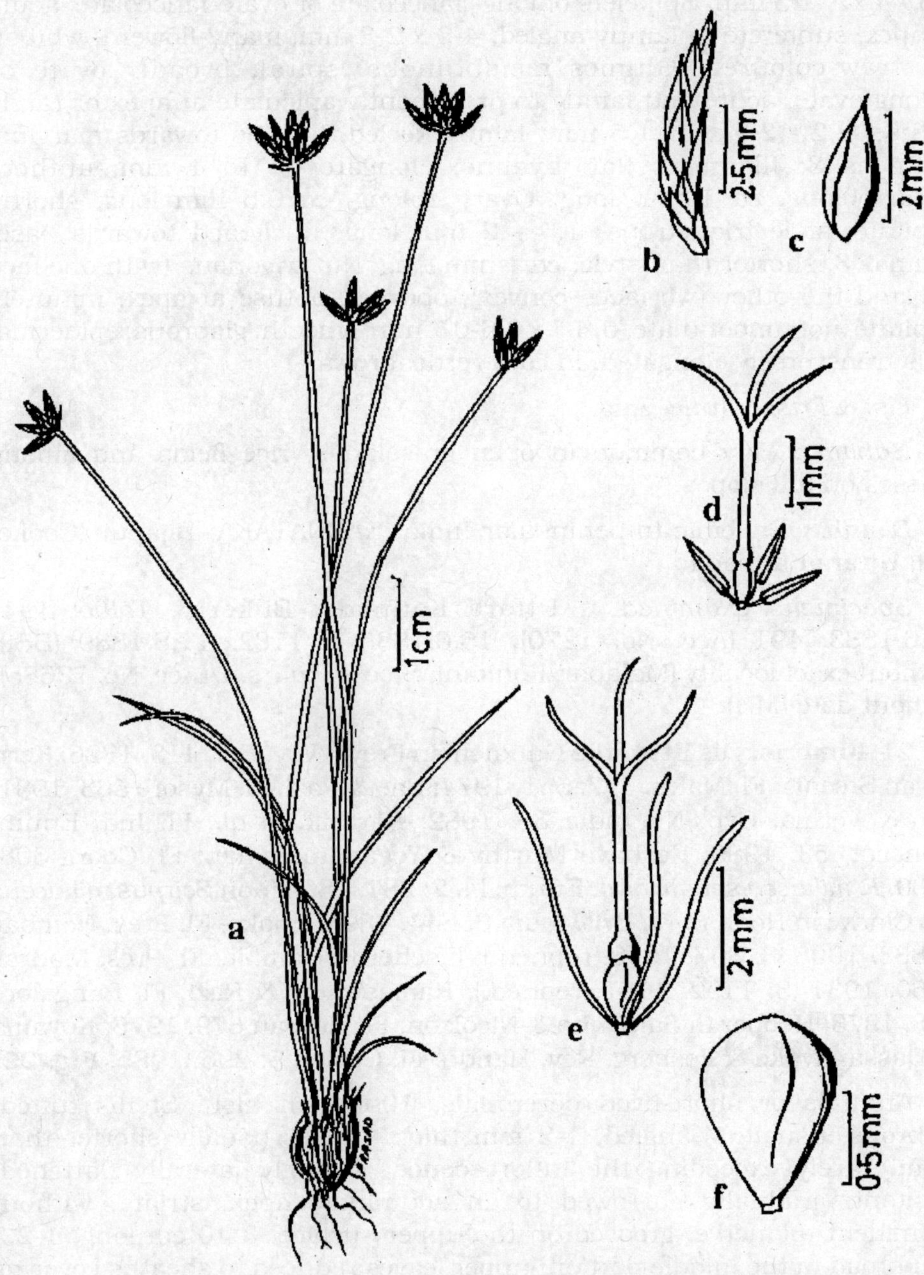

Fig. 28. *Fimbristylis lawiana* (Boeck.) Kern
a. Habit, **b**. Spikelet, **c**. Glume, **d**. Flower (early stage), **e**. Flower (later stage), **f**. Nut

much smaller than spikelets, ovate to ovate-lanceolate, acuminate at apex, 2.5 - 4 x 1-1.5 mm. Spikelets oblong- lanceolate or ovate-lanceolate, acute at apex, subterete or faintly angled, 4-9 x 2-3 mm, many-flowerd, whitish to straw-coloured. Glumes membranous, spiral, broadly ovate or oblong-ovate, acute and faintly to prominently apiculate at apex or rarely obtuse, 2.2 - 2.5 x *ca* 1.5 mm, faintly keeled, hyaline towards margins. Stamens 3; filaments flat, hyaline, elongate up to 4 mm; anthers linear-oblong, *ca* 1 mm long. Ovary oblong, *ca* 0.5 mm long, shortly stipitate; style triquetrous, 1.5 - 2 mm long, thickened towards base; stigmas 3, shorter than style, *ca* 1 mm long. Nut trigonous (with one face flat and the other two faces convex), obovoid, obtuse at apex, minutely stipitate, not umbonulate, 0.8-1 x 0.6-0.7 mm, smooth, glabrous; epidermal cells transversely elongated, in *ca* 8 vertical rows.

Fls. & Frts. : June- Aug.

Habitat : Very common in open grasslands, rice fields and among grasses on hill slopes.

Distrib. : Endemic to peninsular India. KARNATAKA: Bijapur (Cooke, *l.c.*), Uttara Kannada.

Specimens examined : Uttara Kannada: Bilkerry, *Talbot* 491, 10.6.1883; 491 (Acc. No. 1270), 15.6.1885 & 1182, 17.8.1889 (BSI). Without exact locality (Malabar, Konkan), *Stocks, Law s.n.* (Acc. No. 73685), without date (MH).

21. **Fimbristylis littoralis** Gaudich. in Freyc. Voy. Bot. 413. 1826; Kern in van Steenis, Fl. Males. 1, 7: 551. 1974; Rao & Razi, Fl. Mysore 563. 1981; Rao & Verma, Cyp. NE India 36. 1982; Karthik. *et al.*, Fl. Ind. Enum. Monocot. 53. 1989; Keshava Murthy & Yoganarasimhan, Fl. Coorg 508. 1990. *F. miliacea sensu* Vahl, Enum. Pl. 2: 287. 1806 (non *Scirpus miliaceus* L.); Clarke in Hook.f., Fl. Brit. India 6: 644. 1894; Cooke, Fl. Pres. Bombay 2: 883. 1908 (3: 397. 1958, repr.ed.); Fischer in Gamble, Fl. Pres. Madras 1660. 1931 (3: 1152. 1157, repr.ed.); Ramaswamy & Razi, Fl. Bangalore 108. 1973; Hooper in Saldanha & Nicolson, Fl. Hassan 679. 1976; Koyama in Dassanayake & Fosberg, Rev. Handb. Fl. Ceylon 5: 296. 1985. **Fig. 29.**

Annuals or short-lived perennials, 10-60 cm high. Stems tufted, glabrous, usually 4-angled, 1-2 mm thick. Leaves usually shorter than stem, rarely exceeding the inflorescence, strongly laterally flattened, equitant, gradually narrowed to an acuminate apex, striate, without prominent midnerve, grooved on the upper surface, 3-40 cm long, 1-2.5 mm broad in the middle portion; upper leaves reduced to sheaths covering basal portion of the stem, with a very short appendage-like blade, up to 15 cm long; ligule absent; orifice membranous, smooth, glabrous. Inflorescence compound or decompound, with many to numerous spikelets, 1.5 - 10 cm long, 1-9 cm wide. Involucral bracts 3-5, much shorter than

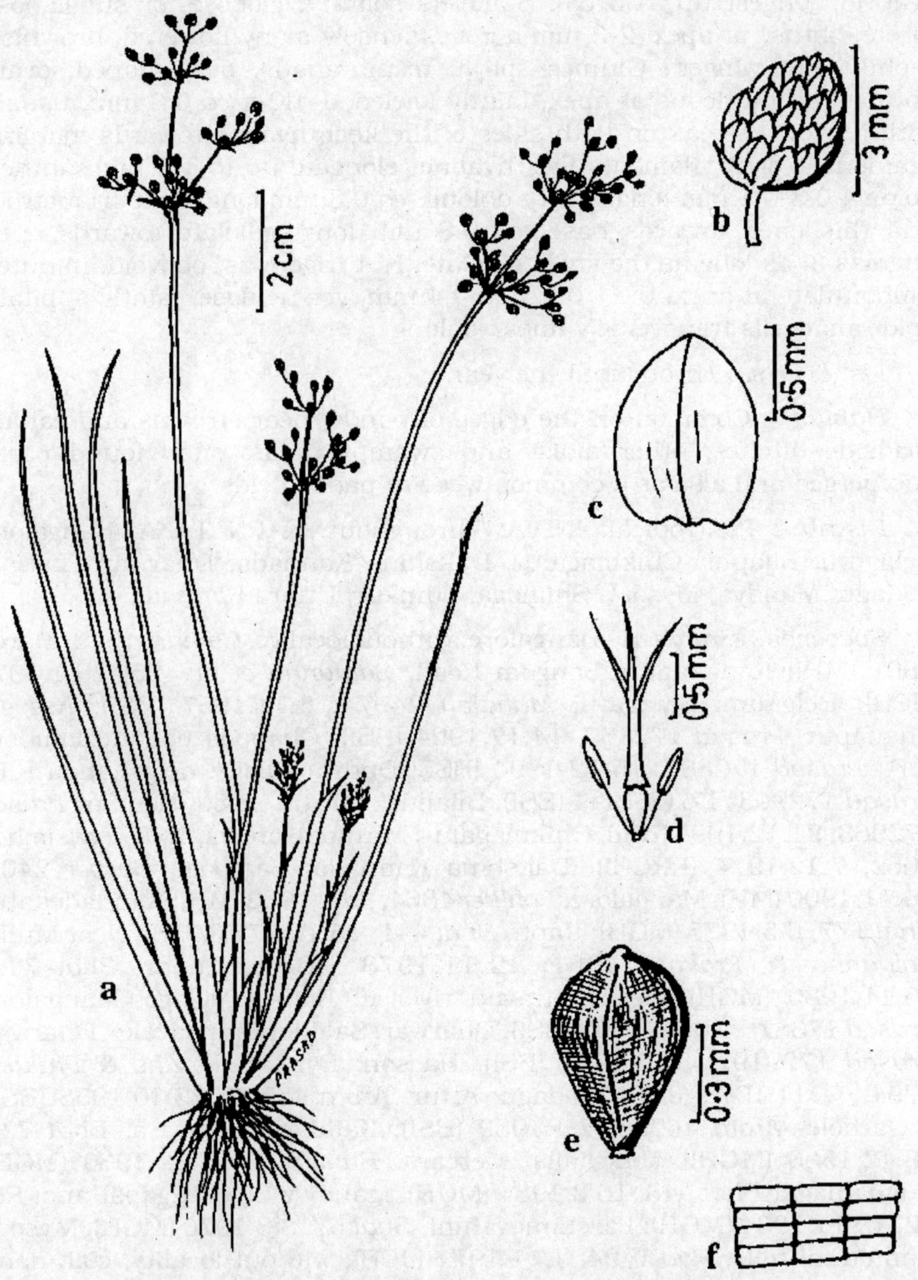

Fig. 29. *Fimbristylis littoralis* Gaudich.
a. Habit, **b**. Spikelet, **c.** Glume, **d.** Flower, **e.** Nut, **f.** Epidermal cells

inflorescence; lowest up to 3 cm long. Primary rays few to several angled, scabrid, longest up to 5 cm. Spikelets solitary, globose, or subglobose, terete, obtuse at apex, 2-3 mm across, densely many-flowered, brownish; rachilla not winged. Glumes spiral, membranous, boat-shaped, ovate, obtuse and muticous at apex, faintly keeled, 1-1.2 x *ca* 0.8 mm, usually with a brown streak on both sides of the keel, hyaline towards margins. Stamens 1 or 2; filaments flat, hyaline, elongate up to 1.1 mm; anthers oblong, 0.3-0.5 mm long. Ovary oblong, *ca* 0.3 mm long; style triquetrous and thickened towards base, *ca* 0.8 mm long, ciliolate towards apex; stigmas 3, as long as the style, ciliolate. Nut trigonous, obovoid, minutely umbonulate at apex, 0.6 - 0.7 x *ca* 0.4 mm, verruculose, faintly stipitate; epidermal cells transversely linear-oblong.

Fls. & Frts. : Throughout the year.

Habitat : Common on the edges of ponds, near streams and canals, roadside ditches, other moist and swampy areas; often found partly submerged and also as a common weed in paddy fields.

Distrib. : Pantropical. INDIA: Throughout. KARNATAKA: Bangalore, Belgaum, Bijapur, Chikmagalur, Dakshina Kannada, Dharwar, Hassan, Kodagu, Mandya, Mysore, Shimoga, Tumkur, Uttara Kannada.

Specimens examined : Bangalore: without locality, *Camaron* 579, March 1891 (MH); Kanakapura-Sangam Road, *Saldanha et al.* 25, 28.1.1978 (JCB). Belgaum: Satagatti, *Mahajan* 24974, 4.11.1957 (BSI); Assoga, Khanapur, *Prasad* 172865, 14.12.1994 (BSI); Chappoli nallah, Jambotti R.F., *Prasad* 172881, 15-12-1994 (BSI); Dudwa village, near Londa R.F., *Prasad* 172903, 17.12.1994 (BSI). Bijapur: Varathi Kavlu, Kendur, *Prasad* 172963, 21.12.1994 (BSI). Chikmagalur: Muthodi forests, *Yoganarasimhan* 0652, 7.12.1974 (RRCBI). Dakshina Kannada: Sampagi, *Barber* 2409, 15.11.1900 (MH); Mangalore, *Foulker* 4864, Jan. 1902 (MH); Kannadekatte, *Bhat* 177, 6.5.1975 (MGH); Kapu, *Bhat* 441, 15.1.1977 (MGH); Near Mulki; *Saldanha & Prakash* 4091, 12.11.1978 (JCB); Venoor, *Bhat* 714, 16.11.1980 (MGH); Near Netravati river in Kadkar village, Mangalore, *Prasad* 173876, 10.11.1995 (BSI). Dharwar: Salikkinikuppa lake, Dharwar, *Prasad* 172919, 18.12.1994 (BSI). Hassan: Byra, *Saldanha & Prakash* 3794, 11.11.1978 (JCB). Kodagu: Attur, *Arora* 45733, 22.10.1958 (BSI); Nagarhole, *Arora* 46296, 7.1.1959 (BSI); Katakare, Mercara, *Bhat* 726, 18.12,1980 (MGH); Abbe falls, Mercara, *Bhat* 873, 23.12.1980 (MGH); Kushalnagar *Bhat* 948, 16.2.1981 (MGH). Mandya: Sreerangapattana, *Rao* 32, 10.11.1964 (MGH); Paschimavahini, *Bhat* 57, 8.8.1970 (MGH). Mysore: Chamundi hills, *Rao* 1494, 1.11.1971 (JCB); without locality, coll. name and number (Acc. No. 73794), without date (MH). Shimoga: Chytrumane, Agumbe, *Raghavan* 81151, 14.5.1962 (BSI); Thirthanuthru Pond, near Agumbe, *Raghavan* 81019, 27.5.1962 (BSI); Varahi falls, Hulical, *Raghavan* 80801, 80802 and 80805, 20.5.1962 (BSI); Sagar-Jog Road, *Raghavan*

74084, 15.6.1961 (BSI). Tumkur: On way to Devarayanadurga, *Saldanha* 2173, 15.8.1978 (JCB). Uttara Kannada: Yellapore, *Talbot s.n.* (Acc. Nos. 763 & 764), 5.10.1884 (BSI); Karwar, *Talbot* 1304, 1885 (BSI); Castle Rock, *Gammie s.n.* (Acc. No. 2751), Oct. 1902 (BSI); Kanchikeri, *Fernandez* J.F. 1553, 27.5.1950 (BLAT); Gersoppa, *Fernandez* J.F. 5, 28.11.1950 (BLAT); Castle Rock, *Almeida* 2702, 14.4.1973 (BNHS); Along Kalinadi, Bagrigadde-Shivpur, *Gurudev Singh & Udayakumar* 14480, 21.5.1982 (JCB); Ullettikeri, Halyal, *Prasad* 172997 & 173000, 29.10.1995 (BSI).

Note: This species can be identified easily by the equitant, laterally flattened leaves and the small globose or subglobose spikelets. In many Floras this species is described as *F.miliacea* (L.) Vahl based on *Scirpus miliaceus* L. But this Linnean specimen was found to be *F.quinquangularis* (Vahl) Kunth (Clarke in *J. Linn. Soc. Bot.* 30: 312. 1894). Consequently S.T. Blake (in *J. Arn. Arb.* 35: 217. 1954) accepted *F. littoralis* Gaudich. as the correct name for this species and the same was followed by many subsequent authors including Kern, *l.c.* Same is followed in the present study also.

22. **Fimbristylis merrillii** Kern in Blumea 8: 135, f.6. 1955 *et* in van Steenis, Fl. Males. 1, 7: 579. 1974; Mistry & Almeida in Journ. Econ. Tax. Bot. 9: 403,f. A-G. 1987; Karthik. *et al.*, Fl. Ind. Enum. Monocot.I (Addenda & Corrigenda). 1989. *F. ligulata* Govind. in Proc. Ind. Acad. Sci.76: 187, t. 3. 1972. **Fig. 30.**

Type : Philippines, *Ramos Phil. Pl.* 1431-Holotype (N.Y.).

Illus. : Kern, *l.c.*; Mistry & Almeida *l.c.;* Govind., *l.c.*

Annuals with fibrous roots, 9-30 cm high. Stems tufted, narrow, compressed but angular, smooth, usually scabrous at top, 0.5 - 1 mm thick. Leaves mostly basal, shorter than stem, flat, linear, abruptly acuminate at apex, 3-20 cm long, 0.7 - 1.5 mm wide, usually scabrous on the margins; ligule a dense fringe of short hairs; sheaths striate, usually densely puberulous on the upper portion. Inflorescence simple to compound, loose, with 3- few spikelets, rarely reduced to a single spikelet, up to 3 cm long and wide. Involucral bracts 1-3, usually shorter than inflorescence, at times overtopping, 1-3.5 cm long, ciliate- puberulous on the dilated base. Primary rays up to 6, smooth, 0.5 - 2 cm long. Spikelets solitary, terete, ovoid or broadly oblong-ovoid, acute at apex, 3-4 x *ca* 2 mm, densely few to several-flowered; rachilla narrowly winged. Glumes spiral, subchartaceous, broadly ovate, acute at apex, slightly apiculate, *ca* 1.6 x 1.5 mm; keel not prominent, 3-nerved; sides without nerves; margins hardly hyaline. Stamens 1 or 2; filaments hyaline, elongate up to 2.5 mm; anthers oblong, *ca* 0.5 mm long. Style flat, slightly dilated at base, *ca* 1 mm long, glabrous or with a few cilia at top; stigmas 2, shorter than style. Nut biconvex, obovate, shortly stipitate, minutely umbonulate, *ca* 1 x 0.8 mm,

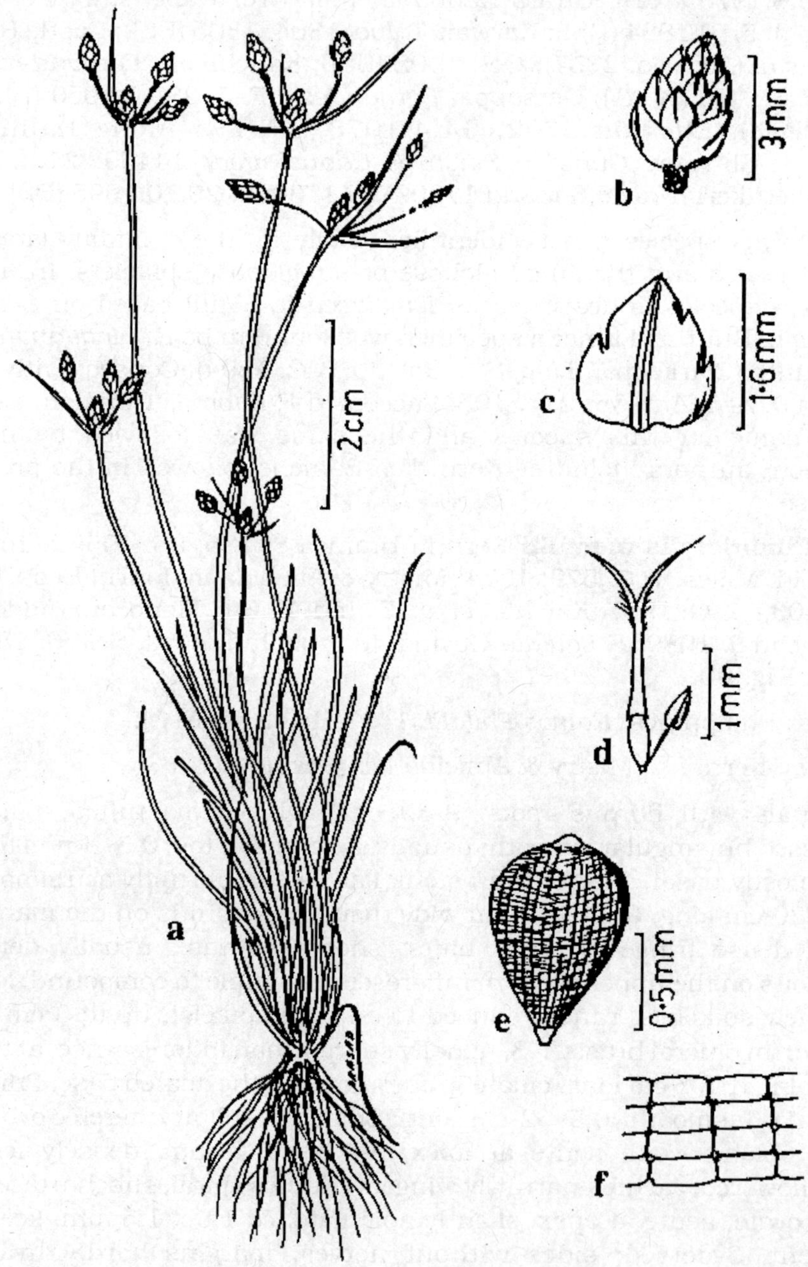

Fig. 30. *Fimbristylis merrillii* Kern
a. Habit, **b.** Spikelet, **c.** Glume, **d.** Flower, **e.** Nut, **f.** Epidermal cells

conspicuously trabeculate by the transversely oblong epidermal cells in (5-) *ca* 9 vertical rows on either face.

Fls. & Frts. : July- Nov.

Habitat : Paddy fields and wet grasslands, especially during rainy season.

Distrib. : South China, Thailand, Malesia and Queensland. INDIA: Western India. KARNATAKA: Bangalore, Dharwar (Mistry & Almeida *l.c.*), Hassan.

Specimens examined : Bangalore: Bannnergatta, *Hooper & Saldanha* 18049, 7.11.1971 (JCB). Hassan: Byra, *Hooper & Gandhi* 2438, 12.11.1971 (JCB).

23. **Fimbristylis microcarya** F.v. Muell., Fragm. Phyt. Austr. 1: 200. 1859; Kern in van Steenis, Fl. Males. 1, 7: 550. 1974; Karthik. *et al.*, Fl. Ind. Enum. Monocot. 53. 1989. *F. complanata* var. *microcarya* (misprinted as *microcarpa*) Clarke in Hook.f., Fl. Brit. India 6: 646. 1893; Cooke, Fl. Pres. Bombay 2: 886. 1908 (3: 400. 1958, repr.ed.). **Fig. 31.**

Glabrous, annuals with fibrous roots, without rhizome, 10-35 cm high. Stems highly tufted, slender, strongly compressed, ridged, usually flattened below inflorescence, 0.75-1.5 mm broad, usually glabrous, rarely scabrous on flattened edges near apex. Leaves mostly basal, shorter than stem, flat, abruptly acuminate at apex, 3.5 - 25 cm long, 1-3 mm wide, often minutely scabrous on margins near apex; sheaths flattened, equitant in early stage, 1.5 - 9 cm long; ligule a fringe of short hairs; orifice membranous, oblique, with few bristles at apex. Inflorescence decompound, open, 3-7 cm long, 1.5 - 6 cm wide, with many to numerous spikelets. Involucral bracts 3-4, shorter than to as long as the inflorescence, acuminate at apex, scabrous on margins near apex; lowest 1-5 cm long. Primary rays few-many, very unequal, compressed, up to 4.5 cm long, scabrous. Spikelets solitary, ovoid or oblong-lanceolate, acute at apex, angular, 3-4 x *ca* 1 mm, few- flowered; rachilla winged. Glumes spiral, membranous, ovate, acute-mucronate at apex, keeled, *ca* 1.25 x 1 mm. Stamen 1; filament flat, hyaline, elongate up to 1.25 mm; anther oblong, *ca* 0.3 mm long. Ovary oblong, *ca* 0.3 mm long, shortly stipitate; style triquetrous, pyramidally thickened at base, 0.6 - 0.7 mm long, glabrous; stigmas 3, slightly shorter than style. Nut trigonous, obovoid, minutely umbonulate, shortly stipitate, 0.6- 0.7 x 0.4-0.5 mm, smooth or sparsely verruculose; epidermal cells transversely linear-oblong, in *ca* 6 vertical rows.

Fls. & Frts. : Dec.

Habitat: Margins of lakes, wet soil in crevices of rocks along canals.

Distrib. : Widely distributed in Eastern Asia, Queensland and in North Australia. INDIA: Peninsular India and Himalaya. KARNATAKA: Belgaum.

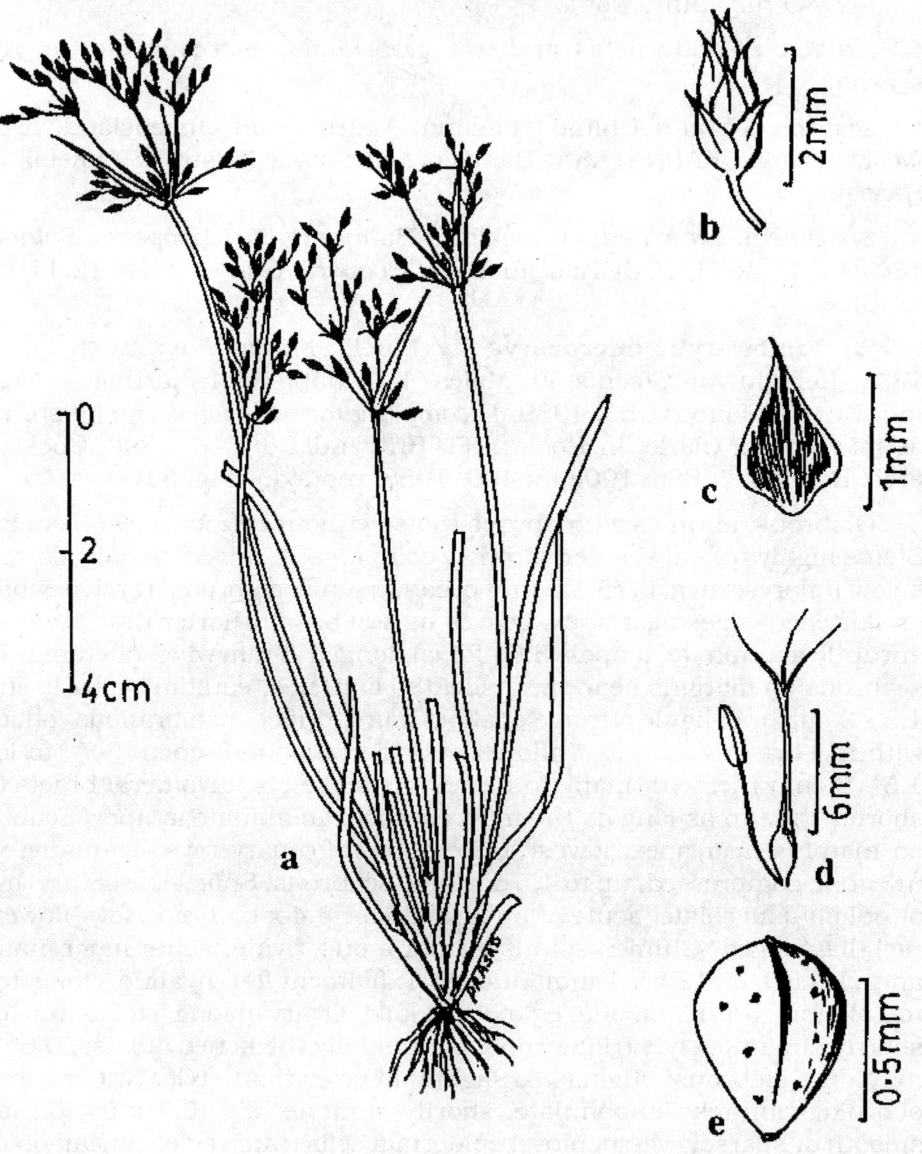

Fig. 31. *Fimbristylis microcarya* F.v. Muell.
a. Habit, **b**. Spikelet, **c.** Glume, **d.** Flower, **e.** Nut

Specimens examined : Belgaum: Dinman Hassur, way to Belgaum from Gokak, *Prasad* 172844, 13.12.1994 (BSI).

Note: This species was reported for the first time from Karnataka as a result of the present study (*Prasad & Singh*, 1996a).

24. **Fimbristylis miliacea** (L.) Vahl, Enum. Pl. 2: 287. 1806; Kern in van Steenis, Fl. Males. 1, 7: 552. 1974; Rao & Verma, Cyp. NE India 36. 1982; Sharma *et al.*, Fl. Karnataka 310. 1984; Karthik. *et al.*, Fl. Ind. Enum. Monocot. 53. 1989. *Scirpus miliaceus* L., Syst. Veg. ed. 10: 868. 1759. *S. quinquangularis* Vahl, Enum. Pl. 2: 279. 1806. *Fimbristylis quinquangularis* (Vahl) Kunth, Enum. Pl. 2: 229. 1807; Clarke in Hook. f, Fl. Brit. India 6: 644. 1894; Cooke, Fl. Pres. Bombay 2: 883. 1908 (3:397. 1958, repr.ed.); Fischer in Gamble, Fl. Pres. Madras 1659. 1931 (3:1151. 1957, repr.ed.); Ramaswamy & Razi, Fl. Bangalore 109. 1973; Koyama in Dassanayake & Fosberg, Rev. Handb. Fl. Ceylon 5: 295. 1985.

Illus.: Clarke, Illus. Cyp. t. 43, f. 4-6.1909.

Glabrous annuals with fibrous roots, 8-55 cm high. stems densely tufted, erect, 4-5-angled, striate, smooth, 0.8 - 1.5 mm thick. Leaves usually basal, smaller than to as long as the stem, often overtopping the inflorescence, flat, gradually tapering to an acute apex, often very minutely serrulate-scabrous on the margins and on the veins on the upper surface, 1-2.5 mm wide in the middle portion; sheaths papery, 1-9 cm long; ligule absent; orifice membranous, glabrous. Upper leaves reduced to sheaths or with very short blades. Inflorescence compound to decompound, with many to numerous spikelets, 2-8 x 2-7 cm. Involucral bracts 3-6, much shorter than inflorescence, lowest 0.6 - 3.5 cm long. Primary rays few to many, compressed, sulcate, up to 5 cm long, usually scabrid. Spikelets solitary, ovoid, acute at apex, angled, 2-4 x 1.5-2 mm, brownish, many- flowered; rachilla prominently winged. Glumes membranous, spiral, broadly ovate, obtuse and apiculate at apex, keeled, *ca* 1.7 x 1.7 mm, with a brown streak on both sides of the keel, hyaline towards margins. Stamen 1; filament flat, hyaline, elongate up to 1.5 mm; anther oblong, acute at apex, *ca* 0.5 mm long. Ovary oblong, *ca* 0.4 mm long; style trigonous, pyramidally thickened at base, 0.5 - 0.7 mm long, minutely ciliate in the upper half; stigmas 3, 0.7 - 1 mm long, minutely ciliate. Nut obtusely trigonous, broadly obovoid to subglobose, unbonulate, 0.7-0.8 x 0.4-0.5 mm, verruculose, minutely stipitate; epidermal cells transversely linear-oblong, in *ca* 6 vertical rows on each face.

Fls. & Frts. : Nov.-Jan.

Chrom. No. : 2n = 10 (*Taxon* 21: 683. 1972).

Habitat : Paddy fields, shallow ponds, swampy grasslands, margins of streams and tanks, wet roadsides and other marshy areas.

Distrib. : Sri Lanka, South China, Formosa, Malesia, tropical Africa, tropical Australia. INDIA: Throughout. KARNATAKA: Bangalore (Ramaswamy & Razi; *l.c.*), Chickmagalur (Rao *et al.*, 2000), Kodagu, Mysore, Shimoga, Tumkur (Gowda *et al.*, 1986), Uttara Kannada.

Specimens examined : Kodagu: Kalbatta, *Arora* 46139, 4.1.1959 (BSI). Mysore: without locality, coll. name and number (Acc. No. 73766), without date (MH). Shimoga: Nalur, Agumbe, *Raghavan* 68039A, 1.11.1960 (BSI). Uttara Kannada: Ulletikeri, Halyal, *Prasad* 172997 A, 29.10.1995 (BSI).

Note: Plant reported from Hassan district (Fl. Hassan 679) as *F. miliacea* is actually *F. littoralis* Gaudich.

25. **Fimbristylis monospicula** Govind. in Proc. Ind. Acad. Sci. 79B: 169, f. 4. 1974; Sharma *et al.*, Fl. Karnataka 311. 1984; Bhargavan in Henry *et al.*, Fl. Tamil Nadu 1,3: 79. 1989; Karthik. *et al.*, Fl. Ind. Enum. Monocot. 53. 1989.

Illus. : Govind., *l.c.*

Type : Karnataka state, Biligirirangans, *Barnes* 628 (PCM).

Annuals, hirsutely hairy, dirty brown. Stems few, erect, tufted, not thickened at base, 5-7-ribbed, sulcate, rigid, 10-35 cm long, 0.2 - 0.3 (-0.4) mm thick. Leaves many 1/3 to as long as the stem, erect or flexuous, flat, acuminate, 10-20 cm long, 0.5 -0.75 mm wide; margins thickened, hairy; sheaths obliquely truncate, 9-11-ribbed, glabrous, chestnut brown. Spikelets solitary or 2-3, terminal, ovoid or elliptic- ovoid, obtuse at apex, 3-6 x 3-4 mm, terete, bracteate, castaneous brown, many-flowered. Glumes densely imbricate, crustaceous, oblong, ovate, or suborbicular, obtuse at apex, 1.75-2 x 1.5-1.8 mm, shining glabrous, castaneous- fuscous brown, not distinctly keeled, with nerveless sides and narrow scarious margins; nerves on the keel white, evanescent. Rachilla shortly winged or wingless, excavated. Stamens 3; anthers linear, apiculate, 0.6 - 0.8 mm long. Style triquetrous, with slightly dilated pyramidal base, 0.7-0.8 mm long, densely fimbriate; stigmas 3, hairy (ultimately papillate), as long as the style. Nut broadly obovoid, with convex sides, obtuse at apex, 0.8 - 0.9 x 0.6-0.7 mm (incl. stipe), distinctly tricostulate, densely tubercled, stipitate, brown; epidermal cells in upper half hexagonal, transversely elongated, much impressed (trabeculate), in 7-8 regular vertical rows on each face.

Fls. & Frts. : Not reported.

Habitat : An alpine species found in high altitude.

Distrib. : Endemic to South Indian hill stations. (Karnataka and Tamilnadu). KARNATAKA: Mysore (Balagiri ranges, Govind., *l.c.*).

Note: There are no specimens of this species in BSI, but is included here on the authority of Govindarajalu, *l.c.* According to him this species differs from *F.kingii*, another high altitude species endemic to South Indian hill stations, in its annual habit, stems without thickened base, flat, acuminate and narrower leaves, smaller ovoid or elliptic-ovoid spikelets which are solitary or 2-3 per stem, smaller glumes, usually with 3- nerved keel, shorter anthers, shorter style with stigmas of same length, smaller, densely tubereled and distinctly tricostulate brown nuts with 7-8 regular vertical rows of transversely elongated, much impressed hexagonal epidermal cells on each face.

26. **Fimbristylis narayanii** Fischer in Kew Bull. 1831: 46. 1831 *et* in Gamble, Fl. Pres. Madras 1660. 1931 (3: 1152. 1957, repr.ed.); Bhargavan in Henry *et al.*, Fl. Tamil Nadu 1, 3: 79. 1989; Karthik. *et al.*, Fl. Ind. Enum. Monocot. 53. 1989. **Fig. 32.**

Glabrous annuals, *ca* 16 cm high. Stems densely tufted, slender, 4-angled, 0.3-0.5 mm thick. Leaves many, shorter than stems, linear, abruptly acuminate at apex, 3-9 cm long, 1-1.5 mm wide, usually sparsely scabrous on the margins near apex; sheaths striate, up to 1.5 cm long; ligule absent. Inflorescence simple, with 3-4 loosely arranged spikelets, 1-2 cm long, 1-2.5 cm broad. Involucral bracts 2-3, very short, inconspicuous; longest 4-8 mm long, scabrous on the margins near apex. Rays (peduncles) 0.5-1.2 cm long, slightly compressed. Spikelets solitary, oblong- lanceolate, acute at apex, flat, *ca* 6 x 2 mm, 10-12-flowered; rachilla winged. Glumes distichous, ovate-lanceolate or oblong-lanceolate, mucronate at apex, *ca* 2.5 x 1.5 mm, keeled, scarious towards margins. Stamens 3; filaments hyaline, elongate up to 2.5 mm; anthers linear-oblong, apiculate at apex, *ca* 0.5 mm long. Ovary oblong, *ca* 0.4 mm long; style *ca* 2 mm long, trigonous and pyramidally thickened towards base, glabrous; stigmas 3, much shorter than style (*ca* 0.5 mm). Nut trigonous, obovoid, umbonulate at apex, shortly stipitate, *ca* 0.9 x 0.5 mm, sparsely and minutely tubercled, whitish; epidermal cells broad, isodiametric, in 7-10 vertical rows.

Fls. & Frts. : Sept.-Nov.

Habitat : Moist rocky slopes.

Distrib. : South India and North-West Himalaya. KARNATAKA: Dakshina Kannada (not common).

Specimens examined : Dakshina Kannada: Belthangady, *Bhat* 76, 20.9.1970 (BSI); Shiradi, *Raghavan* 145828, 6.10.1976 (BSI); Kannadekatte, *Bhat* 387, 9.11.1976 (MGH).

Note : This species was reported for the first time from the state as a result of the present study (*Prasad & Singh,* 1996a).

27. **Fimbristylis ovata** (Burm.f.) Kern in Blumea 15: 126. 1967; Ramaswamy & Razi, Fl. Bangalore 108. 1973; Kern in van Steenis, Fl. Males. 1, 7: 565. 1974; Hooper in Saldanha & Nicolson, Fl. Hassan 679. 1976; Rao & Razi, Fl. Mysore 563. 1981; Arora *et al.*, Bot. S. Kanara 61. 1981; Rao & Verma, Cyp. NE India 39. 1982; Sharma *et al.*, Fl. Karnataka 310. 1984; Koyama in Dassanayake & Fosberg, Rev. Handb. Fl. Ceylon 5:

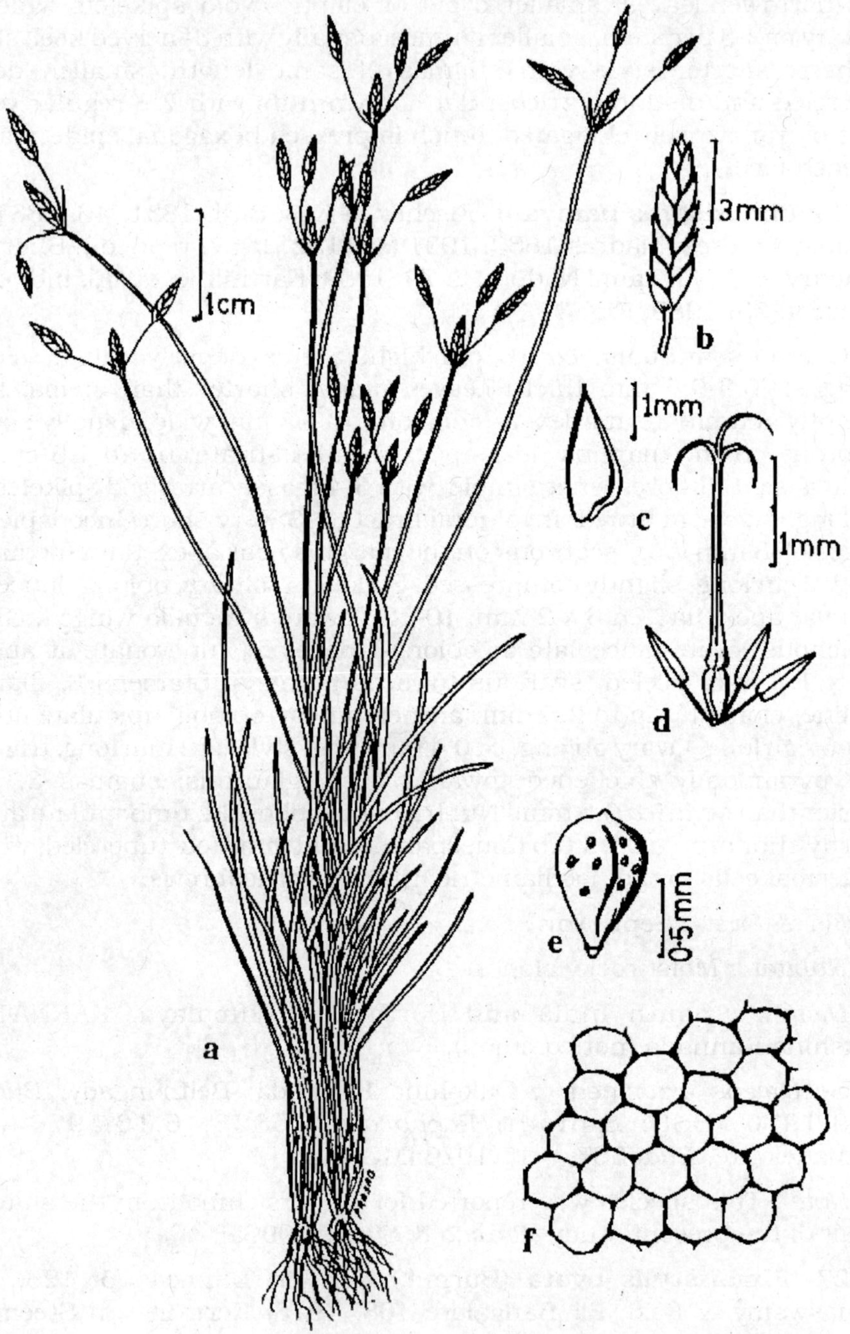

Fig. 32. *Fimbristylis narayanii* Fischer
a. Habit, **b**. Spikelet, **c.** Glume, **d**. Flower, **e**. Nut, **f.** Epidermal cells

273. 1985; Singh, Fl. E. Karnataka 2: 637. 1988; Karthik. *et al.*, Fl. Ind. Enum. Monocot. 54. 1989; Keshava Murthy & Yoganarasimhan, Fl. Coorg 508. 1990. *Carex ovata* Burm. f., Fl. Ind. 194. 1768. *Cyperus monostachyos* L., Mant. 2: 180. 1771. *Fimbristylis monostachyos* Hassk., Pl. Jav. Rar. 61. 1848 (*monostachya*) ; Clarke in Hook.f., Fl. Brit. India 6: 649. 1893; Cooke, Fl. Pres. Bombay 2: 885. 1908 (3:399. 1958, repr.ed.); Fischer in Gamble, Fl. Pres. Madras 1660. 1931 (3: 1152. 1957, repr.ed.). **Fig. 33.**

Type : Java.

Illus. : Mahesh., Illus. Fl. Delhi f. 221. 1966.

Perennials with short rhizome 6-40 cm high. Stems tufted, slender, compressed, usually trigonous, often flattened towards apex and slightly swollen towards base, sulcate, 0.4-1 mm thick, often minutely scabrous on the edges. Leaves very narrow, shorter than stem, flat or canaliculate with incurved margins, abruptly acuminate, 3-20 cm long, 0.6-1 mm wide, often scabrous on the slightly thickened margins; sheaths up to 4.5 cm long; ligule absent; orifice membranous, glabrous. Inflorescence usually a single terminal spikelet, often proliferate into 1 or 2 additional peduncled spikelets. Bracts 2, very small, scale-like, with the mid-nerve elongated into an awn, 4-8 mm long, deciduous. Peduncle of the proliferated spikelets 0.5-1.5 cm long. Spikelets ovate, strongly compressed, acute at apex, 6-11 x 3-5 mm, greenish; rachilla prominently winged. Glumes distichous, subspiral in later stage, coriaceous, broadly ovate, acute and mucronate at apex, strongly keeled, *ca* 5 x 4-5 mm. Stamens 3; filaments flat, hyaline, elongate up to 5 mm; anthers linear oblong, apiculate, *ca* 2 mm long. Ovary oblong, *ca* 1 mm long; style triquetrous, *ca* 3 mm long, pyramidally thickened at base, ciliate throughtout; stigmas 3, shorter than style, 1-1.5 mm long. Nut obtusely trigonous, pyriform, almost truncate with umbonate apex, *ca* 2.5 x 1.5 mm, muricate except on the narrow base, stramineous to greyish; epidermal cells isodiametric, in many vertical rows.

Fls. & Frts.: March-Dec.

Chrom. No. : 2n = 10 (*Taxon* 21: 683. 1972).

Habitat: Sunny or partly shaded grasslands, stony ground, shady places along roadsides, over grazed grasslands, grassy patches on rocky sea coasts, wet grasslands, open moist places, bunds of cultivated fields, near streams, etc.

Distrib. : Pantropical. INDIA: Throughout the warmer parts. KARNATAKA: Bangalore, Bidar, Bijapur, Dakshina Kannada (Arora *et al.*, *l.c.*), Dharwar, Hassan, Kodagu, Kolar, Mandya, Mysore, Raichur (Sharma *et al.*, *l.c.*) , Shimoga, Tumkur, Uttara Kannada.

Specimens examined : Bangalore: Without locality and coll. name, R.D.A., B-258, 1917 (MH); Bannerghatta National Park, *Saldanha* 2145, 15.8.1978 (JCB). Bidar: Birnali R.F., Humnabad, *Singh* 142818, 10.8.1976

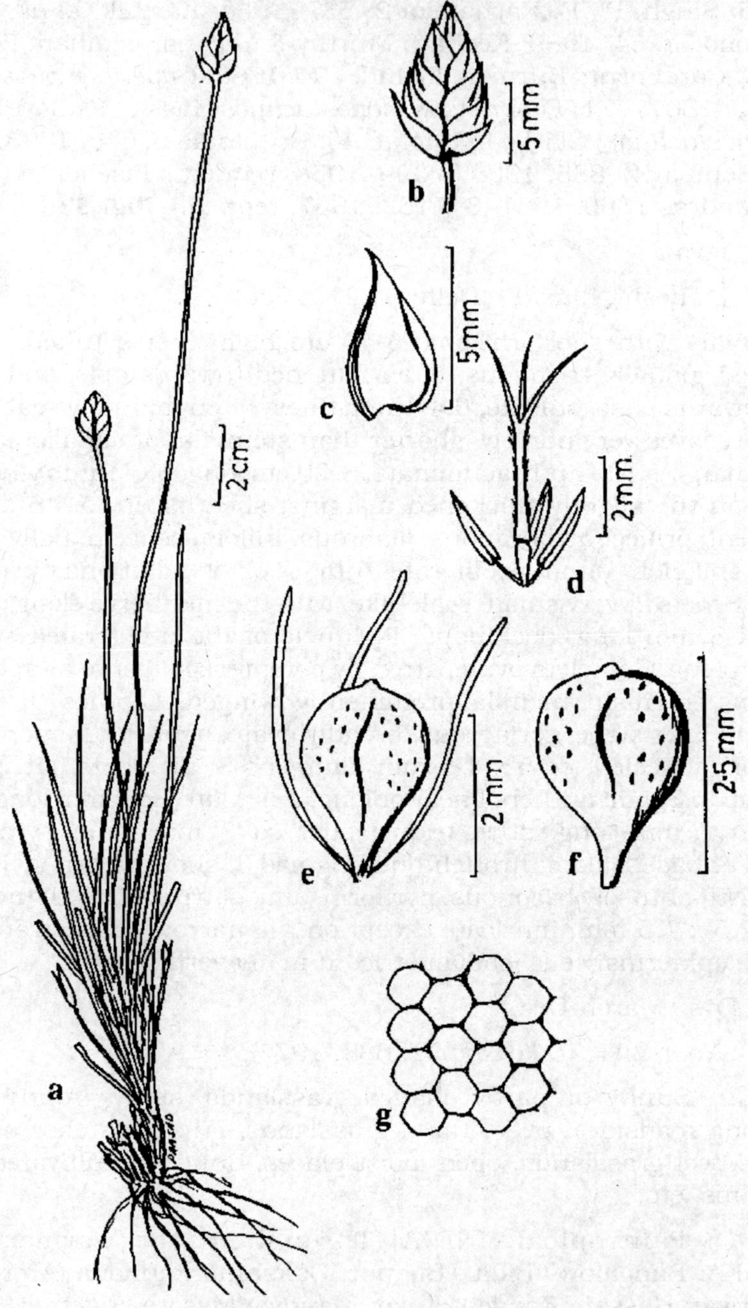

Fig. 33. *Fimbristylis ovata* (Burm.f.) Kern
a. Habit, **b.** Spikelet, **c.** Glume, **d.** Flower, **e.** Nut with filaments, **f.** Nut, **g.** Epidermal cells

(BSI). Bijapur: Badami, without coll. name, *s.n.* (Acc. No. 2704), 27.6.1893 (BSI); Adgal-Bagalkot Road, *Ramesh & Ramesh* 9340, 16.9.1979 (JCB). Dharwar: Dharwar, *Talbot* 2502. 10.7.1890 (BSI). Hassan: Nagpuri Road, *Saldanha* 13844, 20.6.1969 (JCB). Kodagu: Kushalnagar, *Bhat* 1038, 21.9.1981 (MGH). Kolar: Chickballapur - Bangalore Road 10th km, *Singh* 143200, 22.8.1976 (BSI); Maderahalli, *Prakash & Sreenath* 2803, 21.9.1978 (JCB). Mandya: Malavalli-Kollegal Road, *Saldanha & Ramesh* 2406, 5.9.1978 (JCB). Mysore: Bandipur, *Naithani* 21132, 24.8.1964 (MH); Near St. Philomina College, *Bhat* 30, 20.7.1970 (MGH); Chamundi hills, *Rao* 1289, 22.5.1971 (MGH); M.M. hills, *Rao* 2053, 29.6.1972 (MGH). Shimoga: Agumbe, *Raghavan* 83331, 18.10.1962 (BSI). Tumkur: Manchaldore R.F., *Singh* 143245, 23.8.1976 (BSI); Hills on way to Devarayanadurga, *Saldanha* 2185, 15.8.1978 (JCB). Uttara Kannada: Yellapur, *Talbot* 560, 1880 (BSI).

28. **Fimbristylis polytrichoides** (Retz.) R. Br., Prodr. 226. 1810; Clarke in Hook.f., Fl. Brit. India 6: 632. 1893; Cooke, Fl. Pres. Bombay 2: 879. 1908 (3: 393. 1958, repr.ed.); Fischer in Gamble, Fl. Pres. Madras 1658. 1931 (3: 1150. 1957, repr.ed.); Kern in van Steenis, Fl. Males. 1, 7: 586. 1974; Sharma *et al.*, Fl. Karnataka 310. 1984; Koyama in Dassanayake & Fosberg, Rev. Handb. Fl. Ceylon 5: 318. 1985; Karthik. *et al.*, Fl. Ind. Enum. Monocot. 54. 1989. *Scirpus polytrichoides* Retz., Obs. Bot. 4, 11. 1789. **Fig. 34.**

Type: Sri Lanka, *Koenig.*

Illus. : Clarke, Illus. Cyp. t. 40, f. 8-9. 1909.

Perennials with fibrous roots; rhizome when present very short, clothed with brown scales. Stems densely tufted, slender, erect, glabrous, subterete to trigonous, sulcate, 3-35 cm long, up to 1 mm thick. Leaves much smaller than stem, 2-15 cm long, canaliculate; sheaths 0.5-5.5 cm long; ligule a row of short hairs. Inflorescence a solitary terminal spikelet. Bracts usually glume-like, papery, oblong-lanceolate, *ca* 3 x 1 mm, mucronate, with 3 very prominent midveins; often leafy, slender, 0.5 - 1.8 cm long. Spikelets terete, ovoid to elliptic, acute to obtuse at apex, 5-9 x 2-4 mm, densely many-flowered, pale brown; rachilla narrowly winged. Glumes membranous, densely imbricate, boat-shaped, oblong-ovate, acute to obtuse at apex, 2.5 - 3 x 1- 1.25 mm, minutely mucronate at apex, faintly keeled. Stamen 1 or 2; filaments 0.5 - 2 mm, unequal when 2; anthers linear, oblong- lanceolate, *ca* 1 mm long, minutely apiculate at apex. Ovary oblong to obovate, *ca* 0.6 mm long; style *ca* 1 mm long, not prominently dilated at base; stigmas 2, rarely 3, *ca* 1 mm long, tapering towards apex, minutely ciliate. Nut obovate, biconvex, rounded at apex, cuneate towards base, *ca* 1 x 0.8 mm, greyish to black, verruculose, minutely stipitate.

Fls. & Frts. : Aug.- Dec.

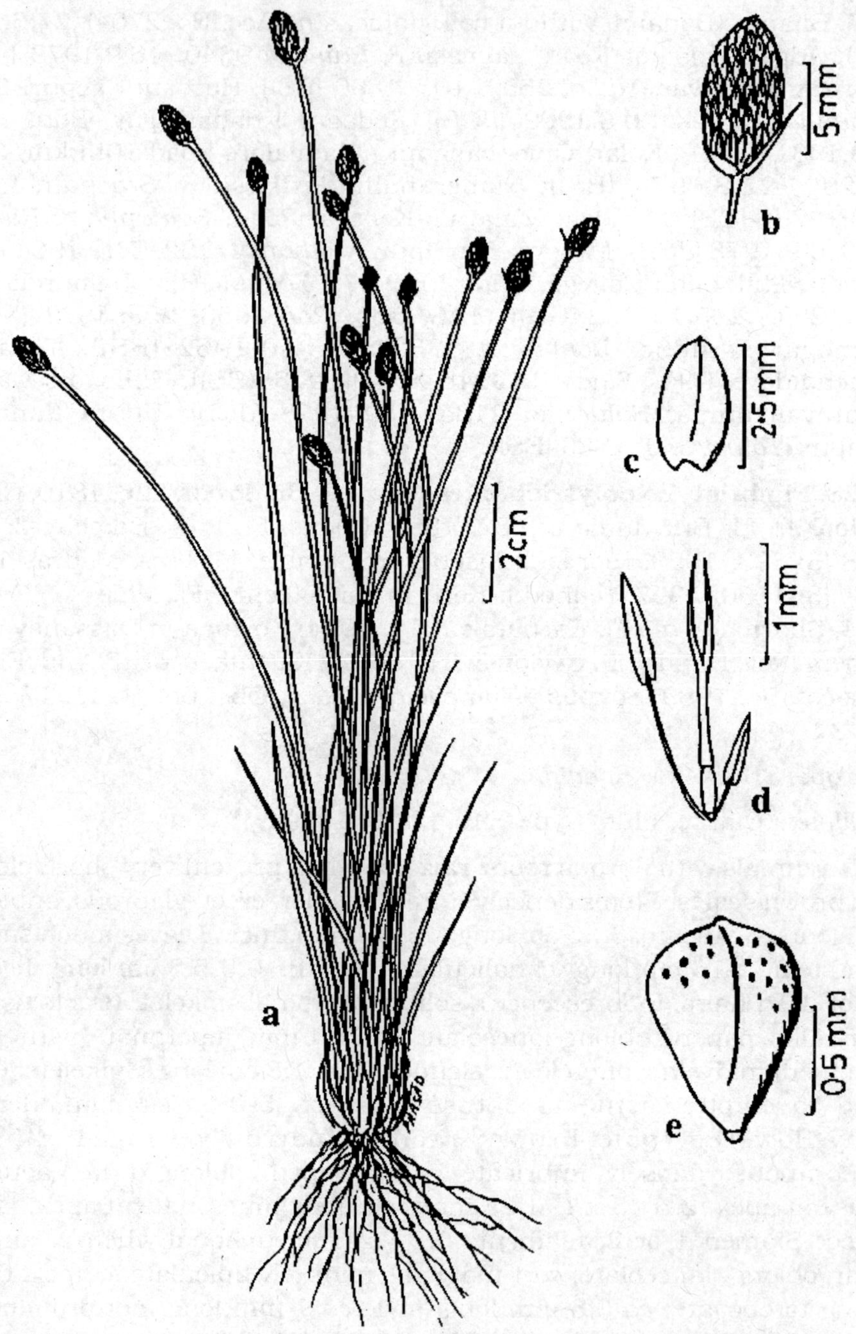

Fig. 34. *Fimbristylis polytrichoides* (Retz.) R. Br.
a. Habit, **b.** Spikelet, **c**. Glume, **d**. Flower, **e.** Nut

Habitat : Wet saline areas near sea shore and crevices of rocks near sea.

Distrib. : Tropical Asia, Africa and Australia. INDIA: Western Peninsula, Central and Eastern India. KARNATAKA: Dakshina Kannada, Uttara Kannada. Not common. Confined to coastal areas.

Specimens examined: Dakshina Kannada: Byndoor (Beach), *Raju* 500, 26.12.1938 (MH); St. Mary's Island, opposite Malpe, *Raghavan* 146139, 18.10.1976 (BSI); Badergudda, Malpe, *Bhat* 440, 18.12.1976 (MGH). Uttara Kannada: Karwar, *Talbot s.n.* (Acc. No. 740), 26.8.1885 (BSI).

29. **Fimbristylis pseudomicrocarya** Govind. in Proc. Ind. Acad. Sci. 100: 77, f. 1. 1990.

Type : India, Karnataka state, Dakshina Kannada (South Kanara) dt., Someshwar, Bachapu, *Govindarajalu* 13141-Holotype (CAL).

Illus.: Govind., *l.c.*

Annuals. Stems few to many, caespitose, filiform, 4-5 angled, rigid, erect, glabrous, ribbed, sulcate, leafy at base, (4-) 5-6 (-7) cm long, 0.4 - 0.5 mm thick. Leaves few to many, filiform, flat, acute at apex, 2-4 cm long, 0.3 - 0.5 (-0.8) mm wide, with smooth margins, ligulate; uppermost sheaths leaf bearing, obliquly erect; ciliate at mouth; 2-3 basal sheaths bladeless, glabrous. Inflorescence simple, obliquely erect, contracted, usually with 3-9 (-15) spikelets, 6-10 cm long. Involucral bracts somewhat leafy, shorter than to as long as the inflorescence, stiff, erect, ovate-lanceolate, 4-10 mm long, with smooth margins, 3-5-nerved. Spikelets usually paired or in threes, sessile, erect, linear-oblong, obtuse or subacute at apex, angular, very small, 3-4 x 0.8-1 mm, cinnamomeous brown, many-flowered. Glumes distichous in lower half, spiral in upper half, chartaceous, deltoid-ovate, subacute at apex, mucronate, 1.2 - 1.3 x *ca* 1 mm, keeled, shining glabrous throughout, nerveless in each half, with distinctly scarious margins, somewhat inflated by the nut; mucro erect or recurved, 0.1-0.2 mm long; keel 3-nerved; nerves excurrent into mucro. Rachilla winged, excavated. Stamen 1; anther linear-oblong, obtuse at both ends, minute, 0.2 - 0.3 mm long, spurred at base. Style triquetrous, slightly pyramidal at base, 0.4 - 0.5 mm long, glabrous; stigmas 3, 0.2-0.3 mm long, glabrous. Nut trigonous, tricostulate with convex sides, obovoid, umbonulate, minutely stipitate, 0.5-0.6 x 0.3-0.4 mm, usually crystalline, smooth; epidermal cells distinct in the upper half, transversely elongated-hexagonal, in 4- 5 regular rows on each face giving transversely lineolate appearance.

Fls. & Frts. : Dec.

Habitat : Common in grasslands.

Distrib. : Endemic to South Western peninsular India (Karnataka and

Kerala). KARNATAKA: Dakshina Kannada (Someshwar and Kudremukh, Govindarajalu, *l.c.*).

Note : Specimens of this species were not seen; but included here on the authority of Govindarajalu,*l.c.* and the description provided here is as given in the protologue. According to him it closely resembles *F. microcarya* F.v. Muell. and can be easily mistaken for the latter.

30. **Fimbristylis schoenoides** (Retz.) Vahl, Enum. Pl. 2: 286. 1806; Clarke in Hook.f., Fl. Brit. India 6: 634. 1893; Cooke, Fl. Pres. Bombay 2: 880. 1908 (3: 394. 1958, repr.ed.); Fischer in Gamble, Fl. Prs. Madras 1658. 1931 (3: 1150. 1957, repr.ed.); Ramaswamy & Razi, Fl. Bangalore 109. 1973; Kern in van Steenis, Fl. Males, 1, 7: 573. 1974; Hooper in Saldanha & Nicolson, Fl. Hassan 680. 1976; Rao & Razi, Fl. Mysore 563, 1981; Rao & Verma, Cyp. NE India 30. 1982; Sharma *et al.*, Fl. Karnataka 310. 1984; Koyama in Dassanayake & Fosberg, Rev. Handb. Fl.Ceylon 5: 315. 1985; Singh, Fl. E. Karnataka 637. 1988; Karthik. *et al.*, Fl. Ind. Enum. Monocot. 54. 1989; Keshava Murthy & Yoganarasimhan, Fl. Coorg 509. 1990. *Scirpus schoenoides* Retz., Obs. Bot. 5: 14. 1788. *Kadu gundu hullu, Gedde gundu hullu.* **Fig. 35.**

Type : India, *Koenig.*

Annuals or perennials with or without a short rhizome. Stems densely tufted, smooth, very slender, striate-sulcate, 3-30 cm long, 0.3 - 0.8 mm thick. Leaves filiform, shorter than stem, 6-15 cm long, 0.7-1 mm wide; sheaths 2-5 cm long; ligule a fringe of short hairs. Inflorescence usually a solitary, terminal spikelet, often proliferates into 1-2 peduncled spikelets. Involucral bracts usually glume-like, ovate or triangular, *ca* 2 x 1 mm, apiculate, spinular-scabrous on the margins, often leafy, up to 6 mm long. Spikelets ovoid, 4-6 x 3-5 mm, many-flowered. Glumes spiral, membranous, broadly ovate, apiculate at apex, prominently keeled, *ca* 2 x 2 mm, with brown spots towards centre. Stamens 3; anthers oblong, *ca* 1 mm long, mucronate. Ovary obovate, *ca* 0.5 mm long; style *ca* 1 mm long, dilated at base, ciliate in the upper half; stigmas 2, *ca* 0.7 mm long, brownish. Nut biconvex, obovate 1.5-1.75 x *ca* 1 mm, whitish to brownish, stipitate; gynophore (stipe) *ca* 0.25 mm long; epidermal cells isodiametric.

Fls. & Frts. : Aug. - Jan; April.

Chrom. No. : 2n = 10 (*Taxon* 21: 683. 1972).

Habitat : Open grasslands, open areas in forests, fallow rice fields, dried up dykes, and edges of rice fields, tanks and canals.

Distrib. : Sri Lanka, China, Malesia, Thailand, Formosa, and tropical Australia. INDIA: Throughout. KARNATAKA: Bangalore, Belgaum,

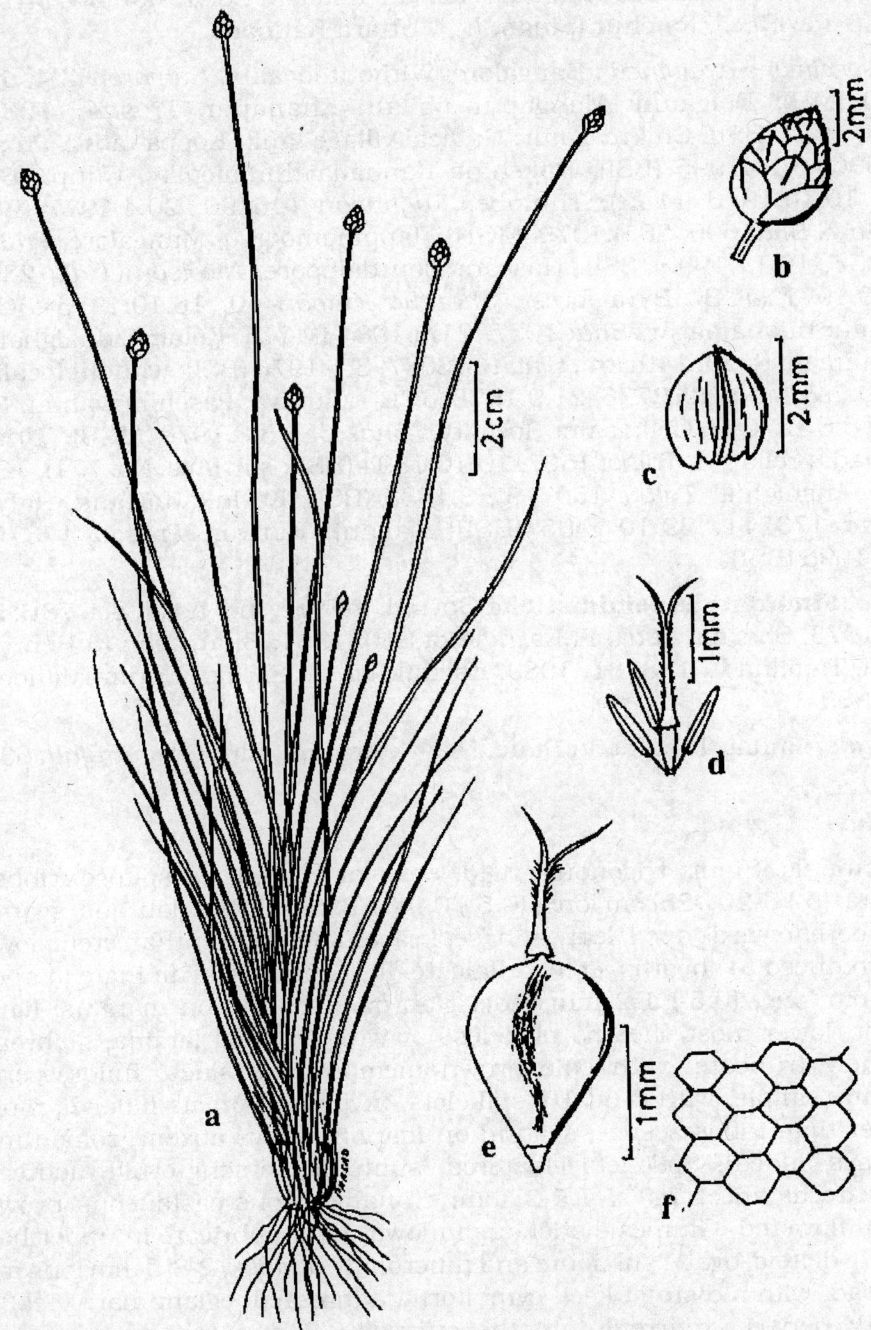

Fig. 35. *Fimbristylis schoenoides* (Retz.) Vahl
a. Habit, **b.** Spikelet, **c.** Glume, **d.** Flower, **e.** Nut, **f.** Epidermal cells

Chikmagalur, Dakshina Kannada, Hassan, Kodagu, Kolar, Mandya, Mysore (Rao & Razi, *l.c.*), Raichur (Singh, *l.c.*), Uttara Kannada.

Specimens examined : Bangalore: Without locality, *Camaron* 424, Jan. 1890 (MH); Belgaum: Mangapur nallah, Khanapur, *Prasad* 172859, 14.12.1994 (BSI). Chikmagalur: Gadical village tank, Koppa taluk, *Prasad* 173836, 8.11.1995 (BSI). Dakshina Kannada: Kinimoolky, Udupi, *Bhat* 301, 10.10.1976 (MGH); Thumbe, *Raghavan* 146490, 20.4.1977 (BSI); Manipal, *Bhat* 605, 28.9.1979 (MGH); Jappinamogeru, Mangalore, *Prasad* 173853, 10.11.1995 (BSI). Hassan: Bourdalboore, *Nicolson et al.*, 2350, 26.10.1971 (JCB); Byra forest, *Shantha Kumari* 40, 16.10.1975 (JCB). Kodagu: Kushalnagar, *Bhat* 1037, 21.9.1981 (MGH). Kolar: Siddalghatta-Chintamani Road, 14th km, *Singh* 142037, 3.1.1976 (BSI); without locality, *Prakash & Sreenath* 2776, 21.9.1978 (JCB). Mandya: Paschimavahini, *Rao* 816, 8.8.1970 (MGH); same locality, *Bhat* 53, 8.8.1970 (JCB). Uttara Kannada: Yellapur, *Talbot* 1507, 10.10.1894 (BSI); *s.n.* (Acc. No. 771), 1883 (BSI); Anshighat, *Talbot* 1507, 12.8.1885 (BSI); Karlukatta tank, Halyal, *Prasad* 173711. 29.10.1995 (BSI); Kaliguj, Karwar, *Prasad* 173759, 1.11.1995 (BSI).

31. **Fimbristylis semidisticha** Govind. in Proc. Ind. Acad. Sci. 78B: 51, f. 3. 1973; Sharma *et al.*, Fl.Karnataka 310. 1984; Bhargavan in Henry *et al.*, Fl. Tamilnadu 1, 3: 81. 1989; Karthik. *et al.*, Fl. Ind. Enum. Monocot. 54. 1989.

Type : India, Tamilnadu State, Nilgiri, Aravankadu, *Govindrajalu* 6307 (PCM).

Illus. : Govind., *l.c.*

Annuals. Stems trigonous, rigid, erect or curved, caespitose, ribbed, sulcate, (5-) 8-20 (-25) cm long, (0.5-) 0.9 - 1.3 mm thick, glabrous, scabrid or smooth towards apex, leafy at base. Leaves few, shorter than stem, lower most reduced to sheaths, erect or falcate, flat, abruptly acuminate at apex, 5-15 cm long, (0.5-) 1-2 mm wide, glabrous, scabrid on margins; ligule absent; lower most sheaths bladeless or with reduced lamina, glabrous, obliquely truncate, with pale brown membranous sides. Inflorescence capitate, simple, with 3-6 (-10) spikelets. Involucral bracts filiform, much shorter than inflorescence, scabrid on margins. Rays absent (some times 1-2 rays added). Spikelets clustered, subterete, oblong-ovoid, acute or subacute at apex, 7-8 x 1.5-2 mm, cinnamomeous-castaneous brown, 10-20- flowered. Glumes distichous in lower half, imbricate in upper half, broadly deltoid-ovate, subacute and mucronate at apex, 2.8-3 mm long and as broad, with a distinct keel, cymbiform, adpressed, eglandular; keel 3-4 nerved; nerves excurrent into the curved, 0.2 mm long mucro; sides nerveless, cinnamomeous, with distinct scarious glabrous margins. Rachilla winged, excavated. Stamens 3; anthers ellipsoid, apiculate, 0.3 - 0.6 mm long, lobed at base. Style triquetrous, with slightly dilated pyramidal base, 1.2 - 1.3 mm long, sparsely fimbriately hairy throughout; stigmas 3, slender, as long as the style, fimbriately hairy. Nut trigonous, globose-obovoid, tricostulate, with convex sides, umbonulate, abruptly attenuating at base, 1-1.1 x 0.75 - 0.9 mm, minutely stipitate, verruculose, dark brown; epidermal cells transversely hexagonal in the upper half, in 10-14 more or less regular vertical rows.

Fls. & Frts. : Nov.

Habitat : High ranges with rather heavy rain fall.

Distrib. : Endemic to South Indian hill stations (Karnataka and Tamilnadu). KARNATAKA: Chikmagalur (Bababudan hills, Govind. *l.c.*).

Note : Specimens of this species were not seen; but included here on the authority of Govindarajalu *l.c.* According to him it is related to *F. merguensis* Clarke, but differs in having caespitose habit, shorter stems, fewer leaves per stem, lower most sheaths bladeless or with reduced blades, simple capitate inflorescence, filiform bracts, smaller spikelets, broadly ovoid-deltoid glumes arranged distichously in the lower half with strongly curved base and glabrous margins, shorter anthers and style, longer stigmas, dark brown larger verruculose nuts with transversely elongated hexagonal epidermal cells.

32. **Fimbristylis sieberiana** Kunth, Enum. Pl. 2: 237. 1837; Kern in van Steenis, Fl. Males. 1, 7: 572. 1974; Hooper in Saldanha & Nicolson, Fl. Hassan 680. 1976; Sharma *et al.*, Fl. Karnataka 30. 1984; Karthik. *et al.*, Fl. Ind. Enum. Monocot. 55. 1989. *F. ferruginea* auct. non Vahl, 1806; Decne, Nouv. Ann. Mus. Hist. Nat. Paris 3: 352. 1834. *F. ferruginea* var. *sieberiana* Boeck. in Linnaea 37: 17. 1871. **Fig. 36.**

Perennials with short woody rhizome, 25-60 cm high. Stems usually tufted, rather stiff, glabrous, striate, 1-1.5 mm thick. Cauline leaves shorter than stem, but compratively longer than that of *F. ferrugenia*, up to 40 cm long, 1.5-2 mm wide; basal sheaths membranous to papery, not shining brown; ligule a fringe of short hairs. Inflorescence simple to decompound, 2-7.5 cm long. Involucral bracts usually 3, lower 1 or 2 usually overtopping the inflorescence, up to 13 cm long. Primary rays up to 4.5 cm long. Spikelets few, terete, ovoid, acute at apex, 5-10 x 2.5- 4 mm, densely many-flowered, brownish. Glumes spiral, broadly oblong-ovate, obtuse and apiculate at apex, 3-3.5 x *ca* 3 mm, often as broad as long, keeled, puberulous towards apex. Stamens 3, filaments elongate up to 4 mm; anthers linear-oblong, *ca* 1 mm long. Ovary oblong, *ca* 1 mm long, distinctly stipitate; style flat, *ca* 2 mm long, ciliate at margin; stigmas 2, *ca* 1.5 mm long. Nut biconvex, broadly oblong-obovate, *ca* 1.5 x 1 mm, smooth, whitish, distinctly stipitate; stipe *ca* 0.25 mm long.

Fls. & Frts. : June - March.

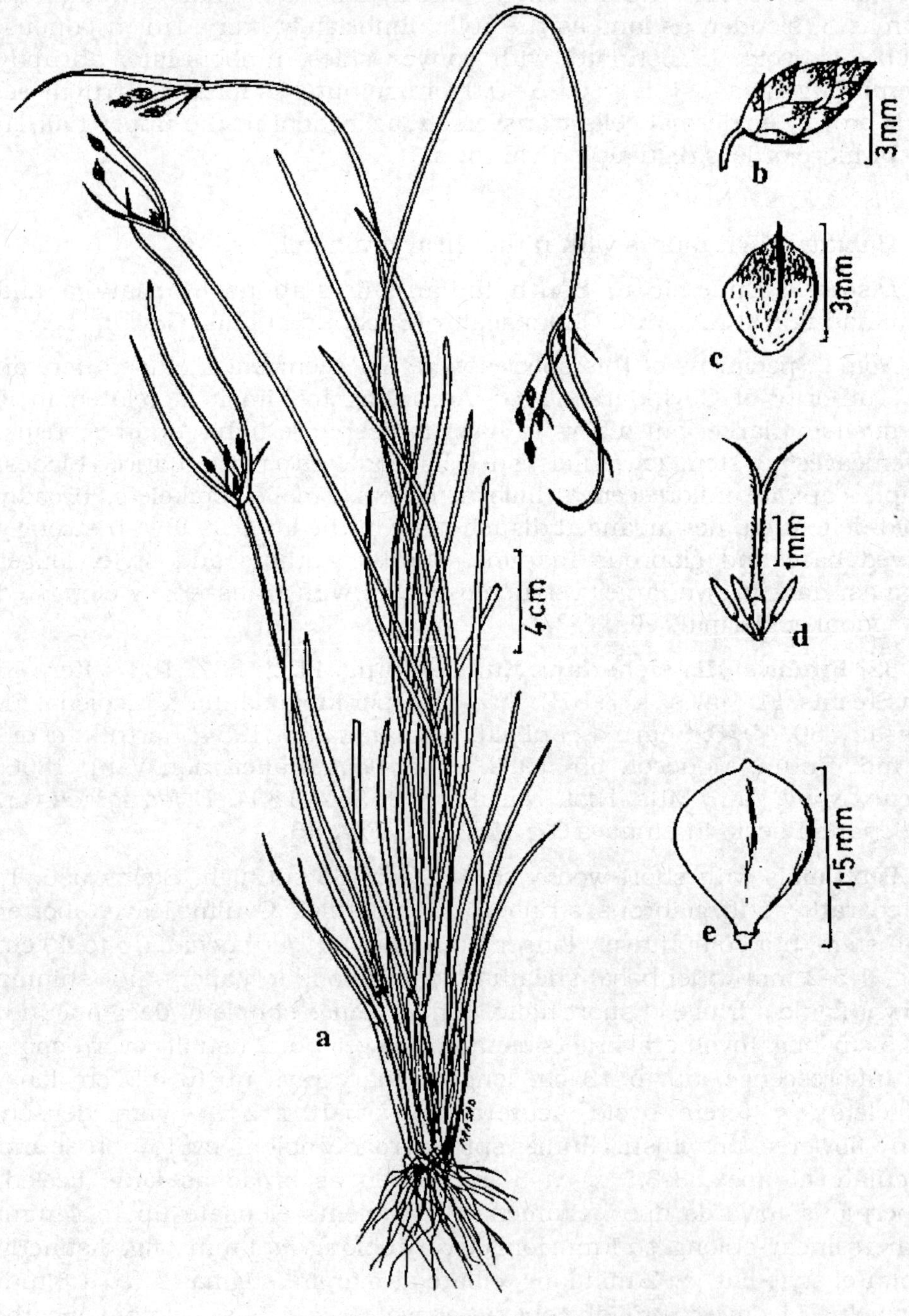

Fig. 36. *Fimbristylis sieberiana* Kunth
a. Habit, **b**. Spikelet, **c**. Glume, **d**. Flower, **e**. Nut

Habitat : Saline marshy areas, dry muddy areas near puddles, muddy shores of lakes and canals, wet gravelly soil in gardens, marshy roadsides, irrigated areas; as a weed in rice fields; comparatively less halophytic than *F. ferruginea* and more confined to inland areas.

Distrib. : Asia (Syria, Arabia, Iran, Malesia) and Africa (South Africa, Madagascar, Mauritius). INDIA: Peninsular India, North & North-West India. KARNATAKA: Bangalore, Belgaum, Bellary, Bijapur, Chitradurga, Dharwar, Gulbarga, Hassan, Mandya, Mysore, Raichur.

Specimens examined : Bangalore: Ghate, *Ramesh* 614, 29.3.1978 (JCB). Belgaum: Near Ghataprabha dam, *Prasad* 172811, 10.12.1994 (BSI); Near Gokak falls, *Prasad* 172824, 11.12.1994 (BSI); Along Ghataprabha-Gokak road, *Prasad* 172835, 12.12.1994 (BSI). Bellary: Kudligi-Hospet Road, *Manohar & Ramesh* 5895, 29.1.1979 (JCB). Bijapur: Hebballi seemi, Badami, *Prasad* 172973, 22.12.1994 (BSI); Hossur Forest Research Nursery, Badami, *Prasad* 172987, 23.12.1994 (BSI). Chitradurga: Huliyur-Hiriyur road 43rd km, *Singh* 141201, 13.10.1975 (BSI); Donnehalli-Chikka Ujjani Road, *Murthy & Manohar* 7769, 16.6.1979 (JCB). Dharwar: Kottur, near Kilthur, *Saldanha & Prakash* 3429, 24.10.1978 (JCB); Sadankeri, Alnavar road, *Prasad* 172939, 19.12.1994 (BSI). Gulbarga: Kamalapur-Sulahpet, *Ramesh & Manohar* 10068, 27.10.1979 (JCB). Hassan: Hassan, *Gandhi* 2119, 22.9.1971 (JCB). Mandya: Malavalli-Kolligal Road, *Saldanha & Ramesh* 2404, 5.9.1978 (JCB). Mysore: Without locality, *Suma* 2, 12.8.1968 (MGH); Kushalnagar- Hussur Road, *Ramesh* 2388, 24.8.1978 (JCB). Raichur: Gangavati on Kanakgiri Road 8th km, *Singh* 147017, 8.5.1977 (BSI); Lingsugur- Sindhnur Road 40th km, *Singh* 143045, 18.6.1976 (BSI).

Note : Very similar to *F. ferruginea* and hence often treated as its synonym. But this species can be distinguished by the longer, less stiff cauline leaves, herbaceous sheaths and lowest involucral bracts usually overtopping the inflorescence. Specimens having intermediate characters of both the species were also found.

33. **Fimbristylis simpsonii** Prasad & N.P. Singh in Journ. Bombay nat. Hist. Soc. 96 (3): 456, f. 2. 1999. **Fig. 37.**

Type : India, Karnataka state, Shimoga dt., Thirthahalli, Kanagalgudda, *Sundara Raghavan* 90025 - Holotype (CAL) and 90025 A-Isotype (BSI).

Illus. : Prasad & Singh, *l.c.*

Glabrous annuals with fibrous roots, *ca* 9 cm high. Stems tufted, slender, trigonous below the inflorescence, deeply furrowed below, 0.2-0.5 mm thick. Leaves numerous, half to 3/4 length of the stem, flat, uniformly linear, abruptly acuminate at apex, 2.5 - 6.5 cm long, 0.7 - 1.2 mm wide, with slightly thickened margins; sheaths chartaceous, up to 1.5 cm long;

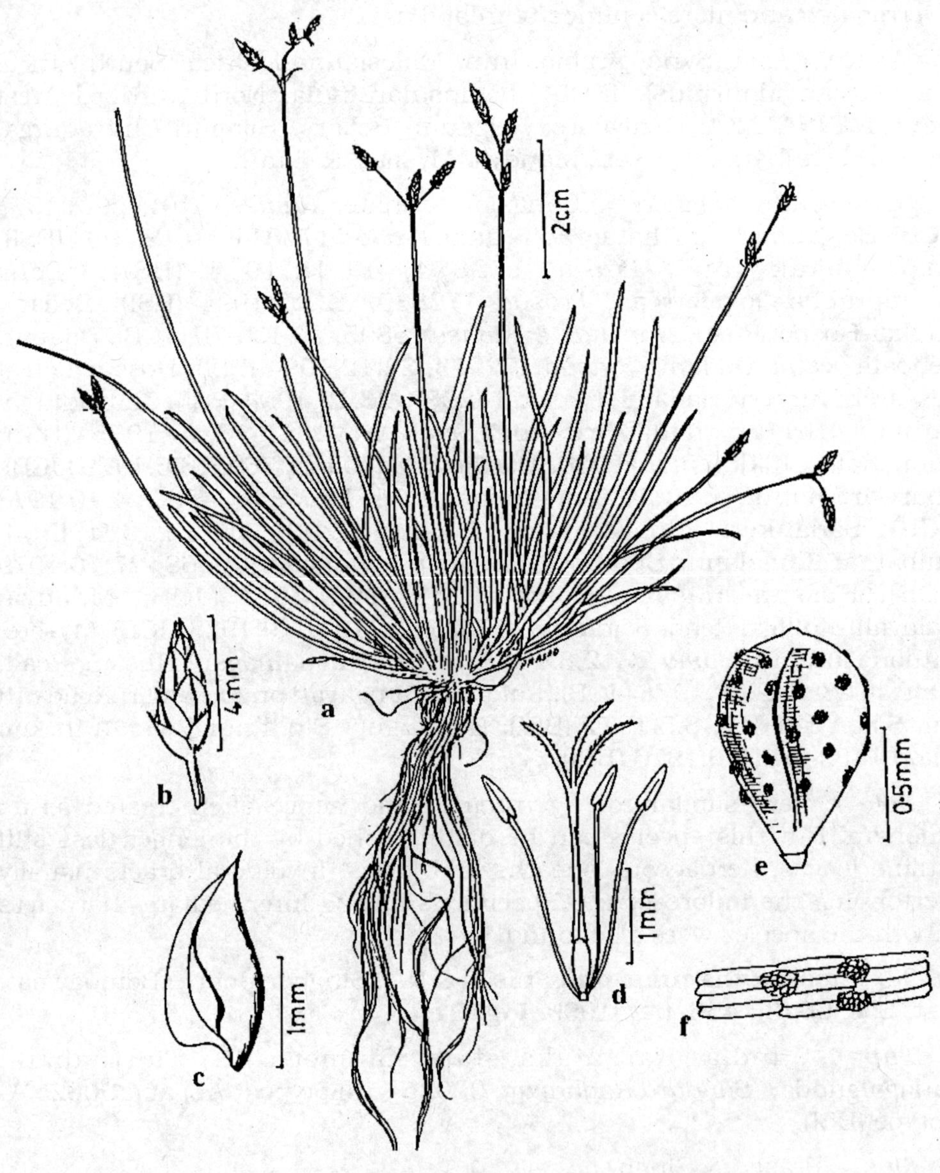

Fig. 37. *Fimbristylis simpsonii* Prasad & N.P. Singh
a. Habit, **b.** Spikelet, **c.** Glume, **d**. Flower, **e.** Nut, **f.** Epidermal cells.
"Reproduced from *J. Bombay nat. Hist. Soc.*" Vol. 96, p. 457, 1999 with permission of BNHS.

ligule a fringe of short hairs. Orifice membranous, ciliate. Inflorescence simple, with 1-3 spikelets, 5-8 mm long, 6-13 mm wide. Involucral bracts 1 or 2, much shorter than to as long as the inflorescence, short- laminate or glume-like with an awn, 2-7 mm long. Rays if present trigonous, striate, 2-4 mm long. Spikelets solitary, ovoid to oblong-lanceolate, acute at apex, slightly angled, 3-4 x 1-1.5 mm, brown, few-flowered. Rachilla winged. Glumes spiral, ovate, acute-mucronate at apex, strongly keeled, 1.8 - 2 x 1.2-1.5 mm, hyaline towards margins but brown-lineolate. Stamens 3; filaments hyaline, elongate up to 2 mm; anthers linear-oblong, obtuse-acute at apex, *ca* 0.6 mm long. Ovary linear-oblong, *ca* 0.5 mm long; style triquetrous, slightly thickened towards base, *ca* 1.2 mm long, brownish, glabrous; stigmas 3, about half the length of the style, scabrous. Nut trigonous with 2 convex faces and 1 flat face, obovoid, minutely umbonulate, shortly stipitate, 0.8-1 x 0.6-0.7 mm, smooth or verruculose, creamish-white; epidermal cells transversely elongated, in *ca* 4 vertical rows on each face.

Fls. & Frts. : Aug.

Habitat : Rocky slopes near riverlets.

Distrib. : Endemic to Karnataka (Shimoga dt.).

Specimens examined : As mentioned above under Type.

34. **Fimbristylis squarrosa** Vahl var. **esquarrosa** Makino, Bot. Mag. Tokyo 17: 47. 1903; Kern in Blumea 8: 143. 1955, in Reinwardtia 6: 48. 1961 *et* in van Steenis, Fl. Males. 1, 7: 585. 1974; Hooper in Saldanha & Nicolson, Fl. Hassan 681. 1976; Sharma *et al.*, Fl. Karnataka 310. 1984. *F. velata* R. Br., Prodr. 227. 1810; Govind. in Reinwardtia 8: 511. 1974; Karthik. *et al.*, Fl. Ind. Enum. Monocot. 560. 1989. *F. squarrosa* var. *velata* (R. Br.) Clarke ex Cheesem., Man. New Zeal. Fl. 770. 1906. *F. dichotoma* var. *villosa* Fischer in Gamble, Fl. Pres. Madras 1658. 1931. *F. bisumbellata* var. *hirtistyla* Fischer in Kew Bull. 1935: 150.1935, in Gamble, Fl. Pres. Madras 1898. 1936 (Corrig.) *et* 3: 1151. 1957, repr.ed.). *F. aestivalis* var. *esquarrosa* Koyama in Journ. Fac. Sci. Univ. Tokyo 3, 8:116. 1961.

Annuals with fibrous roots. Stems densely tufted, slender, setaceous, compressed-subtrigonous, 10-25 cm long, 0.5 - 1 mm thick, smooth. Leaves shorter than stem, very narrow, flat or canaliculate, acute at apex, 0.3 - 1 mm wide, more or less soft- hairy at least on the sheaths; ligule absent. Inflorescence compound or decompound, loose, 2-5 cm long, with several to many spikelets. Involucral bracts 3-7, dilated at base, soft-hairy; lowest shorter than to about as long as the inflorescence. Primary rays up to 6, slender, filiform, up to 2.5 cm, smooth. Spikelets solitary, lanceolate, angular, acute at apex, 4-7 x 1- 1.5 mm, stramineous or ferrugineous, densely many-flowered; rachilla narrowly winged. Glumes spiral, thinly membranous, suberect, ovate or oblong-ovate, acute at apex, acutely keeled, 1.5 - 2 x *ca* 1 mm, 3-nerved, with nerveless sides, glabrous or pubescent; the strong midnerve excurrent into a strait or slightly excurved, up to 0.3 mm long mucro. Stamen 1; anther oblong, 0.25-0.3 mm long. Style slender, flat, dilated at base, not with hyaline margin, 0.75-1 mm long, sparsely ciliate at top; lower margins fringed with a whorl of long, pendent hairs, closely appressed to the nut covering its half to three-fourth; stigmas 2, shorter than style. Nut biconvex, with acute edges, obovate, shortly stipitate, minutely umbonulate, smooth, 0.6-0.75 x 0.5 mm, obscurely reticulate, shining stramineous; epidermal cells not impressed, hexagonal.

Fls. & Frts. : Feb; May.

Habitat : Marshy areas, wet sandy soil and shores of lakes in high altitudes (1200-1300 m in S India).

Distrib. : The typical var. *squarrosa* is widely distributed in the tropical and warm temperate regions of Asia, South Europe, Africa and South America. In India this variety is found in the North and North-East.

The var. *esquarrosa* is found in Malesia, Thailand, Indo-China, North and North-East China, South Korea, Japan, Polynesia, Australia and New Zealand. INDIA: Confined to South Indian hill stations. KARNATAKA: Hassan, Mysore (Fischer, *l.c.*), Uttara Kannada.

In western extreme of the global distributional range only var. *squarrosa* occurs and in the eastern extreme var. *esquarrosa* . But in Eastern Asia there is considerable overlapping of both the varieties. Hence a clear-cut distributional demarcation is not possible between the two.

Specimen examined : Hassan: Hebbasale, *Saldanha* 16450, 24.12.1970 (JCB). Uttara Kannada: Along Kargatte river, *Ahmed* 1159, 29.5.1978 (JCB).

Note : *F. squarrosa* is very similar to *F. aestivalis* and F. v. Muller merged these two (Fragm. 9. 1875). Koyama also followed this. But typical *F. squarrosa* is clearly distinct from *F. aestivalis* by the much longer, strongly recurved mucros of the glumes which give the spikelets a squarrose appearance and by the long pendent hairs of the style base. The mucros of var. *esquarrosa* do not differ from those of *F. aestivalis.* So the trichomes of the style base which are absent or very short in *F. aestivalis,* the slightly longer style and the often somewhat larger nuts are the only characters to distinguish *F. squarrosa* var. *esquarrosa* and *F. aestivalis.*

35. **Fimbristylis tenera** Schult., Syst. Veg. 2, Mant. 57. 1824; Clarke in Hook.f., Fl. Brit. India, 6: 642. 1894; Fischer in Gamble, Fl. Pres. Madras 1660. 1931 (3: 1152. 1957, repr.ed.); Rao & Razi, Fl. Mysore 563. 1981; Koyama in Dassanayake & Fosberg, Rev. Handb. Fl. Ceylon 5: 293. 1985; Singh, Fl. E. Karnataka 2: 637. 1988; Karthik. *et al.*, Fl. Ind. Enum. Monocot. 55. 1989. *Scirpus tenellus* Roxb., Fl. Ind. 1, 1: 227. 1820 & 2, 1: 224. 1832. *F. oxylepis* Steud., Syn. Pl. Glum. 2: 110. 1855. *F. tenera* Schult. var. *oxylepis* (Steud.) Clarke in Hook.f., Fl. Brit. India 6: 642. 1894; Hooper

in Saldanha & Nicolson, Fl. Hassan 681. 1976; Rao & Verma, Cyp. NE India 35. 1982; Sharma *et al.*, Fl. Karnataka 311. 1984. **Fig. 38.**

Type: India, Coramandal, *Roxburgh.*

Glabrous annuals with fibrous roots, 10-30 cm high. Stems tufted, slender, trigonous, 0.2-0.5 mm thick, smooth. Leaves usually up to half the length of the stem, flat, slender, abruptly acuminate at apex, 0.5 - 1 mm wide, often spinulose scabrous on the margins towards top; sheaths papery, up to 4 cm long; ligule absent; orifice oblique, smooth at margin. Leaves on the flowering stem often reduced to an appendage. Inflorescence simple or subcompound, 1-3 cm long, 0.8 - 2 cm wide, with few- many spikelets. Involucral bracts 3-4, usually shorter than inflorescence, dilated at base, at times the lowest overtopping the inflorescence, 2-15 mm long. Primary rays up to 2 cm long. Spikelets solitary, ovate-lanceolate, acute at apex, faintly angled, 4-6 x 1.5 - 2 mm, many-flowered, brownish; rachilla winged. Glumes spiral, broadly triangular-ovate, acute-mucronate at apex, keeled, *ca* 2.5 x 2 mm, membranous towards margins, minutely puberulous towards apex. Stamen 1 (or 2); filament flat, elongate up to 2 mm; anther linear-oblong, acute at apex, *ca* 0.7 mm long. Ovary oblong-obovoid, *ca* 0.5 mm long, shortly stipitate; style *ca* 1.2 mm long, glabrous, angled towards the base, dilated and pyramidal at base; stigmas 3, *ca* 0.5 mm long, minutely scabrous. Nut trigonous, with convex faces, obovoid, *ca* 0.9 x 0.6 mm, verruculose, often sub-tuberculate, whitish, with *ca* 10 series of transversely hexagonal epidermal cells.

Fls. & Frts. : July - Nov.

Habitat : Among grasses in sandy soil, near puddles and along roadside.

Distrib. : Sri Lanka and tropical Africa. INDIA: Western Peninsula, Central and North India. KARNATAKA: Bangalore, Bellary, Bijapur, Dakshina Kannada (Fischer, *l.c.*), Hassan, Mandya, Mysore (Rao & Razi, *l.c.*), Raichur (Singh, *l.c.*), Uttara Kannada.

Specimens examined : Bangalore: Nandi hills, *Hooper & Saldanha* 18099, 4.12.1971 (JCB). Bellary: Without locality, *Gamble* 17763, July 1886 (MH). Bijapur: Badami hills, *Paranjpye s.n.* (Acc. No. 2770), 1.9.1992 (BSI). Hassan: Byra, *Nicolson et al.* 2251, 23.10.1977 (JCB). Mandya: Paschimavahini, *Bhat* 55, 8.8.1970 (JCB). Uttara Kannada: Castle Rock, *Almeida* 2704, 14.4.1973 (BNHS).

Note: Absence of ligule is an important character to identify this species.

36. **Fimbristylis tetragona** R. Br., Prodr. 226. 1810; Clarke in Hook.f., Fl. Brit. India 6: 631. 1893; Cooke, Fl. Pres. Bombay 2: 679. 1908 (3: 393. 1958, repr.ed.); Fischer in Gamble, Fl. Pres. Madras. 1658. 1931 (3: 1150. 1957, repr.ed.); Ramaswamy & Razi, Fl. Bangalore 110. 1973; Kern in van Steenis, Fl. Males. 1, 7: 590. 1974; Hooper in Saldanha & Nicolson, Fl. Hassan 681. 1976; Rao & Razi, Fl. Mysore 563. 1981; Rao & Verma, Cyp.

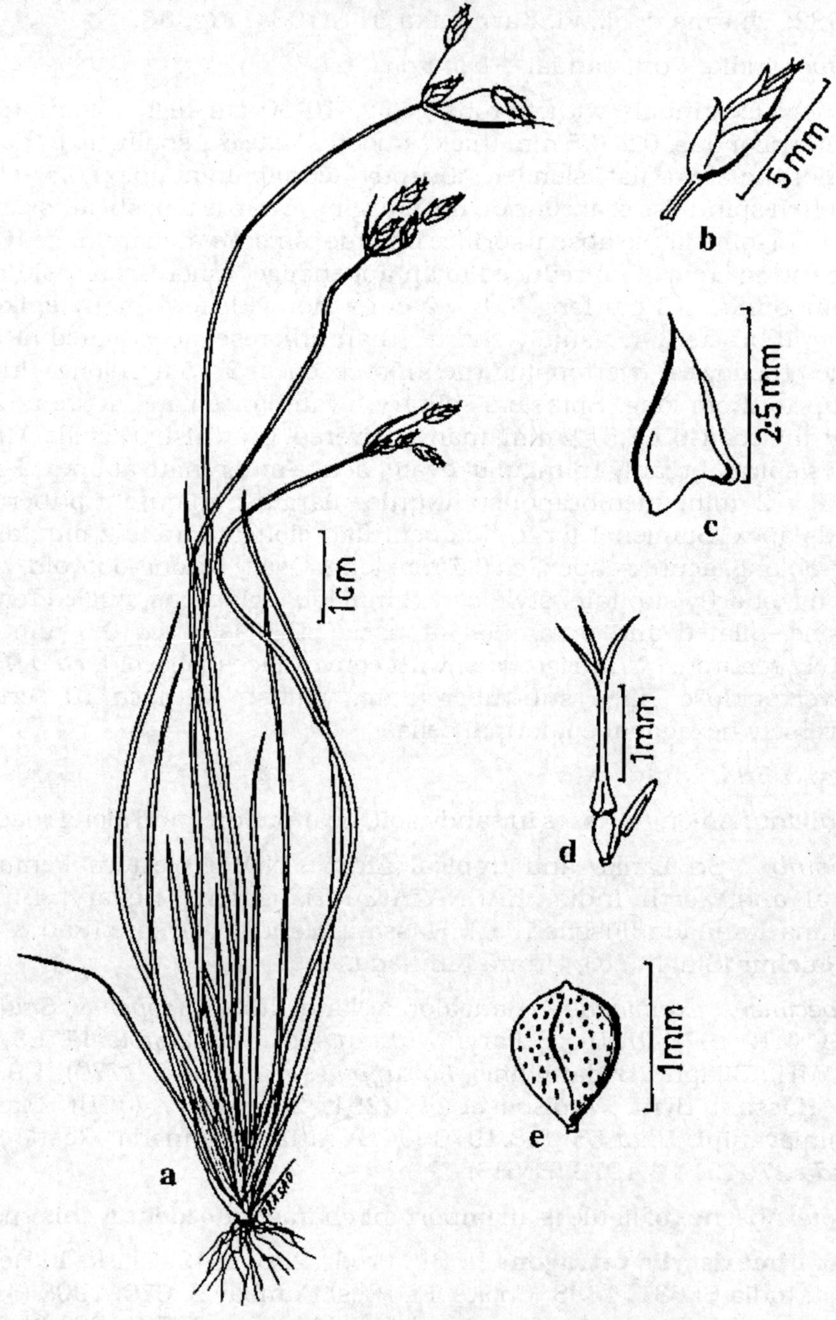

Fig. 38. *Fimbristylis tenera* Schult.
a. Habit, **b**. Spikelet, **c.** Glume, **d.** Flower, **e.** Nut

NE India 29. 1982; Sharma *et al.*, Fl. Karnataka 311. 1984. Koyama in Dassanayake & Fosberg, Rev. Handb. Fl. Ceylon 5: 323, f. 25. 1985; Singh, Fl. E. Karnataka 638. 1988; Karthik. *et al.*, Fl. Ind. Enum. Monocot. 55. 1989; Keshava Murthy & Yoganarasimhan, Fl. Coorg 509. 1990.

Type : Australia.

Illus. : Clarke, Illus. Cyp. t. 40, f. 1-4. 1909; Koyama, *l.c.*

Perennials, with short rhizome. Stems densely tufted, slender, smooth, erect, quadrangular, 7-40 cm long, 0.7- 1 mm thick. Leaf sheaths 2-3, 2.5-7 cm long, uppermost often with a small lanceolate blade, acute-acuminate at apex. Inflorescence a single spikelet. Spikelet ebracteate, ovoid to spherical, 5-8 x 4-6 mm, densely many-flowered. Glumes membranous, spiral, closely imbricate, oblong or ovate, obtuse or acute-obtuse at apex, *ca* 3 x 2.5 mm, usually with brownish lines in the centre, hyaline towards margins, lower few empty. stamen usually 1, at times up to 3; filament elongate while fruiting; anther linear-oblong, *ca* 1 mm long, apiculate. Ovary oblong, *ca* 1 mm long; style *ca* 1.5 mm long, slightly dilated and dark brown at base, minutely ciliate; stigmas 3, *ca* 0.5 mm long; gynophore *ca* 0.5 mm long. Nut oblong-cylindrical, obtuse at apex, slightly curved, *ca* 2 x 0.5 mm, trabeculate, light yellow, with persistent style and staminal filaments; stipe very prominent.

Fls. & Frts. : Aug. - Feb.

Chrom. No. : 2n = 10 (*Taxon* 21: 683. 1972).

Habitat : Open wet places, swampy grasslands, wet harvested rice fields and on its edges, near streams, tanks, marshes and swamps, muddy base of rocky river beds, along with short grasses in wet soil. Once collected from crevices of rocks in sandy area.

Distrib. : Sri Lanka, Nepal, South China, Malesia, Thailand, Indo- China, Formosa, tropical North Australia. INDIA: Throughout. KARNATAKA: Bangalore, Belgaum, Bijapur, Dakshina Kannada, Dharwar, Hassan, Kodagu, Mandya, Mysore, Shimoga, Uttara Kannada.

Specimen examined : Bangalore: Bannergatta, *Hooper & Saldanha* 18034, 7.11.1971 (JCB). Belgaum: Londa, *Gammie s.n.* (Acc. No. 2780), 6.10.1899 (BSI); 15839, 2.11.1902 (BSI); without locality and coll. name, 2779, 16.10.1891 (BSI). Bijapur: Badami, *Bhide s.n.* (Acc. No. 2778), 8.9.1911 (BSI). Dakshina Kannada: Kannadekatti, *Bhat* 388, 9.11.1976 (MGH). Dharwar: Sadankeri on Alnavar Road, *Prasad* 172937, 19.12.1994 (BSI). Hassan: Nagpuri, *Saldanha* 15211, 8.10.1969 (JCB). Kodagu: Hardoor Sangam, *Rao* 75045, 10.10.1961 (BSI); Bhagamandala, *Bhat* 803, 19.12.1980 (MGH). Mandya: Paschimavahini, *Bhat* 51, 8.8.1990 (MGH). Mysore: Sivasamudram, without coll. name, 10410, 12.5.1914 (MH); Hirakulguda state forest, *Wadhwa* 44551, 23.11.1957 (BSI); Hirakulguda state forest, Arsikere, *Rao* 73843, 11.9.1961 (BSI). Shimoga: Hosuru, near Gubbiga, *Raghavan* 83003, 5.10.1962 (BSI); Sharavati river, above Jog falls, *Prasad* 173774 and 173782, 3.11.1995 (BSI); Gaurikeri, Thalaguppa, *Prasad* 173809, 5.11.1995 (BSI). Uttara Kannada: Karwar, *Talbot* 1306, 25.8.1885 (BSI); Ulletikeri, Halyal, *Prasad* 173701, 29.10.1995 (BSI); Karlukatta tank, Halyal, *Prasad* 173708, 29.10.1995 (BSI).

37. **Fimbristylis tomentosa** Vahl, Enum. Pl. 2: 290. 1806; Kern in van Steenis, Fl. Males. 1, 7: 576, f. 45. 1974; Rao & Verma, Cyp. NE India 34. 1982; Karthik. *et al.*, Fl. Ind. Enum. Monocot. 55. 1989. *F. podocarpa* Nees *et* Meyer ex Nees in Wight, Contr. Bot. India 98. 1834, *p.p.* typ.; Clarke in Hook.f., Fl. Brit. India 6: 638. 1893; Hooper in Saldanha & Nicolson, Fl. Hassan 680. 1976; Sharma *et al.*, Fl. Karnataka. 310. 1984. **Fig. 39.**

Illus.: Clarke, Illus. Cyp. t. 42, f. 5-6.1909; Kern, *l.c.*

Glabrous annuals with fibrous roots, 15-30 cm high. Stems tufted, compressed, angular, smooth or often with few bristles below the inflorescence, 0.5- 1 mm thick. Leaves many, shorter than to as long as the stem, flat, abruptly acuminate at apex, 5- 20 cm long, 0.7 - 2 mm wide, with bristles on the slightly thickened margins in the upper half; sheaths 1-4.5 cm long, glabrous or densely clothed with white bristles on the upper half; ligule a dense fringe of short hairs; orifice membranous, usually with bristles at apex. Inflorescence simple or compound, loose, wider than long, 1.5 - 4 cm wide, with 3-8 spikelets, very often reduced to a single spikelet. Involucral bracts 3-5, lowest much longer than inflorescence, acuminate at apex, 2-5 cm long, with bristles on the margins. Primary rays 2-5, unequal, compressed, longest *ca* 2 cm long. Spikelets solitary, ovoid or oblong- ovoid, acute at apex, terete, 4-6 x 3-4 mm, many-flowered; rachilla narrowly winged. Glumes spiral, chartaceous, glabrous, broadly ovate, mucronate at apex, *ca* 2.5 x 2-2.5 mm, faintly keeled, with a strong prominent midnerve and faint nerves on both sides. Stamens 2; filaments flat, hyaline, elongate up to 3 mm; anthers linear - oblong, acute at apex, *ca* 0.8 mm long. Ovary oblong, *ca* 7 mm long, stipitate; style flat, *ca* 1.5 mm long, ciliate on the margins, slightly dilated towards base; stigmas 2, *ca* 1 mm long, scabrous. Nut biconvex, broadly obovoid or orbicular, umbonulate, prominently stipitate, 1.5 - 1.7 x 1-1.1 mm (including the stipe), sparsely tubercled in the upper half towards apex; margins obtuse, thikened; stipe obpyramidal, 0.2-0.3 mm long; epidermal cells transversely oblong, in *ca* 18 vertical rows.

Fls. & Frts. : Aug. - Nov.

Habitat : Open or shady areas, moist grasslands, grassy roadsides, in and along the rice fields, forest clearings and marshy areas in semi-evergreen forests.

Distrib. : Bangladesh, China, Malesia, tropical Africa, Madagascar, Mauritius. INDIA: Western Peninsular India, Chotanagpur, Western Himalaya, Khasia, Upper Assam. KARNATAKA: Dakshina Kannada, Hassan.

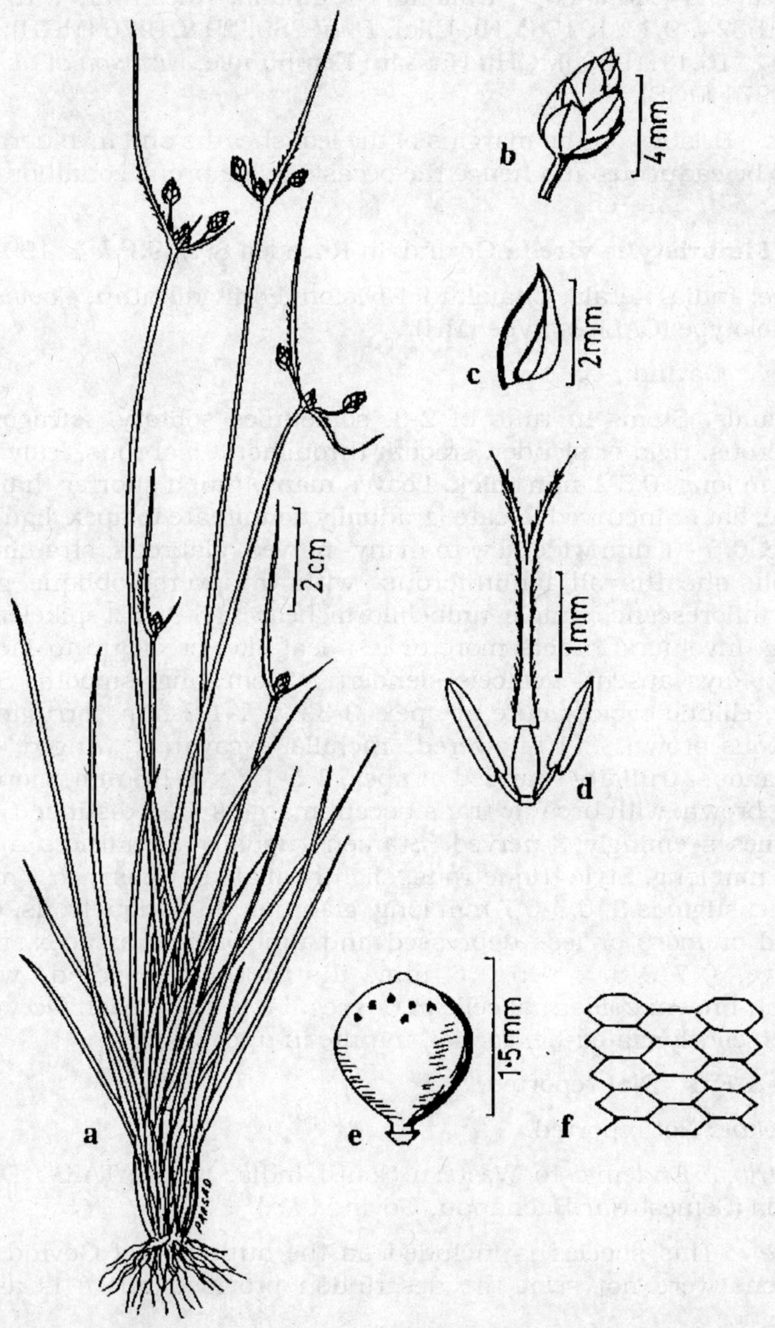

Fig. 39. *Fimbristylis tomentosa* Vahl
a. Habit, **b**. Spikelet, **c**. Glume, **d**. Flower, **e**. Nut, **f**. Epidermal cells

Specimens examined : Dakshina Kannada: Nileshwar, without coll. name, 15324, 9.11.1917 (MH); Ullal, *Bhat* 289, 29.8.1976 (MGH); Venoor, *Bhat* 712, 16.11.1980 (MGH). Hassan: Kempuhole, *Nicolson et al.* 2336 A, 25.10.1971 (JCB).

Note : Bristles on the margins of the leaf sheaths and involucral bracts seem to be caducous and hence the occasional glabrous condition of these parts.

38. **Fimbristylis virella** Govind. in Rheedea 8(1): 93, f. 4. 1998.

Type: India, Kerala, Chalakudi Division, Poringalkuthu, *Govindarajalu* 6072-Holotype (CAL); Isotype (MH).

Illus. : Govind., *l.c.*

Annuals. Stems in tufts of 2-3, sometimes solitary, tetragonous or tetrapterous, rigid or slender, smooth throughout, glabrous, leafy at base, 10-16 cm long, 0.5-1 mm thick. Leaves many, much shorter than stems, rosulate, flat or incurved falcate, gradually acuminate to apex, ligulate, 5-8 cm long, 0.5 - 1 mm wide, few to many- nerved, glabrous, stramineous or greenish; sheaths all laminiferous, with transverse-oblique glabrous mouth. Inflorescence simple, umbelliform, bearing 6-9(-12) spikelets, 1- 2.5 cm long. Involucral bracts more or less leaf-like, erect, up to 1 cm long, smooth. Rays absent. Pedicels slender, 1-2 cm long, smooth. Spikelets solitary, elliptic-ovoid, acute at apex, 3-3.2 x 1-1.2 mm, ferrugineous or castaneous-brown, 8-10-flowered; rachilla excavated, winged. Glumes membranous, trullate, rounded at apex, 1.5-1.7 x 1-1.5 mm, more or less shining brown, with broadly transluscent margins; keel distinct, 5-nerved, sometimes seemingly 3-nerved. Stamens usually 3; anthers apiculate, 0.2-0.3 mm long. Style triquetrous, slightly dilated at base, *ca* 1 mm long, glabrous; stigmas 3, 0.5-0.7 mm long, glabrous. Nut triquetrous, obovoid, rounded or more or less depressed and umbonulate at apex, minutely stipitate, 0.7-0.8 x 0.5-0.6 mm, distinctly tubercled, white or yellowish-brown; epidermal cells in 6-8 regular rows on each face, distinct, transversely elongated-hexagonal, minute in upper half.

Fls. & Frts. : Not reported.

Habitat : Not reported.

Distrib. : Endemic to Western South India. KARNATAKA: Dakshina Kannada (Someshwar Bachappu, Govind., *l.c.*).

Note : This species is included on the authority of Govind., *l.c.* As specimens were not seen the description provided above is as in the protologue.

39. **Fimbristylis woodrowii** Clarke in Bull. Misc. Inf. 227. 1898 *et* in Journ. Linn. Soc. 34: 68. 1898; Cooke, Fl. Pres. Bombay 2: 884. 1908 (3: 398. 1958, repr.ed.); Hooper in Saldanha & Nicolson, Fl. Hassan 682. 1976;

Sharma *et al.*, Fl. Karnataka 311. 1984; Karthik. *et al.*, Fl. Ind. Enum. Monocot. 56. 1989. **Fig. 40.**

Glabrous annuals, with fibrous roots, 8-20 cm high. Stems highly tufted, very slender, 3-5-angled, compressed below the inflorescene, 0.2 - 0.8 mm thick. Leaves shorter than stem, flat, abruptly acuminate at apex, 2-12 cm long, 0.5 - 1.5 mm wide; margins slightly thickened on the upper surface; sheaths striate, 0.5 - 2.5 cm long; ligule a fringe of minute hairs. Inflorescence simple or compound umbel of spikelets, 0.5- 2 cm long, 0.5 - 2.2 cm wide, with few to many spikelets. Involucral bracts 1-3 ; lowest smaller than to overtopping the inflorescence, up to 2.5 cm long. Rays 2-7, angular, compressed. Spikelets paired or in clusters of 3, of which one sessile and the other two short-peduncled, at time solitary, ovoid-lanceolate, acute at apex, faintly angled, 2-6 x 1-2 mm, few-flowered; rachilla winged. Glumes spiral, membranous, broadly triangular-ovate, acute and mucronulate at apex, keeled, 1.2 -1.8 x 1-1.3 mm, hyaline towards margins; keel 3- nerved. Stamen 1; filament hyaline, elongate up to 1.2 mm; anther oblong, 0.3 - 0.5 mm long, minutely apiculate at apex. Ovary oblong, *ca* 0.3 mm long ; style glabrous 0.7 - 0.8 mm long; triquetrous and pyramidally thickened at base; stigmas 3, shorter than to as long as the style. Nut trigonous with prominent angles, obovoid, umbonulate, shortly stipitate, 0.7 - 0.8 x 0.4 - 0.7 mm, smooth, at times sparsely verruculose, creamish; epidermal cells transversely linear-oblong.

Fls. & Frts. : Aug.-Dec.

Habitat: Among grasses in open areas, open, seasonally wet sandy places in upper ghats, river beds, rocky crevices, cultivated fields and roadsides.

Distrib. : Endemic to Western Peninsular India. KARNATAKA: Dakshina Kannada, Hassan, Kodagu, Uttara Kannada.

Specimens examined : Dakshina Kannada: Manipal, *Bhat* 254, 2.8.1976 (MGH); Udyavara, *Bhat* 269, 8.8.1976 (MGH). Hassan: Yettenahalla, *Hooper & Gandhi* 2442, 12.11.1971 (JCB). Kodagu: Talakaveri, *Bhat* 1073, 27.9.1981 (MGH); Lakshmana Thirtha, *Murthy & Yoganarasimhan* 5055, 2.12.1984 (RRCBI). Uttara Kannada: Karwar, *Talbot* 999, Aug. 1883 (BSI).

SPECIES EXCLUDED

Fimbristylis glabra Hochst. ex Steud., Syn. Pl. Glum. 111. 1854.

Sharma *et al., l.c.* reported this species from Shimoga district, most probably based on misidentified specimens in BSI. During the present study these specimens were found to be interesting and confirmed in the herbarium at Kew that these belong to a new species and later on named

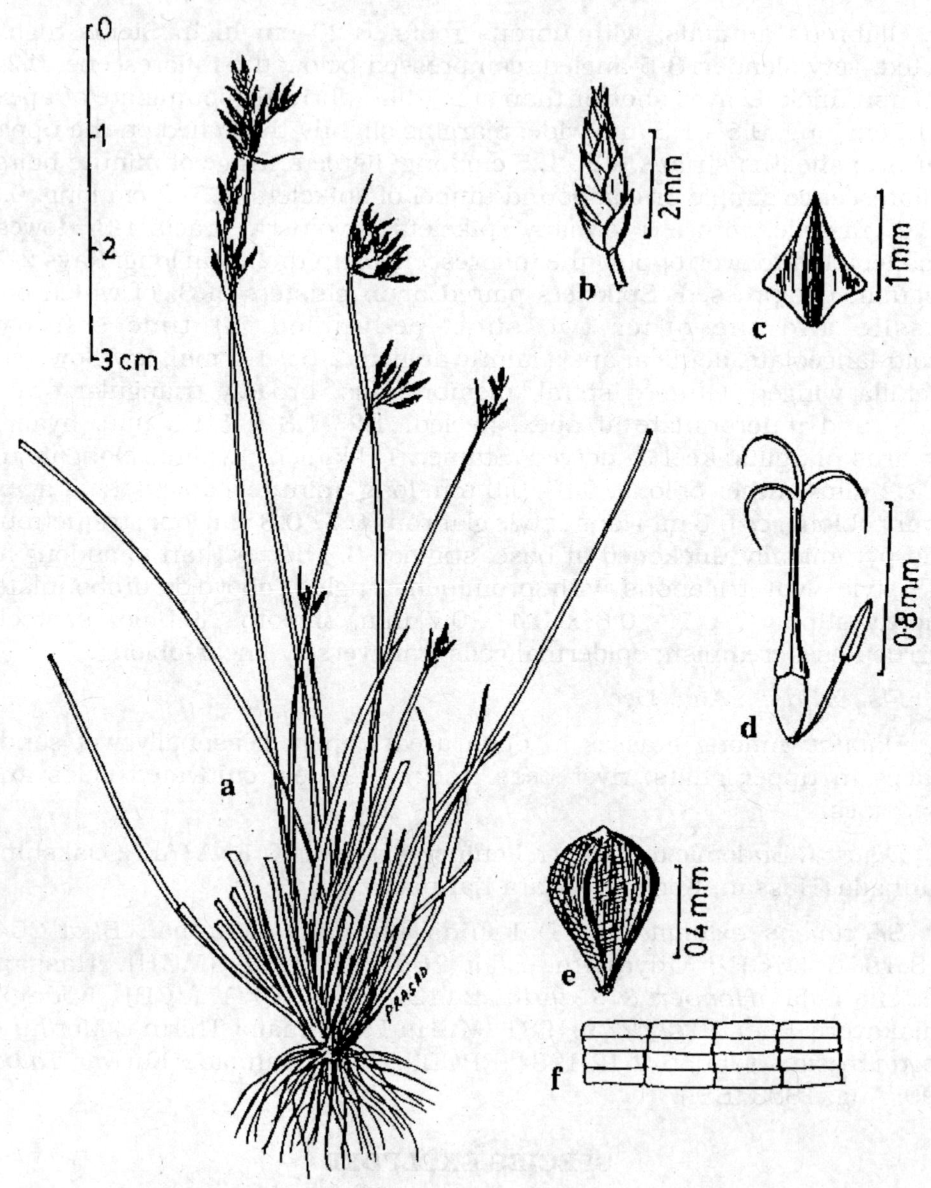

Fig. 40. *Fimbristylis woodrowii* Clarke
a. Habit, **b.** Spikelet, **c**. Glume, **d**. Flower, **e.** Nut, **f.** Epidermal cells

as *F. simpsonii* Prasad & N.P. Singh. Hence, *F. glabra* Hochst. ex Steud. is excluded from the State Flora of Karnataka (Prasad & Singh, 2001).

Further, according to Clarke (1893) Steudel's *F. glabra* was founded on a Hohenacker specimen which is actually *Abildgaardia cinnamometorum* (Vahl) Thw. [*F. cinnamometorum* (Vahl) Kunth] mixed with *F. tenera* var. *oxylepis* (Steud.) Clarke; also Steudel's description regards mostly of *F. cinnamometorum.* Hence, existence of a species namely *F. glabra* is also doubtful.

10. FUIRENA

Rottb., Descr. & Ic. 70. 1773; Kunth, Enum. Pl. 2: 180. 1837; Boeck. in Linnaea 37: 98. 1871; Benth. & Hook., Gen. Pl. 3: 1053. 1883; Clarke in Kew Bull. add. Ser. 8: 115. 1908.

Type : *Fuirena umbellata* Rottb.

Annual or perennial herbs, at times with short rhizome. Stems erect, angular, noded, leafy throughout the length. Radical leaves reduced to cataphylls. Cauline leaves usually with elongate, linear or lanceolate blades and closed sheaths; ligule membranous. Inflorescence paniculate, with a terminal and few to several axillary clusters of sessile spikelets, often much reduced. Bracts leafy, sheathing at base. Spikelets terete, ovoid or ellipsoid, usually greyish-green and pubescent, many-flowered. Rachilla persistent, not winged, with diamond-shaped scars. Glumes spirally imbricate, acropetally caducous, not keeled, awned, 3-nerved, hairy on the back at least in the upper half, basal 1 or 2 empty. Flowers hermaphrodite. Perianth (rarely absent) with 3 outer bristles (some times absent) and 3 inner scales, rarely with 6 bristles. Stamens (2-) 3. Style not dilated at base, continuous with the ovary; stigmas 3. Nut small, triquetrous, obovate or ovate, smooth or trabeculate, beaked.

About 30 species in the warm regions of the world, mostly in the tropical Africa and tropical America. 9 species in India; 5 in Karnataka.

Literature: **GOVINDARAJALU, E.** (1969) The systematic anatomy of South Indian Cyperaceae: *Fuirena* Rottb. : in *Bot. Journ. Linn. Soc.* 67: 27-40. **NIJALINGAPPA, B.H.M.**(1977) Cytological studies in *Fuirena* Rottb. (Cyperaceae), in *Curr. Sci.* 46(4): 121-122. **RATH. S.P.** & **PATNAIK, S.N.** (1974) A note on the Cytotaxonomy of East Indian species of the genus *Fuirena* Rottb., in *Bot. Mag. Tokyo* 87 (1008): 333-336. **SABNIS S.D.** & **BEDI S.J.** (1971) The genus *Fuirena* (Cyperaceae) in Gujarat, in *Journ. Bombay nat. Hist. Soc.* 68: 857-858. **VARTAK, V.D.** (1971) Taxonomic study of the genus *Fuirena* Rottb. from India, in *Journ. Univ. Poona. Sci. Tech. Sect.* No. 40: 185-203.

Key to the species

1a. Perianth often absent, if present uniseriate (scales strictly absent); nut with 3-5 vertical lines on the faces .... **3. F. cuspidata**

1b. Perianth always present, biseriate; nut without 3-5 vertical lines on the faces 2

2a. Annuals; perianth scales subquadrate or anchor shaped 3

2b. Annuals or perennials; perianth scales oblong-ovate, broadly elliptic or broadly obovate 4

3a. Perianth scales subquadrate .... **2. F. ciliaris**

3b. Perianth scales anchor-shaped .... **4. F. trilobites**

4a. Annuals, or short-lived perennials, without thick rhizome, 4-20 cm high; stems not stout, 1-2 mm thick, densely pilose; leaves 1.5 - 10 cm long, 2-5 mm wide, densely pilose on both surfaces; staminal filaments elongate up to 2 mm; anthers oblong, *ca* 0.6 mm long .... **1. F. capitata**

4b. Perennials with short thick rhizome, 50-90 cm high; stems somewhat stout, 4-5 mm thick, usually glabrous, but pubescent below the inflorescence; leaves up to 25 cm long, 5-9 mm wide, glabrous or sparsely scabrous on the nerves below; staminal filaments elongate up to 4.5 mm; anthers linear-oblong, up to 1.5 mm long .... **5. F. umbellata**

1. **Fuirena capitata** (Burm.f.) Koyama in Dassanayake & Forsberg, Rev. Handb. Fl. Ceylon 5: 151. 1985. *Scirpus capitatus* Burm.f., Fl. Ind. 21. 1768, *non* L. 1753. *S. uncinatus* Willd., Sp. Pl. 1: 300. 1798; Vahl, Enum. Pl. 2: 259. 1806. *Isolepis uncinata* (Willd.) Roem. & Schult., Syst. Veg. 2: 111. 1817. *Fuirena ciliaris* auct. non Roxb. 1814; Nees in Wight, Contr. Bot. India 93. 1834. *F. uncinata* (Willd.) Kunth, Enum. Pl. 2: 184. 1837; Clarke in Hook.f., Fl. Brit. India 6: 666. 1893; Cooke, Fl. Pres. Bombay 2: 898. 1908 (3: 413. 1958, repr.ed.); Fischer in Gamble, Fl. Pres. Madras 1669. 1931 (3: 1158. 1957, repr.ed.); Vartak in Journ. Univ. Poona Sci. & Tech. Sect. 40: 200. 1971; Ramaswamy & Razi, Fl. Bangalore 97. 1973; Hooper in Saldanha & Nicolson, Fl. Hassan 684. 1976; Rao & Razi, Fl. Mysore 563. 1981; Sharma *et al.*, Fl. Karnataka 311. 1984; Singh, Fl. E. Karnataka 2: 639. 1988; Karthik. *et al.*, Fl. Ind. Enum. Monocot. 56. 1989; Keshava Murthy & Yoganarasimhan, Fl. Coorg 510. 1990.

Type : India.

Illus. : Vartak, *l.c.*, p. 201; Matthew, Furth. Illus. Fl. Tamilnadu Carnatic Pl. 679. 1988.

Tufted annuals or short-lived perennials with fibrous roots, 4-20 cm high. Stems angular, sulcate, 1-2 mm thick near the base, usually densely pilose. Leaves linear-lanceolate, acute at apex, 1.5-10 cm long, 2-5 mm wide, striate, almost densely pilose on both surfaces; sheaths striate, pilose; orifice membranous, ferrugineous; ligule *ca* 1 mm long; 1 or 2 basal leaves reduced to bladeless sheaths or with very short lamina. Inflorescence a dense cluster of terminal spikelets, usually with 1 or 2 axillary, peduncled clusters; clusters 1-1.5 cm across, bearing few to many spikelets; peduncles pilose. Spikelets sessile, squarrose, broadly ovoid or ellipsoid, 3-6 mm long, 2-3 mm wide, few- flowered. Glumes obovate or elliptic-obovate, obtuse at apex, *ca* 2.3 x 1.2-1.5 mm (excluding the awn), densely pubescent on the upper surface, with a prominent keel; sides with 3 strong yellowish nerves, two of which extending to the 1-1.5 mm long, recurved awn; margins pilose. Perianth biseriate; bristles 3, 0.3 - 0.4 mm long, retrorsely scabrous; scales broadly oblong- ovate or broadly elliptic, truncate-obtuse and short-awned at apex, short-stipitate at base, *ca* 1 x 0.6 mm (including the stipe and awn), ciliate on the margins towards apex. Stamens 3; filaments elongate up to 2 mm; anthers oblong, *ca* 0.6 mm long. Style *ca* 1 mm long; stigmas 3, as long as the style. Nut triquetrous, broadly obovoid or broadly ellipsoid, acute at apex and base, *ca* 0.8 x 0.6 mm, minutely stipitate, smooth, light brown.

Fls. & Frts. : Aug.-Feb.

Habitat : Seasonal wet areas, wet sandy soil in open grasslands; at low altitudes.

Distrib.: Restricted to Sri Lanka and Peninsular India. KARNATAKA: Bangalore (Ramaswamy & Razi, *l.c.*), Hassan, Kodagu, Kolar (Singh, *l.c.*), Mandya, Mysore (Rao & Razi, *l.c.*), Shimoga, Uttara Kannada.

Chrom. No. : 2 n = 36 (Cytologia 36: 13. 1972).

Specimens examined : Hassan: 6 miles before Arsikeri on Tiptur-Arsikerei Road, *Nicolson et al.* 2228, 22.10.1971 (JCB). Kodagu: Kerugoor, *Bhat* 923, 26.1.1981 (MGH). Mandya: Paschimavahini, *Bhat* 59, 8.8.1970 (MGH); Narayanasurya, *Dinesh* 714, 4.12.1983 (MGH). Shimoga: Varahi falls, Hulical, *Raghavan* 80804, 20.5.1962 (BSI). Uttara Kannada: Karwar, *Talbot* 1312, 20.9.1885 (BSI); Jugglepet, *Talbot* 1570, 15.11.1885 (BSI). Without exact locality, *Vartak s.n.* (S. No. 10128 & 10129), Nov. 1971 (MACS).

2. **Fuirena ciliaris** (L.) Roxb., Hort. Beng. 81. 1814 *et* Fl., Ind. 1: 184. 1820; Vartak in Journ. Univ. Poona Sci. & Tech. Sec. 40: 198. 1971; Ramaswamy & Razi, Fl. Bangalore 97. 1973; Kern in van Steenis, Fl. Males. 1, 7: 519, f. 32. 1974; Hooper in Saldanha & Nicolson, Fl. Hassan 683. 1976; Rao & Razi, Fl. Mysore 563. 1981; Rao & Verma, Cyp. NE India 47. 1982; Sharma *et al.*, Fl. Karnataka 311. 1984; Koyama in Dassanayake & Fosberg, Rev. Handb. Fl. Ceylon 5: 150. 1985; Karthik. *et al.*, Fl. Ind. Enum. Monocot. 56. 1989; Keshava Murthy & Yoganarasimhan, Fl. Coorg 509. 1990. *Scirpus ciliaris* L., Mant. 2: 182. 1771. *Fuirena glomerata* auct. non

Lamk. 1791; Clarke in Hook.f., Fl. Brit. India 6: 666. 1894; Cooke, Fl. Pres. Bombay 2: 898. 1908 (3: 412. 1958, repr.ed.); Fischer in Gamble, Fl. Pres. Madras 1669. 1931 (3:1158.1957, repr.ed.).

Illus. : Clarke, Illus. Cyp. t. 59, f.6. 1909; Kern, *l.c.;* Matthew, Furth. Illus. Fl. Tamilnadu Carnatic Pl. 680. 1988.

Annuals with somewhat thick fibrous roots, 10-40 cm high. Stems tufted, obtuse-angular, striate-sulcate, pubescent towards apex, 1-1.5 mm thick. Leaves linear-lanceolate, acute at apex, 2.5-15 cm long, 2-7 mm wide, pubescent throughout, ciliate on the margins; ligule very prominent, 2-3 mm long; sheaths striate, pubescent. Inflorescence usually a terminal cluster along with 1 or 2 peduncled clusters from the upper axils, often a terminal cluster only; each cluster, 0.6-3 cm across, bearing 3-13 spikelets; peduncles pubescent. Bracts leaf-like, shorter than or overtopping the inflorescence. Spikelets ovoid or oblong-ellipsoid, subacute at apex, 5-10 x 3-5 mm, squarrose, densely pubescent. Glumes membranous, oblong-obovate or obovate, obtuse at apex, *ca* 3 x 1 mm (including a 1 mm long, recurved, bristly awn), pubscent throughout outside, pilose in the centre; costa 3- nerved. Perianth biseriate; bristles 3, usually shorter than nut, at times as long as the nut, scaberulous to smooth; scales subquadrate, cordate at base, 3-dentate at apex (with the central teeth largest), distinctly clawed, long-stipitate at base, *ca* 1 x 0.6 mm (including the claw and the stipe), glabrous or minutely hairy at apex. Stamens 3; filaments elongate up to 2 mm; anthers oblong, *ca* 0.3 mm long. Style *ca* 1 mm long; stigmas 3, slightly shorter than style. Nut triquetrous, obovate in outline, cuneate towards base, narrowed to a beak at apex, *ca* 1 x 0.6 mm (including the beak), smooth, stramineous.

Fls. & Frts. : Aug.- April.

Habitat : Open wetlands, swampy grasslands, water loged areas, river banks, along the margin of paddy-fields, wet sandy soil and other marshy areas at low altitudes.

Distrib. : Tropics of the old world; throughout South-East Asia, tropical Africa and Australia. INDIA: Throughout the warmer parts. KARNATAKA: Bangalore (Ramaswamy & Razi, *l.c.*), Belgaum, Chikmagalur (Sharma *et al.,l.c.*) , Dakshina Kannada, Hassan, Kodagu, Mandya, Mysore (Rao & Razi, *l.c.*), Uttara Kannada.

Specimens examined : Bangalore: Bangalore, *Camaron* 327, Feb. 1889 (MH). Belgaum: Londa, without coll. name *s.n.* (Acc. No. 2876), 5.11.1891 (BSI); Londa, along Belgaum Road, *Santapau* 10849, 21.4.1950 (BLAT); Khanapur, *Ahuja* 47740, 5.1.1959 (BSI). Dakshina Kannada: Sampagi, without coll. name 2412, 15.11.1900 (MH); Sullia, *Barber* 2108, 25.10.1900 (MH); Mangalore, *Foulker* 4862, Jan. 1902 (MH); Udyavara, *Bhat* 321, 24.10.1976 (MGH); Panamboor, *Bhat* 480, 5.11.1977 (MGH); Venoor, *Bhat*

711, 16.11.1980 (MGH). Hassan: Magodi estate, *Saldanha* 12245, 15.1.1969 (JCB). Kodagu: Kothamudi, along Cauvery river, Napoklu, *Rao* 74949, 5.10.1961 (BSI); Betoly, Virajpet, *Bhat* 898, 24.12.1980 (MGH); Kirugoor, *Bhat* 937, 26.1.1981 (MGH). Mandya: Paschimavahini, *Padma Rani* 25, 8.8.1970 (MGH); Ranganathittu, *Dinesh* 800, 1.2.1984 (MGH). Uttara Kannada: Yellapur, *Talbot s.n.* 1883; 779, 1883; *s.n.* 1884 (all BSI); Karwar, *Talbot* 1532, 25.9.1885 (BSI); Castle rock, *Gammie s.n.* 1902 (BSI).

3. **Fuirena cuspidata** (Roth) Kunth, Enum. 2: 187. 1837; Singh, Fl. E. Karnataka 2: 638. 1988. *Scirpus cuspidatus* Roth, Nov. Pl. Sp. 31. 1821. *Fuirena wallichiana* Kunth, Enum. 2: 182. 1837; Clarke in Hook.f., Fl. Brit. India 6: 665. 1894; Cooke, Fl. Pres. Bombay 2: 898. 1908 (3: 412. 1958, repr.ed.); Fischer in Gamble, Fl. Pres. Madras 1669. 1931 (3: 1158. 1957 repr.ed.); Vartak in Journ. Univ. Poona Sci. Tech. Sect. 40: 192. 1971; Ramaswamy & Razi, Fl. Bangalore 97. 1973; Hooper in Saldanha & Nicolson, Fl. Hassan 684. 1976; Rao & Razi, Fl. Mysore 563. 1981; Sharma *et al.*, Fl. Karnataka 311. 1984; Karthik. *et al.*, Fl. Ind. Enum. Monocot. 56. 1989.

Illus.: Clarke, Illus. Cyp. t. 58, f. 8-9. 1909; Vartak, *l.c.*, p. 193; Matthew, Furth. Illus. Fl. Tamilnadu Carnatic Pl. 681. 1988.

Perennials with creeping rhizome, 10-90 cm high. Stems robust, many, closely arising from the rhizome, triquetrous, 1.5- 3 mm thick, pubescent towards apex. Lowest leaves reduced to bladeless sheaths or with very small lamina. Upper leaves well developed, linear, narrowed to an acute apex, 4-13 cm long, 2-5 mm wide in the middle portion, with a prominent midrib and recurved margins below, ciliolate towards apex on the margins and the midrib below; sheaths striate and glabrous to puberulous towards mouth. Inflorescence usually a terminal cluster of 3 - few spikelets, often with 1 or 2 axillary peduncled clusters. Bracts 2-3; lowest usually slightly overtopping the inflorescence. Spikelets terete, oblong or oblong-ovoid, subacute at apex, 6-10 x 4-5 mm, many-flowered. Glumes oblong, obtuse at apex, with a subapical awn, 3.5-4 x *ca* 1.5 mm (including the 1-1.5 mm long awn), puberulous on the outer surface towards apex, ciliolate on the margins near apex. Perianth bristles often absent, when present shorter than nut; scales absent. Stamens 3; filaments elongate up to 3 mm; anthers oblong, minutely apiculate at apex, 1-1.5 mm long. Style 1.5 - 1.8 mm long; stigmas 3, as long as the style or slightly shorter. Nut trigonous, broadly obovoid to suborbicular, with a pubescent conical apex, shortly stalked, *ca* 1 x 0.6 mm, with prominent angles and 3-5 vertical lines on the faces, ultimately dark brown; epidermal cells transversely oblong, in *ca* 9 vertical rows on each face.

Fls. & Frts. : Oct.- Aug.

Chrom. No. : 2 n = 38 (*Cytologia* 36: 13. 1972).

Habitat : Seasonal water-logged areas, banks of streams and marshy areas near ponds.

Distrib.: Endemic to India. Western Peninsular India, Central and North-West India. KARNATAKA: Bangalore, Bijapur, Chickmagalur, Dharwar, Hassan, Mysore.

Specimens examined : Bangalore: Bangalore, *Govindu* COA. 162, March 1943 (MH); Doddabadlapur, *Ramesh* 616, 29.3.1978 (JCB). Bijapur: Sindgi-Jivargi Road, 25th km, *Singh* 142628, 19.1.1976 (BSI). Chickmagalur: Sakrepatna, *Ahuja* 59353, 18.5.1959 (BSI). Dharwar: Mugad Khan, *Bhide s.n.* (Acc. No. 2891 & 2892), 8.12.1910 (BSI). Hassan: Sakalaspur - Hassan Road, *Puri* 19933B, 6.6.1957 (BSI); Arsikeri, *Saldanha* & *Ramesh* 1624, 26.6.1978 (JCB). Mysore: Eichlakolame betta, North of Ketedevargudi, *Rao* 80235, 22.4.1962 (BSI); Gundlapet, *Naithani* 23266, 19.4.1965 (MH); without exact locality, *Vartak s.n.* (S.Nos. 10684, 10685 & 10686), Nov. 1971 (MACS); Bhogadi, *Suma* 11, 31.8.1968 (MGH); St. Philomina College, Mysore, *Bhat* 32, 20.7.1970 (JCB).

Note : *Scirpus cuspidatus* Roth (1821) is the earliest valid name for this species and when it is treated under the genus *Fuirena* the correct name is *F. cuspidata* (Roth) Kunth as followed by Singh (1988) and the name *F. wallichiana* Kunth (1837) becomes a taxonomical synonym of the former as it is based on a different type.

4. **Fuirena trilobites** Clarke in Hook.f., Fl. Brit. India 6: 666. 1893; Vartak in Journ. Univ. Poona Sci.& Tech. Sect. 40: 196. 1971; Hooper in Saldanha & Nicolson, Fl. Hassan 683. 1976; Sharma *et al.*, Fl. Karnataka 311.1984; Singh, Fl. E. Karnataka 2: 639. 1988; Karthik. *et al.*, Fl. Ind. Enum. Monocot. 56. 1989. **Fig. 41.**

Illus. : Clarke, Illus. Cyp. t. 59, f. 5. 1909; Vartak, *l.c.*, P. 197; Matthew, Illus. Fl. Tamilnadu Carnatic 790. 1982.

Annuals, tufted, with somewhat thick fibrous roots, 25-50 cm high. Stems compressed, sulcate, densely scabrous on the stem below the inflorescence, 1-2 mm thick. Leaves linear, narrowed to an acute apex, 6-15 cm long, 2-6 mm wide, densely scabrous on both surfaces, with 5-9 prominent veins below; sheaths striate-sulcate, densely scabrous. Inflorescence usually a terminal cluster of 3-13 spikelets along with 1 or 2 peduncled cluster from the upper axils; peduncles densely scabrous. Bracts shorter than to slightly overtopping the inflorescence. Spikelets oblong-ovoid, subacute at apex, 4-8 x 2-3 mm, squarrose, pubescent. Glumes membranous, broadly oblong or oblong-obovate, obtuse at apex, with a long, recurved bristly subapical awn, 2- 2.5 x 1-1.2 mm (including the 1-1.2 mm long awn), pubescent outside, black tinged towards apex; costa 3-nerved, with bristles. Perianth biseriate; bristles 3, shorter than nut, retrorsely scaberulous; scales anchor shaped, with a long stalk, as long

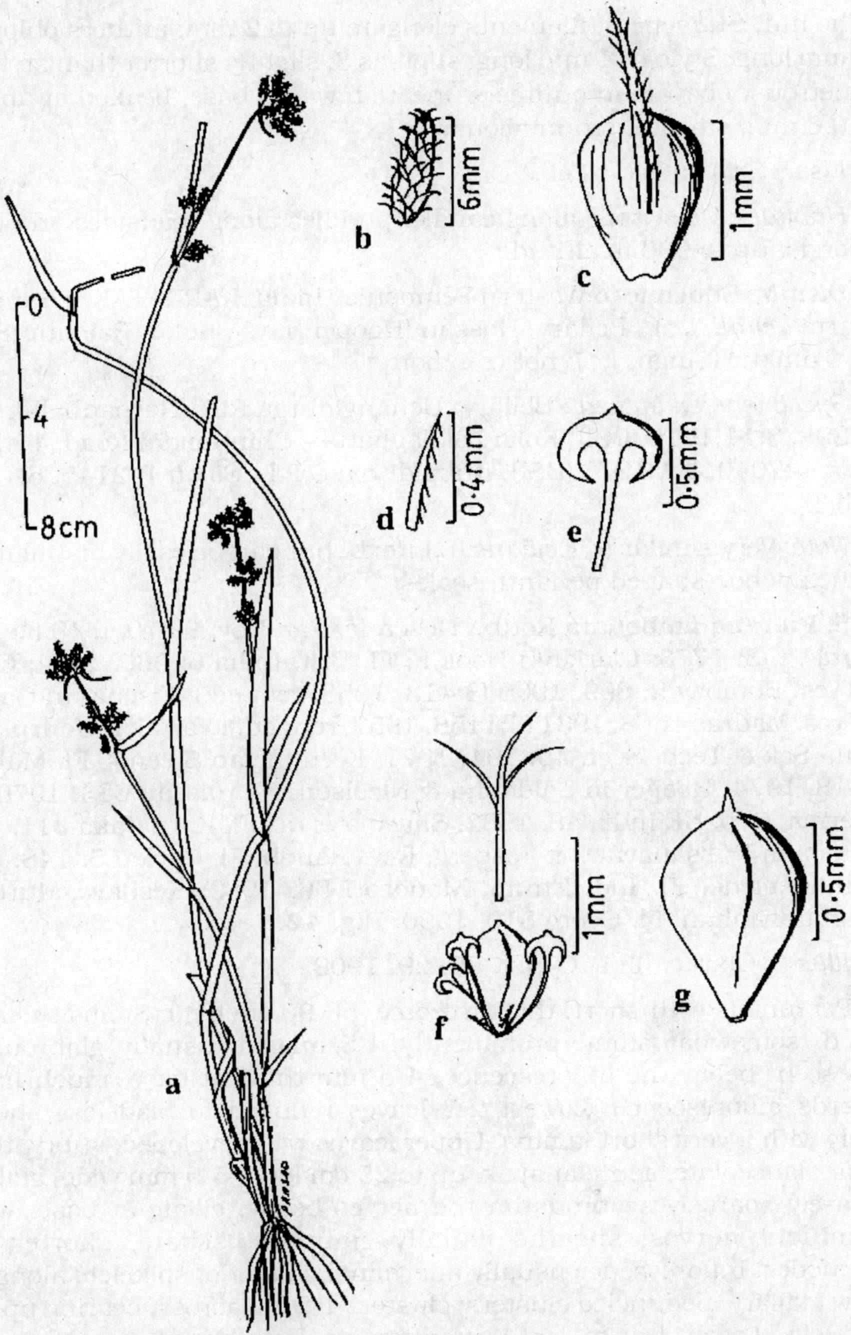

Fig. 41. *Fuirena trilobites* Clarke
a. Habit, **b**. Spikelet, **c**. Glume, **d**. Perianth bristle, **e**. Perianth scale, **f**. Nut with perianth bristles and scales, **g**. Nut

as the nut. Stamens 3; filaments elongate up to 2 mm; anthers oblong, *ca* 0.3 mm long. Style *ca* 1 mm long; stigmas 3; slightly shorter than style. Nut triquetrous, obovate in outline, cuneate towards base, beaked at apex, *ca* 1 x 0.6 mm, smooth, stramineous.

Fls. & Frts. : Oct. - Feb.

Habitat : Moist soils along canals, puddles along roadsides and banks of ponds; up to 900 m altitude.

Distrib.: Endemic to Western Peninsular India. KARNATAKA: Bangalore (Sharma *et al., l.c.*) , Bellary, Hassan (Hooper, *l.c.*) , Kolar, Raichur (Singh, *l.c.*), Tumkur (Singh, *l.c.*); not common.

Specimens examined : Bellary: Uchangidurga R.F. (Harpanhalli), *Singh* 141298, 3.11.1975 (BSI). Kolar: Sidlaghatta - Chintamani Road, 14th km, *Singh* 142040, 3.1.1976 (BSI); Ittikaldurga S.F.I., *Singh* 142145, 6.1.1976 (BSI).

Note: Very similar to *F.ciliaris* (L.) Roxb. but can be easily distinguished by the anchor-shaped perianth scales.

5. **Fuirena umbellata** Rottb., Descr. Ic. Rar. Nov. Pl. 70, t. 19 (i.e. t. 18 *altera*) , f. 3. 1773; Clarke in Hook.f., Fl. Brit. India 6: 666. 1894; Cooke, Fl. Pres. Bombay 2: 899. 1908 (3:413. 1958, repr.ed.); Fischer in Gamble, Fl. Pres. Madras 1668. 1931 (3:1158. 1957, repr.ed.); Vartak in Journ. Univ. Poona Sci. & Tech. Sect. 40: 194. 1971; Kern in van Steenis, Fl. Males. 1, 7: 518. 1974; Hooper in Saldanha & Nicolson, Fl. Hassan 683. 1976; Rao & Verma, Cyp. NE India 48. 1982; Sharma *et al.*, Fl. Karnataka 311. 1984; Koyama in Dassanayake & Fosberg, Rev. Handb. Fl. Ceylon 5: 148. 1985; Karthik. *et al.*, Fl. Ind. Enum. Monocot. 56. 1989; Keshava Murthy & Yoganarasimhan, Fl. Coorg 510. 1990. **Fig. 42.**

Illus. : Clarke, Illus. Cyp. t. 59, f.9. 1909.

Perennials with short, thick rhizome, 50-90 cm high. Stems solitary or tufted, somewhat stout, prominently 4-5 angled, usually glabrous but pubescent below the inflorescence, 4-5 mm thick at base, much narrow towards inflorescence. Lowest 2-3 leaves reduced to bladeless sheaths, rarely with a very short lamina. Upper leaves well developed, subrigid, flat, linear- lanceolate, acute at apex, up to 25 cm long, 5-9 mm wide, glabrous or rarely sparsely scabrous on the nerves below, ciliate at base, with 5 prominent nerves; sheaths usually glabrous, striate, shorter than internodes. Inflorescence usually a terminal cluster of spikelets along with 1-few axillary, peduncled clusters; clusters 1-2 cm across, bearing up to 12 spikelets. Lower bracts leaf-like, often as long as or overtopping the inflorescence, upper ones much shorter. Spikelets ovoid to oblong- ovoid, acute at apex, 5-10 x *ca* 3 mm, pubescent, more or less squarrose. Glumes membranous, obovate or oblong-obovate, obtuse at apex, 2.5-3 x 1.5-2 mm, with a recurved awn of *ca* 1.2 mm, pubescent on the outer surface, ciliate

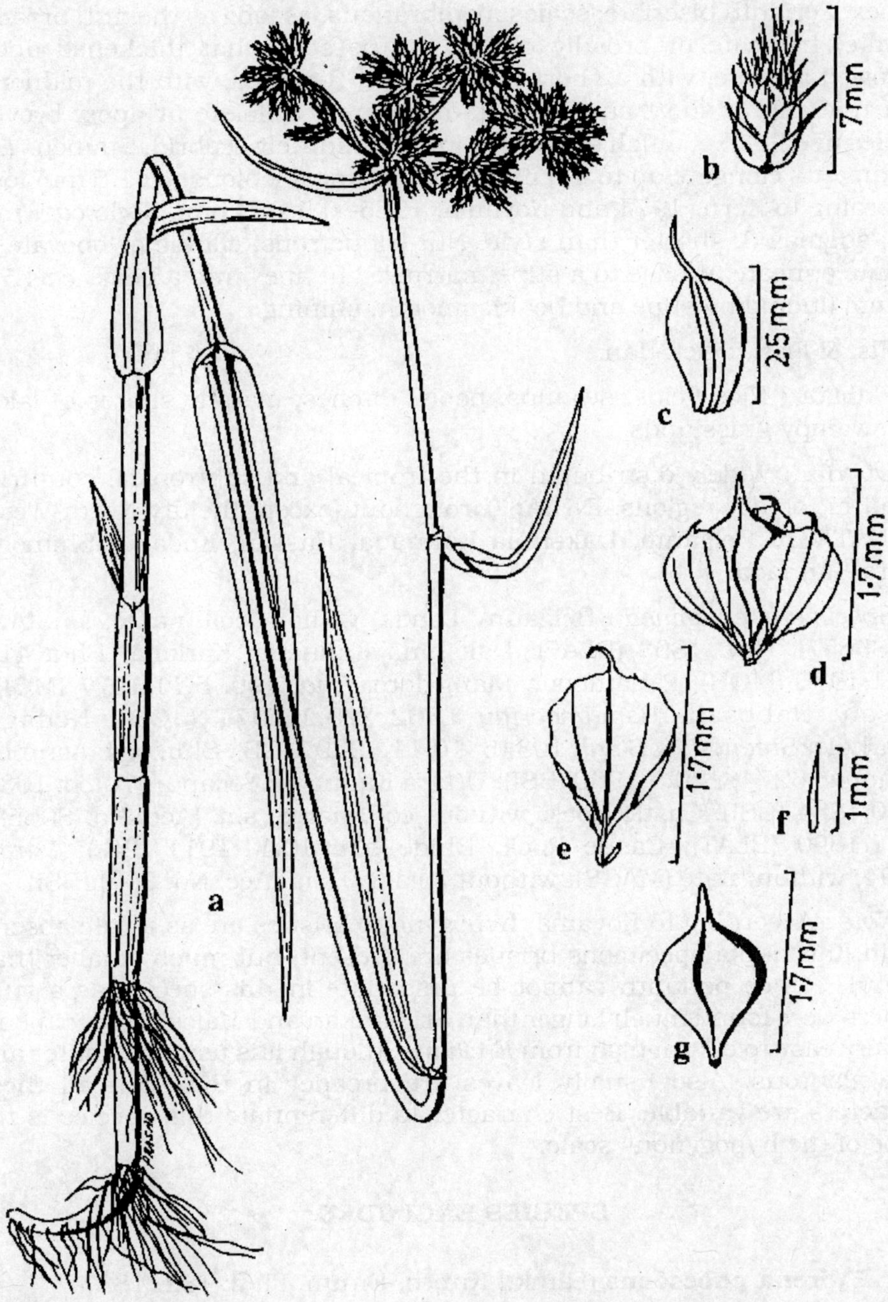

Fig. 42. *Fuirena umbellata* Rottb.
a. Habit, **b.** Spikelet, **c.** Glume, **d.** Nut with perianth scales,
e. Perianth scale, **f.** Perianth bristle, **g.** Nut

at apex. Perianth biseriate; scales membranous, as long as the nut, broadly obovate, truncate or broadly obtuse at the somewhat thickened apex, narrowed at base, with a short curved claw, 3-nerved, with the midnerve excurrent into a short curved mucro, minutely ciliolate at apex, brown; bristles needle-like, much shorter than nut, minutely scabrid. Stamens (2-) 3; filaments elongate up to 4.5 mm; anthers linear-oblong, *ca* 1.5 mm long (according to Kern, 1974 and Koyama, 1985- 0.5-0.7 mm). Style *ca* 1 mm long; stigmas 3, shorter than style. Nut triquetrous, elliptic or obovate in outline, cuneate at base to a stipe, narrowed at apex into a beak, *ca* 1.7 x 0.8 mm (including stipe and beak), smooth, shining.

Fls. & Frts. : Oct.-Jan.

Habitat : Rice fields, swamps, pools, ditches, marshy shores of lakes and swampy grasslands.

Distrib. : Widely distributed in the tropical and subtropical countries except in too dry regions. INDIA: Throughout (except the dry North-West). KARNATAKA: Belgaum, Dakshnia Kannada, Hassan, Kodagu, Shimoga, Uttara Kannada.

Specimens examined : Belgaum: Londa, without coll. name, *s.n.* (Acc. No. 80657), Nov. 1893 (BLAT). Dakshina Kannada: Karkala, *Bhat* 411, 20.11.1976 (MGH); Panamboor, Mangalore, *Bhat* 490, 5.11.1977 (MGH). Hassan: Habbsali, *Ramamoorthy* 1362, 28.1.1971 (JCB). Kodagu: Nagarhole, *Sreenath & Singh* 10345, 10.11.1979 (JCB). Shimoga: Agumbe, *Raghavan* 62746, 20.5.1960 (BSI). Uttara Kannada: Yellapur, *Talbot* 1032, 10.10.1884 (BSI); Castle Rock, without coll. name, *s.n.* (Acc. No. 80658), 14.11.1890 (BLAT); Castle Rock, Bhide *s.n.* 10.11.1911 (BSI); *Vartak* 23094, without date (MACS); without locality, *s.n.* (Acc. No. 803) (BSI).

Note : According to Koyama, hypogynous bristles are as a rule absent. But in Karnataka specimens bristles are present, but much smaller than the nut. Hence perianth cannot be uniseriate in this species as a rule. Anthers were found much longer than Sri Lankan and Malesian specimens. Not very easy to distinguish from *F. ciliaris*, though it is usually stouter and more glabrous. Also usually leaves are broader in this. But all these characters are variable. Best character to differentiate this species is the shape of the hypogynous scale.

SPECIES EXCLUDED

1. **Fuirena pubescens** (Lamk.) Kunth, Enum. Pl. 2: 182. 1837.

Singh (1983 & 1988) reported this species as a new record from Karnataka based on specimens collected from Raichur district (Coll. No. 143045). But a critical study of the specimens revealed that these are *Fimbristylis sieberiana* Kunth. Hence this species is excluded from the State Flora of Karnartaka (*Prasad & Singh*, 1996a).

2. **Fuirena tuwensis** Deshpande & Shah in Bull. Bot. Surv. India 10: 239. 1968 (1969).

Sharma *et al.* included this species in the Flora of Karnataka, probably based on a wrongly identified specimen in BSI (Acc. No. 2876). A study of this specimen revealed that it is nothing but *Fuirena ciliaris* (L.) Roxb. Hence this species is also excluded from the State Flora of Karnataka (*Prasad & Singh, l.c.*).

11. HEMICARPHA

Nees in Linnaea 9: 287. 1834.

Small tufted annuals. Stems slender, leafy at base. Inflorescence a pseudolateral cluster of 1- several spikelets. Lower bract erect, stem-like. Flowers bisexual, stamens 1-2. Ovary partly surrounded by a membraneous scale, sometimes much reduced. Stigmas 2. Nut narrowly to broadly obovate, minutely papillate, mucronate.

6 species distributed in the tropical and subtropical countries; 1 in India.

Hemicarpha isolepis Nees in Arnold, Edinb. New Phil. Journ. 17: 263. 1834; Hooper in Saldanha & Nicolson, Fl. Hassan 685. 1976; Sharma *et al.*, Fl. Karnataka 311. 1984; Karthik. *et al.*, Fl. Ind. Enum. Monocot. 57. 1989. *Scirpus isolepis* (Nees) Boeck. in Linnaea 36: 498. 1870; Clarke in Hook.f., Fl. Brit. India 6: 663. 1893; Fischer in Gamble, Fl. Pres. Madras 1666. 1931 (3: 1156. 1957, repr.ed.); Rao & Razi, Fl. Mysore 564. 1981. *Lipocarpha isolepis* (Nees) Haines in Bot. Not. 124: 476. 1971. *Sanna gundu hullu.*

Illus.: Clarke, Illus. Cyp. t. 52, f. 14-17. 1909.

Annuals with short rhizome. Stems filiform, tufted, 2.5 - 15 cm long, smooth. Leaves 2-3, shorter than stem, filiform, obtuse at apex; sheaths reddish-brown. Inflorescence a solitary, pseudolateral spikelet at right angle to the stem near the apex. Bract single, erect, as if continuation of the stem, 6-30 mm long. Spikelet oblong, obtuse at apex, 2.5 - 4.2 mm long. Glumes densely imbricate, obovate, truncate and mucronulate at apex, *ca* 0.6 mm long, reddish-brown. Hypogynous bristles replaced by 1 or 2 membranous scales, as long as the nut, hyaline. Stamen 1. Style very short; stigmas 2. Nut biconvex (according to Hooper, *l.c.* planoconvex to almost terete), narrowly obovoid, slightly shorter than the glume, minutely muricate, blackish.

Fls. & Frts. : Nov.

Habitat : Moist sandy soil, with bright sunlight.

Distrib. : Thailand, tropical Africa. INDIA: South, Central and East India. KARNATAKA: Hassan, Mysore (Fischer, *l.c.*).

Specimen examined : Hassan: Tank near Dandiganahalli, *Hooper & Gandhi* 2407, 11.11.1971 (JCB).

Note : Species of this genus is variously placed under *Scirpus, Cyperus* and *Lipocarpha* by different workers. As no specimens were available in B S I, description provided above is compiled from Clarke, *l.c.;* Fischer, *l.c.* and Hooper, *l.c.*

12. HYPOLYTRUM

L.C. Rich. in Pres. Syn. Pl. 1: 70. 1805. *Hypaelyptum* Vahl, Enum. Pl. 2: 283. 1806, *p.p. Beera* Lestib., Essai Fam. Cyp. 43. 1819. *Tringa* Roxb., Hort. Beng. 81. 1814 *nom. nud. Tunga* Roxb., Fl. Ind. 1: 187. 1820, *p.p. Albikia* Presl, Rel. Haenk. 1: 184. 1828. *Pandanophyllum* Hassk., Tijid. Nat. Gesch. Phys. 10: 118. 1843, *p.p.*

Type: Hypolytrum latifolium L.C. Rich.

Perennial herbs of medium or large size with short, stout, rhizome. Stems arising from centre of a tuft of radical or subradical leaves, erect, trigonous. Cauline leaves sheathing the stem. Inflorescence a terminal panicle, more or less corymbose, sometimes contracted into a head-like cluster. Bracts leafy, not sheathing. Spikelets few to numerous, terete, ovoid to ellipsoid, often globose while fruiting, many-flowered. Glumes membranous or chartaceous, spirally imbricate, smooth, 1-nerved, muticous or with slightly protruding midrib, basal few empty. Flowers hermaphrodite. Hypogynous scales 2, hyaline, opposite, boat-shaped, ciliate on the acute keel, often dilicately connate on the adaxial side in early stage. Stamens 2. Ovary terminal; style slender, continuous with ovary; stigmas 2, long. Nut biconvex, ovate, elliptic or suborbicular, hard, smooth or irregularly wrinkled, with a long spongy beak adnate to the nut proper.

About 40 species distributed in the tropical regions. 3 in India ; 1 in Karnataka.

Literature: **UITTIEN, H.** (1936) Studies in Cyperaceae - Mapanieae, in *Rec. Trav. Bot. Neerl.* 33: 133-155.

Hypolytrum nemorum (Vahl) Spreng., Syst. 1: 233. 1825; Kern in Blumea 9: 218. 1958, in Back. & Bakh. f., Fl. Java 3: 456. 1968 *et* in van Steenis, Fl. Males. 1, 7: 490. 1974; Hooper in Saldanha & Nicolson, Fl. Hassan 685. 1976; Arora *et al.*, Bot. S. Kanara 61. 1981; Rao & Verma, Cyp. NE India 53. 1982; Sharma *et al.*, Fl. Karnataka 312. 1984; Koyama in Dassanayake & Fosberg, Rev. Handb. Fl. Ceylon 5: 138. 1985; Karthik. *et al.*, Fl. Ind. Enum. Monocot. 57. 1989. *Schoenus nemorum* Vahl, Symb. Bot. 3: 8. 1794. *Hypolytrum latifolium* L.C. Rich. in Pers. Syn. 1: 70. 1805; Clarke in Hook. f., Fl. Brit. India 6: 678. 1894; Fischer in Gamble, Fl. Pres. Madras 1673. 1931 (3: 1161. 1957, repr.ed.). *H. wightianum* Boeck. in Linnaea 37:

130. 1871; Clarke in Hook.f., Fl. Brit. India 6: 678. 1894; Cooke, Fl. Pres. Bombay 2: 899. 1908 (3: 414. 1958, repr.ed.). **Fig. 43.**

Type: India.

Illus. : Clarke, Illus. Cyp. t. 106, f.1-14. 1909.

Rhizome woody, with many thick, wiry roots. Stems usually solitary, stout, triquetrous, with 1 or 2 well developed cauline leaves, 50-120 cm high, 2-6 mm thick. Leaves mostly crowded at base, longer than stem, broadly linear, gradually narrowed to an acute apex, up to 130 cm long, (6-) 13-20 mm wide, with 3 prominent nerves, antrorsely serrulate on margins in the upper half. Lower most leaves reduced to bladeless sheaths. Inflorescence paniculate-corymbose, open or dense, hemispherical or broadly ovoid in shape, (2-) 4-10 cm across, with few to numerous spikelets; primary rays spreading, 1-4 cm long, rigid, smooth, base covered with a bifid prophyll. Bracts 2-4; lowest 1 or 2 leaf-like, much overtopping the inflorescence, up to 65 cm long. Spikelets oblong-ovoid or ellpsoid before fruiting, acute to subobtuse at apex, 4-7 x 2-3 mm, densely many-flowered; subglobose or ovoid while fruiting. Glumes membranous, obovate-orbicular or oblong-orbicular, obtuse at apex, 2-2.5 x 1.2-1.5 mm, densely brown-lineolate throughout; midnerve prominent. Hypogynous scales not connate, up to 2 mm long, with a spinulose keel. Staminal filaments up to 4 mm long; anthers up to 1 mm long. Nut slightly compressed, ovoid to subglobose, 2-4 x 1.5-2 mm (including the conical beak), smooth or rugulose, light brown to chestnut brown, often brown-spotted; at times with a constriction between the nut proper and the conspicuous beak.

Fls. & Frts. : Almost throughout the year.

Habitat : Wet or swampy areas in forests, usually in shady areas, along streams, near ponds, along canals, at times half submerged; 100-1000 m altitude.

Distrib. : South China, Formosa, Polynesia, Malesia, Fiji and Australia. INDIA: South, North-East India and Andaman & Nicobar Islands. KARNATAKA: Belgaum, Chikmagalur, Dakshina Kannada, Hassan, Shimoga, Uttara Kannada.

Specimens examined : Belgaum: Castle Rock - Anmod Road, *Saldanha* 7832, 22.5.1979 (JCB). Chikmaglur: Kemmangudi, *Bharati s.n.,* without date (MGH). Dakshina Kannada: Varavi R.F., without coll. name, 018139, 20.11.1927 (MH); Bisle ghat, without coll. name, 018250, 1.12.1927 (MH); Kervashe, *Raghavan* 143556, 9.4.1977 (BSI); Mundaje-Kudrmukha foot hills, *Saldanha & Sreenath* 4965, 10.12.1978 (JCB). Hassan: Kempuhole, Shiradi ghat, *Saldanha* 14445, 7.8.1969 (JCB). Shimoga: Agumbe, State reserve - Vanakaeppa falls reserve, *Rao* 1302, 10.3.1960 (BSI); Agumbe-Someshwar Road, near Forest nursery, *Raghavan* 62643, 18.5.1960 (BSI); Ghatibagh, Agumbe, *Raghavan* 80594, 14.5.1962 (BSI);

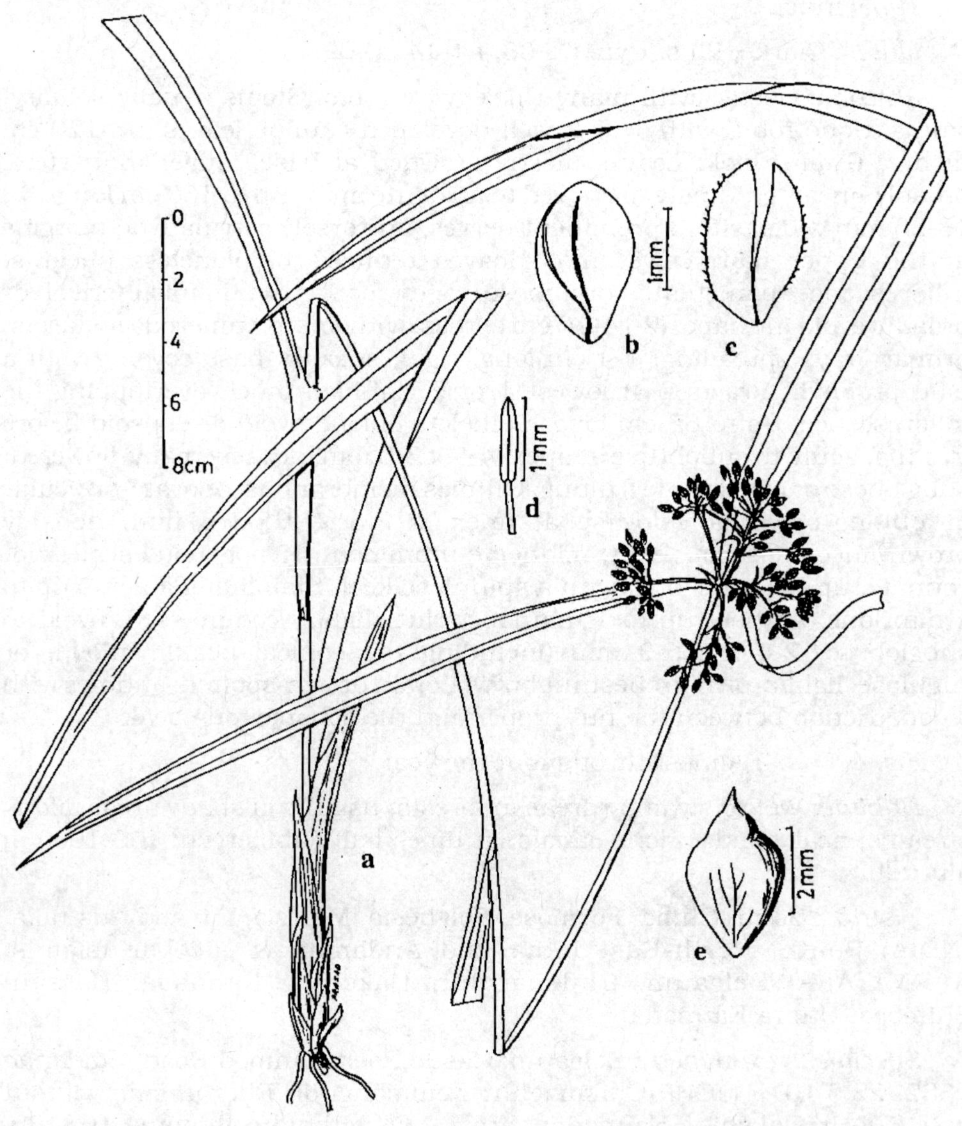

Fig. 43. *Hypolytrum nemorum* (Vahl) Spreng
a. Habit, **b**. Glume, **c**. Hypogynous scales, **d**. Stamen, **e.** Nut

Kolegar, 14 km from Murdeshwar, *Raghavan* 79561, 19.2.1962 (BSI); Gubbia, near Yadur, *Raghavan* 80837, 21.5.1962 (BSI); Varahi falls near Hulical, *Raghavan* 90248, 26.8.1963 (BSI); Jog, *Karanth* 101, 16.12.1965 (MGH); Agumbe, *Bhat* 717, 7.12.1980 (MGH). Uttara Kannada: Gersoppa, *Talbot* 1075, 18.11.1884 (BSI); Falls of Gersoppa, without coll. name, 29, 22.11.1884 (MH); Yellapur, *Talbot* 1075, Dec. 1884 (BSI); Bhatgal, without Coll. name, *s.n.* (Acc. No. 80666), Feb. 1893 (BLAT); Castle Rock, Chibber *s.n.* (Acc. No. 2898), 23.3.1908 (BSI); Yellapur Road, *Bell* 2128, 3.7.1920 (BLAT); Castle Rock, *Vartak* 22935, 27.12.1962 (MACS); Gersoppa Road, *Chibber s.n.* (Acc. No. 2897), 16.5.1911 (BSI); Katlekan, *Almeida* 1212, 8.5.1969 (BLAT); Castle Rock, *Almeida* MRA-1178, 21.2.1981 (BNHS); Near Nagarajeddinala, *Ramesh & Udayakumar* 13571, 17.9.1981 (JCB). Kanara Coast without exact locality, coll. name, number and date (Acc. No. 2901) (BSI). Without locality, coll. name, number and date (Acc. No. 73971) (MH).

Note : A polymorphic species showing high degree of variation in nut morphology which resulted in the seggregation of several other species from it (see Prasad & Singh, 1996b).

13. KYLLINGA

Rottb., Descr. Ic. Rar. Nov. Pl. 12. 1773. *Cyperus* subgenus *Kyllinga* (Rottb.) Valck. Sur. Gesl. Cyp. Mal. Arch. 42. 1898; Kern in van Steenis, Fl. Males. 1, 7: 653. 1974 (excl. sect. *Queenslandiella*). *Cyperus* sect. *Eu-kyllinga* Kukenth. in Engl. Pflanzenr. 4 (20), Heft 101: 576. 1936.

Type : *Kyllinga brevifolia* Rottb.

Perennials with short or horizontally creeping rhizome. Stems trigonous. Leaves basal, narrow, elongated, at times reduced to bladeless sheaths. Inflorescence terminal, capitate, with 1-few sessile spikes. Spikes cylindrical, ellipsoid or globose. Spikelets many to numerous, densely packed on a short rachis, laterally compressed, usually with 1-2 (rarely up to 5) flowers, falling off as a whole; rachilla short, disarticulating above the prophyll. Glumes several (usually 4), 2-ranked, folded, membranous or hyaline; lowest two smaller than the upper ones, empty; third and fourth longer and flower bearing; distal glume empty or bearing staminate flower; keel prominent, sometimes winged. Flowers bisexual and staminate; hypogynous bristles absent; stamens 2 or 3; pistil digynous; style not jointed at base, bifid into 2 stigmas. Nut bilaterally flattened, with one angle facing rachilla.

About 40 species distributed in temperate, subtropical and tropical regions of both the hemispheres. 10 species in India; 7 in Karnataka.

Literature : **PADHYE, M.D.** (1966) The pollen grains of *Kyllinga* Rottb. in *Palynological Bull. Lucknow II & III: 101-103.*

Key to the species

1a.Rhizome well developed, creeping 2

1b.Rhizome very short or insignificant, never creeping 5

2a.Keel of the nut-bearing glumes distinctly winged 3

2b.Keel of the glumes wingless 4

3a.Rhizome horizontally creeping; stems triquetrous. ....**5. K. nemoralis**

3b.Rhizome vertically (obliquely) descending; stems 3-winged. **3. K. eglandulosa**

4a.Rhizome slender, 1-2 (-3) mm thick; stems usually very distant (at times closely arranged), usually less than 30 cm long, 1-1.5 mm thick; leaves usually well developed; nut 1-1.5 mm long ....**1. K. brevifolia**

4b.Rhizome stout, 2-4 mm thick; stems always closely arranged, usually 30-170 cm long (less than 30 cm in ssp. *bifolia*), 2-4 mm thick; leaves usually reduced to sheaths or the upper one or two short-laminate (upper leaves fairly well developed in ssp. *bifolia*); nut 1.5-2 mm long ....**4. K. melanosperma**

5a.Keel of the glumes winged with a broad, hyaline, toothed crest .... **7. K. squamulata**

5b.Keel of the glumes wingless 6

6a.Central spike globose to ovoid-globose, 5-9 x 5-6 mm; nut oblong, not truncate at apex, 1.2-1.5 x 0.5-0.7 mm .... **2. K. bulbosa**

6b.Central spike cylindrical, 10-12 x 4-5 mm; nut obovate or elliptic-obovate, truncate at apex, *ca* 1.5 x 1 mm .... **6. K. odorata**

1. **Kyllinga brevifolia** Rottb., Descr. Ic. Rar. Nov. Pl. 13, t. 4, f. 3. 1773; Clarke in Hook. f., Fl. Brit. India 6: 588. 1893; Fischer in Gamble, Fl. Pres. Madras 1624. 1931 (3: 1130. 1957, repr.ed.); Hooper in Saldanha & Nicolson, Fl. Hassan 686. 1976; Sharma *et al.*, Fl. Karnataka 312. 1984; Koyama in Dassanayake & Fosberg, Rev. Handb. Fl. Ceylon 5: 248. 1985; Singh, Fl. E. Karnataka 2: 640. 1988; Karthik. *et al.*, Fl. Ind. Enum. Monocot. 60. 1989; Keshava Murthy & Yoganarasimhan, Fl. Coorg 510. 1990. *Cyperus brevifolius* (Rottb.) Hassk., Cat. Hort. Bogr. 24. 1884; Ramaswamy & Razi, Fl. Bangalore 81. 1973; Kern in van Steenis, Fl. Males., 1, 7: 656. 1974; Rao & Razi, Fl. Mysore 558. 1981; Rao & Verma, Cyp. NE India 6. 1982.

Type: India, *Koenig.*

Illus. : Clarke, Illus. Cyp. t. 1, f. 1-4. 1909; Matthew, Furth. Illus. Fl. Tamilnadu Carnatic Pl. 682. 1988.

Key to the varieties

1a. Glumes 2.5-3.5 mm long, with spinulose keel... .... var. **brevifolia**

1b. Glumes *ca* 2 mm long, with glabrous keel ....var. **stellulata**

var. **brevifolia**

Perennials. Rhizome horizontally creeping, subterranean or epigeal, covered by brownish scales, 1-3 mm thick. Stems distantly or closely arranged in a row, triquetrous, 2-30 (-45) cm long, 0.5 - 1.5 (-2) mm thick, smooth. Leaves usually well developed, radical or subradical, usually shorter than stems, at times up to the inflorescence, narrowly linear, gradually or abruptly acuminate at apex, 1-3 (-4) mm wide; sheaths membranous, brownish or purplish, lowest 1 or 2 (-4) bladeless. Inflorescence a terminal, globose or ovoid head, 4-10 x 5-8 mm, greenish, ultimately stramineous, rarely with 1 or 2 smaller heads from the base, densely bearing numerous spikelets. Involucral bracts (2-)3-4, leafy, unequal in size; lowest 2-10 (-20) cm long; often erect. Spikelets strongly compressed, elliptic-oblong or oblong-lanceolate, 2-3.5 x *ca* 1 mm, 1 (-2)-flowered. Glumes elliptic-ovate to boat-shaped, membranous; first glume 0.5 - 1 mm long; second *ca* 1 mm; third and fourth mucronate at apex, 2.5- 3.5 mm long, strongly compressed, acutely keeled, with 2-3-nerved sides; keel usually spinulose. Stamens 1-2 (rarely 3); filaments up to 3 mm long; anthers linear, *ca* 1 mm long. Style *ca* 1 mm long; stigmas 2, *ca* 1 mm long. Nut biconvex, laterally compressed, obovate or elliptic, apiculate at apex, 1-1.5 x 0.5-0.8 mm, ultimately brownish.

Fls. & Frts.: Throughout the year.

Chrom.No. : 2n=120 (*Cytologia* 37: 13.1972).

Habitat : Moist soil near canals, streams, rivers, tanks, lakes etc.; muddy bottom of rocky river beds and small puddles in rocky area; moist margin of paddy fields; also along sea shores. Common.

Distrib. : Pantropic. INDIA: Throughout. KARNATAKA: Bangalore, Belgaum, Bijapur, Chikmagalur, Dakshina Kannada, Dharwar, Hassan, Kodagu, Mandya, Mysore, Shimoga, Tumkur, Uttara Kannada.

Specimens examined : Bangalore: Hulimavu, *Hooper & Saldanha* 18016, 7.11.1971 (JCB). Belgaum: Gokak falls, *Prasad* 172812 & 172817, 11.12.1994 (BSI); Gottni Nallah, Jambotti R.F, *Prasad* 172874, 15.12.1994 (BSI); Dudwawada R.F. Londa, *Prasad* 172895, 17.12.94 (BSI). Bijapur: Varathikavlu (Canal), Kendur, *Prasad* 172959, 21.12.1994 (BSI). Chikmagalur: Near Forest Office, Koppa, *Prasad* 173812, 7.11.1995 (BSI); Gadical village tank, Koppa taluk, *Prasad* 173840, 8.11.1995 (BSI). Dakshina Kannada: Kannadekatte, *Bhat* 180, 6.5.1975 (MGH); Suratkal,

Saldanha, Ramesh & Ravindra 2013, 29.7.98 (JCB); Jappinamogeru village, Mangalore, *Prasad* 173857, 10.11.95 (BSI). Dharwar: Sadankeri, along Alnavar Road, *Prasad* 172940, 19.12.1994 (BSI). Hassan: Bisle forest, *Mahajan* 19829A, 5.6.1957 (BSI); Byra, *Nicolson et al.*, 2267, 23.10.1971 (JCB); Halbag, *Hooper & Saldanha* 2504, 26.11.1971 (JCB). Kodagu: Katakeri, *Puri* 31736 (A), 7.3.1958 (BSI); Karavala Badiga, West of College, *Rao* 4518, 28.9.1961 (BSI); Kambile, Santicoppa, *Rao* 8611, 3.3.1963 (BSI); Gorikoppal, *Bhat* 850, 22.12.1980 (MGH); Kushalnagar, *Bhat* 953, 16.2.1981 (MGH); Abbe falls, *Yoganarasimhan & Murthy* 3829, 21.5.1983. Mandya: Ranganathittu, *Dinesh* 738, 1.2.1984 (MGH). Mysore: Kottathuhalla, Biligirirangan hill ranges, *Rao* 80175, 21.4.1962 (BSI) & 80386, 27.4.1962 (BSI). Shimoga: Thirthahalli, *Raghavan* 82829, 30.9.1962 (BSI), 82973, 4.10.1962 (BSI); Gaurikeri, Thalaguppa, Sagar taluk, *Prasad* 173798, 5.11.95 (BSI); Jog falls, Forest Guest house campus, *Prasad* 173810, 5.11.1995 (BSI). Tumkur: Namadachilume, near Forest Rest house, *Rao* 73271, 28.8.1961 (BSI); Namadachilume, *Singh* 132681, 26.2.1975 (BSI). Uttara Kannada: Yellapur, *Talbot s.n.*, 10.10.1884 (BSI); Karwar, *Talbot s.n.*, 1.8.1885 (BSI); Castle Rock, *Gammie s.n.*, without date (BSI); Castle Rock, *Almeida* 2703, 14.4.1973 (BNHS); Karlukatta tank, Halyal, *Prasad* 173713, 29.10.1995 (BSI); Karka river bed, Dandeli R.F., *Prasad* 173721, 30.10.1995 (BSI); Javalli village, Halyal range, *Prasad* 173741, 30.10.1995 (BSI); Karwar beach, *Prasad* 173750, 31.10.1995 (BSI).

var. **stellulata** (Valck. Sur.) Ohwi, Bot. Mag. Tokyo 56: 199. 1942; Hooper in Saldanha & Nicolson, Fl. Hassan 687. 1976; Karthik. *et al.*, Fl. Ind. Enum. Monocot. 60.1989. *Cyperus brevifolius* f. *stellulatus* Valck. Sur., Gesl. Cyp. Males. Arch.48,t. 2, f. 5. 1898. *C. brevifolius* var. *stellulatus* (Valck. Sur.) Kukenth. in Engl., Pflanzenr. 4 (20), Heft 101: 603. 1936; Kern in van Steenis, Fl. Males. 1, 7: 658. 1974.

Plants dwarf and inflorescence small with fewer spikelets (Kern, *l.c.*). According to Hooper, *l.c.* lowermost bract usually erect.

Fls. & Frts. : Oct. (Hooper, *l.c.*).

Habitat : Roadside in dry deciduous forest zone (Hooper, *l.c.*).

Distrib. : Malesia, Philippines, Australia. INDIA: South India. KARNATAKA: Hassan (Hooper, *l.c.*).

Note : This variety is included on the authority of Hooper (*l.c.*).

2. **Kyllinga bulbosa** Beauv., Fl. d'Oware & Benin 1: 11, t. 8, f.1. 1804; Koyama in Dassanayake & Fosberg, Rev. Handb. Fl. Ceylon 5: 245. 1985. *K. triceps* Rottb., Descr. Ic. 14, t. 4, f. 6. 1773, *nom. illeg*; Clarke in Hook.f., Fl. Brit. India 6: 587. 1893; Cooke, Fl. Pres. Bombay 2: 877. 1908 (3: 391. 1958, repr.ed.); Fischer in Gamble, Fl. Pres. Madras 1623. 1931 (3: 1130. 1957, repr.ed.). *Cyperus triceps* (Rottb.) Endl., Cat. Hort. Acad. Vindb. 1:

94. 1842; Kunkenth. in Engl., Pflanzenr. 4 (20) Heft. 101: 578. 1936; Ramaswamy & Razi, Fl. Bangalore 95. 1973; Kern in van Steenis, Fl. Males. 1, 7: 659. 1974; Rao & Razi, Fl. Mysore 561. 1981. *Kyllinga tenuifolia* Steud., Syn. Pl. Glum. 2: 69. 1855; Hooper in Saldanha & Nicolson, Fl. Hassan 688. 1976; Yoganarasimhan *et al.*, Fl. Chikmagalur 359. 1981; Sharma *et al.*, Fl. Karnataka 312. 1984; Singh, Fl. E. Karnataka 2: 641. 1988; Karthik. *et al.*, Fl. Ind. Enum. Monocot. 61. 1989; Keshava Murthy & Yoganarasimhan, Fl. Coorg 511. 1990. *K. cylindrica* auct. non Nees, 1834; Clarke in Hook. f., Fl. Brit. India 6: 588. 1893. *Ananta gonde hullu.*

Illus. : Beauv., *l.c.*; Rottb., *l.c.*; Matthew, Illus. Fl. Tamilnadu Carnatic Pl. 785. 1982.

Tufted perennials with short rhizome, 15-25 cm high; stolons absent. Stems slender, obtusely trigonous, 0.5-1.5 mm thick, smooth; base forming an ovoid enlargement covered by brownish fibres of disintegrating sheaths. Leaves shorter or longer than the stems, linear; gradually acuminate, 1-3 mm wide; sheaths somewhat stout, often brownish at base. Inflorescence a head of (1-) 3 (-5) sessile spikes; central spike globose to ovoid-globose, obtuse at apex, 5-9 x 5-6 mm; lateral spikes globose, smaller. Involucral bracts 3-4, spreading to reflexed, up to 10 cm long. Spikelets numerous, densely packed, strongly compressed, oblong to oblong-lanceolate, acute at apex, 1.7-2.5 x 0.6-1 mm, pale green to whitish, 1-flowered; rachilla short, disarticulating at base. Glumes mostly 4, ovate-oblong to oblong-lanceolate, acute to submucronulate at apex, with a smooth keel, hyaline; lowest two glumes very small, 0.3-0.5 mm long, 1 or 3-nerved; third glume 1.5-2 mm long, 5 or 7-nerved; fourth glume, 1.7-2.5 mm long, 5-nerved. Stamens 2; filaments elongate up to 2.5 mm; anthers linear-oblong. Style *ca* 0.5 mm long; stigmas 2, up to 0.8 mm long. Nut laterally compressed, oblong, apiculate at apex, 1.2-1.5 x 0.5 - 0.7 mm, brownish.

Fls. & Frts. : July - Oct.

Chrom. No. : n = 48 (*Taxon* 20: 612. 1971).

Habitat : In moist sandy soils, on hilly areas during monsoon, among grasses in swampy areas, grasslands and as a weed in lawns.

Distrib. : Sri Lanka, China, Myanmar, Indo-China, Malesia, North Australia and tropical Africa. INDIA. Throughout (except North-East). KARNATAKA: Bangalore, Bidar, Bijapur, Chikmagalur, Dakshina Kannada, Gulburga, Hassan, Kodagu, Mandya, Raichur, Shimoga, Tumkur.

Specimens examined : Bangalore: Bangalore, *Anstead* 83381, Nov. 1917 (MH); Lalbag Botanic garden, *Rao* 73489, 31.8.1961 (BSI); Bannargatta, *Saldanha* 19305, 13.8.1977 (JCB); Bangalore, *R.D.A.* B. 218, Dec. 16. without year (MH). Bidar: Birnali R.F., *Singh* 142814, 10.8.1976 (BSI) & 140892, 23.10.1975 (BSI). Bijapur: Badami hills, *Paranjpye s.n.* (Acc. No. 2508), 1.9.1912 (BSI). Chikmagalur: Basavani tank, Koppa taluk, *Prasad* 173848, 8.11.1995 (BSI). Dakshina Kannada: Manipal, *Bhat* 255,

2.8.1976 (MGH). Gulbarga: Narayanpur Road, *Singh* 142999, 16.8.1976 (BSI). Hassan: 6 miles before Arsikere, on Tiptur-Arsikere Road, *Nicolson et al.* 2235, 22.10.1971 (JCB). Kodagu: Kushalnagar, *Bhat* 978, 15.7.1981 (MGH). Mandya: Paschimavahini, *Bhat* 95, 20.11.1970 (MGH); Belluru, *Nicolson et al.* 2354, 26.10.1971 (JCB). Mysore: Kuntugudi, Biligirirangan temple vicinity, *Rao* 73677, 5.9.1961 (BSI); Chamundi hills, *Bhat* 10, 10.7.1970 (MGH); Mysore, *Thomson s.n.* (Acc. No. 73408 B), without date (MH). Raichur: Maliyabab, *Wadhwa* 44724, 27.9.1958 (BSI). Shimoga: Heggarguda, Thirthahalli, *Raghavan* 90047, 20.8.1963 (BSI). Tumkur: Siddarabetta, Kollikullu state forest, *Yoganarasimhan* 488, 20.8.1974 (RRCBI); Devarayadurga forest, *Ramesh & Murthy* 2723, 15.9.1998 (JCB).

3. **Kyllinga eglandulosa** Govind. & Ramani in J. Econ. Tax. Bot. 18 (2): 335, f. 1.1994.

Type : India, Karnataka State, Bababudan Hills, *Govindarajalu* 8726 (CAL).

Illus. : Govind. & Ramani, *l.c.*

Perennials with thick, woody, vertically (obliquely) decending rhizome. Roots thick, pale brown, sometimes woolly. Stems 1-2 (-3), approximate, rigid, erect, 10-15 cm long, 3-winged; basal portion covered by 4-6, broadly ovate, brown, membranous sheaths. Leaves as long as stems, flat, acuminate at apex, 2-4 mm wide, scabrid on upper surface; sheaths tubular, with purple dotted, membranous sides and transversely truncate mouth. Spikes solitary, ovate-subglobose, 6-8 mm broad, pale brown. Spikelets numerous, compressed, ovate-elliptic, 2-2.5 mm long, usually 1-flowered. First glume linear-oblong or subulate, nerveless, *ca* 1.5 mm long; second glume broadly ovate, 2-nerved on each side, 1x 1.2- 1.4 mm; third glume nut-bearing, distinctly setulose in the middle, pale brown or stramineous, 3(-4)-nerved, abruptly constricted at base, with winged keel, 2-2.1 x 2 mm; wing moderately and uniformly broad from base to apex, eglandular; mucro short, excurved or erect; fourth glume similar to third glume, 2.6 x 2 mm, smooth. Stamens 2; anthers linear-oblong, 0.6-0.7 mm long, with red apiculus. Nut biconvex, ellipsoid, narrowed towards both ends or ovate-ellipsoid, apiculate at apex, stipitate, 1.5 x 0.9 mm, with granular surface, dark castaneous brown.

Fls. & Frts. : Nov.

Distrib.: Endemic to Karnataka (Bababudan hills, Chickmagalur).

Note: It seems there is not much difference between this and *K. nemoralis* (Forst.) Dandy, though it is included here on the authority of Govindarajulu & Ramani, *l.c.* As no specimens were available in BSI, description provided above is as given in the protologue.

4. **Kyllinga melanosperma** Nees in Wight, Contr. Bot. India 91. 1834; Clarke in Hook. f., Fl. Brit. India 6: 588. 1893; Fischer in Gamble, Fl. Pres.

Madras 1624. 1931 (3:1130. 1957, repr.ed.); Hooper in Saldanha & Nicolson, Fl. Hassan 687. 1976; Sharma *et al.*, Fl. Karnataka 312. 1984; Koyama in Dassanayake & Fosberg, Rev. Handb. Fl. Ceylon 5: 246. 1985; Karthik. *et al.*, Fl. Ind. Enum. Monocot. 60. 1989; Keshava Murthy & Yoganarasimhan, Fl. Coorg 511. 1990. *Cyperus melanospermus* (Nees) Valck. Sur., Gesl. Cyp. Mal. Archip. 50, t. 2, f. 8. 1898; Kukenth. in Engl., Pflanzenr. 4 (20), Heft. 101: 583. 1936; Ramaswamy & Razi, Fl. Bangalore 90. 1973; Kern in Fl. Males. 1, 7: 655. 1974.

Type: India, *Rottler*.

Illus.: Valck. Sur., l.c.; Matthew, Furth. Illus. Fl. Tamilnadu Carnatic 683. 1988.

Key to the subspecies

1a. Stems 30-170 cm long, 2-4 mm thick; leaves all reduced to sheaths or the upper 1 or 2 short laminate .... subsp. **melanosperma**

1b. Stems up to 70 cm long, *ca* 2 mm thick; leaves fairly well developed except the basal ones .... subsp. **bifolia**

subsp. **melanosperma**

Perennials, 30-170 cm high; rhizome creeping, 2-4 mm thick, covered by ovate-lanceolate brownish to blackish sheaths, very aromatic. Stems arranged closely in a row along the rhizome, erect, triquetrous, almost winged, 2-4 mm thick, smooth. Leaves reduced to bladeless sheaths or the uppermost 1 or 2 short-bladed; blades if present, up to 3 cm long, 3-4 mm wide, abruptly acuminate at apex; sheaths membranous, transversely corrugate on the anterior side, purplish to reddish-brown tinged. Inflorescence a single globose or ovoid to oblong-ovoid head, 8-14 x 7-9 mm, greenish. Involucral bracts 3, spreading or reflexed, up to 20 cm long. Spikelets numerous, densely arranged, flattened, elliptic-oblong to elliptic-ovate, 3-3.5 (-4.5) x 1-1.2 mm, 1 (rarely -2)-flowered. Glumes 4, ovate-lanceolate or elliptic- ovate, mucronulate, with a smooth or spinulose keel; lower two 2.5-3 mm long; third 2.5- 3.5 mm long, with 3-4 nerved sides; fourth up to 4 mm long, with 2-nerved sides. Stamens 3; anthers linear, *ca* 1 mm long. Style 1.5-2 mm long; stigmas 2, shorter than style. Nut laterally compressed, oblong or elliptic-oblong, apiculate at apex, 1.5-2 x 0.6 - 0.7 mm, brown to blackish.

Fls. & Frts. : July-May.

Habitat : Wet slippery rocks near water falls, shady areas, rice fields, open wet places and wet grasslands.

Distrib. : South Asia, Malesia and tropical Africa. INDIA: Peninsular India, Central and Eastern India. KARNATAKA: Bangalore (Ramaswamy &

Razi, *l.c.*), Belgaum, Chikmagalur (Sharma *et al., l.c.*) , Dakshina Kannada, Hassan, Kodagu, Mandya, Mysore, Shimoga, Uttara Kannada.

Specimens examined : Belgaum: Jamboti, *Ahuja* 53766, 10.3.1959 (BSI). Dakshina Kannada: Kodachadri, *Bhat* 592, 17.11.1978 (MGH). Hassan: Near Cardamom estate, Hassan-Saklespur Road, *Nicolson et al.* 2279, 23.10.1971 (JCB). Kodagu: Abbe falls, *Bhat* 880, 23.12.1980 (MGH); Ponnampet, *Bhat* 914, 26.1.1981 (MGH); Nagarhole, *Bhat* 967, 7.3.1931 (MGH); Bhagamandala-Karike Road, *Yoganarasimhan et al.* 3876, 23.5.1983 (RRCBI). Mandya: Ranganathittu, *Bhat* 66, 28.8.1970 (MGH). Mysore: Sampagee, Koynad Range, *Wadhwa* 40211, 6.3.1959 (BSI); Biligiri Rangan hills, *Ramesh & Manohar* 8679, 1.8.1979 (JCB). Shimoga: Varahi, *Raghavan* 97161, 24.3.1969 (BSI); Jog falls, *Prasad* 173784, 4.11.1995 (BSI). Uttara Kannada: Arabial ghat, *Ramesh & Sivaprakash* 13358, 20.7.1981 (JCB); without exact locality, *Talbot s.n.* (Acc.no.718), without date (BSI).

subsp. **bifolia** (Miq.) Karthik. in Karthik. *et al*., Fl. Ind. Enum. Monocot. 60. 1989. *Kyllinga bifolia* Miq., Fl. Ind. Bat. 3: 293; 1856; Sharma *et al.,* Fl. Karnataka 312. 1984. *Cyperus brevifolius f. vaginatus* Valck. Sur., Gesl. Cyp. Mal. Arch. 47, t. 2, f. 4. 1898, *p.p. Cyperus melanospermus* (Nees) Valck. Sur. subsp. *bifolius* (Miq.) Kern in Reinwardtia 3: 62, f. 12. 1954; Govind. in Bull. Bot. Surv. India 8: 352 (1966); Kern in van Steenis, Fl. Males. 1, 7: 656, f. 69. 1974.

Illus.: Valck. Sur., *l.c.*; Kern, *l.c.*

Plants much smaller in size. Stems up to 70 cm long, *ca* 2 mm thick. Upper most leaves more developed, 7-18 cm long, 3-5 mm wide. Spikes 6-7 x 5-7 mm. Spikelets 2.5 - 3 x *ca* 1 mm. Upper glumes 2.5 - 3 mm long; with spinulose keel. Nut elliptic - oblong, *ca* 1.5 x 0.7 mm.

Fls. & Frts. : Sept., Jan.

Habitat : Open, water logged areas near streams.

Distrib. : Malesia. INDIA: South and North India. KARNATAKA: Balgaum, Kodagu, Uttara Kannada.

Specimens examined : Belgaum: Khanapur, *Ahuja* 47741, 5.1.1959 (BSI). Kodagu: Behind 'Roshanara', *Rao* 74712, 27.9.1961 (BSI). Uttara Kannada: Yellapur, *Talbot s.n.* (Acc No. 721), Sept. 1884 (BSI).

Notes: Floral characters of BSI specimens do not agree exactly with the characters given by Kern, *l.c.* According to him nut is broadly elliptic to obovate-ovoid, *ca* 1.5 x 1-1.2 mm. But in the above specimens nuts are elliptic-oblong and *ca* 1.5 x *ca* 0.7 mm. Probably it is a smaller form of *K. melanosperma* with fairly well developed upper leaves.

5. **Kyllinga nemoralis** (Forst.) Dandy ex Hutchins. & Dalziel, Fl. W. Trop. Africa 2: 487. 1936; Hooper in Saldanha & Nicolson, Fl. Hassan 687. 1976; Koyama in Gard. Bull. Singapore 30: 163. 1977; Sharma *et al.,* Fl.

Karnataka 312. 1984; Koyama in Dassanayake & Fosberg, Rev. Handb. Fl. Ceylon 5: 249. 1985; Singh, Fl. E. Karnataka 2: 640. 1988; Karthik. *et al.*, Fl. Ind. Enum. Monocot. 60. 1989. *K. monocephala* Rottb., Descr. Ic. Rar. Nov. Pl. 13, t. 4, f. 4. 1773, *nom. illeg.*; Clarke in Hook.f., Fl. Brit. India 6: 588. 1893; Cooke, Fl. Pres. Bombay 2: 876. 1908 (3: 390. 1958, repr.ed.); Fischer in Gamble, Fl. Pres. Madras 1624. 1931 (3: 1130. 1957, repr.ed.); Satyanarayan & Shankaranarayan in Ann. Arid Zone 2: 146. 1963. *Thryocephalon nemoralis* J.R. & G. Forst., Char. Gen. Pl. 130. 1776. *Cyperus kyllinga* Endl., Cat. Hort. Acad. Vindb. 1: 94. 1842; Ramaswamy & Razi, Fl. Bangalore 89. 1973; Kern in van Steenis, Fl. Males. 1, 7: 659. 1974; Rao & Razi, Fl. Mysore 560. 1981; Rao & Verma, Cyp. NE India 7. 1982.

Type : Society Islands, *J.R. & G. Forster.*

Illus. : Rottb., *l.c.* ; Clarke, Illus. Cyp. t. 2, f. 1-2. 1909; Matthew, Furth. Illus. Fl. Tamilnadu Carnatic Pl. 684. 1988.

Perennials with horizontally creeping rhizome, 10-45 cm high; rhizome slender, 1-2 mm thick, covered by brownish scales. Stems approximate to distantly arranged along the rhizome, triquetrous, 1-1.5 mm thick, smooth. Leaves well developed, basal and subbasal, linear 2-4 mm wide; sheaths brown to purplish. Inflorescence a head of one terminal spike, usually along with 1-3 lateral spikes; terminal spike subglobose or broadly ovoid, 5-10 x 4-8 mm; lateral ones much smaller. Involucral bracts 3-4, spreading or reflexed, lowest up to 30 cm long. Spikelets many, densely packed, strongly compressed, ovate-elliptic or elliptic-lanceolate, 2.5 - 3.5 x 1-1.5 mm, 1 or 2-flowered. Glumes 4 or 5, lowest two smaller, narrow, 1-1.5 mm long; remaining glumes strongly compressed, boat-shaped, apiculate or mucronulate at apex, broadly winged on the acute keel; third glume 2.5 - 3 mm long, enclosing the female flowers; fourth glume 3-3.5 mm long; fifth glume if present, small, sterile. Stamens 3; filaments elongate up to 3 mm; anthers linear, 0.5-1mm long. Style up to 1 mm long; stigmas 2, longer than style. Nut laterally compressed, biconvex, oblong or oblong- obovate, 1.2 - 1.5 x 0.7- 1 mm, yellowish to brownish.

Fls. & Frts. : May - Jan.

Chrom. No. : n = 60 (*Reg. Veg.* 77: 29. 1971).

Habitat : Open moist grasslands, wet shady areas near water, moist humous soils in shady areas of forests; waste lands and fallow fields.

Distrib. : Pantropic, especially in warmer parts of Asia, relatively less in tropical Africa and Australia, rare in tropical America. INDIA: Throughout. KARNATAKA: Bangalore, Belgaum, Bellary (Satyanarayan & Shankaranarayan, *l.c.*), Chikmagalur, Dakshina Kannada, Dharwar (Sharma *et al.*, *l.c.*), Hassan, Mandya, Mysore (Rao & Razi, *l.c.*), Shimoga (Sharma *et al.*, *l.c.*), Tumkur, Uttara Kannada.

Specimens examined : Bangalore: Bangalore, *R.D.A.* B 256, 1917 (MH); Lalbag Botanic Garden, *Rao* 73481, 31.8.1961 (BSI). Belgaum: Londa, *Cooke s.n.* (Acc. Nos. 2481 & 2482), Oct. 1897 (BSI). Chikmagalur: Basavani tank, Koppa taluk, *Prasad* 173850, 8.11.1995 (BSI). Dakshina Kannada: Sampagi, *Barber* 2197, 10.11.1900 (MH); Kannadekatte, *Bhat* 175, 6.5.1975 (MGH). Hassan: Near Balvathally state forest, *Saldanha* 14140, 16.7.1969 (JCB). Mandya: Ranganathittu, *Bhat* 67, 28.7.1970 (JCB). Tumkur: Banglibetta-Devarayadurga, *Rao* 73359, 28.8.1961 (BSI); Siddarabetta, Kollikallu state forest, *Yoganarasimhan* 0504, 20.8.1974 (RRCBI). Uttara Kannada: Castle rock, *Gammie s.n.* (Acc. No. 2404 A.), Oct. 1902 (BSI); Sirsi, *Fernandez* J.F.1726, 9.11.1950 (BLAT.); Karka R.F., Dandeli range, Halyal, *Prasad* 173717, 30.10.1995 (BSI).

Note. : Though Sharma *et al. (l.c.)* reported this species from Kodagu district, specimens from this district in BSI are nothing but *Kyllinga brevifolia* Rottb.

6. **Kyllinga odorata** Vahl, subsp. **cylindrica** (Nees ex Wight) Koyama, Gard. Bull. Singapore 30: 161. 1977 *et* in Dassanayake & Fosberg, Rev. Handb. Fl. Ceylon 5: 244. 1985. *K. cylindrica* Nees ex Wight, Contr. Bot. India 91. 1834; Clarke in Hook.f., Fl. Brit. India 6: 588. 1893 (incl. var. *subtriceps* Nees); Fischer in Gamble, Fl. Pres. Madras 1624. 1931 (3: 1130. 1957, repr. ed.). *K. odorata* Vahl var. *cylindrica* (Nees) Kukenth. ex Merr., J. Straits, Branch Roy. Asiat. Soc. 76: 80. 1917. *Cyperus sesquiflorus* (Torr.) Mattfeld & Kukenth. var. *cylindricus* (Nees) Kukenth. in Engl., Pflanzenr. 4 (20), Heft 101: 593. 1936; Kern in Reinwardtia 6:66. 1961; Ramaswamy & Razi, Fl. Bangalore 94. 1973. *C. sesquiflorus* (Torr.) Mattfeld & Kukenth. subsp. *cylindricus* (Nees) Koyama, Bot. Mag. 83: 187. 1970. *C. sesquiflorus* Mattfeld & Kukenth. var. *subtriceps* (Nees) Koyama in Quart. Jour. Taiwan Mus. 14: 191. 1961; Kern in van Steenis, Fl. Males. 1, 7: 659. 1974; Rao & Verma, Cyp. NE India 6. 1982. *Kyllinga odorata* sensu Hooper in Saldanha & Nicolson, Fl. Hassan 687. 1976; Sharma *et al.*, Fl. Karnataka 312. 1984; Karthik. *et al.*, Fl. Ind. Enum. Monocot. 61. 1989. **Fig. 44.**

Illus.: Clarke, Illus. Cyp. t. 2, f. 3-6. 1909 (*Kyllinga odorata*).

Tufted perennials, 10-30 cm high. Rhizome very short, clothed with brown fibres; stolons absent. Stems slender, triquetrous, *ca* 1 mm thick, smooth, somewhat thickened at base. Leaves few, narrowly linear, gradually acuminate at apex, 2-4 mm wide, subrigid; sheaths pale brownish. Inflorescence a head of 1-3 spikes, whitish, ultimately straw-coloured; central spike cylindrical, 10-12 x 4-5 mm; lateral spikes subglobose, much smaller than the terminal one. Involucral bracts 3-5, reflexed; lowest up to 10 cm long. Spikelets numerous, flattened though slightly turgid, elliptic, 2-2.5 x 1-1.3 mm, 1 or 2-flowered, maturing one fruit; rachilla cylindric. Glumes 4; lowest two small, *ca* 2 mm long, 9-11-nerved; upper two broadly ovate, acute or apiculate at apex, *ca* 2.5 mm long, with a smooth or hardly spinulose keel, 5- 7-nerved. Stamens 2; anthers oblong, 0.5-0.7 mm long. Style 0.6 - 0.8 mm; stigmas 2. Nut laterally compressed, biconvex, obovate or elliptic-obovate, truncate at apex, *ca* 1.5 x 1 mm, yellowish, ultimately dark brown.

Fls. & Frts. : June-Oct.

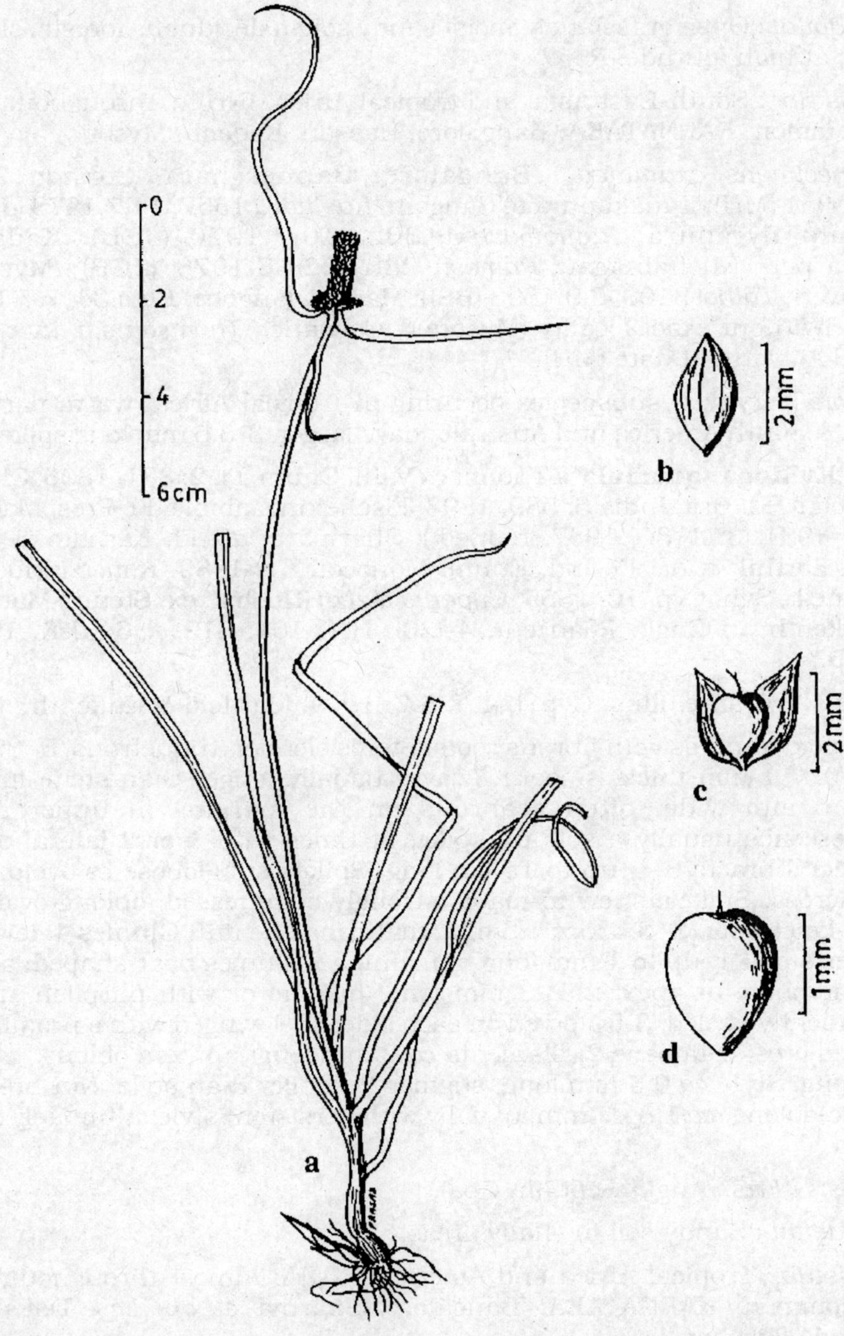

Fig. 44. *Kyllinga odorata* subsp. *cylindrica* (Nees ex Wight) Koyama
a. Habit, **b**. Spikelet, **c**. Nut with glumes, **d**. Nut

Habitat : Open grasslands, moist sandy soil in deciduous forests, shady areas; at high altitude. Rare.

Distrib. : South-East Asia, and tropical Africa. INDIA: Throughout, but not common. KARNATAKA: Bangalore, Hassan, Kodagu, Mysore.

Specimens examined : Bangalore: Bannerghata, *Govindu* 281, 29.7.1951 (MH); Kanakapura to Sangam, *Ravindra* 1557, 14.7.1978 (JCB). Hassan: Byrapura, *Ramamoorthy* 362, 10.7.1970 (JCB). Kodagu: Somvarpet- Kushalnagar, *Ramesh* 2382, 24.8.1978 (JCB). Mysore: Santaveri, *Talbot* 3108, 6.9.1893 (BSI); Manasagangotri, *Bhat* 34, 7.7.1970 (JCB). Without exact locality (Mysore & Carnatic), *Thomson s.n.* (Acc. No. 73408 A), without date (MH).

Note : Typical subspecies occuring in tropical Africa, warm parts of North & South America and Australia have larger, 3-3.5 mm long spikelets.

7. **Kyllinga squamulata** Thonn. ex Vahl, Enum. Pl. 2: 381. 1806; Clarke in Hook.f., Fl. Brit. India 6: 589. 1893; Fischer in Gamble, Fl. Pres. Madras 1624. 1931 (3: 1130. 1957, repr.ed.); Sharma *et al.*, Fl. Karnataka 312. 1984; Karthik.*et al.*, Fl. Ind. Enum. Monocot. 61. 1989. *K.metzii* Hochst. ex. Stued., Syn. Cyp. 70. 1855. *Cyperus metzii* (Hochst. ex. Steud.) Mattfeld & Kukenth. in Engl., Pflanzenr. 4 (20), Heft 101: 612, f. 64 J-K. 1936. **Fig.45.**

Illus. : Clarke, Illus. Cyp. t. 1, f. 5-7. 1909; Mattfeld & Kukenth., *l.c.*

Tufted annuals with fibrous roots. Stems slender, triquetrous, 5- 30 cm long, 0.7- 1 mm thick, smooth. Leaves usually longer than stem, linear, 1.5-2.5 mm wide, often scabrous on the margins in upper part. Inflorescence usually a solitary spike, at times with 1 or 2 lateral ones. Involucral bracts 3-4, up to 15 cm long. Spikes subglobose or ovoid, 4-5 mm across. Spikelets few to many, strongly compressed, oblong-ovate to ovate-lanceolate, 3- 3.5 x *ca* 1.3 mm, maturing one nut. Glumes 4; lower 2 glumes smaller, up to 1 mm long; remaining 2 glumes boat-shaped, acute or mucronate at apex, 2.5 -3 mm long, hyaline or with purplish spots, prominently keeled, 3-5-nerved on each side; keel winged with a prominent toothed crest. Stamens 2; filaments *ca* 2 mm long; anthers oblong, *ca* 0.5 mm long. Style *ca* 0.5 mm long; stigmas 2, longer than style. Nut broadly elliptic-oblong, *ca* 1.8 x 1 mm, usually with persistent style, ultimately dark brown.

Fls. & Frts.: Aug. - Sept. (in Goa).

Habitat : Sandy soil in shady areas.

Distrib.: Tropical Africa and America. INDIA: Almost throughout, but not common. KARNATAKA: Bangalore (Sharma *et al., l.c.),* Dakshina Kannada (Fischer, *l.c.*).

Note: This species is included based on earlier reports (Fischer, *l.c.* and Sharma *et al., l.c.*). As specimens are not available from Karnataka the above description is based on 2 specimens available in BSI from Goa.

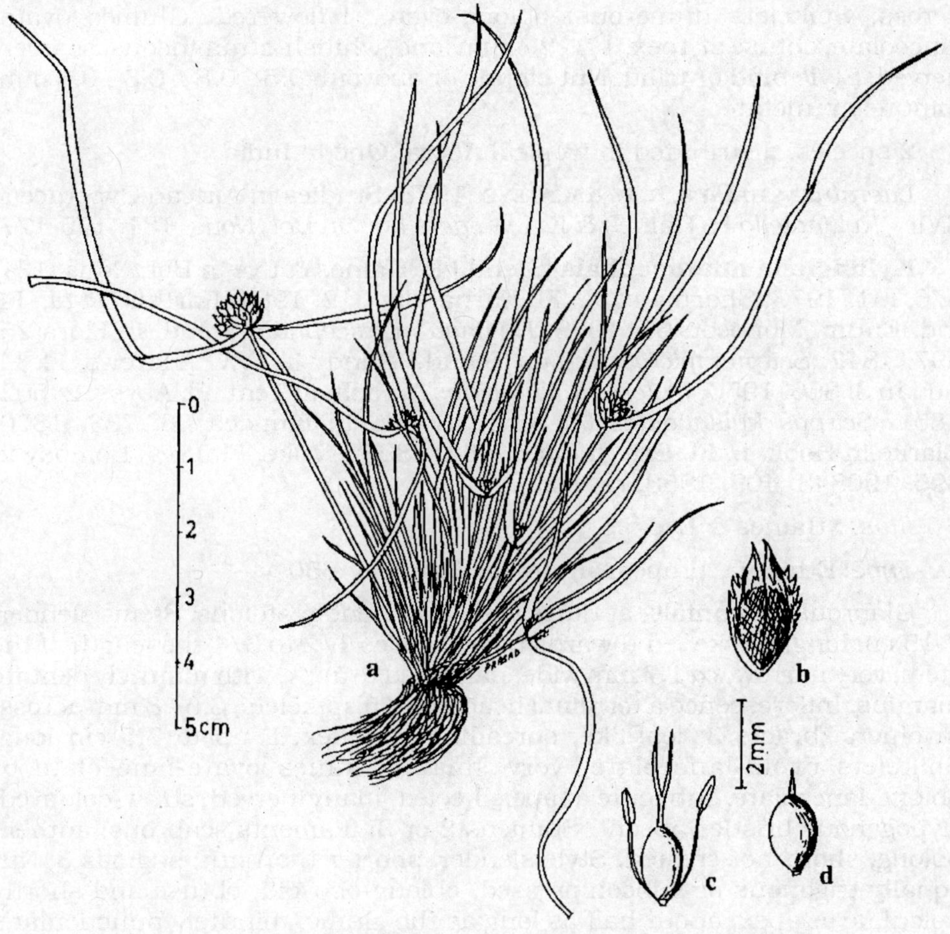

Fig. 45. *Kyllinga squamulata* Thonn. ex Vahl
a. Habit, **b**. Spikelet, **c.** Flower (later stage), **d.** Nut

14. KYLLINGIELLA

R. Haines & Lye in Bot. Notis. 131: 176. 1978.

Type : *Kyllingiella microcephala* (Steud.) R. Haines & Lye

Perennials with short rhizome. Stems slender, several, tufted, 5-40 cm high, 0.2 - 0.5 mm thick. Leaves 1-7, 3-15 cm long, 1-2 mm wide; sheaths long, brownish or greyish. Inflorescence crowded, very dense, 2-8 mm across. Spikelets numerous, oblong-ovate, 1-flowered. Glumes ovate-lanceolate, obtuse at apex, 1.2 - 2.4 mm long, whitish-stramineous, scarcely nerved. Style bifid or trifid. Nut elliptic or obovoid, 0.5 - 0.8 x 0.3 - 0.6 mm, minutely punctate.

2 species, distributed in tropical Africa. One in India.

Literature : HAINES, R. W. & K.A. LYE (1978) Studies in African Cyperaceae XVII - *Kyllingiella* R. Haines & K. Lye, *gen. nov. in Bot. Notis.* 131: 175-177.

Kyllingiella microcephala (Steud.) R. Haines & Lye in Bot. Notis. 131: 176, f. 1. 1978; Sharma *et al.*, Fl. Karnataka 312. 1984; Karthik. *et al.*, Fl. Ind. Enum. Monocot. 61. 1989. *Kyllinga microcephala* Steud. in Flora 25: 597. 1842. *Scirpus microcephalus* (Steud.) Dandy in F.W. Andrews, Fl. Pl. Sudan 3: 366. 1956. *Isolepis kyllingoides* A. Rich. in Tent. Fl. Abyss. 2: 502. 1851. *Scirpus kyllingoides* (A. Rich.) Boeck. in Linnaea 36: 733. 1870; Clarke in Hook. f., Fl. Brit. India 6: 662. 1893; Cooke, Fl. Pres. Bombay 2: 895. 1908 (3: 409. 1958, repr.ed.).

Illus. : Haines & Lye, *l.c.*

Type: Ethiopia, Prope Gon Amba, *Schimper* 650.

Glabrous perennials, at times with long slender stolons. Stems slender, 7-15 cm long, thickened towards base. Leaves 1/2 to 3/4 the length of the stem, very narrow, *ca* 1.7 mm wide, narrowed to apex, with minutely dentate margins. Inflorescence a terminal head of 5-15 spikelets, 3.5 - 8 mm across. Involucral bracts 3, leaf-like, spreading or reflexed, up to 7.5 cm long. Spikelets ovoid-lanceolate, very small. Glumes ovate-lanceolate or oblong-lanceolate, subacute at apex, keeled, many-nerved, straw-coloured. Hypogynous bristles absent. Stamens 2 or 3; filaments scabrous; anthers oblong, short, not crested. Style slender, shorter than nut; stigmas 3. Nut equally trigonous or subcompressed, oblong-obovoid, obtuse and shortly apiculate at apex, about half as long as the glume, minutely puncticulate, yellowish- black or reddish.

Fls. & Frts. : Not reported.

Habitat : Not reported.

Distrib. : Tropical Africa. INDIA: Peninsular India, Central, North and North-West India. KARNATAKA: Canara (Clarke, *l.c.*).

Note : This species is included on the authority of Clarke, *l.c.* and Cooke, *l.c.* Specimens were not found in BSI. Description given above is compiled from Clarke, *l.c.* and Cooke, *l.c.*.

15. LIPOCARPHA

R. Br. in Tucky, Narr. Exped. Congo, Append. 459. 1818, *nom. cons.*; Kunth, Enum. Pl. 2: 266. 1837; Steud., Syn. 2: 129. 1855; Boeck. in Linnaea 37: 114. 1871; Clarke in Kew Bull. add. Ser. 8: 116. 1908. *Hypaelyptum* Vahl, Enum. Pl. 2: 283. 1806. *Tunga* Roxb., Fl. Ind. 1: 187. 1820, *p.p.*

Type: *Lipocarpha argentea* (Vahl) R. Br. [=*Hypaelyptum argenteum* Vahl, *L. chinensis* (Osb.) Kern]

Annual or perennial, glabrous herbs, Stems tufted, erect, smooth, leafy at base. Leaves with incurved margins, eligulate. Inflorescence terminal, capitate, with 1-8 sessile spikelets. Involucral bracts leafy. Spikelets terete, densely many-flowered. Rachilla persistent, not winged. Glumes spirally imbricate, acropetally caducous, not keeled. Flowers hermaphrodite. Perianth represented by 2 hyaline, nerved scales; the posticous scale embracing the anticous one. Stamens 1-2; anthers small, linear-oblong, with a shortly produced connective. Style short, not dilated at base, continuous with the ovary, glabrous; stigmas 2-3. Nut dorsiventrally compressed, trigonous or planoconvex, oblong-obovate to narrowly oblong, smooth, reticulate, often slightly curved, enclosed by the hypogynous scales; epidermal cells isodiametric.

About 154 species, mostly in tropical Africa. Also in South and South East Asia, tropical America and South Africa; 3 in India and 2 in Karnataka.

Literature: **DATAR, R.** & **V.D. VARTAK** (1976) Taxonomic study of the genera *Lipocarpha* R. Br. and *Bulbostylis* Kunth from India - 2, in *Journ. Univ. Poona* 48: 19-32. **KOYAMA, T.** (1960) Some transfers of names related to Cyperaceae, in *Bot. Mag. Tokyo* 73: 438.

Key to the species

1a. Stems 1-2 mm thick; inflorescence whitish; glumes 2-3 mm long; hypogynous scales *ca* 2 mm long, much longer than nut; staminal filament elongate up to 2 mm; anther 0.8-1.2 mm long; style 0.5-1 mm long; nut apiculate at apex .... **1. L. chinensis**

1b. Stems 0.6-1 mm thick; Inflorescence purplish-brown; glumes 1.7 - 2 mm long; hypogynous scales *ca* 1.3 mm, not much longer than nut; staminal filament elongate up to 1.5 mm; anther *ca* 0.4 mm long; style *ca* 0.2 mm long; nut with a conical beak at apex .... **2. L. sphacelata**

1. **Lipocarpha chinensis** (Osb.) Kern in Blumea Suppl. 4: 167. 1958 *et* in Back. & Bakh. f., Fl. Java 3: 457. 1968; Ramaswamy & Razi, Fl. Bangalore 76. 1973; Kern in van Steenis, Fl. Males. 1, 7: 521, t. 33. 1974; Datar & Vartak in Journ. Univ. Poona, Sci. Tech. 48: 20, t. 1. 1976; Hooper

in Saldanha & Nicolson, Fl. Hassan 688. 1976; Rao & Razi, Fl. Mysore 563 1981; Rao & Verma, Cyp. NE India 48. 1982; Sharma *et al.*, Fl. Karnataka 312. 1984; Koyama in Dassanayake & Fosberg, Rev. Handb. Fl. Ceylon 5: 251. 1985; Karthik. *et al.*, Fl. Ind. Enum. Monocot. 61. 1989; Keshava Murthy & Yoganarasimhan, Fl. Coorg 511. 1990. *Scirpus chinensis* Osb., Dugb. Ostind. Resa 220. 1757. *Lipocarpha argentea* R. Br. ex. Nees in Linnaea 9: 287. 1835; Clarke in Hook.f., Fl. Brit. India 6: 667. 1893; Fischer in Gamble, Fl. Pres. Madras 1670. 1931 (3: 1158. 1957, repr.ed.).

Illus.: Clarke, Illus. Cyp. t. 60, f. 4-7.1909; Kern, *l.c.;* Datar & Vartak, *l.c.*; Matthew, Furth. Illus. Fl. Tamilnadu Carnatic Pl. 685. 1988.

Annuals or short-lived perennials with or without a short rhizome, 12-55 cm high. Stems usually tufted, obtusely trigonous, striate-sulcate, uniformly (1-2 mm) thick throughout the length. Leaves shorter than stem (rarely up to the inflorescence), flat or canaliculate, slightly thick, gradually narrowed to a subobtuse apex, loosely surrounding the stem. Inflorescence a terminal head of (1-) 2-6 (-7) spikelets, 6-13 x 7-18 mm. Involucral bracts 2-3, quite unequal, overtopping the inflorescence, horizontal or reflexed in later stage, dilated at base, longest up to 9 cm long. Spikelets terete, ovoid to oblong-ovoid, obtuse at apex, 5-8 x 4-6 mm, whitish, densely many-flowered. Glumes membranous, spathulate to oblong-obovate, triangular at apex, cuneate towards base, 2-3 x *ca* 1 mm, with a strong midnerve and purplish-lineolate sides. Hypogynous scales oblong or oblong-lanceolate, *ca* 2 x 0.5 mm, 5-7-nerved, much longer than the nut. Stamen 1; filament hyaline, elongate up to 2 mm; anther linear, 0.8-1.2 mm long. Style 0.5-1 mm long; stigmas (2-) 3, as long as the style. Nut trigonous, oblong or oblong-obovate, at times slightly curved, apiculate at apex, with a short discoid stipe at base, 1-1.5 x *ca* 0.4 mm, light brown; epidermal cells very minute, in many vertical rows on each face.

Fls. & Frts.: July-May.

Habitat: Open water logged areas, marshes, swamps; along streams, canals and ponds; on the banks of rivers and lakes, in wet rice fields, muddy crevices of rocks in river beds and moist grasslands; often found along with other sedges and *Eriocaulon* sp., up to 200 m.

Distrib.: South & South-East Asia, tropical and South Africa, tropical Australia. INDIA: Throughout. KARNATAKA: Bangalore (Ramaswamy & Razi, *l.c.*), Belgaum, Chickmagalur, Dakshina Kannada, Hassan, Kodagu, Mandya, Mysore (Rao & Razi, *l.c.*), Shimoga, Uttara Kannada.

Specimens examined: Belgaum: Khanapur, *Ahuja* 47743, 5.1.1959 (BSI). Chikmagalur: Hosakere village forests, on way to Byrapura forests, *Yoganarasimhan* 1385, 16.11.1972 (RRCBI); Charmudi ghat, *Kulkarni* 15597, 25.5.1979 (BSI). Dakshina Kannada: Kannadekatte, *Bhat* 165, 6.5.1975 (MGH). Hassan: Byra, *Saldanha & Prakash* 3797, 11.11.1978 (JCB). Kodagu: Katakeri, *Puri* 31736, 7.3.1958 (BSI); Mercara, *Arora* 47564,

10.1.1959 (BSI); Behind 'Roshanara', *Rao* 74709, 27.9.1961 (BSI); Kottamudi, along cauvery river at Napok, *Rao* 74951, 5.10.1961 (BSI); on way to Makut, *Ramesh & Prakash* 3161, 10.10.1978 (JCB); Mercara, *Bhat* 788, 18.12.1980 (MGH); Bhagamandala, *Bhat* 801, 19.12.1980 (MGH); Kirugur, *Bhat* 925, 26.1.1981 (MGH). Mandya: Ranganathittu, *Bhat* 65, 28.8.1970 (MGH); Srirangapatanam, *Bhat* 65, 28.7.1970 (MGH). Shimoga: Galinguda, Agumbe, *Raghavan* 80607, 14.5.1962 (BSI); Gubbiga, near Yedur, *Raghavan* 83004, 5.10.1962 (BSI); East of Tunga river, Gajnur-Shimoga Road, *Saldanha & Ramesh* 7456, 24.4.1979 (MGH); Sharavathi river bed, above Jog falls, *Prasad* 173779, 3.11.1995 (BSI); Gaurikeri, Thalaguppa, *Prasad* 173808, 5.11.1995 (BSI). Uttara Kannada: Yellapur, *Talbot s.n.*, 1.9.1884 (BSI). Without exact locality, *Vartak s.n.* (Sr. Nos. 10160, 10161, 10162), Nov. 1971 (MACS).

2.**Lipocarpha sphacelata** (Vahl) Kunth, Enum. Pl. 2: 267. 1837; Clarke in Hook. f., Fl. Brit. India 6: 667. 1893; Kern in Reinwardtia 6: 32. 1962; Ramaswamy & Razi, Fl. Bangalore 77. 1973; Datar & Vartak in Journ. Univ. Poona Sci. & Tech. 48: 22, t. 2. 1976; Hooper in Saldanha & Nicolson, Fl. Hassan 688. 1976; Rao & Razi, Fl. Mysore 564. 1981; Rao & Verma, Cyp. NE India 49. 1982; Sharma *et al.*, Fl. Karnataka 312. 1984; Koyama in Dassanayake & Fosberg, Rev. Handb. Fl. Ceylon 5: 252. 1985; Singh, Fl. E. Karnataka 2: 641. 1988; Karthik. *et al.*, Fl. Ind. Enum. Monocot. 62. 1989; Keshava Murthy & Yognarasimhan, Fl. Coorg 512. 1990. *Hypaelyptum sphacelatum* Vahl, Enum. Pl. 2: 283. 1806. *Lipocarpha triceps* (Roxb.) Nees in Wight, Contr. Bot. India 92. 1834; Fischer in Gamble, Fl. Pres. Madras 1670. 1931 (3: 1158. 1957, repr.ed.). **Fig. 46.**

Type: India, Tranqebar, *Koenig, Rottler.*

Illus. Datar & Vartak, *l.c.;* Matthew, Illus. Fl. Tamilnadu Carnatic Pl. 791. 1981.

Tufted annuals with fibrous roots, up to 35 cm high. Stems slender, obtusely trigonous or subterete, 0.6-1 mm thick, smooth. Leaves shorter than stem, linear, canaliculate, slightly thick, gradually narrowed to a subobtuse apex, 7-19 cm long, 1-2 mm wide, smooth; sheaths 1.5-5 cm long, subloosely surrounding the stem. Inflorescence a terminal head of (1-) 2-3 spikelets, 4-8 x 6-12 mm. Involucral bracts (1-) 2, quite unequal, overtopping the inflorescence, horizontal or reflexed in later stage, dilated at base; lowest 3-12 cm long. Spikelets terete, broadly ovoid to oblong-ovoid, obtuse at apex, 4-7 x 3-4 mm, purplish-brown, densely many-flowered. Glumes membranous, spathulate or oblong- obovate, triangular near apex, slightly narrowed towards base, incurved above, 1.7-2 x *ca* 1 mm, with a prominent midnerve and purplish-lineolate sides. Hypogynous scales oblong, subacute at apex, *ca* 1.3 x 0.6 mm, transluscent, with 5-7 nerves, not much longer than nut. Stamen 1; filament elongate up to 1.5 mm; anther linear-oblong, *ca* 0.4 mm long. Style short, *ca* 0.2 mm long; stigmas

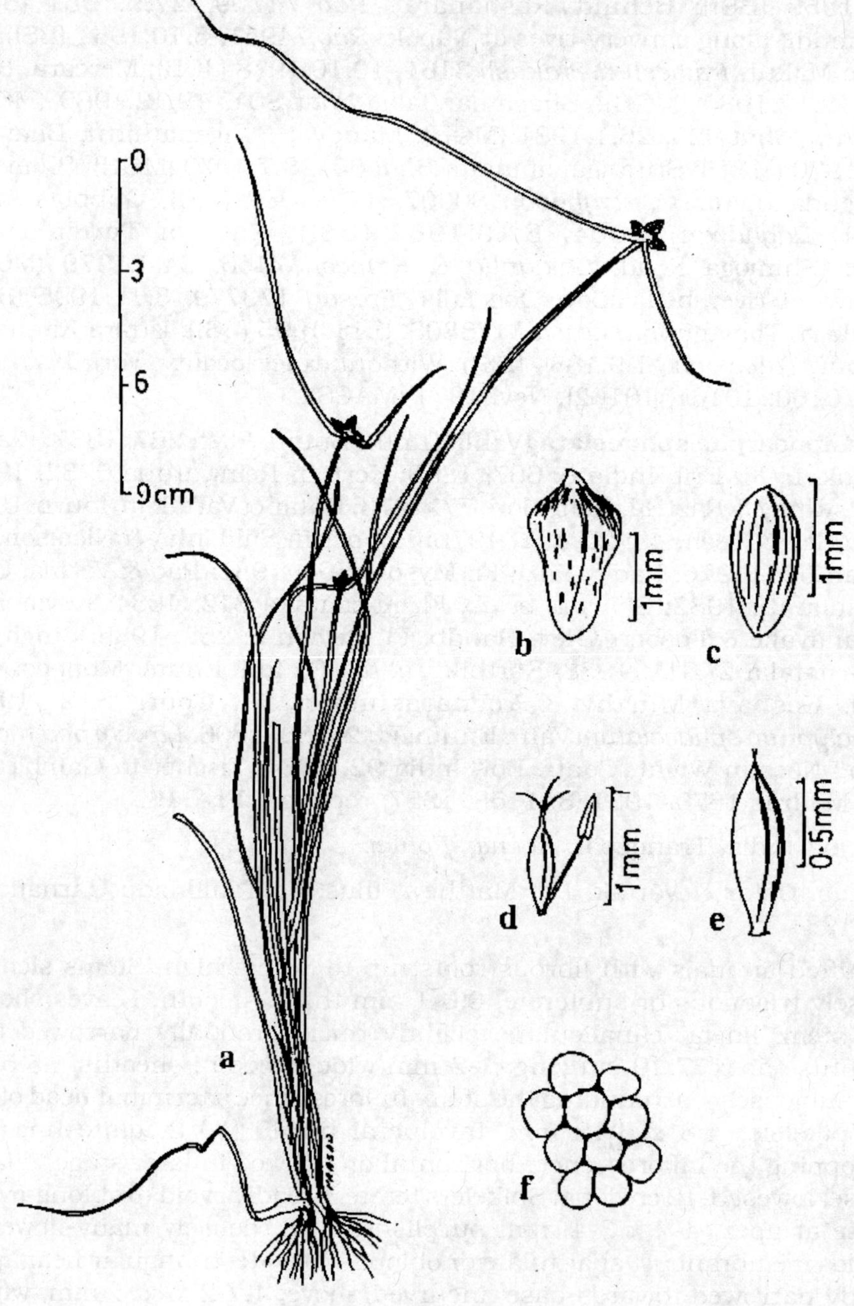

Fig. 46. *Lipocarpha sphacelata* (Vahl) Kunth
a. Habit, **b**. Glume, **c**. Hypogynous scale, **d**. Flower, **e**. Nut, **f**. epidermal cells

3, longer than style, *ca* 0.3 mm long. Nut trigonous, oblong, with a conical beak at apex, with a minute discoid stipe at base, 1-1.2 x 0.3 - 0.4 mm, light brown; epidermal cells very minute, in many vertical rows on each face.

Fls. & Frts. : Aug. - Jan.

Habitat: Open wet areas, margins of swamps and paddy fields, damp grassy areas, along the margin of tanks, marshy places in forests and along roadsides, wet paddy fields.

Distrib.: Sri Lanka, Nepal, Thailand, tropical Africa, tropical America. INDIA: Throughout. KARNATAKA: Bangalore, Belgaum (Sharma *et al.,l.c.*), Bellary, Chikmagalur, Hassan, Kodagu, Kolar, Mandya, Mysore, Shimoga (Sharma *et al.,l.c.*), Uttara Kannada.

Specimens examined: Bangalore: Bannergatta National Park, *Saldanha* 18746, 2.10.1975 (JCB). Bellary: Kudilegi-Sandhur Road, *Ramesh & Sreenath* 4287, 14.11.1978 (JCB). Chikmagalur: Hilikeri village, along Sringeri Road, *Prasad* 173816, 173821 & 173831, 7.11.1995 (BSI); Gadical village tank, Koppa taluk, *Prasad* 173842, 8.11.1995 (BSI). Hassan: 3 km before Hassan from Dudda, *Hooper & Gandhi* 2384, 10.11.1971 (JCB). Kolar: Nandi S.F., *Singh* 142034, 3.1.1976 (BSI). Kodagu: Bitoly, Virajpet, *Bhat* 897, 24.12.1980 (MGH); Kirugoor, *Bhat* 926, 26.1.1981 (MGH). Mandya: Ranganathittu, *Rao* 866, 28.8.1970 (MGH); Paschimavahini, *Rao* 1555, 3.12.1971 (MGH); Sangam, *Murthy & Prakash* 4585, 25.11.1978 (JCB); Narayana Surya, *Dinesh* 611, 14.10.1983 (MGH). Mysore: Mysore, *Vartak* 1524, Nov. 1971 (MACS). Uttara Kannada: Karwar, *Chibbar s.n.* Acc. Nos. 2895 & 2896), Nov. 1910 (BSI).

16. MARISCUS

Vahl, Enum. Pl. 2:372.1806 *nom. cons. Cyperus* sect. *Mariscus* (Vahl) Benth. Fl. Austr. 7: 288. 1878. *Cyperus* sect. *Umbellati* Clarke, Fl. Brit. India 6:620. 1893.

Type : *Mariscus capillaris* Vahl

Herbs with vegetative characters as in the genus *Cyperus*, usually perennials. Inflorescence a single head to simple or compound umbels. Spikelets with few to many distichously disposed glumes, subterete, subtetragonous or laterally flattened; rachilla jointed at base and hence spikelets falling entire. Flowers bisexual; stamens 1-3; style not jointed at base; stigmas 3. Nut trigonous, linear-oblong or elliptical, with one side facing rachilla.

About 200 species distributed in the temperate, subtropical and tropical regions of the whole world, but mostly in tropical America; 14 species in India and 9 in Karnataka.

Literature: **FERNALD, M.L.** (1923) Nomenclatural transfers in *Mariscus.* in *Rhodora* 25: 49-54.

Key to the species

1a.Annuals with fibrous roots 2

1b.Perennials with short or long rhizome 3

2a.Spikes hemispheric; glumes narrowly oblong, 2.5 - 3 mm long; nut oblong, 1.2-1.5 mm long . . . .**6. M. maderaspatanus**

2b.Spikes oblong-ovoid to subglobose; glumes oblong-ovate or oblong-elliptic, 1-2 mm long; nut broadly to narrowly oblong-obovoid, 0.7-0.9 mm long. . . . .**8. M. squarrosus**

3a.Stem base bulbous, surrounded by inflated basal sheaths 4

3b.Stem base not bulbous 5

4a.Rhizome emitting slender stolons; stems solitary; spikelets obtusely trigonous, elliptic-lanceolate, 0.8-1 mm broad, 1-flowered . . . . **1. M. clarkei**

4b.Rhizome without stolons; stems tufted; spikelets subterete or turgid-flattened, ovate to oblong-lanceolate, 1.5-2.2 mm broad, 3-6-flowered . . . . **4. M. dubius**

5a.Leaves septate-nodulose; inflorescence compound to decompound 6

5b.Leaves not septate-nodulose; inflorescence simple (rarely subcompound in *M.sumatrensis*) 7

6a.Spikes globose or subglobose; spikelets stellately arranged, linear-lanceolate, 0.8-1 mm broad; glumes oblong- lanceolate; nut linear-oblong, 1.8 - 2.2 x *ca* 0.5 mm . . . .**2. M. compactus**

6b.Spikes cylindrical or oblong; spikelets arranged at right angles to the rachis or slightly reflexed, oblong to oblong-lanceolate, 1.8 - 2.5 mm broad; glumes ovate to broadly ovate; nut obovoid to subellipsoid, 1.2 - 1.5 x 0.7-0.9 mm . . . .**5. M. javanicus**

7a.Rhizome emitting long, very slender (*ca* 0.5 mm thick) stolons; stems *ca* 0.5 mm thick; leaves 1-2 mm wide; spikelets prominently horizontally spreading, maturing a single nut . . . . **7. M. paniceus**

7b.Rhizome without stolons (if present short and stout, *ca* 1.5 mm thick); stems 1-3 mm thick; leaves 2-6 mm wide. Spikelets obliquely erect or horizontally spreading at right angles to the rachis, at times reflexed, maturing 1- 3 nuts 8

8a.Inflorescence usually with well developed rays; rays 3-15, up to 8 cm long; spikes always cylindrical, 2-4 cm long; spikelets spreading at right angles to the rachis; nut linear-oblong, *ca* 0.5 mm broad . . . . **9. M. sumatrensis**

8b. Inflorescence usually contracted; rays usually absent, if present less than 1 cm long; spikes broadly cylindrical to obovoid, usually attenuate towards base, 8-20 mm long; spikelets obliquely erect to horizontally divergent; nut oblong- ellipsoid, 0.7 - 0.9 mm broad . . . . **3. M. cyperinus**

1. **Mariscus clarkei** (Cooke) Koyama, Journ. Jap. Bot. 51 (10): 313. 1976, in Gard. Bull. Singapore 30: 158, f. 9. 1977 *et* in Dassanayake & Fosberg, Rev. Handb. Fl. Ceylon 5. 236, f. 16. 1985; Karthik. *et al.*, Fl. Ind. Enum. Monocot. 62. 1989. *Cyperus clarkei* Cooke, Fl. Pres. Bombay 2: 873. 1908 (3: 387. 1958, repr.ed.); Kukenth. in Engl., Pflanzenr. 4 (20), Heft 101: 540, f. 58. 1936; Rao & Razi, Fl. Mysore 558. 1981. *Mariscus pictus* Nees in Wight, Contr. Bot. India 90. 1834 (in part, excluding type). *M. bulbosus* (Vahl) Clarke in Hook.f., Fl. Brit. India 6: 620. 1893; Fischer in Gamble, Fl. Pres. Madras 1644. 1931 (3: 1142. 1957, repr.ed.); Sharma *et al.*, Fl. Karnataka 313. 1984; Singh, Fl. E. Karnataka 642. 1988. *Kyllinga bulbosa* Vahl, Enum. 2: 376. 1806 (in note).

Illus.: Kukenth., *l.c.*; Koyama, *l.c.*; Matthew, Illus., Fl. Tamilnadu Carnatic Pl. 792. 1982.

Perennials with short, tuberous rhizome; stolons slender, 5-10 cm long, *ca* 1 mm thick, clothed with oblong-lanceolate, striate, pale-brown scales. Stems solitary, with an enlarged bulbous base, triquetrous, 10-20 cm long, *ca* 1.5 mm thick, smooth; bulbous base enclosed with inflated, membranous, brownish basal sheaths. Leaves shorter than or equalling the stem, linear, gradually acuminate at apex, 2-4 mm wide, often with incurved blades. Inflorescence simple, with very short rays or capitate, bearing 5-8 spikes. Involucral bracts 5-7, leafy, reflexed, up to 15 cm long. Spikes usually subsessile, cylindrical, obtuse at apex, 6-15 x 5-7 mm. Spikelets many, spreading, obtusely trigonous, elliptic-lanceolate, 4-5 x 0.8-1 mm, bearing 1 nut. Lower glume ovate or ovate-lanceolate, acute at apex, *ca* 3 mm long; slightly keeled, membranous towards margins. Nut-bearing glume lanceolate, acute at apex, *ca* 4.5 x 1.2 mm, many-striate. Stamens 3; filaments *ca* 2 mm long; anthers linear-oblong, *ca* 1 mm long. Style *ca* 1.5 mm long; stigmas 3. Nut trigonous, ellipsoid or elliptic- oblong, *ca* 2.5 x 0.8 mm, dark brown when mature.

Fls. & Frts. : Sept.-Oct.

Habitat : Found among grasses

Distrib. : Sri Lanka. INDIA: Confined to western peninsular India and North-West India. KARNATAKA: Bijapur (Badami, Cooke, *l.c.*), Dharwar, Mysore, Uttara Kannada (N. Kanara, Cooke, *l.c.*) .

Specimens examined: Dharwar: Dharwar golf ground, *Talbot* 2907, Oct. 1893 (BSI). Mysore: Chamundi hills, *Bhat* 43, 29.7.1970 (MGH); Same locality, *Rao* 1413, 11.9.1971 (JCB).

2. **Mariscus compactus** (Retz.) Boldingh, Zakfl. Landb. Java 77. 1916; Fischer in Gamble, Fl. Pres. Madras 1645. 1931 (3: 1143. 1957, repr.ed.);

Koyama in Gard. Bull. Singapore 30:153, f.8. 1977 *et* in Dassanayake & Fosberg, Rev. Handb. Fl. Ceylon 5: 228. 1985; Karthik. *et al.*, Fl. Ind. Enum. Monocot. 62. 1989. *Cyperus compactus* Retz., Obs. Bot. 5: 10. 1789; Kukenth. in Engl., Pflanzenr. 4 (20), Heft 101: 423. 1936; Kern in van Steenis, Fl. Males. 1, 7: 638. 1974; Rao & Verma, Cyp. NE India 21. 1982. *C. dilutus* Vahl, Enum. Pl. 2: 357. 1806; Cooke, Fl. Pres. Bombay 2: 875. 1908 (3: 389. 1958, repr.ed.). *Mariscus microcephalus* Pres., Reliq. Haenk. 1: 182. 1828; Clarke in Hook. f., Fl. Brit. India 6: 624. 1893; Sharma *et al.*, Fl. Karnataka 313. 1984.

Type: China, *Osbeck*.

Illus.: Koyama, *l.c.;* Matthew, Furth. Illus. Fl. Tamilnadu Carnatic Pl. 686. 1988.

Perennials with short rhizome, 30-100 cm high; stolons absent. Stems solitary or a few together, obtusely trigonous to subterete, 2-5 (-6) mm thick, smooth. Leaves longer or shorter than stem, rigid, linear, canaliculate, 4-8 (-12) mm wide, septate-nodulose, scabrous on the margins and midrib above; lower sheaths strongly septate-nodulose, reddish-brown. Inflorescence usually large, compound or decompound, 6-20 (-30) cm across, often reddish-brown. Involucral bracts 4-8, patent to reflexed; lower ones much overtopping the inflorescence, longest up to 90 cm long. Primary rays 6-9 (-12), spreading, up to 18 cm long; secondary rays 4-9, up to 3 cm long. Spikes globose or subglobose, dense, 10-20 mm across, with numerous stellately arranged spikelets; rachis short, up to 5 mm long. Spikelets subterete or slightly compressed, linear-lanceolate, 5-10 (-15) x 0.8 -1 (- 1.5) mm, 4-8-flowered; rachilla flexuous, distinctly winged. Glumes membranous, oblong-lanceolate, obtuse or subobtuse at apex; 3-4.5 x 1-1.5 mm, reddish-brown or rarely pale, with 2-3-nerved sides; keel not prominent. Stamens 3; filaments up to 4 mm long; anthers linear-oblong, *ca* 1 mm long. Style *ca* 1.5 mm long; stigmas 3, *ca* 1.5 mm long. Nut trigonous, linear-oblong, apiculate at apex, slightly convex on one side, 1.8-2.2 x *ca* 0.5 mm, yellowish-brown.

Fls. & Frts.: Sept. - Dec.

Habitat: Swampy grassfields, ditches in forests, wet rice fields, along muddly lake beds and river banks; up to 1000 m.

Distrib.: Sri Lanka, South China, Taiwan, Formosa, Malesia and Mauritius (introduced). INDIA: Throughout. KARNATAKA: Chikmagalur (Sharma *et al., l.c.*), Dharwar, Mysore (Fischer, *l.c.*), Uttara Kannada.

Specimens examined : Dharwar: Halligari Thalab, adjacent to Reserve Forest, *Prasad* 172929, 18-12-1994 (BSI). Uttara Kannada: Yellapur, *Talbot* 1024, 20.9.1884 (BSI); Supa-Dandeli Road, *Saldanha* & *Prakash* 3758, 28.10.1978 (JCB); Berchi R.F., Dandeli, *Prasad* 173736, 30.10.1995 (BSI); Javalli village, Halyal forest range, *Prasad* 173742, 30.10.1995 (BSI).

Without exact locality (Mysore & Carnatic), *Thomson s.n.* (Acc. No. 73660), without date (MH).

Note: This species can be easily recognized in the field by the large, prominent reddish-brown inflorescence composed of globose spikes.

3. **Mariscus cyperinus** (Retz.) Vahl, Enum. Pl. 2: 377. 1806; Clarke in Hook. f., Fl. Brit. India 6: 621. 1893; Fischer in Gamble, Fl. Pres. Madras 1644. 1931 (3: 1143. 1957, repr.ed.); Hooper in Saldanha & Nicolson, Fl. Hassan 689. 1976; Yoganarasimhan *et al.*, Fl. Chikmagalur 361. 1981; Sharma *et al.*, Fl. Karnataka 313. 1984; Koyama in Dassanayake & Fosberg, Rev. Handb. Fl. Ceylon 5: 233. 1985; Karthik. *et al.*, Fl. Ind. Enum. Monocot. 62. 1989. *Kyllinga cyperina* Retz., Obs. Bot. 6: 21. 1791. *Cyperus cyperinus* (Retz.) Valck. Sur., Gesl.Cyp. Mal. Archip. 154, t. 6, f. 10. 1898; Kukenth. in Engl., Pflanzenr. 4 (20), Heft 101: 518. 1936; Kern in van Steenis, Fl. Males. 1, 7: 641. 1974.

Type : India.

Illus. : Valak. Sur., *l.c.*; Clarke, Illus. Cyp. t. 22, f. 3-4. 1909.

Key to the varieties

1a. Spikelets 2-3 (usually 2)- flowered; rachilla well developed .... var. **cyperinus**

1b. Spikelets 1-flowered; rachilla minute .... var. **pictus**

var. **cyperinus**

Perennials with short woody rhizome clothed with brown fibres, 15-50 cm high; stolons absent (very rarely with short and somewhat thick stolons). Stems solitary or a few tufted, triquetrous, 1-2 (-3) mm thick, smooth. Leaves usually shorter than stem, flat or canaliculate, linear, long-attenuate towards apex, 2-3 (-6) mm wide, scabrid on the margins and midrib towards apex; lower sheaths reddish to dark brown. Inflorescence simple, usually contracted and head-like, at times with short rays. Involucral bracts 4-5 (-10), much overtopping the inflorescence. Rays usually absent, when developed less than 1 cm long. Spikes broadly cylindrical to obovoid, usually attenuate towards base, 8-20 x 6-8 mm, densely bearing numerous spikelets. Spikelets obliquely erect to horizontally divergent, subterete, linear- oblong, 3-7 x *ca* 1 mm, usually 2-flowered. Glumes 4; lower 2 ovate, 1-1.3 mm long; third glume elliptic to ovate-lanceolate, acute at apex, 3-4 x *ca* 1.5 mm, with membranous sides; keel greenish, 3-nerved. Stamens 3; filaments up to 3 mm long; anthers *ca* 1 mm long. Style very short, *ca* 0.5 mm long; stigmas 3, much longer than style. Nut trigonous, oblong-ellipsoid, shortly apiculate, slightly curved, 2-2.5 x 0.7-0.9 mm, brown.

Fls. & Frts.: May-Oct.

Habitat: Moist soil near streams, rivers, canals etc., moist rocky slopes.

Distrib.: South-East Asia, Malesia, Polynesia and Australia. INDIA: Peninsular India, North-West, Eastern India and Andaman & Nicobar Islands. KARNATAKA: Bangalore. Chikmagalur (Yoganarasimhan *et al.,l.c.*), Hassan, Tumkur.

Specimens examined: Bangalore: Bangalore, *Camaron* 536A, 4.9.1890 (MH). Hassan: Bisle ghat, *Saldanha* 13623, 27.5.1969 (JCB). Tumkur: On way to Devarayadurga, *Ravindra & Ramesh* 1541, 14.7.1978 (JCB).

var. **pictus** (Nees) Karthik. in Karthik. *et al.*, Fl. Ind. Enum. Monocot. 63. 1989. *Mariscus pictus* Nees in Wight, Contr. Bot. India 90. 1834; Clarke in Hook.f., Fl. Brit. India 6: 621. 1893; Fischer in Gamble, Fl. Pres. Madras 1645. 1931 (3: 1143. 1957, repr.ed.); Sharma *et al.*, Fl. Karnataka 313.1984. *Cyperus cyperinus* (Retz.) Valck. Sur. var. *pictus* (Nees) Kukenth. in Engl., Pflanzenr. 4 (20), Heft 101: 519. 1936; Ramaswamy & Razi, Fl. Bangalore 83. 1973.

Illus. : Clarke, Illus. Cyp. t. 23, f. 3-4.1909.

This variety can be distinguished only on the basis of presence of a single nut per spikelet. All other characters like size of the stems, leaves, bracts, spikes, rays etc. are variable. Hence probably it is only a form of *M.cyperinus* bearing a single fertile glume per spikelet.

Fls. & Frts.: May - Nov.

Habitat : Marshy areas.

Distrib.: Endemic to peninsular India. KARNATAKA: Bangalore, Dakshina Kannada, Kodagu, Hassan, Shimoga, Uttara Kannada.

Specimens examined : Bangalore: Bangalore, *Camaron* 536 B, 4.9.1890 (MH); Lalbag botanic garden, *Rao* 73482 A, 31.8.1961 (BSI). Dakshina Kannada: Keelar *Rao* 74774, 19.9.1961 (BSI). Hassan: Bisle forest, *Mahajan* 19829, 5.6.1957 (BSI). Kodagu: Sahdudu area, Makut, *Arora* 55475, 21.5.1959 (BSI). Mysore: Bandipur, *Naithani* 2112, 24.8.1964 (MH). Shimoga: Yellakibante area, Agumbe, *Raghavan* 62511, 15.5.1960 (BSI); Karkala-Barakana valley, Agumbe, *Raghavan* 80753, 18.5.1962 (BSI); Agumbe, *Raghavan* 74178, 18.6 1961 (BSI); Ghatibagh, *Raghavan* 80539, 14.5.1962 (BSI); Ulikappa, near Hulical, *Raghavan* 80979, 25.5.1962 (BSI); Thirthamuthur, Agumbe, *Raghavan* 81003, 27.5.1962 (BSI); Kavaledurga, Agumbe, *Raghavan* 82923, 2.10.1962 (BSI); Yedur, *Raghavan* 82948, 4.10.1962 (BSI); Nishanigudd, Agumbe, *Raghavan* 83309, 18.10.1962 (BSI); Begar, Agumbe, *Raghavan* 90385, 2.9.1963 (BSI); Varahi falls, Hulical, *Raghavan* 90246, 26.8.1963 (BSI); Jog falls, *Prasad* 173794, 4.11.1995 (BSI). Uttara Kannada: Near Katgal, *Puri* 1938, 2.5.1956 (BSI); Katlekan, *Almeida* 1218, 8.5.1969 (BLAT).

Note: It is worth mentioning that Kern, *l.c.* opined that Malesian specimens cited by Kukenthal under *Cyperus cyperinus* var. *pictus* are typical *Cyperus cyperinus* (*Mariscus cyperinus*) . It seems most of the South Indian specimens of this species bear a single nut per spikelet and hence is more common than the typical variety which usually bears 2 nuts. The variety *pictus* is often wrongly identified as *Mariscus paniceus*, most probably due to the presence of a single nut per spikelet. But *M.paniceus* can be identified by the very slender habit with long and 0.5 - 1 mm thick stolons, *ca* 0.5 mm thick stems, 1-2 mm wide leaves and horizontally spreading spikelets. At times stolons are found in *M. cyperinus* also, but are comparatively shorter and more stout (*ca* 1.5 mm thick).

4. **Mariscus dubius** (Rottb.) Kukenth. ex Fischer in Gamble, Fl. Pres. Madras 1644. 1931 (3: 1142. 1957, repr.ed.); Hooper in Saldanha & Nicolson, Fl. Hassan 689. 1976; Koyama in Gard. Bull. Singapore 30: 158. 1977; Sharma *et al.*, Fl. Karnataka 313. 1984; Koyama in Dassanayake & Forsberg, Rev. Handb. Fl. Ceylon 5: 237. 1985; Singh, Fl. E. Karnataka 2: 642. 1988; Karthik. *et al.*, Fl. Ind. Enum. Monocot. 63. 1989; Keshava Murthy & Yoganarasimhan, Fl. Coorg 512. 1990. *Cyperus dubius* Rottb., Descr. Ic. Rar. Pl. 20, t. 4, f. 5. 1773; Kukenth. in Engl., Pflanzenr. 4 (20), Heft 101: 563. 1936; Ramaswamy & Razi, Fl. Bangalore 85. 1973; Kern in van Steenis, Fl. Males. 1, 7: 643. 1974; Rao & Razi, Fl. Mysore 559. 1981. *Mariscus dregeanus* Kunth, Enum. Pl. 2: 120. 1837; Clarke in Hook.f., Fl. Brit. India 6: 620. 1893.

Illus. : Rottb. *l.c.*; Clarke, Illus Cyp. t. 21, f. 6. 1909; Matthew, Furth. Illus. Fl. Tamilnadu Carnatic Pl. 688. 1988.

Type: India, Malabar, *Koenig*.

Tufted perennials with short rhizome; stolons absent. Stems erect, slender, triquetrous, 15-45 cm long, 1-2 (-3) mm thick, smooth; base bulbous with thickened basal sheaths and surrounded by their fibrous remains. Leaves much shorter than to nearly equalling the stem, flat or with slightly incurved margins, linear, gradually acuminate to apex, 2-5 mm wide, scabrid on the margins towards apex. Inflorescence a dense, lobed head consisting of 1-6 confluent spikes, broadly ovoid, subglobose or suboblong, 1-2 cm across. Involucral bracts 3-5, spreading to reflexed up to 30 cm long. Spikelets numerous, patent, subterete, or turgid-flattened, ovate to oblong- lanceolate, acute at apex, oblique at base, 4-7 x 1.5-2.2 mm, 3-6-flowered; rachilla nearly straight, conspicuously winged; wings lanceolate, whitish-hyaline, persistent. Glumes herbaceous to subcoriaceous, broadly ovate to lanceolate, obtusish at apex, 3-4 x 1.5-3 mm, 15-19- nerved, slightly keeled, whitish to stramineous, with hyaline margins. Stamens 3; anthers linear-oblong, 1-1.5 mm long. Stigmas 3. Nut trigonous, ellipsoid to slightly oblong-ellipsoid, apiculate at apex, 1.6-2.5 x 0.6-0.7mm.

Fls. & Frts.: May-Oct.

Habitat: Moist grasslands, rocky slopes, sandy soil and waste places.

Distrib.: Sri Lanka, China, Malesia and tropical Africa. INDIA: Southern and Eastern India. KARNATAKA: Bangalore, Chitradurga, Hassan, Kodagu (Keshava Murthy & Yoganarasimhan, *l.c.*), Kolar, Mysore.

Specimens examined: Bangalore: Bangalore, *Camaron* 621, Sept. 1891 (MH); Bannergatta state forest, *Saldanha* 18602, 2.8.1975 (JCB). Chitradurga: Joginath S.F., *Singh* 133011, 15.9.1974 (BSI). Hassan: Kunigal- Channarayapattana Road, South of Channarayapattana, *Saldanha* 14250, 4.8.1969 (JCB). Kolar: Kendatti hill, *Prakash & Sreenath* 2771, 21.9.1978 (JCB). Mysore: Bandipur area, *Naithani* 21121, 24.8.1964 (MH); Yelwala, *Bhat* 47, 7.8.1970 (MGH); B.R. Hills, *Manohar & Ramesh* 7523, 14.5.1979 (JCB).

5. **Mariscus javanicus** (Houtt.) Merr. & Metcalfe, Lingnan Sci. J. 21: 4. 1945; Koyama in Gard. Bull. Singapore 30: 153. 1979 *et* in Dassanayake & Forsberg, Rev. Handb. Fl. Ceylon 5: 227. 1985; Karthik. *et al.*, Fl. Ind. Enum. Monocot. 63. 1989. *Cyperus javanicus* Houtt., Nat. Hist. II. 13. Aanw. Pl. (1), t. 88, f. 1. 1782; Kern in van Steenis, Fl. Males. 1, 7: 635, f. 63. 1974. *C. pennatus* Lamk. Illus. 1: 144. 1791; Kukenth. in Engl., Pflanzenr. 4 (20), Heft 101: 476, f. 53A-G. 1936; Cooke, Fl. Pres. Bombay 2: 875. 1908 (3: 389. 1958, repr.ed.). *Mariscus albescens* Gaud. in Freyc., Voy. Bot. 415. 1826; Clarke in Hook. f., Fl. Brit. India 6: 623. 1893. *M. pennatus* (Lamk.) Dom., Bibl. Bot. 85: 440. 1915; Fischer in Gamble, Fl.Pres. Madras 1645. 1931 (3: 1143. 1957, repr.ed.). **Fig. 47.**

Illus.: Kukenth., *l.c.*; Kern, *l.c.*

Tufted perennials with short rhizome, 40-100 cm high; stolons absent. Stems stout, obtusely trigonous or subterete, 3-5 mm thick, densely papillose. Leaves many, mostly exceeding the stem, rigid, coriaceous, slightly spongy, septate-nodulose, canaliculate, linear, gradually acuminate to apex, with drooping top, 5-10 mm wide, very scabrous on the margins and keel, greyish-green; sheaths castaneous to purplish- brown, often shiny, cylindrical. Infloresence large, compound to decompound, 10-17 cm across. Involucral bracts 5-7, widely spreading, with drooping top, lower ones much overtopping the inflorescence. Primary rays 6-12, rigid, up to 10 cm long, papillose; secondary rays short. Spikes cylindrical or oblong, obtuse at apex, 1.5 - 3 x 1- 1.5 cm, bearing many spikelets; rachis 1-2 cm long, glabrous. Spikelets arranged at right angles to the rachis or slightly reflexed, somewhat flattened and turgid, oblong to oblong-lanceolate, acute at apex, 5-10 x 1.8-2.5 mm, 6-10-flowered; rachilla broadly winged; wings ovate, pale. Glumes subcoriaceous, ovate to broadly ovate, acute at apex, 2.5-3.5 x 2-2.5 mm, hardly keeled, 7-9-nerved, pale, ultimately shining brown or purplish-lineolate; keel greenish; margins whitish-hyaline. Stamens 3; filaments up to 3.6 mm long; anthers linear-oblong, 0.8-1.2 mm long. Style *ca* 1 mm long; stigmas 3, slightly

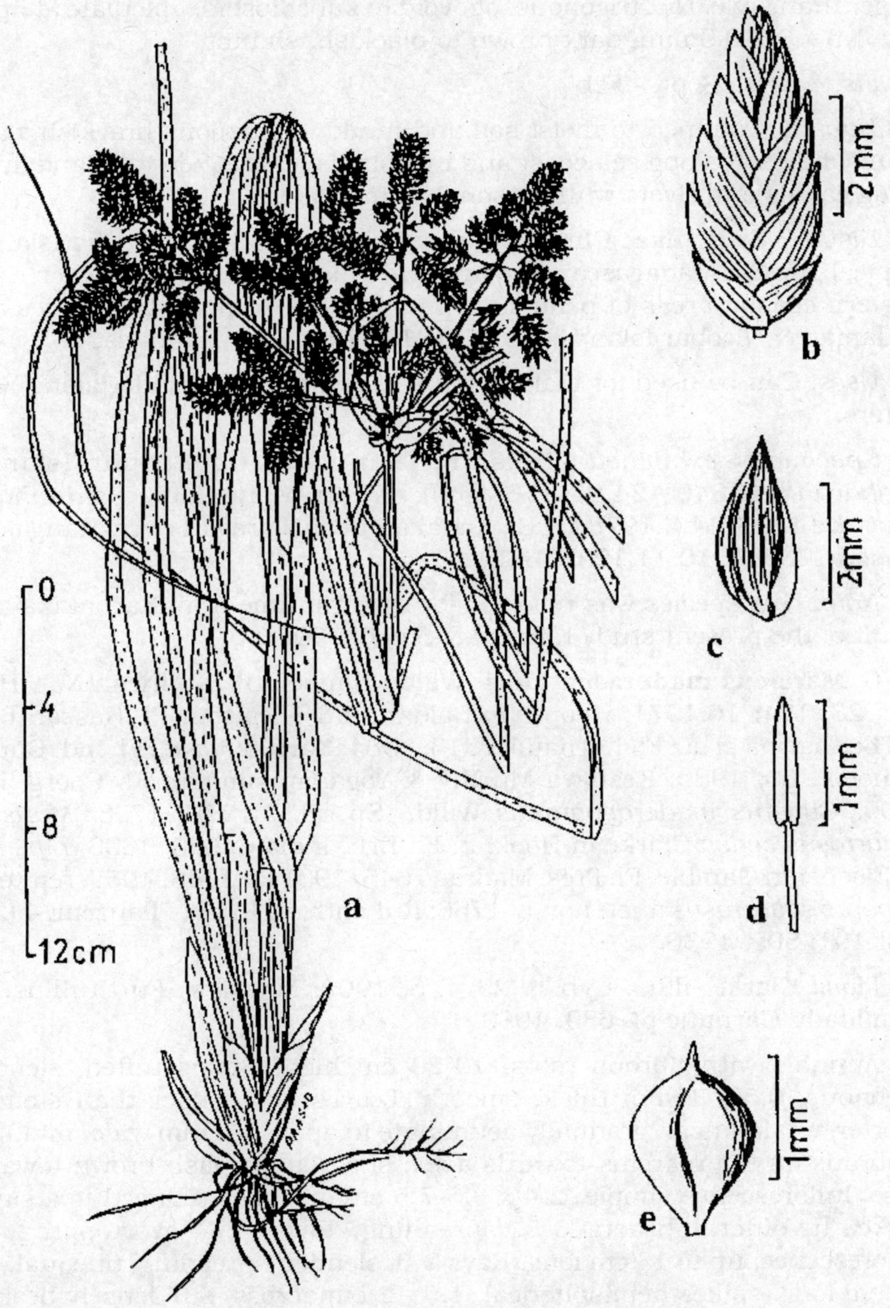

Fig. 47. *Mariscus javanicus* (Houtt.) Merr. & Metcalfe
a. Habit, **b**. Spikelet, **c.** Glume, **d.** Stamen, **e.** Nut
"Reproduced with permission from *J. Econ. Taxon. Bot.* Vol. 21, p. 669. 1997".

longer than style. Nut trigonous, obovoid to subellipsoid, apiculate at apex, 1.2- 1.5 x 0.7-0.9 mm, dark brown to blackish, shining.

Fls. & Frts.: Sept. - Feb.

Habitat: Confined to moist soil and muddy flats along brackish water or in sandy soil along sea coast and in small islands; often in inland areas along canals and rivers with seasonal saline water.

Distrib.: Sri Lanka, China, Ryukue Islands, Malesia, Pacific Islands, tropical Africa, Madagascar and tropical Australia. INDIA: Western and Eastern coastal areas of peninsular India, Eastern India, Laccadives and Andaman & Nicobar Islands. KARNATAKA: Dakshina Kannada.

Uses: Can be used for making mats. It is also a soil binder along back waters.

Specimens examined: Dakshina Kannada: Coondapur Islands, *Raghavan* 145340, 24.9.1976 (BSI); Kulur-Mangalore, *Saldanha & Ravindra* 1326, 14.6.1978 (JCB); Kadakar, near Netravati river, Mangalore, *Prasad* 173869, 10.11.1995 (BSI).

Note: This species was reported for the first time from Karnataka as a result of the present study (*Prasad & Singh,* 1997c).

6. **Mariscus maderaspatanus** (Willd.) Napper, J. E. African Nat. Hist. Soc. 28, 124: 10.1971; Hooper in Saldanha & Nicolson, Fl. Hassan 690. 1976; Sharma *et al.*, Fl. Karnataka 313. 1984; Karthik. *et al.*, Fl. Ind. Enum. Monocot. 63. 1989; Keshava Murthy & Yoganarasimhan, Fl. Coorg 513. 1990. *Cyperus maderaspatanus* Willd., Sp. Pl. 1: 278. 1798. *Mariscus squarrosus sensu* Clarke in Hook. f., Fl. Brit. India 6: 623. 1893, *p.p., non* L.; Fischer in Gamble, Fl. Pres. Madras 1645. 1931 (3: 1143. 1957, repr.ed.). *Cyperus squarrosus* auct. non L. 1756; Kukenth. in Engl., Pflanzenr. 4 (20), Heft 101: 505. 1936.

Illus.: Clarke, Illus. Cyp. t. 29, f. 5. 1909; Matthew, Furth. Illus. Fl. Tamilnadu Carnatic pl. 689. 1988.

Annuals with fibrous roots, 10-20 cm high. Stems tufted, slender, trigonous, 0.5 - 1 mm thick, smooth. Leaves few, longer than stem or shorter, weak, linear, gradually acuminate to apex, 1-2 mm wide, minutely scabrous on the margins towards apex; sheaths purplish-brown towards base. Inflorescence simple, 2-4 x 3.5-7.5 cm, at times reduced to a single spike. Involucral bracts 3-5, spreading, lower 2-3 overtopping the inflorescence, up to 13 cm long. Rays 2-6, slender, spreading, unequal, 1.5 - 3 cm long, spikes hemispherical, 1.5 - 2 cm across, subdensely bearing 6-18 spikelets. Spikelets compressed, linear, 6-16 mm long, 2-3 mm wide (including the mucro), 8-22-flowered; rachilla flexuous, broadly winged, with 0.9- 1 mm long internodes. Glumes narrowly oblong, 2.5-3 x *ca* 1 mm (including the long arista); keel strong, green, excurrent into a 0.6-1 mm long recurved arista, prominently 3-nerved; sides stramineous, nerveless. Stamen 1; filament elongate up to 2 mm; anther linear-oblong, *ca* 0.5 mm long. Style *ca* 1 mm long; stigmas 3, slightly shorter or as long as the style. Nut trigonous, oblong, slightly curved, minutely apiculate, 1.2 - 1.5 x *ca* 0.3 mm, puncticulate, reddish-brown.

Fls. & Frts.: July-Jan.

Habitat: Moist shady areas along river banks and damp sandy slopes in semi-evergreen forests of lower ghats.

Distrib.: South-East Asia to tropical East Africa. INDIA: Peninsular India. KARNATAKA: Bangalore, Dakshina Kannada, Hassan, Kodagu (Keshava Murthy & Yoganarasimhan, *l.c.*), Shimoga.

Specimens examined : Bangalore: Bannergatta, *Hooper & Saldanha* 18048, 7.11.1971 (JCB). Dakshina Kannada: Kapu, *Bhat* 445, 15.1.1977 (MGH); Kunjargiri, *Bhat* 626, 21.7.1980 (MGH). Hassan: Kanchankumri, *Hooper & Gandhi* 2451, 12.11.1971 (JCB). Shimoga: Varahi falls near Hulical, *Raghavan* 90245, 26.8.1963 (BSI).

7. **Mariscus paniceus** (Rottb.) Vahl, Eum. Pl. 2: 373. 1806; Clarke in Hook.f., Fl. Brit. India 6: 620. 1893; Fischer in Gamble. Fl. Pres. Madras 1644. 1931 (3: 1143. 1957, repr.ed.); Hooper in Saldanha & Nicolson, Fl. Hassan 690. 1976; Sharma *et al.*, Fl. Karnataka 313. 1984; Koyama in Dassanayake & Fosberg, Rev. Handb. Fl. Ceylon 5: 235. 1985, *p.p.;* Singh, Fl. E. Karnataka 2: 642. 1988; Karthik. *et al.*, Fl. Ind. Enum. Monocot. 63. 1989; Keshava Murthy & Yoganarasimhan, Fl. Coorg 513. 1990. *Schaenoides paniceus* Rottb., Descr. Pl. Rar. Progr. 15. 1772 (*nom. Provis.*). *Kyllinga panicea* Rottb., Descr. & Ic. Rar. Nov. Pl. 15, t. 4, f. 1. 1773. *Cyperus paniceus* (Rottb.) Boeck. in Linnaea 36: 381. 1870, *p.p.* ; Kukenth. in Engl., Pflanzenr. 4 (20), Heft 101: 526. 1936; Kern in van Steenis, Fl. Males. 1, 7: 643. 1974; Rao & Verma, Cyp. NE India 20. 1982. *C. cyperoides* auct. non (L.) Kuntze; Ramaswamy & Razi, Fl. Bangalore 83. 1973, *p.p.* ; Rao & Razi, Fl. Mysore 558. 1981, *p.p.* **Fig. 48.**

Type : India, Malabar.

Illus.: Clarke, Illus. Cyp. t. 22, f. 1-2, 1909.

Perennials, 10-30 cm high. Rhizome short, 2-3 mm thick, clothed with dark brown fibres; stolons long, slender, *ca* 0.5 mm thick, covered with reddish- brown, membranous scales. Stems very slender, trigonous, *ca* 0.5 mm thick, smooth. Leaves shorter or equalling the stem, narrow, gradually narrowed to an acute apex. Inflorescence simple, up to 1 x 1.5 cm. Bracts 3-5, longest up to 13 cm or more. Spikes sessile or subsessile, cylindrical, up to 10 x 6 mm, bearing many spikelets. Spikelets horizontally spreading, lanceolate, slightly curved, 2.5-3.5 x *ca* 0.5 mm, bearing one nut. Glumes 4; lower 2 ovate, *ca* 1 mm long; third glume ovate-lanceolate, acute at apex, up to 3 mm long, nut bearing; fourth glume much reduced, sterile. Stamens

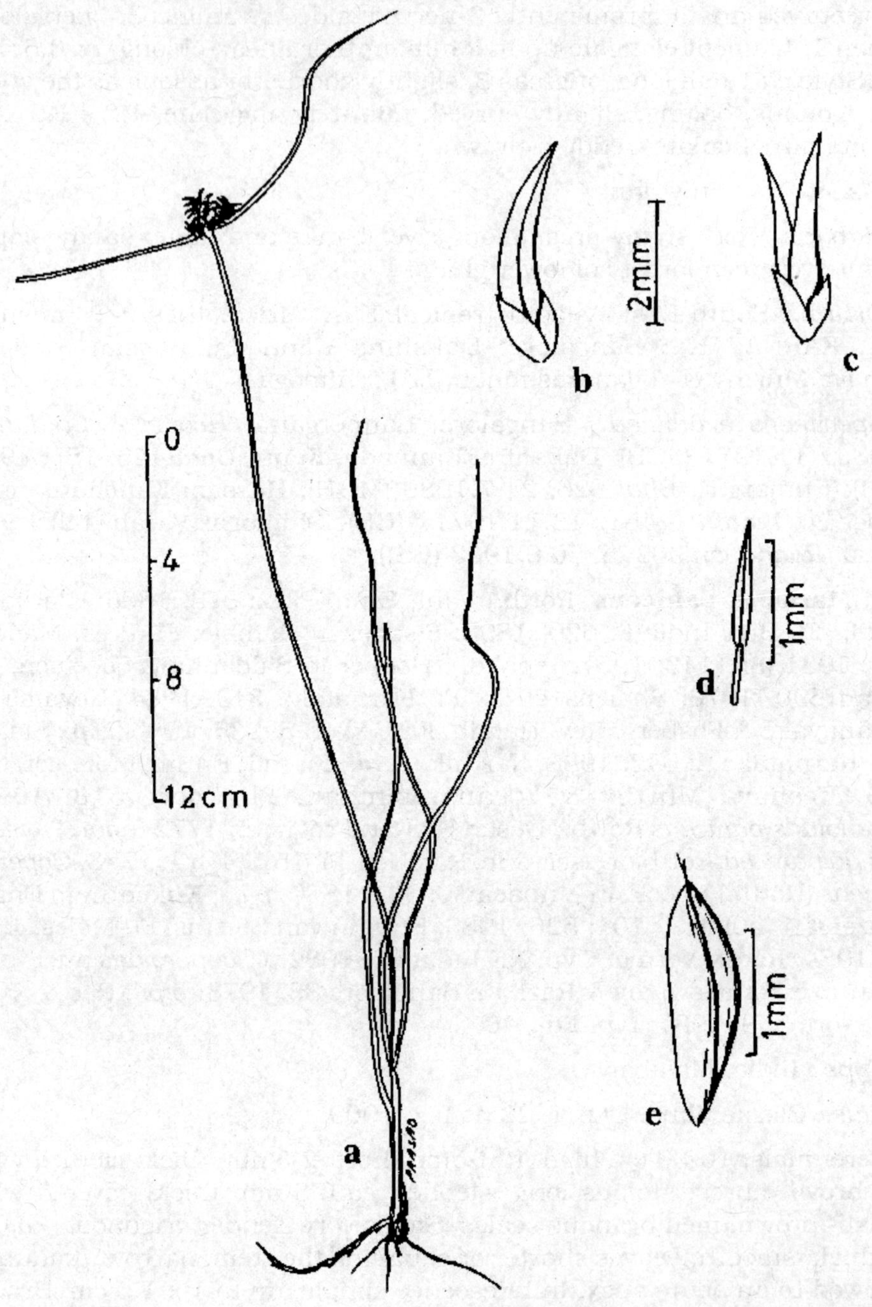

Fig. 48. *Mariscus paniceus* (Rottb.) Vahl
a. Habit, **b.** Spikelet, **c.** Glumes, **d.** Stamen, **e.** Nut

3; anthers linear, *ca* 1 mm long. Style very short; stigmas long, reflexed. Nut trigonous, oblong, slightly curved, *ca* 2 x 0.7 mm.

Fls. & Frts.: May - Nov.

Habitat : Moist shady areas, semi-evergreen forests of upper ghats. Not common.

Distrib.: Sri Lanka to Vietnam. INDIA: Peninsular India, Central, East and North- East India. KARNATAKA: Bangalore (Ramaswamy & Razi, *l.c.*), Bijapur, Chikmagalur (Sharma *et al.*, *l.c.*), Dakshina Kannada, Hassan, Mysore (Rao & Razi, *l.c.*), Uttara Kannada.

Specimens examined : Bijapur: Badami, *Bhide s.n.* (Acc. Nos. 2600 & 2601), 8.9.1911 (BSI). Dakshina Kannada: Kannadekatte, *Bhat* 172, 6.5.1975 (MGH); Subramanya, *Bhat* 576, 26.11.1977 (MGH). Hassan: Kagenari, Shiradi ghat, *Hooper & Gandhi* 2452, 12.11.1971 (JCB); Aglatta Coffee estate, Hanbal Road, *Hooper & Saldanha* 2478, 13.11.1971 (JCB). Uttara Kannada: Karwar, *Talbot s.n.* (Acc. No. 700), July 1883 & *Talbot s.n.* (Acc. No. 699), Aug. 1889 (BSI); Yellapur - Sirsi bridge side, *Ramesh & Udayakumar* 13091, 23.6.1981 (JCB).

8. **Mariscus squarrosus** (L.) Clarke in Hook.f., Fl. Brit. India 6: 623. 1893; Hooper in Saldanha & Nicolson, Fl. Hassan 690. 1976; Sharma *et al.*, Fl. Karnataka 313. 1984; Koyama in Dassanayake & Fosberg, Rev. Handb. Fl. Ceylon 229. 1985; Singh, Fl. E. Karnataka 2: 643. 1988; Karthik. *et al.*, Fl. Ind. Enum. Monocot. 64. 1989. *Cyperus squarrosus* L., Cent. Pl. 2: 6. 1756; Ramaswamy & Razi, Fl. Bangalore 94. 1973; Kern in van Steenis, Fl. Males. 1, 7: 631. 1974; Rao & Razi, Fl. Mysore 561. 1981. *C. aristatus* Rottb., Descr. Pl. Rar. Progr. 22. 1772; Clarke in Hook. f., Fl. Brit. India 6: 606. 1893; Cooke, Fl. Pres. Bombay 2: 866. 1908 (3: 380. 1958, repr.ed.); Fischer in Gamble, Fl. Pres. Madras 1640. 1931 (3: 1140. 1957, repr.ed.); Kukenth. in Engl., Pflanzenr. 4 (20), Heft 101: 502, f. 55, F-J. 1936. *Mariscus aristatus* (Rottb.) Cherm. in Bull. Soc. Bot. France 85: 366. 1938.

Illus.: Kukenth., *l.c.*; Matthew, Furth. Illus. Fl. Tamilnadu Carnatic Pl. 691. 1988.

Tufted, low annuals with fibrous roots, 3-20 cm high. Stems triquetrous, almost 3-winged, 0.5-1.8 mm thick, smooth. Leaves shorter than to nearly equalling the stem, flat, linear, gradually acuminate to apex; sheaths membranous, basal ones purplish. Inflorescence simple, often reduced to a single, hemispherical spike. Involucral bracts 2-4, lowest 1 or 2 much overtopping the inflorescence, up to 8 cm long. Rays when developed up to 3, slender, up to 3 cm long, smooth. Spikes oblong-ovoid to subglobose, 5-20 x 5-10 mm, more or less echinate; rachis 1-5 mm long. Spikelets many to numerous, densely spicate, spreading, flattened, oblong, 5-10 x 2.5-4 mm, 6-25-flowered; rachilla straight, wingless. Glumes membranous, oblong-ovate or elliptic-oblong, tapering into a strong recurved awn, 1-2 x 0.5-1 mm (excluding the awn), strongly 7-9-nerved, ferrugineous to reddish-brown; awn 0.5 - 1.2 mm long. Stamen 1; filament up to 1.5 mm long; anther 0.5 - 0.7 mm long. Style *ca* 0.7 mm long; stigmas 3, shorter than style. Nut variable, trigonous, broadly to narrowly oblong-obovoid, obtuse and minutely apiculate at apex, 0.7-0.9 x 0.2-0.4 mm, brownish.

Fls. & Frts.: July - Nov.

Habitat: Moist sandy soil, in crevices on rocks, gravelly soil, marshy areas along roadsides, wet grasslands.

Distrib.: Widely distributed in tropical and subtropical regions of Asia, Africa, America and Australia; also extending to temperate regions in Canada, Chile and Argentina. INDIA: Throughout. KARNATAKA: Bangalore, Bellary, Bijapur, Chikmagalur (Sharma *et al., l.c.*), Chitradurga, Dakshina Kannada, Dharwar, Hassan, Kodagu, Kolar, Mandya, Mysore, Raichur, Shimoga, Tumkur, Uttara Kannada.

Specimens examined : Bangalore: Bangalore, *Camaron* 629, Sept. 1891 (MH); Bangalore, *Anstead* 83386, Nov. 1917 (MH); Lalbag Botanic Garden, *Rao* 73493 A, 31.8.1961 (BSI); Bannergatta National Park, *Saldanha* 19335, 20.8.1977 (JCB). Bellary: Bellary Fort, *Gamble* 17754, July 1886 (MH). Bijapur: Badami hills, *Paranjpye s.n.*, 1.9.1912 (BSI); Madhavi Pasture plot, *Singh* 142948, 14.8.1976 (BSI). Chitradurga: Kamarkawal S.F., *Singh* 141515, 6.11.1975 (BSI). Dakshina Kannada: Manipal, *Bhat* 258, 2.8.1976 (MGH); Ullal, Mangalore, *Bhat* 293, 29.8.1976 (MGH); Panamboor, Mangalore, *Bhat* 483, 5.11.1977 (MGH); Charmadi ghat, *Bhat* 519, 24.11.1977 (MGH). Dharwar: without exact locality, *Talbot* 2620, 15.8.1891 (BSI). Hassan: Nagpuri, *Hooper & Saldanha* 2550, 26.11.1971 (JCB). Kodagu: Kottamudi along Cauveri river, Napoka, *Rao* 74955, 5.10.1961 (BSI); Somawarpeth-Kushalnagar, *Ramesh* 2385, 24.8.1978 (JCB). Kolar: Kamsandra S.F., *Singh* 133315, 22.9.1974 (BSI); Kendatti hills, *Prakash & Sreenath* 2775, 21.9.1978 (JCB). Mandya: Paschimavahini, *Padma Rani* 20, 8.8.1970 (MGH); Melukotte, *Dinesh* 660, 7.11.1983 (JCB). Mysore: Bandipur, *Naithani* 21130, 24.8.1964 (MH); Chamundi hills, along roadside near temple, *Rao* 73517, 1.9.1961 (BSI); Beduguli estate surroundings, *Kammathy* 73803, 8.9.1961 (BSI); Yelwala *Bhat* 48, 7.8.1970 (MGH). Raichur: Kasagi, *Wadhwa* 44865A, 29.9.1958 (BSI). Shimoga: Barakana, Agumbe, *Raghavan* 74205A, 19.6.1961 (BSI); Kanagulgudda, Agumbe, *Raghavan* 90006, 19.8.1963 (BSI). Tumkur: Namdachilume Forest Rest House vicinity, *Rao* 73252, 28.8.1961 (BSI); Jalangkhatejare, back side of Devarayadurga temple, *Rao* 73391, 29.8.1961 (BSI); Ippadi S.F., *Singh* 140890, 22.10.1975 (BSI); Devarayadurga, *Saldanha* 2210, 15.8.1978 (JCB). Uttara Kannada: Karka bridge, Dandeli, *Prasad* 173726, 30.10.1995 (BSI).

9. **Mariscus sumatrensis** (Retz.) J. Raynal in Adansonia 15: 110. 1975; Hooper in Saldanha & Nicolson, Fl. Hassan 690. 1976; Koyama in Gard. Bull. Singapore 30: 154. 1977; Sharma *et al.*, Fl. Karnataka 313. 1984; Koyama in Dassanayake & Fosberg, Rev. Handb. Fl. Ceylon 5. 232. 1985; Singh, Fl. E. Karnataka 2: 643. 1988; Karthik. *et al.*, Fl. Ind. Enum. Monocot. 64. 1989; Keshava Murthy & Yoganarasimhan, Fl. Coorg 513. 1990. *Kyllinga sumatrensis* Retz., Obs. Bot. 4: 13. 1786. *Scirpus cyperoides* L., Mant. 2: 181. 1771. *Kyllinga umbellata* Rottb., Descr. Ic. Rar. Nov. Pl. 15. 1773, excluding t. 4, f. 2 (*nom. illeg.*). *Mariscus sieberianus* Nees (Linnaea 9: 286. 1835, *nom. nud.*) ex Clarke in Hook.f., Fl. Brit. India 6: 622. 1893; Fischer in Gamble, Fl. Pres. Madras 1645. 1931 (3: 1143. 1957, repr.ed.). *Cyperus cyperoides* (L.) Kuntze, Rev. Gen. Pl. 3 (2). 333. 1898; Kukenth. in Engl., Pflanzenr. 4 (20), Heft 101: 514. 1936; Ramaswamy & Razi, Fl. Bangalore 83. 1973, *p.p.* (excluding syn. *Mariscus paniceus* Vahl); Kern in van Steenis, Fl. Males. 1, 7: 642. 1974; Rao & Razi, Fl. Mysore 558. 1981, *p.p.* (excluding syn. *M. paniceus* Vahl); Rao & Verma, Cyp. NE India 21. 1984.

Type: Sumatra, *Wennerberg.*

Illus.: Clarke, Illus. Cyp. t. 23, f. 5-6.1909.

Perennials with short woody rhizome 15-75 cm high. Rhizome 6-10 cm thick, clothed with brown fibres; stolons absent. Stems solitary or up to a few closely arranged, erect, rigid, triquetrous, 1-3 mm thick, smooth. Leaves shorter than to equalling the stem, linear, gradually long-attenuate at apex, 3-6 mm wide; sheaths reddish-brown at base. Inflorescence simple, rarely subcompound (not found in Karnataka specimens). Involucral bracts 5-10, lower ones much overtopping the inflorescence; lowest up to 20 (-30) cm long. Rays 3-15, slender, patent, straight, up to 8 cm long, smooth. Spikes cylindrical, not attenuate to base, (1.5 -) 2-4 cm long, 6-8 (-10) mm broad, densely bearing numerous spikelets. Spikelets obliquely erect in early stage, later spreading at right angle to the rachis, lower ones often reflexed, suberect, linear, or linear-lanceolate, acute at apex, 3-4 x 0.5-0.8 mm, maturing 1 or 2 nuts (usually 1 in Karnataka specimens); rachilla with broad hyaline wings. Glumes 4-5, membranous, oblong-ovate to oblong-lanceolate, subacute at apex, 3-3.5 x *ca* 1 mm, faintly many-nerved; keel 3- nerved. Stamens 3; filaments up to 3.5 mm long; anthers linear- oblong, up to 1 mm long. Style 0.5 - 1 mm long, stigmas 3, longer than style. Nut trigonous, linear-oblong, slightly curved, apiculate at apex, 1.8 - 2.2 x *ca* 0.5 mm, straw-coloured, minutely puncticulate.

Fls. & Frts.: June - Nov.

Habitat: Wet sandy soil in open fields, moist open grasslands, forest clearings, roadsides, grassy slopes and wet crevices of rocks.

Distrib.: Tropical and subtropical regions of Asia, Africa and Australia.

INDIA: Throughout. KARNATAKA: Bangalore (Ramaswamy & Razi, *l.c.*), Bidar, Chikmagalur, Dharwar, Hassan (Hooper, *l.c.*), Kodagu, Mandya, Mysore, Shimoga, Uttara Kannada.

Specimens examined: Bidar: Bohrampalli R.F., *Singh* 142840, 10.8.1976 (BSI). Chikmagalur: Sankar falls, Bababuddan hills, *Saldanha & Sreenath* 8579, 28.7.1979 (JCB); Dharwar: Bandur, *Arora* 37653, 2.6.1958 (BSI). Kodagu: Hardoor estate, *Rao* 75067, 11.10.1961 (BSI); Kushalnagar, *Bhat* 1014, 21.9.1981 (MGH); Malambibetta, *Bhat* 1133, 7.11.1981 (MGH); Manchalli State Forest, Nagarhole, *Yoganarasimhan & Murthy* 6246, 10.1.1989 (RRCBI). Mandya: Srirangapattanam, *Bhat* 36, 26.7.1970 (MGH). Mysore: Bogadi, *Bhat* 49, 7.8.1970 (JCB); M.M.Hills, *Rao* 2082, 30.6.1972 (MGH). Shimoga: Kyasdur forests, *Raghavan* 74044, 14.6.1961 (BSI); Kavaledurga, *Raghavan* 82923, 2.10.1962 (BSI); Jog falls, *Prasad* 173793, 4.11.1995 (BSI). Uttara Kannada: Halyal, *Talbot* 1884, 2.8. 1889 (BSI); Yellapur, *Talbot s.n.* (Acc. No. 701), 10.8.1894 (BSI); Dandeli, banks of Kalinadi, *Santapau* 18750, 29.5.1954 (BLAT); Yellapur-Hubli Road, *Arora* 37969, 6.6.1958 (BSI); Kalsur village, *Udayakumar & Ramesh* 13091, 25.6.1981 (JCB). Without locality *Arora*, 38171, without date (BSI).

Note : The name *Scirpus cyperoides* L. cannot be transferred under *Mariscus* due to the existence of the name *Mariscus cyperoides* Dietrich. *Kyllinga umbellata* is an illegitimate name and *Kyllinga sumatrensis* is the next available name to transfer this species under *Mariscus*.

It is a variable species usually having a single nut per spikelet in Karnataka specimens. Hence many old specimens are wrongly identified as *M. paniceus*.

17. PYCREUS

Beauv., Fl. Oware & Benin 2: 48, t. 86. 1807. *Cyperus* L. subgen. *Pycreus* (Beauv.) Miq., Fl. Ind. Bot. 3: 254. 1856; Clarke in Journ. Linn. Soc. 21:33. 1884.

Type: Pycreus polystachyos (Rottb.) Beauv.

Annuals or perennials with vegetative characters as in *Cyperus*. Spikelets digitately or spicately arranged, laterally flattened, with few to many glumes distichously arranged; rachilla winged or wingless, persistent. Glumes usually with nerveless sides, bearing a hermaphrodite flower. Hypogynous bristles absent. Stamens 1-3. Pistil digynous; stigmas 2. Nut laterally flattened, with one angle facing the rachilla.

About 90 species distributed in the temperate, subtropical and tropical parts of the world, mainly in Africa. 33 species in India; 13 in Karnataka.

Literature: Given under genus *Cyperus*.

Key to the species

1a. Nut with longitudinally oblong epidermal cells, on magnification wrinkled, wavy, often with white broken transverse lines 2

1b. Nut with isodiametric epidermal cells, on magnification puncticulate or finely reticulate 7

2a. Spikelets suberect, or divergent while fruiting; glumes acute and mucronate at apex 3

2b. Spikelets spreading; glumes obtuse and muticous at apex 5

3a. Stems 1-3 noded at base; leaves flat; spikelets divergent while fruiting; glumes 1.6 -1.8 mm long; nut subglobose, transversely zonate .... **1. P. atroglumosus**

3b. Stems without nodes; leaves canaliculate; spikelets suberect; glumes *ca* 2 mm long; nut laterally compressed, transversely rugulose 4

4a. Spikelets *ca* 2.5 mm broad; glumes reddish-brown to blackish- brown .... **7. P. malabaricus**

4b. Spikelets *ca* 2 mm broad; glumes stramineus .... **13. P. stramineus**

5a. Stems *ca* 0.5 mm thick; spikelets oblong-ovate to oblong- lanceolate, subacute at apex .... **4. P. kanarensis**

5b. Stems 1-2 mm thick; spikelets oblong or narrowly elliptic, often with parallel margins, obtuse or subobtuse at apex 6

6a. Stems trigonous; spikelets 3-5 mm wide (*ca* 2.5 mm in var. *gracilescens*); rachilla wingless; glumes 2.5 - 3.5 mm long (*ca* 1.8 mm in var. *gracilescens*); anthers 0.7 - 1 mm long; style *ca* 1.7 mm long; nut broadly ovoid .... **2. P. diaphanus**

6b. Stems triquetrous; spikelets 2.5 mm wide; rachilla narrowly winged; glumes *ca* 2 mm long; anthers *ca* 0.25 mm long; style *ca* 0.25 mm long; nut suborbiculoid or elliptic orbiculoid .... **6. P. mahadevanii**

7a. Glumes prominently mucronate; anthers *ca* 0.2 mm long . . . **10. P. pumilus**

7b. Glumes muticous or at the most mucronulate; anthers 0.5 mm or more long 8

8a. Stems with 1-4 nodes at the base 9

8b. Stems without nodes 10

9a. Rachilla of the spikelets with membranous wings; glumes acute and mucronulate at apex; nut *ca* 0.5 mm broad .... **8. P. plurinodosus**

9b. Rachilla of the spikelets wingless; glumes obtuse and muticous at apex; nut 0.8-1 mm broad .... **12. P. sanguinolentus**

10a. Glumes 2.5- 3 mm long; stamens 3; nut 1.8-2.5 mm long .... **5. P. macrostachyos**

10b. Glumes 1.5 - 2.5 mm long; stamens 2; nut 0.8 - 1.5 mm long 11

11a. Spikelets linear-lanceolate, 1.5-2 mm wide; nut oblong .... **9. P. polystachyos**

11b. Spikelets linear-oblong to oblong-lanceolate, 2-3 mm wide; nut obovate-orbicular, oblong-obovate or oblong-elliptic 12

12a. Stems 1-2 mm thick; leaves shorter than stem, 1-2 mm wide; inflorescence simple or subcompound; rays 3-6; spikelets up to 60-flowered ; glumes oblong-ovate, obtuse and muticous at apex; anthers *ca* 0.5 mm long; nut oblong-obovate to oblong-elliptic, ultimately dark brown .....**3. P. flavidus**

12b. Stems 2.5 - 4 mm thick; leaves slightly overtopping the stem, 4-8 mm wide; inflorescence compound; rays 4-11; spikelets 12-34-flowered; glumes ovate-orbicular, mucronulate at apex; anthers 0.8 - 1 mm long; nut obovate-orbicular, orange-brown .... **11. P. puncticulatus**

1. **Pycreus atroglumosus** (Govind.) P. Singh & V. Singh in Journ. Econ. Taxon. Bot. 5: 467. 1984. *Cyperus atroglumosa* Govind. in Proc. Ind. Acad. Sci. 81B, (5): 187, f.1. 1975; Karthik. *et al.*, Fl. Ind. Enum. Monocot. 43. 1989.

Type: India, Karnataka State, Shimoga Dist., Megaravalli to Nalur, *Govindarajalu* 4826 (PCM).

Illus. : Govind., *l.c.*.

Annuals, without stolons. Stems tufted, slender, 1-3-noded at base, 10-25 cm long, 0.5-0.7 mm thick, smooth. Leaves 1-3 per stem, as long as the stem in smaller specimens, flat, acuminate at apex, 0.5 - 1.5 mm wide, scabrid on upper margins; sheaths purplish. Infloresence simple, somewhat congested, 1-3-rayed, bearing 3-10 spikelets. Involucral bracts 2-3, suberect or divergent, longer than inflorescence, up to 10 cm long. Spikelets spicately arranged, strongly compressed, linear-oblong, acute at apex, divergent in fruit, 6-7.5 (-20) x 1.75-2 mm; rachilla straight, excavated, wingless. Glumes triangular-ovate, acute and mucronulate at apex, keeled, 1.6 - 1.8 x 1.2-1.6 mm; keel 3- nerved, green; sides nerveless, atrosanguineous, with broad hyaline wavy margins. Stamens 2; anthers linear, apiculate at apex, 0.3 - 0.75 mm long, yellow. Style flat, *ca* 1 mm long, glabrous; stigmas 2, shorter than style. Nut turgid, subglobose, asymmetric, elliptic, apiculate, 0.75-1 x 0.5-0.6 mm, conspicuously transversely zonate, dark brown to blackish, shining; epidermal cells longitudinally oblong.

Fls. & Frts.: Oct.

Habitat: Marshy places at higher altitudes (1000-1500 m) receiving heavy rainfall.

Distrib.: Endemic to Karnataka (Shimoga dt.).

Note: This species is included on the authotiry of Govindarajalu, *l.c.* As no specimens available in BSI, description provided above is as in the protologue.

2. **Pycreus diaphanus** (Roem. & Schult.) Hooper & Koyama in Journ. Jap. Bot. 51: 316. 1976; Hooper in Saldanha & Nicolson, Fl. Hassan 691. 1976; Sharma, *et al.*, Fl. Karnataka 313. 1984; Karthik. *et al.*, Fl. Ind. Enum. Monocot. 65. 1989; Keshava Murthy & Yoganarasimhan, Fl. Coorg 514. 1990. *Cyperus diaphanus* Schrad. ex Roem. & Schult. Mant. 2: 477. 1824; Kern in Blumea 10: 644. 1960 *et* in van Steenis, Fl. Males. 1, 7: 651. 1974. *C. latespicatus* Boeck. in Flora 42: 433 (441). 1859; Cooke, Fl. Pres. Bombay 2: 855. 1908 (3: 368. 1958, repr.ed.); Kukunth. in Engl., Pflanzenr. 4 (20), Heft 101: 392. 1936; Kern in Reinwardtia 2: 124, f. 13. 1952. *Pycreus latespicatus* (Boeck.) Clarke in Hook.f., Fl. Brit. India 6: 590. 1893.

Illus.: Clarke, Illus. Cyp. t. 3, f. 1-6. 1909; Kern, *l.c.*

Key to the varieties

1a. Spikelets 10-20 x 3-5 mm; glumes 2.5 - 3.5 mm long; nut 1- 1.2 mm long. .... var. **diaphanus**

1b. Spikelets *ca* 9 x 2.5 mm; glumes *ca* 1.8 mm long; nut *ca* 0.9 mm long .... var. **gracilescens**

var. **diaphanus**

Tufted annuals, 10-35 cm high. Stems trigonous, 1-2 mm thick, smooth. Leaves gradually narrowed to apex, 1-3 mm wide, minutely scabrid on margins towards apex. Inflorescence usually simple, often contracted and reduced to a single head, at times with 2-3 rays. Involucral bracts 2-3, patent, lowest much longer than inflorescence. Spikelets strongly compressed, oblong with almost parallel margins, subobtuse at apex, 10-20 x 3-5 mm; rachilla wingless, persistent. Glumes membranous, obtuse and muticous at apex, 2.5- 3.5 x *ca* 2 mm, with somewhat undulate margins; keel 3-nerved, green. Stamens 2; filaments up to 3.5 mm long; anthers oblong to linear-oblong, 0.7-1 mm long. Style *ca* 1.7 mm long; stigmas 2, slightly shorter than style. Nut laterally compressed, biconvex, broadly obovate, apiculate at apex, 1- 1.2 x *ca* 1 mm, rugulose with transverse wavy lines, ultimately black; epidermal cells longitudinally oblong.

Fls. & Frts. : Sept. - Dec.

Habitat : Wet rice fields, moist grasslands and other marshy areas.

Distrib. : Central Asia, Malesia. INDIA: Peninsular India, East and North East India. KARNATAKA: Chikmagalur (Sharma *et al., l.c.*), Hassan, Kodagu, Shimoga, Uttara Kannada.

Specimens examined: Hassan: 3 km before Hassan from Dudda, *Hooper*

& *Gandhi* 2501, 10.11.1971 (JCB). Kogadu: Talakaveri, *Bhat* 1079, 27.9.1981 (MGH). Shimoga: Agumbe, *Raghavan* 62618 A, 17.5.1960 (BSI). Uttara Kannada: Yellapur, *Talbot* 1023, 20-9.1884 (BSI).

var. **gracilescens** (Kukenth.) Hooper in Saldanha & Nicolson, Fl. Hassan 692. 1976; Sharma *et al.*, Fl. Karnataka 313. 1984; Karthik. *et al.*, Fl. Ind. Enum. Monocot. 64. 1989. *Cyperus latispicatus* Boeck. var. *gracilescens* Kukenth. in Engl., Pflanzenr. 4 (20), Heft. 101: 393. 1936.

This variety is differentiated on the basis of smaller floral parts and nut as shown in the key.

Fls. & Frts.: Oct.

Habitat: Marshy hollow in dry deciduous forests.

Distrib.: Indomalaysia. INDIA: Peninsular India and Central India. KARNATAKA: Hassan (Hooper, *l.c.*) .

Note: This variety is included on the authority of Hooper, *l.c.* Specimens are not available in BSI.

3. **Pycreus flavidus** (Retz.) Koyama in Journ. Jap. Bot. 51 (10): 313. 1976 *et* in Dassanayake & Fosberg, Rev. Handb. Fl. Ceylon 5: 222. 1985; Karthik. *et al.*, Fl. Ind. Enum. Monocot. 65. 1989; Keshava Murthy & Yoganarasimhan, Fl. Coorg 514. 1990. *Cyperus flavidus* Retz., Obs. Bot. 5: 13. 1788; Kern in van Steenis, Fl. Males. 1, 7: 648. 1974; Rao & Verma, Cyp. NE India 10. 1982. *C. globosus* All., Auct. Fl. Pedem. 49. 1789, *non* Forsk.; Cooke, Fl. Pres. Bombay 2: 857. 1908 (3: 370. 1958, repr.ed.); Ramaswamy & Razi, Fl. Bangalore 87. 1973; Rao & Razi, Fl. Mysore 559. 1981 (including var. *nilagiricus* Clarke in Journ. Linn. Soc. 21: 49. 1884 & var. *stricta* Clarke, *l.c.*). *C. strictus* Roxb., Fl. Ind. ed. 1, 1: 203. 1820. *C. capillaris* Koenig. ex. Roxb., *l.c.* 1: 198. 1820; Nees in Wight, Contr. Bot. Ind. 76. 1834. *Pycreus globosus* (All.) Reichb., Fl. Germ. Excurs. 140. 1830; Fischer in Gamble, Fl. Pres. Madras 1627. 1931 (3: 1132. 1957, repr.ed.); Hooper in Saldanha & Nicolson, Fl. Hassan 692. 1976 (var. *nilgiricus*) ; Sharma *et al.*, Fl. Karnataka 314. 1984; Singh, Fl. E. Karnataka 2: 644. 1988. *Cyperus nilagiricus* Hochst. ex Steud., Syn. Pl. Glum. 2: 2. 1854. *P. capillaris* (Koenig ex Roxb.) Nees ex Clarke in Hook. f., Fl. Brit. India 6: 591. 1893.

Type: India, Tranquebar, *Koenig.*

Tufted annuals with fibrous roots, rarely perennials with a short rhizome, 15-60 cm high. Stems slender, trigonous, 1-2 mm thick, smooth. Leaves few, shorter than stem, narrow, canaliculate, gradually narrowed to apex, 1-2 mm wide, smooth or slightly scabrid at top; sheaths reddish-brown towards base. Inflorescence simple or subcompound, at times congested to a single head-like cluster. Involucral bracts 2-4, suberect to patent; lower 1 or 2 much overtopping the inflorescence, lower most at times looks like continuation of the stem, up to 20 cm long. Primary rays 3-6, very slender, up to 5 cm long, smooth; secondary rays if present up to 1.5 cm long. Spikes ovoid to subglobose, comprised of 5-20 spikelets; rachis short. Spikelets spreading, strongly compressed, linear-oblong, parallel-sided, 6-20 (-30) x 2-3 mm, purplish- brown to dark brown, shining, up to 60-flowered; rachilla straight, wingless, persistent. Glumes chartaceous, oblong-ovate, obtuse and muticous at apex, 1.5 - 2.5 x 1- 1.5 mm; keel 3-nerved, green; sides pale brown to dark reddish- brown, without nerves, shiny, with hyaline margins. Stamens 2; filaments up to 1.5 mm long; anthers oblong, *ca* 0.5 mm long. Style 0.6-0.8 mm long; stigmas 2, longer than style. Nut laterally compressed, biconvex, oblong-obovate to oblong-elliptic, apiculate at apex, 0.8-1.2 x 0.4 - 0.6 mm, ultimately dark brown; epidermal cells minute, isodiametric.

Fls. & Frts. : Throughout the year.

Habitat: Marshy areas along riverbanks, canals and lakes; in and around ponds and puddles with muddy bottom; wet muddy soil with rocky bottom in running water; muddy crevices of rocks in river beds; paddy fields, dried up muddy areas and puddles along roadsides.

Distrib. : China, Japan, Malesia, South Europe and Africa. INDIA: Throughout. KARNATAKA: Bangalore, Belgaum, Bijapur, Chikmagalur (Sharma *et al., l.c.*), Dakshina Kannada, Dharwar, Gulbarga, Hassan, Kodagu, Kolar, Mandya, Mysore, Raichur, Shimoga, Tumkur, Uttara Kannada.

Specimens examined: Bangalore: Domlur Tank, *Hooper* & *Saldanha* 18081, 18.11.1971 (JCB). Belgaum: Ghataprabha Dam side, *Prasad* 172803, 10.12.1994 (BSI); Gokak falls, Gokak, *Prasad* 172818, 11.12.1994 (BSI); Ghataprabha river, near the bridge, *Prasad* 172826, 12.12.1994 (BSI); on way to Ghataprabha from Gokak, *Prasad* 172833, 12.12.1994 (BSI); Dinman Hassur, Belgaum-Gokak Road, *Prasad* 172843, 13.12.1994 (BSI); Gottni nallah, Jambotti, R.F., *Prasad* 172871, 15.12.1994 (BSI). Bijapur: Bagalkot, *Patwardhan s.n.* (Acc. Nos. 2538, 2539 & 2545), 17.4.1906 (BSI); Sholapur - Bijapur Road, *Singh* 138742, 4.9.1974 (BSI); Varathikavlu, Kendur, *Prasad* 172961, 21.12.1994 (BSI); Hebballi Seemi, Badami, *Prasad* 172974, 22.12.1994 (BSI). Dakshina Kannada: Kodachadri, *Bhat* 593, 17.11.1978 (MGH). Dharwar: Salikkinikuppa lake, Dharwar, *Prasad* 172924, 18.12.1994 (BSI); Sadankeri, on Alnawar Road, *Prasad* 172936 & 172944, 17.12.1994 (BSI). Gulbarga: Tank, Gulbarga, *Singh* 138657, 24.3.1974 (BSI); Bidar-Chincholi Road, *Singh* 128607, 9.2.1975 (BSI); Sindgi-Jevargi Road, *Singh* 142026, 19.1.1976 (BSI). Hassan: 3 miles south of Channarayapattana, *Saldanha* 14275, 4.8.1969 (JCB); Byra, *Nicolson et al.* 2262, 23.10.1971 (JCB). Kodagu: Mercara, *Ahuja* 47531, 10.1.1959 (BSI); Kauble, Sounticoppa, *Rao* 86110, 3.3.1963 (BSI); Bhagamandala, *Bhat* 805, 19.12.1980 (MGH). Kolar: Nandi S.F.,

Kolar, *Singh* 142033, 3.1.1976 (BSI). Mandya: Ranganathittu, *Dinesh* 739, 1.2.1984 (MGH); Hemagiri, *Dinesh* 1034, 12.7.1984 (MGH). Mysore: Kottathuhalla, Biligrirangan ranges, *Rao* 80177, 21.4.1962 (BSI); Srirangapatana, *Sebastin* 18731, 11.3.1964 (MH); Oogharkeeta- Bandipur, *Naithani* 23187, 27.1.1965 (MH); Bilikeri, *Bhat* 3, 6.7.1970 (JCB); St. Philomina College, *Padma Rani* 12, 19.7.1970 (MGH). Raichur: Sindhnur-Gangavati Road, *Singh* 132810, 9.9.1974 (BSI); Kushtagi-Ron Road, *Singh* 141568, 11.11.1975 (BSI). Shimoga: Sharavathi river, above Jog falls, *Prasad* 173776, 3.11.1995 (BSI); Jog falls, *Prasad* 173788, 4.11.1995 (BSI). Tumkur: Namdachilume, *Singh* 132672, 26.2.1975 (BSI). Uttara Kannada: Yellapur, *Talbot* 2950, 28.9.1884 (BSI).

Note: A polymorphic species showing variation mainly in the width and colour of spikelets. In one collection (*Prasad* 172818) the narrow spikelets were found curved in circular shape. This seems to be Clarke's var. *stricta* of *P. capillaris* Nees.

4. **Pycreus kanarensis** Prasad & N.P. Singh in Journ. Econ. & Taxon. Bot. 21 (3): 667, f.1. 1997. **Fig. 49.**

Type: India, Karnataka state, Kanara (without exact locality), *Chibber s.n.,* (Acc. No. 2521), Nov. 1910 (CAL).

Illus.: Prasad & N.P. Singh, *l.c.*

Annuals with fibrous roots. Stems tufted, slender, trigonous, 3-13 cm long, *ca* 0.5 mm thick, smooth, usually with two leaves from the base. Leaves shorter than stem, flat or canaliculate, linear, gradually narrowed to apex, *ca* 1 mm wide, smooth. Inflorescence usually a spike of 3-5 spikelets arranged spicately or subspicately, broadly ovoid, 1-2 cm long, 1-2.5 cm broad; rachis short, *ca* 3 mm long, with internodes of *ca* 1 mm. Rays (found only one), if present *ca* 1.6 cm long. Involucral bracts 1 or 2, reflexed; lower one overtopping the inflorescence, up to 4 cm long. Spikelets spreading, strongly compressed, oblong-ovate to oblong- lanceolate, subacute at apex, 5-9 x 2.5-3 mm, brownish, up to 20-flowered; rachilla wingless, persistent, slightly zig-zag and somewhat canaliculate on both surfaces, with about 0.5 mm long internodes. Glumes membranous, finally obliquely patent, keeled, broadly ovate, obtuse and muticous at apex, 1.8 - 2 x 1-1.4 mm; keel slightly curved, 3-nerved; sides membranous, nervless, brownish, not hyaline at margins. Stamens 2; filaments up to 1 mm long; anthers oblong, *ca* 0.3 mm long, with a purple tip of connective appendage. Style *ca* 1 mm long; stigmas 2, almost as long as the style. Nut biconvex, broadly obovate to orbicular-obovate, minutely apiculate at the obtuse apex, 0.8 - 1 x 0.6-0.7 mm, yellowish-brown to blackish, with longitudinal irregular striations on the faces; epidermal cells longitudinally oblong.

Fls. & Frts.: Nov.

Distrib.: Endemic to Western Peninsular India.

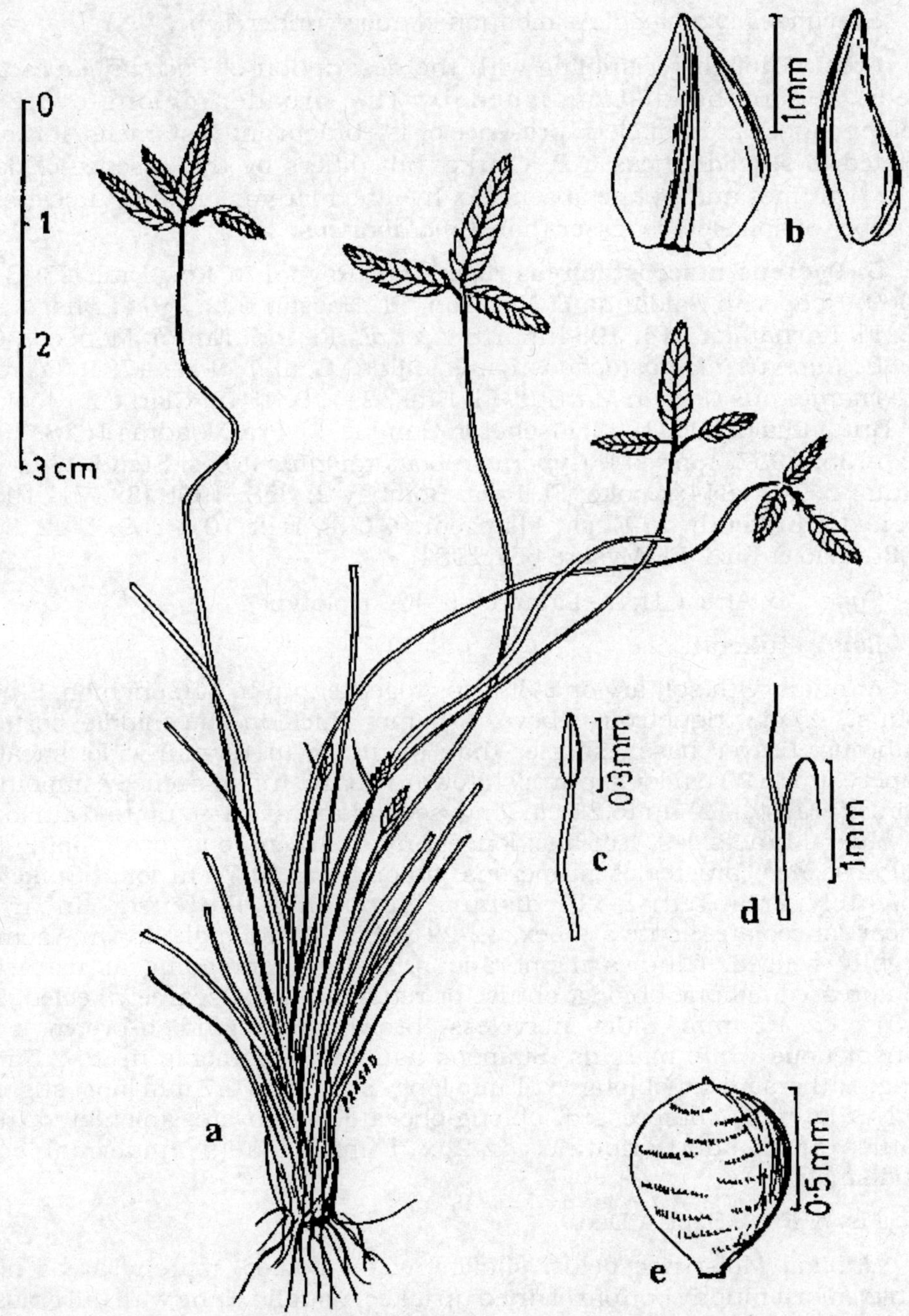

Fig. 49. *Pycreus kanarensis* Prasad & N.P. Singh
a. Habit, **b.** Glumes, **c.** Stamen, **d.** Style, **e.** Nut,
"Reproduced with permission from *J. Econ. Taxon. Bot.* Vol. 21. p. 668. 1997".

Specimens examined: As mentioned above under Type.

Note : Though resembling with the description of *Pycreus flavescens* Nees, it can be distinguished by the broader oblong-ovate to oblong-lanceolate spikelets, presence of 1-2 bracts and 2 stamens. It is also related to *P.malabaricus* C.B. Clarke, but differs by the absence of dark brown glumes and by the absence of hyaline margins on the glumes. Also number of spikelets is lesser than *P.malabaricus*.

5. **Pycreus macrostachyos** (Lamk.) J. Raynal in Kew Bull. 23: 314. 1969; Hooper in Saldanha & Nicolson, Fl. Hassan 692. 1976; Sharma *et al.*, Fl. Karnataka 314. 1984; Karthik. *et al.*, Fl. Ind. Enum. Monocot. 65. 1989. *Cyperus macrostachyos* Lamk., Illus. Gen. 1: 147. 1791. *Pycreus albomarginatus* Nees in Martius, Fl. Bras. 2, 1: 9. 1842; Clarke in Hook.f., Fl. Brit. India 6: 594. 1893; Fischer in Gamble, Fl. Pres. Madras 1628. 1931 (3: 1133. 1957, repr.ed.). *Cyperus albomarginatus* (Nees) Steud., Syn. Pl. Glum. 2: 10. 1854; Cooke, Fl. Pres. Bombay 2: 858. 1908 (3: 371. 1958, repr.ed.); Kukenth. in Engl., Pflanzenr. 4 (20), Heft 101: 359, f. 42 E-H. 1936; Rao & Razi, Fl. Mysore 557.1981.

Type : 'ex Africa' Herb. Lamarck (P-LA, Holotype).

Illus. : Kukenth., *l.c.*.

Annuals with solitary or 2-3 stems together, up to 110 cm high. Stems robust, erect, triquetrous above, 3-6 mm thick in the middle portion, glabrous. Leaves basal, shorter than stem, up to 12 mm wide; sheaths papery, up to 20 cm long, purplish towards base. Inflorescence compound; primary rays 4-12, up to 23 cm long; secondary rays 1-9, up to 6 cm long. Involucral bracts 4-9, leaf-like; longest up to 40 cm long, overtopping the inflorescence; bracteoles numerous, filiform, up to 7 cm long. Spikelets spicately arranged, 8-20, distant, spreading, flattened, linear or linear-lanceolate, acute at apex, 12-22 x 2.5 - 4 mm, pale brown; rachilla slightly winged. Glumes membranous, distant, spreading at maturity, compressed, elliptic-oblong, obtuse or retuse at apex, scarcely keeled, 2.5 - 3 x *ca* 1.5 mm; sides nerveless, brownish to reddish-brown, with conspicuous white margins. Stamens usually 3; filaments up to 2.5 mm long; anthers linear-oblong, *ca* 1 mm long. Style 0.5 - 0.7 mm long; stigmas 2. Nut laterally compressed, oblong-obovate to obovate, apiculate at the obtuse apex, short-stipitate, *ca* 2 x 1 mm, black; epidermal cells isodiametric.

Fls. & Frts.: Sept. - Dec.

Habitat: Moist rice fields, shallow water bodies, marshy areas near tanks, moist muddy bottom of dried up lakes; usually along with tall grasses and other sedges. Often as a weed in rice fields.

Distrib.: Pantropic. INDIA: Peninsular India. KARNATAKA: Bangalore (Sharma *et al.*, *l.c.*), Belgaum, Chikmagalur, Dakshina Kannada, Dharwar, Hassan (Hooper, *l.c.*), Mandya, Mysore (Rao & Razi, *l.c.*), Uttara Kannada.

Specimens examined: Belgaum: Londa, *Gammie s.n.* (Acc.No. 2390), Oct. 1899 (BSI). Chikmagalur: Hilikeri village, Sringeri Road, *Prasad* 173825, 7.11.1995 (BSI). Dakshina Kannada: Venoor, *Bhat* 709, 16.11.1980 (MGH). Dharwar: Hallikeri, near R.F., *Prasad* 172932, 18.12.1994 (BSI). Mandya: Paschimavahini, *Bhat* 97, 20.11.1970 (JCB); T.M. Hossur, *Dinesh* 740, 8.12.1983. Uttara Kannada: Castle Rock, *Gammie s.n.* (Acc No. 2389), Oct. 1902 (BSI); Alurkeri, Dandeli range, Halyal, *Prasad* 173732, 30.10.1995 (BSI).

6. **Pycreus mahadevanii** Govind. in Journ. Econ. Taxon. Bot. 20 (2): 299, f.1. 1996.

Type: India, Karnataka State, Shimoga district, Barkana, *Govindarajalu* 5118 (CAL).

Illus.: Govind., *l.c.*

Annuals with brown or yellowish roots. Stems many, caespitose, spreading, rigid, triquetrous, 10-15 cm long, *ca* 1 mm thick. Leaves shorter than stem, flat, broad, gradually acuminate to apex, 6-10 cm long, *ca* 1.5 mm wide, scabrid towards apex; sheaths membranous, brown with purplish-brown or red nerves, with tannin dotted sides and oblique mouth. Inflorescence compound, pyramidal, distinctly spicate at ends of rays, bearing (6-) 10-30 spikelets. Rays 1-3, stiff, obliquely erect, 1-2 cm long, each consisting of 5-6 spikelets at right angles to rays. Spikelets sessile, oblong or narrowly elliptic-ovate, obtuse at apex, 6-8 x *ca* 2.5 mm, 10-14-flowered; rachilla flexuose, scaly, tannin dotted, narrowly winged. Glumes ovate, obtuse at apex, *ca* 2 x 1.8 mm, brown or cinnamomeous, without hyaline margins; keel green, tannin dotted, 3-nerved. Stamens 2; filaments membranous, persistent; anthers oblong, obtuse at both ends, *ca* 0.25 mm long, yellowish. Style short, *ca* 0.25 mm long; stigmas as long as the style. Nut biconvex, suborbicular or elliptic-orbicular, apiculate at apex, subsessile, *ca* 1 x 0.9 mm, zonate, transversely faintly tuberculate; epidermal cells finely longitudinally striate.

Fls. & Frts. : Oct.

Habitat : Not reported.

Distrib. : Endemic to Karnataka State (Shimoga district).

Note : This species is included on the authority of Govindarajalu, *l.c.* and the description given above is based on the protologue. Specimens were not seen.

7. **Pycreus malabaricus** Clarke in Journ. Linn. Soc. 34: 12. 1898; Woodr. in Journ. Bombay nat. Hist. Soc. 13: 430. 1901; Karthik. *et al.*, Fl. Ind. Enum. Monocot. 65. 1989. *Cyperus malabaricus* (Clarke) Cooke, Fl. Pres. Bombay 2: 856. 1908 (3: 369.1958, repr.ed.); Sharma *et al.*, Fl. Karnataka 306. 1984.

Annuals with fibrous roots, 13-35 cm high. Stems tufted, slender, compressed-trigonous, *ca* 1 mm thick. Leaves up to 2/3 of the stem, gradually narrowed to apex, 1-1.5 mm wide; sheaths up to 5 cm long, purplish-brown tinged. Inflorescence simple, with 5-12 spikelets, 2-2.7 cm long, 1.5-2.7 cm wide. Involucral bracts 3; lowest much overtopping the inflorescence, up to 12 cm long. Spikelets somewhat scattered, strongly compressed, linear-oblong, subacute at apex, 5-10 x *ca* 2.5 mm, 14-20-flowered; rachilla straight, wingless. Glumes membranous, densely imbricate, broadly ovate, acute at apex, *ca* 2 x 1.5 mm, dark and shining reddish-brown, with white-hyaline margins; keel green. Stamens 2; filaments up to 2 mm long; anthers linear-oblong, *ca* 0.5 mm long. Style 1-1.2 mm long; stigmas 2, slightly shorter than style. Nut biconvex, globose-obovate, minutely apiculate at the obtuse apex, often slightly asymmetric near base, *ca* 1 x 0.8 mm, ultimately dark brown, faintly transversely undulate.

Fls. & Frts.: Aug.-Dec.

Habitat: Wet rocky areas, moist rocky slopes, wet gravelly soil along roadsides, along canals and in coastal areas.

Distrib.: Myanmar. INDIA: Western Peninsular India. KARNATAKA: Chikmagalur, Dakshina Kannada, Kodagu, Shimoga, Uttara Kannada (Sharma *et al.*, *l.c.*) .

Specimens examined : Chikmagalur: Koppa, roadsides near the Forest Office, *Prasad* 173811, 6.11.1995 (BSI) & 173813, 7.11.1995 (BSI). Dakshina Kannada: Uppilangady, *Hooper & Saldanha* 2525, 25.11.1971 (JCB). Kodagu: Mercara, *Bhat* 755, 18.12.1980 (MGH); Talakaveri, *Bhat* 1077, 27.9.1981 (MGH). Shimoga: Karodi, near Thirthahalli, *Raghavan* 82765, 28.9.1962 (BSI); Kanagalgudda, near Thirthahalli, *Raghavan* 90045, 19.8.1963 (BSI); Kudadagudda-Nalur Road, *Raghavan* 90432, 3.9.1963 (BSI); Gourikeri, Thalaguppa, Sagar Taluk, *Prasad* 173795, 5.11.1995 (BSI).

8. **Pycreus plurinodosus** (Govind.) P. Singh & V. Singh in Journ. Econ. Taxon. Bot. 5: 467. 1984. *Cyperus plurinodosa* Govind. in Proc. Ind. Acad. Sci. 81 B(5): 192, f. .3. 1975; Karthik. *et al.*, Fl. Ind. Enum. Monocot. 47. 1989.

Type: India, Karnataka State, Shimoga Dist., Megaravalli to Nalur, *Govindarajalu* 4825 (PCM).

Illus.: Govind., *l.c.*

Stolons present. Stems few, tufted, with 2-3 nodes at base, 8-10 cm long, 0.6 - 0.8 mm thick, smooth. Leaves 2-3, longer than stem, flat, rigid, falcate, acuminate at apex, 10-12 cm long, 1-1.2 mm wide; sheaths purple. Inflorescence simple, 1-3- rayed, subspiciform, 15-20 x 5-7.5 mm, bearing 5-8 spikelets. Involucral bracts 3-4, longer than inflorescence, obliquely erect-falcate, up to 6 cm long. Spikelets erect to suberect, compressed, oblong-lanceolate, acute at apex, 5-5.5 x 2-2.2 mm, 10-12-flowered; rachilla straight, membranously winged, excavated. Glumes densely distichous, ovate, acute and mucronulate at apex; keel 3-nerved, green; sides nerveless, atrobrunneous, with broad hyaline margins. Stamens 2; anthers linear-oblong, apiculate, 0.6 - 0.75 mm long. Style flat, 1-1.2 mm long; stigmas 2, shorter than style, 0.5 mm long, glabrous. Nut biconvex, turgid, obovate, asymmetric, 0.8-1 x *ca* 0.5 mm, yellow becoming brown; epidermal cells faint, isodiametric.

Fls. & Frts.: Oct.

Habitat: Higher altitude (1000-1500 m) with very heavy rain fall.

Distrib. : Endemic to Karnataka (Shimoga dt.).

Note: This species is included on the authority of Govindarajalu, *l.c.* As no specimens available in BSI, description provided above is as in the protologue.

9. **Pycreus polystachyos** (Rottb.) Beauv., Fl. Oware & Benin 2: 48, t. 86, f. 2. 1807; Clarke in Hook. f., Fl. Brit. India 6: 592. 1893; Hooper in Saldanha & Nicolson, Fl. Hassan 693. 1976; Arora *et al.*, Bot. S. Kanara 61. 1981; Sharma *et al.*, Fl. Karnataka 314. 1984; Koyama in Dassanayake & Fosberg, Rev. Handb. Fl. Ceylon 5: 219. 1985; Singh, Fl. E. Karnataka 2: 644. 1988; Karthik. *et al.*, Fl. Ind. Enum. Monocot. 65. 1989; Keshava Murthy & Yoganarasimhan, Fl. Coorg. 514. 1990. *Cyperus polystachyos* Rottb., Descr. Ic. Rar. Nov. Pl. 39, t. 11, f. 1. 1773; Nees in Wight, Contr. Bot. India 75. 1834; Kukenth. in Engl., Pflanzenr. 4 (20), Heft. 101. 367. 1936; Ramaswamy & Razi, Fl. Bangalore 91. 1973; Kern in van Steenis, Fl. Males. 1, 7: 649. 1974; Rao & Verma, Cyp. NE India 10. 1982. *C. paniculatus* Rottb., Descr. Ic. Rar. Nov. Pl. 40. 1773. *C. polystachyos* Rottb. var. *laxiflorus* Benth., Fl. Austral. 7: 261. 1878. *Pycreus polystachyos* (Rottb.) Beauv. var. *laxiflorus* (Benth.) Clarke in Hook.f., Fl. Brit. India 6: 592. 1893; Hooper in Saldanha & Nicolson, Fl. Hassan 693. 1976. *P. odoratus* Urb., Symb. Ant. 2: 164. 1900; Fischer in Gamble, Fl. Pres. Madras 1627. 1931 (3: 1132. 1957, repr.ed.). *Cyperus odoratus* auct. non L. 1753; Cooke, Fl. Pres. Bombay 2: 859. 1908 (3:372. 1958, repr.ed.).

Illus. : Rottb., *l.c.*

Tufted annuals or perennials with short rhizome, 20-70 (- 100) cm high. Stems erect, slender, trigonous, 1-3 mm thick, smooth. Leaves shorter than stem, flat or canaliculate, linear, gradually narrowed to apex, 1.5-4 mm wide; sheaths membranous, purplish towards the base. Inflorescence simple or subcompound, at times reduced to a head-like cluster of spikelets. Involucral bracts 3-5, usually the lowest overtopping the inflorescence. Primary rays 3-8, longest up to 7 cm long. Spikelets up to 15, digitately arranged, strongly compressed, linear-lanceolate, gradually narrowed to an acute apex, 10-25 x 1.5 - 2 mm, up to 50-flowered; rachilla flexuos, narrowly winged, persistent. Glumes thinly chartaceous, oblong-ovate, subobtuse and muticous or minutely mucronulate at apex, 1.7-2.5 x *ca* 1 mm; keel 3-nerved, green; sides nerveless ferrugineous to reddish-brown, with white hyaline margins. Stamens 2; filaments up to 2 mm long; anthers linear-oblong, 0.5-1 mm long. Nut laterally compressed, oblong, obtuse-truncate and shortly apiculate at apex, 1-1.5 x *ca* 0.5 mm, ultimately dark brown; epidermal cells isodiametric.

Fls. & Frts.: May-Feb.

Habitat : Common among grasses in moist soil near canals, tanks and rice fields; moist soil in rocky river beds, bunds of back waters and rice fields; in rocky river beds with running water; sandy sea shores and in sandy islands.

Distrib.: Throughout the warmer parts of the whole world. In Eastern Asia extending to warm-temperate regions up to Central Japan and Mediterranean region. INDIA: Peninsular India, Central, East and North-East India, Andaman & Nicobar Islands. KARNATAKA: Bangalore, Balgaum, Bijapur, Chikmagalur, Chitradurga (Sharma *et al., l.c.*), Dakshina Kannada, Hassan, Kodagu, Kolar, Mandya, Mysore, Shimoga, Tumkur, Uttara Kannada.

Specimens examined: Bangalore: Savandurga, *Bhaskar* 147, 16.2.1978 (JCB); Channapatanam, *Ramesh* 277, 23.2.1978 (JCB). Belgaum: Dudwa village, near Londa R.F., *Prasad* 172909 & 172913, 17.12.1994 (BSI). Bijapur: Varathi Kavalu, Kendur, *Prasad* 172958, 21.12.1994 (BSI). Chikmagalur: Charmudi ghat, *Kulkarni* 155186, 25.5.1979 (BSI). Dakshina Kannada: Mangalore, *Foulker* 4854, Jan. 1902 (MH); Kudallu, without Coll. name, 16747, 30.8.1920 (MH); Coondapur Islands, *Raghavan* 145394, 25.9.1976 (BSI); Badergadda, Malpe, *Bhat* 437, 18.12.1976 (MGH); Mulki, *Saldanha* & *Sreenath* 4917, 9.12.1978 (JCB); Surathkal, *Bhat* 608, 2.10.1979 (JCB); Jappinamogeru village, Mangalore, *Prasad* 173852 & 173856, 10.11.1995 (BSI). Hassan: Kenchenkumri, *Hooper & Gandhi* 2449, 12.11.1971 (JCB). Kodagu: Mercara, *Bhat* 735, 18.12.1980; Abbe falls, *Bhat* 886, 23.12.1980 (MGH). Kolar: Gudibanda-Gauribidnur, *Singh* 142171, 7.1.1976 (BSI); Kendatti hill, *Parkash & Sreenath* 4917, 9.12.1978 (JCB). Mandya: Belluru, *Nicolson et al.* 2359, 26.10.1971 (JCB); Sreerangapatanam, *Dinesh* 982, 18.6.1984 (MGH). Mysore: Near Mandakalli, on way to Nanjangud, *Rao* 759, 12.7.1970 (JCB). Shimoga: Balehalli forests, Agumbe, *Raghavan* 90392, 2.9.1963 (BSI); Sharavathi river bed, above Jog falls, *Prasad* 173768, 3.11.1995 (BSI); Gourikeri, Thalaguppa, Sagar taluk, *Prasad* 173802, 5.11.1995 (BSI). Tumkur: Namdachilume forest Rest house vicinity, *Rao* 73261, 28.8.1961 and *Rao* 73428 A, 30.8.1961 (BSI); Namdachilume, *Singh* 132674, 26.2.1975 (BSI). Uttara Kannada: Karwar, *Talbot* 1255A, July 1885 (BSI); Sirsi, *Puri* 2055, 6.5.1956 (BSI); Karwar beach, *Prasad* 173751, 31.10.1995 (BSI); Kaliguj,

Karwar, *Prasad* 173757, 1.11.1995 (BSI); Kinnare, a small island near the Railway bridge over Kalinadi, Karwar, *Prasad* 173762, 1.11.1995 (BSI).

Note: A variable species especially in the size and form of inflorescence. It varies from a dense cluster of suberect spikelets to the open inflorescence of spreading spikelets. Intermediate forms of inflorescence are also found with short rays. In Karnataka, plants with open infloresence are common especially towards coastal areas which are often described as var. *laxiflorus* (Benth.) Clarke.

10. **Pycreus pumilus** (L.) Nees [in Linnaea 9: 283. 1835] ex Clarke in Hook. f., Fl. Brit. India 6: 591. 1893, *quoad basion., excl. descr.;* Dom., Bibl. Bot. 85: 417. 1915; Fischer in Gamble, Fl. Pres. Madras 1627. 1931 (3: 1132. 1957, repr.ed.); Hooper in Saldanha & Nicolson, Fl. Hassan 693. 1976; Yoganarasimhan *et al.*, Fl. Chikmagalur 360. 1981; Sharma *et al.*, Fl. Karnataka 314. 1984; Koyama in Dassanayake & Fosbereg, Rev. Handb. Fl. Ceylon 6: 224. 1985; Singh, Fl. E. Karnataka 2: 644. 1988; Karthik. *et al.*, Fl. Ind. Enum. Monocot. 66. 1989; Keshava Murthy & Yoganarasimhan, Fl.Coorg 515. 1990. *Cyperus pumilus* L., Cent. Pl. 2: 6. 1756; Cooke, Fl. Pres. Bombay 2: 857. 1908 (3: 370. 1958, repr.ed.); Kukenth. in Engl., Pflanzenr. 4 (20), Heft 101: 375, t. 44 A-E 1936; Ramaswamy & Razi, Fl. Bangalore 92. 1973; Kern in van Steenis, Fl. Males. 1, 7: 650, f. 66. 1974; Rao & Razi, Fl. Mysore 560. 1981; Rao & Verma, Cyp. NE India 9. 1982. *C. pulvinatus* Nees & Mey. in Wight, Contr. Bot. India 74. 1834. *Pycreus nitens* Nees, Nova Actorum Acad. Caes. Leop. - Carol. Nat. Cur. 19, Suppl. 1: 53. 1843; Clarke in Hook.f., Fl. Brit. India 6: 591. 1893.

Illus.: Kukenth., *l.c.* ; Kern, *l.c.*

Annuals with fibrous roots, 5-20 cm high. Stems tufted, slender or filiform, triquetrous, 0.3-0.8 mm thick, smooth. Leaves shorter or surpassing the stem, canaliculate, gradually acuminate to apex, 1-2 mm wide, scabrid on margins towards apex; sheaths stramineous to reddish-brown towards base. Inflorescence simple to subcompound, frequently contracted to a head-like cluster of spikelets. Involucral bracts 3-5, overtopping the inflorescence. Primary rays 3-6, slender, up to 3 cm long, smooth, at times absent or very short. Spikes ovoid or subglobose, 1-3 cm across, densely bearing a few to many spikelets; rachis very short. Spikelets spreading, strongly compressed, linear-oblong, subacute at apex, 5-16 x 1.5-2.5 mm (including the mucros); rachilla straight, wingless, persistent, with *ca* 0.25 mm long internodes. Glumes membranous, half imbricate, ovate, truncate or retuse at apex, sharply keeled, 1-1.7 mm long (excluding the mucro), 1-1.2 mm wide; keel 3-5-nerved; sides nerveless, hyaline towards margins; mucro erect or slightly recurved, 0.2-0.3 mm long. Stamens 2; filaments up to 1 mm, anthers oblong, *ca* 0.2 mm long. Style *ca* 0.6 mm long; stigmas 2, longer than style. Nut laterally compressed, biconvex, oblong-obovate, truncate to obtuse and apiculate at apex, 0.5-0.8 x 0.3-0.5 mm, brownish; epidermal cells isodiametric.

Fls. & Frts.: June - Feb.

Habitat: Sandy soil near sea shores, in and around paddy fields, wet soil on the bank of canals, in drying ponds and other marshy areas.

Distrib.: Sri Lanka, South China, Indo-China, Formosa, Malesia, tropical Africa and Australia. INDIA: Throughout. KARNATAKA: Bangalore, Belgaum, Chikmagalur, Dakshina Kannada, Hassan, Kodagu, Kolar, Mandya, Mysore, Shimoga, Uttara Kannada.

Specimens examined : Bangalore: Sringarpura, *Hooper & Saldanha* 18068, 18.11.1971 (JCB). Belgaum: Gokak falls, Gokak, *Prasad* 172820, 11.12.1994 (BSI); Dinman Hassur, Belgaum- Gokak road, *Prasad* 172845 & 172846, 13.12.1994 (BSI); Dudwa village, near Londa R.F., *Prasad* 172911, 17.12.1994 (BSI). Chikmagalur: Hilikeri village, along Sringeri road, *Prasad* 173820, 7.11.1995 (BSI). Dakshina Kannada: Jolpad, *Barber* 2376, 13.11.1900 (MH); Udyavara, *Bhat* 268, 8.8.1976 (MGH); Indrali, Udupi, *Bhat* 279, 22.8.1976 (MGH); Kapu, *Bhat* 442, 15.1.1977 (MGH). Hassan: Kempuhole, Nicolson *et al.* 2388, 25.10.1971 (JCB). Kodagu: Abbe falls, Mercara, *Bhat* 8748, 23.12.1980 (MGH); Kirugoor, *Bhat* 934, 26.1.1981 (MGH). Kolar: Tumkur-Chickballapur Road, *Singh* 142014, 3.1.1976 (BSI). Mandya: Hemagiri, *Dinesh* 741, 1.2.1984 (MGH). Mysore: Chamundi, *Bhat* 22.8.1970 (JCB); Yelwal, *Rao* 1158, 2.12.1970 (MGH). Shimoga: Someshwar, base of ghat zone, *Raghavan* 83215, 14.10.1962 (BSI). Uttara Kannada: Yellapur, *Talbot* 983, 10.9.1883 (BSI); Karwar, *Talbot* 1513, Aug. 1885 (BSI); Saurikatta sea shore, Karwar, *Chibber s.n.* (Acc. No. 2558), 1.11.1910 (BSI); Karlukatta tank, Halyal, *Prasad* 173715, 29.10.1995 (BSI); Karwar beach, *Prasad* 173746 & 173748, 31.10.1995 (BSI).

11. **Pycreus puncticulatus** (Vahl) Nees, Fl. Brasil. 2(1): 10, in note. 1842; Clarke in Hook. f., Fl. Brit. India 6: 593. 1893; Fischer in Gamble, Fl. Pres. Madras 1628. 1931 (3: 1133. 1957, repr.ed.); Sharma *et al.*, Fl. Karnataka 314. 1984; Koyama in Dassayanake & Fosberg, Rev. Handb. Fl. Ceylon 5: 218. 1985; Karthik. *et al.*, Fl. Ind. Enum. Monocot. 66. 1989. *Cyperus puncticulatus* Vahl, Enum. pl. 2: 348. 1806; Cooke, Fl. Pres. Bombay 2: 858. 1908 (3: 371. 1958, repr.ed.); Kukenth. in Engl., Pflanzenr. 4 (20), Heft 101: 362. 1936. *C. baccha* Kunth, Enum. Pl. 2: 115. 1837. **Fig. 50.**

Type : India.

Annuals with brownish, fibrous roots. Stems triquetrous, 30-60 cm tall, 2.5 - 4 mm thick, often somewhat enlarged at base, smooth throughout. Leaves herbaceous to coriaceous, slightly overtopping the stem, gradually acuminate at apex, 15-40 cm long, 4-8 (-10) mm wide. Inflorescence lax, compound, 7-24 cm long and as broad. Involucral bracts 3-5, spreading, mostly longer than inflorescence. Rays 4-11, unequal, slender; longer ones

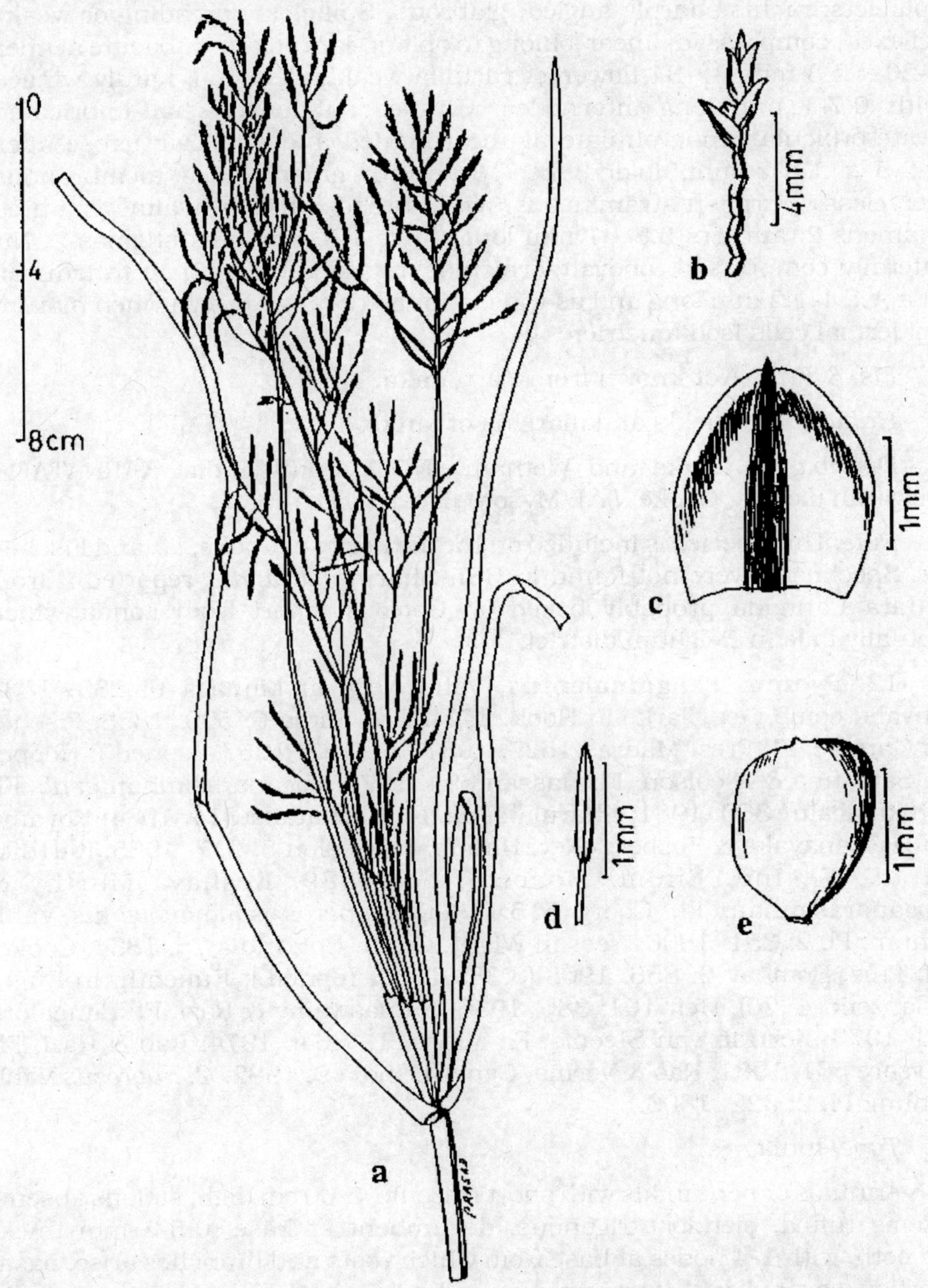

Fig. 50. *Pycreus puncticulatus* (Vahl) Nees
a. Inflorescence (later stage), **b**. Spikelet, **c**. Glume, **d.** Stamen, **e.** Nut

6-14 cm long, smooth. Spikes lax, 2-7 cm long, distantly bearing 7-20 spikelets; rachis sharply angled, glabrous. Spikelets spreading or weakly reflexed, compressed, linear-oblong to oblong-lanceolate, subacute at apex, 8-20 x 2-3 mm, 12-34-flowered; rachilla weakly flexuous, hardly winged, with 0.7-1 mm long internodes. Glumes subdensely half-imbricated, ovate-orbicular, mucronulate at the rounded and shallowly erose apex, 2-2.5 x 1.8-2 mm, folded; keel 3-nerved, green; sides membranous, nerveless, purplish-stramineous, with broad whitish-hyaline margins. Stamens 2; anthers 0.8 - 1 mm long. Style 1.5 mm long; stigmas 2. Nut laterally compressed, obovate-orbicular, rounded-emarginate to truncate at apex, 1-1.2 mm long and as wide, stipitate, orange-brown when mature; epidermal cells isodiametric.

Fls. & Frts.: Not known from Karnataka.

Habitat: Rice fields and margins of tanks.

Distrib.: Sri Lanka and Vietnam. INDIA: South India. KARNATAKA: Belgaum (Londa, Cooke, *l.c.*), Mysore (Fischer, *l.c.*) .

Note: This species is included on the authority of Cooke, *l.c.* and Fischer, *l.c.* Specimens were not found in BSI. Sharma *et al., l.c.* reported it from Uttara Kannada probably based on Cooke's report from Londa which actually falls in Belgaum district.

12. **Pycreus sanguinolentus** (Vahl) Nees in Linnaea 9: 283. 1834 (invalid comb.) ex Clarke in Hook. f., Fl. Brit. India 6: 590. 1893; Fischer in Gamble, Fl. Pres. Madras 1627. 1931 (3: 1132. 1957, repr.ed.); Hooper in Saldanha & Nicolson, Fl. Hassan 694. 1976; Yoganarasimhan *et al.*, Fl. Chikmagalur 360. 1981; Sharma *et al.*, Fl. Karnataka 314. 1984; Koyama in Dassanayake & Fosberg, Rev. Handb. Fl. Ceylon 5: 217. 1985; Karthik. *et al.*, Fl. Ind. Enum. Monocot. 66. 1989. Keshava Murthy & Yoganarasimhan, Fl. Coorg 515. 1990. *Cyperus sanguinolentus* Vahl, Enum. Pl. 2: 351. 1806; Nees in Wight, Contr. Bot. India 75. 1834; Cooke, Fl. Pres. Bombay 2: 856. 1908 (3:369.1958, repr.ed.); Kukenth. in Engl., Pflanzenr. 4 (20), Heft 101: 385. 1936; Ramaswamy & Razi, Fl. Bangalore 93. 1973; Kern in van Steenis, Fl. Males. 1, 7:646.1974; Rao & Razi, Fl. Mysore 561. 1981; Rao & Verma, Cyp. NE India 9. 1982. *C.eragrostis* Vahl, Enum. Pl. 2: 322. 1806.

Type: India.

Annuals or perennials with short rhizome, 5-60 cm high; stolons absent. Stems tufted, slender, trigonous, decumbent at base, 0.5-2 mm thick, smooth, with 1-4 nodes at base from which roots and branches arise; basal portion covered with stramineous or brownish sheaths. Leaves usually shorter than stem, flat or canaliculate, linear, gradually narrowed to an acute tip, 1-4 mm wide, scabrid on the margins towards apex. Inflorescence simple, usually with 1-5 rays, often congested to a cluster of few spikelets.

Involucral bracts 3-5; lower ones overtopping the inflorescence. Spikes ovoid to broadly ovoid, with 3-18 spikelets on a short rachis. Spikelets compressed, ovate to oblong-lanceolate, subacute at apex, 7-15 x 2-3.5 mm, brownish, 10-30-flowered; rachilla wingless, persistent. Glumes membranous, keeled, ovate to broadly ovate, obtuse and muticous at apex, 2.5-3 x 1.5- 2.2 mm; keel 3-5- nerved, green; sides nerveless, ferrugineous to dark brown, usually furrowed in the centre. Stamens 2 or 3; filaments elongate up to 2.5 mm; anthers linear-oblong, 0.7-1 mm long. Style longer than nut, 1.2 - 2 mm long; stigmas 2, shorter than to as long as the style. Nut laterally compressed, biconvex, broadly obovate to orbicular, apiculate at apex, 1-1.3 x 0.8-1 mm, brownish to black when mature; epidermal cells isodiametric.

Fls. & Frts.: Oct.-March

Habitat : Wet soil along canals, tanks, streams etc., wet muddy soil in rocky areas, paddy fields and other marshy areas along with other sedges and grasses.

Distrib.: Widely distributed in warmer parts of Central and Eastern Asia, to Japan, tropical Africa and Australia. INDIA: Throughout. KARNATAKA: Bangalore, Belgaum, Bijapur, Chikmagalur, Hassan, Kodagu, Mysore (Rao & Razi, *l.c.*), Shimoga, Uttara Kannada.

Specimens examined : Bangalore: Halmavu, *Hooper & Saldanha* 18026, 7.11.1971 (JCB). Belgaum: Gokak falls, *Prasad* 172819, 11.12.1994 (BSI); Dudwawada R.F., Londa, *Prasad* 182891, 17.12.1994 (BSI). Bijapur: Varathi Kavlu, Kendur, *Prasad* 172962, 21.12.1995 (BSI). Chikmagalur: Hosakeri village forest, *Yoganarasimhan* 1382, 16.11.1972 (RRCBI); Gadical village tank, Koppa taluk, *Prasad* 173843, 8.11.1995. (BSI). Hassan: Tank near Dandiganahalli, *Hooper & Gandhi* 2401, 11.11.1971 (JCB). Kodagu: Mercara, *Bhat* 775, 18.12.1980 (MGH); Bhagamandala, *Bhat* 802, 19.12.1980 (MGH); Kushalnagar, *Bhat* 945, 16.12.1981 (MGH). Mysore: Mandakalli, *Bhat* 18, 12.7.1970 (MGH). Shimoga: Gourikeri, Talaguppa, Sagar taluk, *Prasad* 173796, 5.11.1995 (BSI). Uttara Kannada: Karwar, without coll. name, *s.n.* (Acc. No. 711), 10.8.1883 (BSI); Yellapur, *Talbot s.n.* (Acc. No. 710), Sept. 1884 (BSI); Karwar, *Talbot* 1616, 10.8.1885 (BSI); Santavery, *Talbot* 3120, 7.9.1893 (BSI); Hangal, *Chibber s.n.* (Acc. Nos. 2561 & 2562), 26.10.1910 (BSI); Karlukatta tank, Halyal, *Prasad* 173706 & 173707, 29.10.1995 (BSI).

13. **Pycreus stramineus** (Nees) Clarke in Hook.f., Fl. Brit. India 6: 589. 1893; Fischer in Gamble, Fl. Pres. Madras 1627. 1931 (3: 1132. 1957, repr.ed.); Hooper in Saldanha & Nicolson, Fl. Hassan 694. 1976; Arora *et al.*, Bot. S. Kanara 62. 1981; Sharma *et al.*, Fl. Karnataka 314. 1984; Koyama in Dassayanake & Fosberg, Rev. Handb. Fl. Ceylon 5: 216. 1985; Karthik. *et al.*, Fl. Ind. Enum. Monocot. 66. 1989; Keshava Murthy & Yoganarasimhan, Fl. Coorg 515. 1990. *Cyperus stramineus* Nees in Wight,

Contr. Bot. India 74. 1834. *C. substramineus* Kukenth. in Engl., Pflanzenr. 4 (20), Heft, 101: 398. 1936; Kern in van Steenis, Fl. Males. 1, 7: 653. 1974; Rao & Verma, Cyp. NE India 7, f. 2-2a. 1982.

Type: India, Silhet, *Wallich* 3320. Not of Desf. ex. Link. 1820.

Illus.: Rao & Verma, *l.c.*

Annuals with fibrous, yellowish roots, 10-40 cm high. Stems tufted, very slender, 0.5 - 1 mm thick, often slightly curved, obtusely trigonous, smooth. Leaves shorter than to equalling the stem, canaliculate, very narrow, gradually acuminate to apex, *ca* 1 mm wide, scaberulous towards apex; sheaths purplish towards base. Inflorescence simple, usually a spike-like cluster of 4-16 spikelets, rarely with 1-3 short rays. Involucral bracts 2(-3), erect- patent or suberect, overtopping the inflorescence, lower one up to 10 cm long. Spikelets suberect, strongly compressed, oblong-lanceolate to linear-oblong, acute at apex, 7-20 x *ca* 2 mm; rachilla straight, wingless; internodes *ca* 0.5 mm long. Glumes membranous, strongly keeled, ovate, acute and mucronulate at apex, *ca* 2 x 1.2 - 1.5 mm; keel 3-nerved, green; sides yellowish, nerveless, with whitish- hyaline margins. Stamens 2; filaments up to 1.5 mm long; anthers linear-oblong, *ca* 0.5 mm long. Style 0.8 - 1 mm long; stigmas 2, as long as the style. Nut laterally compressed, biconvex, orbicular-obovate to elliptic, apiculate at the obtuse apex, often slightly asymmetric, *ca* 1 x 0.8 mm, transversely rugulose, brownish to blackish; epidermal cells longitudinally oblong.

Fls. & Frts.: Aug. - Oct.

Habitat: Moist grasslands, rice fields and along canals.

Distrib.: Sri Lanka, Myanmar, Indo-China and Malesia. INDIA: Throughout (except in drier regions). KARNATAKA: Dakshina Kannada, Hassan (Hooper, *l.c.*) .

Specimens examined : Dakshina Kannada: Sullia, *Barber* 2122, 25.11.1900 (MH); Kudlu, without coll. name, 16740, 28.8.1920 (MH); Udyavara, *Bhat* 275, 8.8.1976 (MGH); Indrali, Udupi, *Bhat* 282, 22.8.1971 (MGH); Kolluru-Nagodi, *Raghavan* 145211, 23.9.1976 (BSI).

SPECIES EXCLUDED

Pycreus flavescens (L.) Beauv. ex Reichb., Fl. Germ. Excurs. 1(1): 72. 1830.

Sharma *et al.* (1984) included this species in the state flora of Karnataka, most probably based on wrongly identified specimens (*Chibber s.n.* Acc. Nos. 2519, 2520 and 2521). A detailed study of these specimens revealed that they do not match with any known species and hence a new species was described based on these specimens (*Prasad & Singh,* 1997c). Therefore, *P. flavescens* (L.) Beauv. ex Reichb. is excluded from the state flora of Karnataka.

18. QUEENSLANDIELLA

Dom., Bibl. Bot. Heft 85. 415. 1915. *Mariscopsis* Cherm. in Bull. Mus. Paris 25: 60. 1919. *Cyperus* sect. *Queenslandiella* (Dom.) Kern in van Steenis, Fl. Males. 1, 7:654.1974.

Type: Queenslandiella mira Dom.

Tufted annuals. Inflorescence umbelliform, once anthelate. Involucral bracts leaf-like. Spikelets strongly flattened, deciduous above the two sterile glumes. Glumes distichous, similar to *Kyllinga* sp., deciduous. Flowers bisexual. Stamens 2. Stigmas 2. Nut laterally compressed, biconvex.

Monotypic genus distributed in paleotropics.

Queenslandiella hyalina (Vahl) Ballard in Hook., Ic. Pl. 33, t. 3208. 1933; Hooper in Saldanha & Nicolson, Fl. Hassan 694. 1976; Sharma *et al.*, Fl. Karnataka 314. 1984; Karthik. *et al.*, Fl. Ind. Enum. Monocot. 67. 1989. *Cyperus hyalinus* Vahl, Enum. Pl. 2:329.1805; Cooke, Fl. Pres. Bombay 2: 857. 1908 (3: 369. 1958, repr.ed.); Ramaswamy & Razi, Fl. Bangalore 88. 1973; Kern in van Steenis, Fl. Males. 1, 7: 655, f. 68. 1974; Rao & Razi, Fl. Mysore 557. 1981. *C. pumilus* Nees in Wight, Contr. Bot. India 74. 1834, *non* L. *Pycreus pumilus* Clarke in Hook.f., Fl. Brit. India 6: 591. 1893, *quoad descr. P. hyalinus* Dom., Bibl. Bot. 85: 417. 1915; Fischer in Gamble, Fl. Pres. Madras 1627. 1931 (3: 1132. 1957, repr.ed.). *Queenslandiella mira* Dom., Bibl. Bot. 85: 416, t. 11, f. 7-13. 1915. *Kyllinga hyalina* (Vahl) Koyama in Journ. Jap. Bot. 51: 313. 1976 *et* in Dassanayake & Fosberg, Rev. Handb. Fl. Ceylon 5: 242, f. 17. 1985. **Fig. 51.**

Illus. : Ballard, *l.c.*; Dom., *l.c.*; Kern, *l.c*; Koyama, *l.c.*

Annuals with fibrous roots, 4-25 (-37) cm high. Stems tufted, triquetrous, 0.5-1.5 mm thick, smooth. Leaves shorter than to overtopping the inflorescence, weak, flat, gradually narrowed to apex, 1-4 mm wide, scabrid on the margins towards apex; sheaths stramineous or reddish-brown tinged. Inflorescence simple, lax, rarely contracted to a head-like cluster. Involucral bracts 3-6, lower ones much overtopping the inflorescence. Rays 3-6, patent, up to 6 cm long; spikes 8-14 x 5-10 mm, loosely to subdensely bearing 5-15 spikelets; rachis 4-winged, glabrous. Spikelets spicately arranged, ultimately spreading, strongly compressed, ovate to oblong-ovate, subacute at apex, 3-7 x 1.5-3 mm, 4-8-flowered, pale greenish; rachilla flexuous, disarticulating at base, broadly winged, with *ca* 1 mm long internodes. Glumes membranous, ovate to broadly ovate, strongly keeled, mucronate at apex, 2-3 x *ca* 2 mm (including the more or less recurved mucro); keel strongly 3-nerved, serrate-scabrous on the back, greenish; sides hyaline, whitish to yellowish, strongly 3-nerved on each side, somewhat reticulate. Stamens 2; filaments up to 2 mm long; anthers linear, oblong-lanceolate, 0.6-1 mm long. Style 0.5 - 1 mm long; stigmas 2,

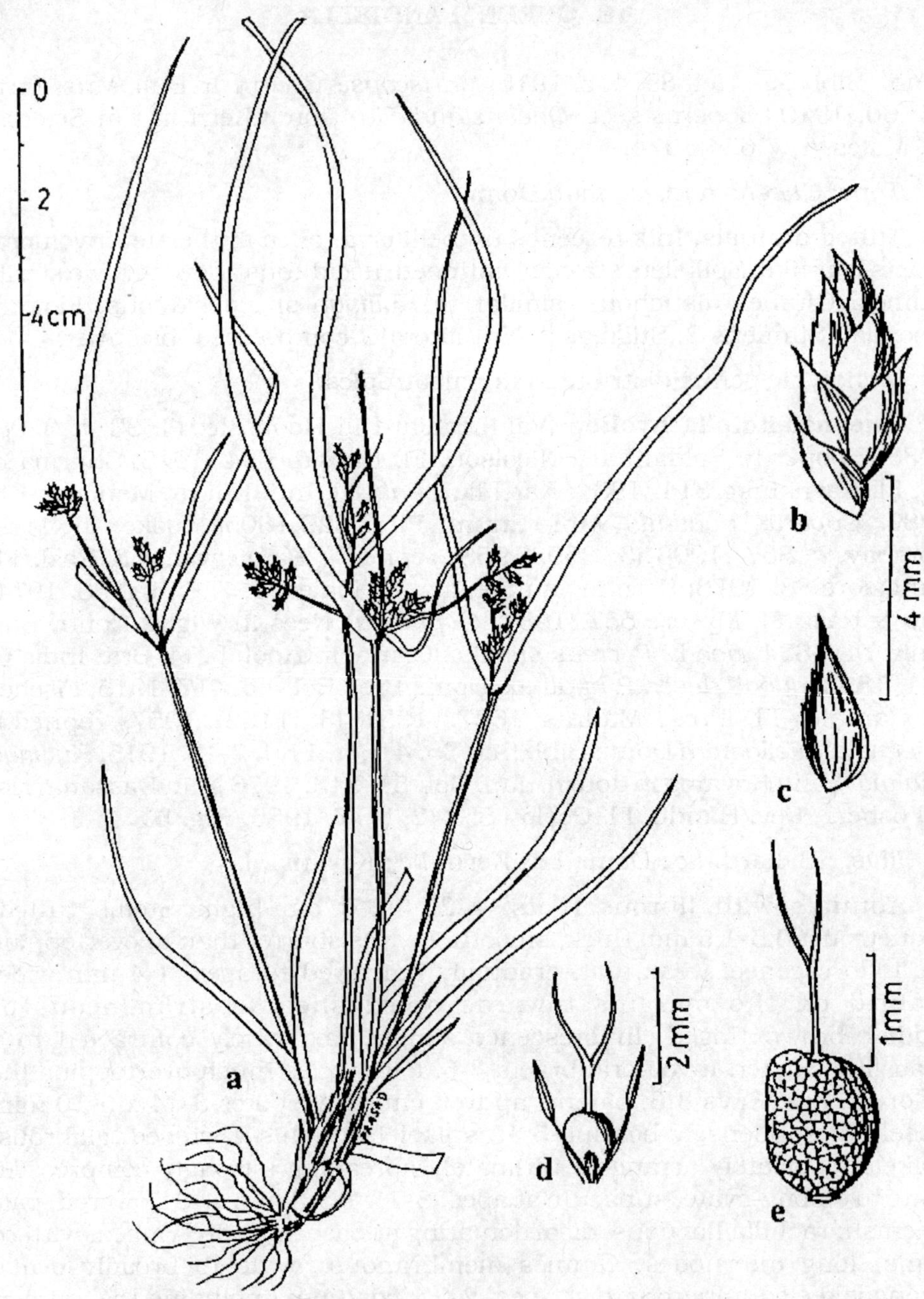

Fig. 51. *Queenslandiella hyalina* (Vahl) Ballard
a. Habit, **b**. Spikelet, **c.** Glume, **d**. Flower (later stage), **e.** Nut

1-1.5 mm long. Nut laterally compressed, biconvex, elliptic to broadly elliptic, usually asymmetric, truncate to emarginate at apex, 1.1-1.3 x 1-1.1 mm, shortly stipitate, brownish to blackish, epidermal cells isodiametric.

Fls. & Frts. : July - Nov.

Habitat : Sandy soil, near sea coast, among mangroves, dry grasslands, marshy areas along the bunds of paddy fields and at times as a weed in cultivated fields.

Distrib.: Sri Lanka, Indo-China, Malesia, tropical East Africa and Australia. INDIA: Peninsular India and Eastern India. KARNATAKA: Bangalore (Ramaswamy & Razi, *l.c.*), Hassan, Mandya, Mysore. Not common.

Specimens examined: Hassan: Near tank, Channarayapattana, *Hooper & Gandhi* 2412, 11.11.1971 (JCB). Mandya: Ranganathittu, *Dinesh* 543, 19.9.1983 (MGH); Melukotte, without coll. name, 657, 7.11.1983 (MGH). Mysore: Saraswathipuram, *Bhat* 62, 21.8.1970 (JCB); St. Joseph College Campus, *Saldanha* 18677A, 13.9.1975 (JCB).

Note: Queenslandiella hyalina shows close relation to *Kyllinga* as well as *Pycreus* and often treated under either of these genera. Laterally compressed, biconvex nuts and bifid styles are found in both these genera. But rachilla with disarticulating base is not found in *Pycreus* but in *Kyllinga*. Glumes are also very similar to that of *Kyllinga* with its prominent nerves on both sides. But by the presence of umbelliform inflorescence it differs from *Kyllinga* which always have a head of sessile spikelets. Hence based on a combination of characters, i.e. disarticulate rachilla and umbelliform inflorescence it is seperated here from *Pycreus* and *Kyllinga* respectively and treated under *Queenslandiella* Dom.

19. REMIREA

Aubl., Hist. Pl. Guian. 44. 1775; Hooper in Hepper, Fl. West. Trop. Africa, ed. 2,3: 297. 1972. *Cyperus* sect. *Remirea* (Aubl.) Koyama in Quart. Journ. Taiwan Mus. 14: 162. 1961; Kern in van Steenis, Fl. Males. 1, 7: 644. 1974.

Perennials with long, creeping rhizome. Stems erect, leafy, branching, rooting from the subterranean nodes. Leaves numerous, crowded on the branches, rigid, spreading and recurved, pungent; sheaths very short, open. Infloresence a terminal, ovoid, solitary or lobed head of crowded spikelets. Spikelets short, 1-flowered. Glumes 4, imbricate, lower 3 empty, suborbicular, concave; upper fertile, very thick, corky, 1-flowered. Hypogynous bristles 0. Stamens 3; anthers linear, acute at apex. Style linear, very short, thickened at base; stigmas 3. Nut trigonous, oblong-ellipsoid, smooth, beaked, closely embraced by the 2 upper glumes.

A monotypic genus, widely distributed in the sandy seashores of the tropics.

Literature: **HOOPER, S.S.** (1983) *Remirea* or *Mariscus* - new support for a monotypic genus in Cyperaceae, in *Kew Bull.* 38: 479-480. **KERN, J.H.** (1958) *Remirea* or *Cyperus*? Florae Malesianae Precursores XXI, in *Act. Bot. Neerl.* 7: 795-798. **OTENG-YEBOAH, A.A.** (1975) Morphology, Anatomy and Taxonomy of the genus *Remirea* Aubl. (Cyperaceae), in *Boissiera* 24: 202.

Remirea maritima Aubl., Hist. Pl. Guian. Franc. 1: 45, t. 16. 1775; Clarke in Hook.f., Fl. Brit. India 6: 677. 1894; Woodr. in Journ. Bombay nat. Hist. Soc. 13: 432. 1901; Cooke, Fl. Pres. Bombay 3: 902. 1908 (3: 417. 1957, repr.ed.); Fischer in Gamble, Fl. Pres. Madras 1673. 1931 (3: 1161. 1958, repr.ed.); Hooper in Kew Bull. 38: 479. 1983; Karthik. *et al.*, Fl. Ind. Enum. Monocot. 67. 1989. *R. pedunculata* R. Br., Prodr. 236. 1810. *Cyperus pedunculatus* (R. Br.) Kern in Act. Bot. Neerl. 7: 798. 1958 *et* in van Steenis, Fl. Males. 1, 7: 644. 1974; Sharma *et al.*, Fl. Karnataka 307. 1984. *Mariscus pedunculatus* (R. Br.) Koyama, Gard. Bull. Singapore 30: 159. 1977 *et* in Dassanayake & Fosberg, Rev. Handb. Fl. Ceylon 5: 240. 1985. **Fig.52.**

Illus. : Clarke, Illus. Cyp. t. 102, f. 7-10. 1909.

Rhizome horizontally creeping, branching, rooting at nodes, clothed with brown sheaths, 1-3 mm thick; internodes 3-6 cm long. Stems solitary or few arising together from the nodes of rhizome, 6-25 cm high, usually basal part subterranean, 1-2 mm thick, trigonous, smooth. Leaves crowded on above ground part of stem, rigid, canaliculate, usually recurved, linear, gradually narrowed to a sharp trigonous apex, 3-8 cm long (including the brown sheath), 4-5 mm wide at base, minutely scabrous on margins towards apex. Infloresence a congested cluster of 3-8 digitately arranged spikes, 1.5-2 cm long, 1.5-3 cm across. Involucral bracts 3-6 (-8), usually 5, patent or reflexed; longest up to 8 cm long, overtopping the inflorescence. Spikes ovoid or ellipsoid, 10-15 x 6-10 mm, with many densely crowded spikelets. Spikelets sessile, ovoid, acute at apex, slightly compressed, with 4 glumes, 1- flowered, 4-5 x 1.5-2 mm. Rachilla jointed above 1 or 2 basal glumes; upper most internode strongly flattened, glume-like (but without nerves), ultimately corky and thickened, *ca* 3 mm long. Glumes ovate or broadly ovate, acute to mucronate at apex, hardly keeled, many-nerved; lower 3 empty, 2-3 mm long; 4th flower-bearing glume 4-4.5 mm long. Stamens 3; filaments elongate up to 5 mm; anthers linear, yellowish, with a short conical connective appendage. Style *ca* 1 mm long; stigmas 3, *ca* 3 mm long. Nut trigonous, oblong, *ca* 2.5 x 0.8 mm, greyish-brown when mature, tightly enclosed in upper internode of the rachilla.

Fls. & Frts.: Aug. - Nov.

Habitat : Common in loose, sandy soil of sea coasts.

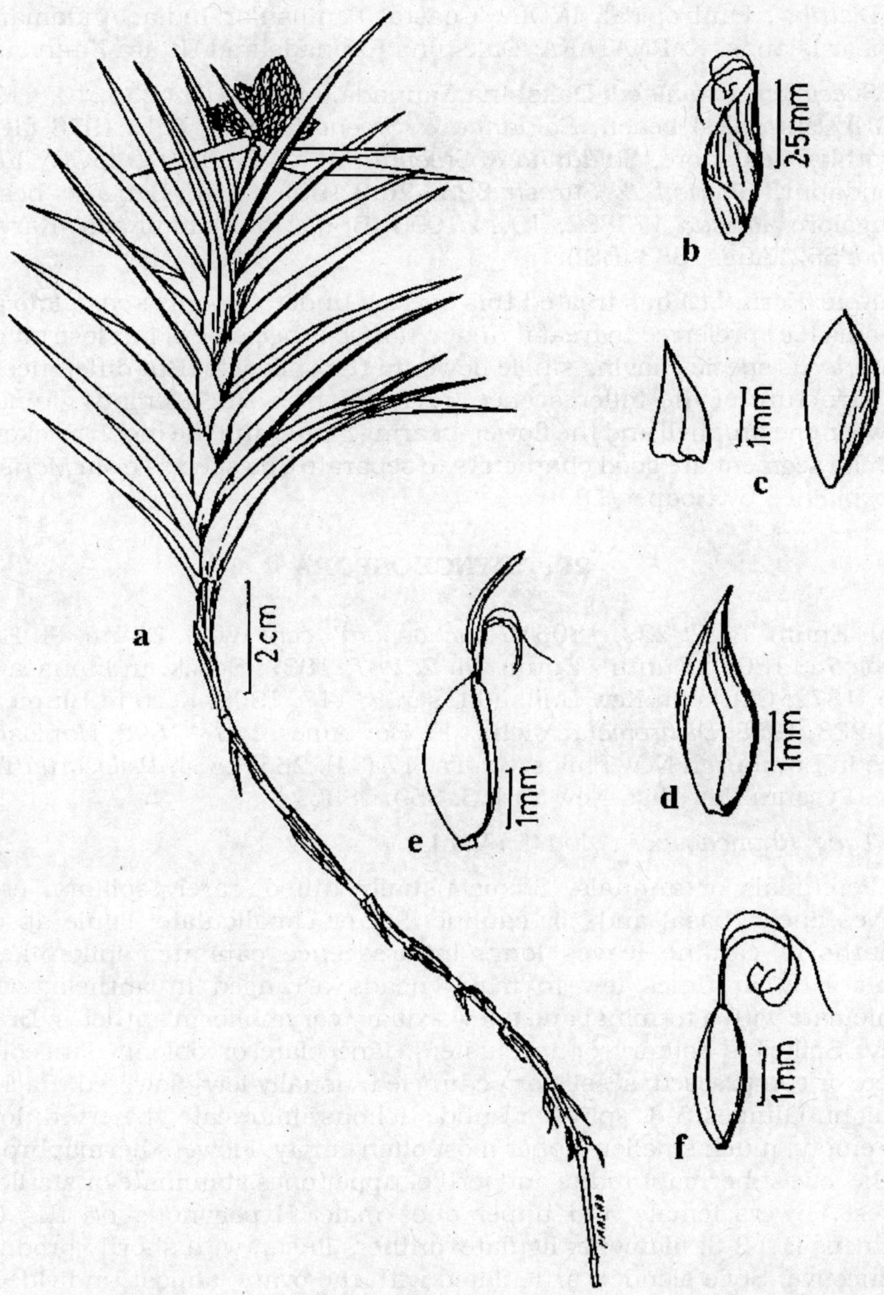

Fig. 52. *Remirea maritima* Aubl.
a. Habit, **b**. Spikelet, **c**. 2 basal glumes, **d**. 4th glume, **e**. Uppermost internode enclosing the nut, slihgtly opened; also shown 1 stamen and the style, **f.** Nut with style

Distrib. : Pantropical. INDIA: Coastal Peninsular India, Andaman & Nicobar Islands. KARNATAKA: Dakshina Kannada and Uttara Kannada.

Specimens examined: Dakshina Kannada: Malpe, *Bhat* 235, 10.8.1975 (MGH); Surathkal beach, *Saldanha & Sreenath* 4891, 9.12.1978 (JCB); Surathkal sea shore, *Saldanha & Sreenath* 9173, 20.9.1979 (JCB); Kodi, Coondapur, *Sherieff & Suresh* 393, 20.2.1985 (MH); Bengare beach, Mangalore, *Prasad* 173882, 10.11.1995 (BSI). Uttara Kannada: Karwar, *Talbot* 552, Aug. 1883 (BSI).

Note: Kern (*l.c.*) has treated this species under *Cyperus sensu lato* and Koyama (*l.c.*) preferred to treat it under *Mariscus* because of its close affinity to *Mariscus* species having single flower in the spikelets. But differences in the structure of the inflorescence, presence of a single empty glume in between the prophyll and the flower-bearing glume and the much thickened rachilla segment are good characters to separate this species from *Mariscus* as explained by Hooper, *l.c.*

20. RHYNCHOSPORA

Vahl, Enum. Pl. 2: 229. 1806 ('*Rynchospora*'; *corr.* Willd. Enum. Pl. Hort. Berol. 71. 1809); Kunth., Enum. Pl. 2: 287. 1837; Boeck. in Linnaea 37: 525. 1872; Clarke in Kew Bull. add. ser. 8: 117. 1908; Kern in Blumea 9: 229-233. 1958. *Dichromena* Michx., Fl. Bor. Amer. 1: 37. 1803. *Haplostylis* Nees in Edinburgh New Philos. Journ. 17 (34): 265. 1834. *Psilocarya* Torr. Ann. Lyceum Nat. Hist. New York 3: 359. 1836.

Type: Rhynchospora alba (L.) Vahl

Perennials or annuals. Stems usually tufted, rarely solitary, erect. Leaves linear, basal and/ or cauline, flat or canaliculate; ligule absent; sheaths of cauline leaves long. Inflorescence capitate, spike-like or paniculate, at times few to many heads arranged in anthela; when paniculate with a terminal and a few axillary corymbiform anthelas. Bracts leafy. Spikelets solitary or in clusters, lanceolate or oblong- lanceolate, terete or compressed, sessile or peduncled, usually few- flowered. Rachilla straight. Glumes 5-8, spiral or subdistichous, imbricate, 1-nerved; lower 3-4 empty, much smaller; upper most often empty. Flowers hermaphrodite or the lowest hermaphrodite and fertile, upper ones staminate or sterile, or lowest flowers female and upper ones male. Hypogynous bristles 0-6. Stamens (1-) 2-3; filaments ligulate; anthers linear, with shortly produced connective. Style slender, articulated with the ovary, almost undivided or deeply bifid, dark brown, dilated at base; style base persistent on the nut, mostly conical. Nut sessile or shortly stipitate, 2 sided, obovate, elliptic or oblong, smooth or transversely rugulose.

More than 200 species, mostly in tropical and subtropical regions of South America. 10 species in India; 3 in Karnataka.

Literature : **KUKENTHAL, G.** (1949-1951) Vorarbeiten Zu einer Monographie der Rhynchosporideae, in *Bot. Jahrb.* 74: 375-509. 1949; *ibid.* 75: 90-195. 1950; *ibid.* 75: 273-314. 1951. **VERMA, D.M.** & **CHANDRA, VEENA** [(1980) 1982] The genus *Rhynchospora* Vahl in India, in *Bull. Bot. Surv. India* 22 (1-4): 126-135.

Key to the species

1a. Annuals; Stems leafy near the base only; inflorescence subglobose or globose heads .... **3. R. wightiana**

1b. Perennials; stems leafy throughout; inflorescence paniculate, with 1-4 corymbiform anthelas 2

2a. Stems 1-1.7 mm thick; leaves 2-3 (-5) mm wide; spikelets with 2-4 hermaphrodite flowers; glumes spiral; style deeply bifid; style base half to as long as the nut, not furrowed; nut 1.5 - 2 mm long .... **2. R. rugosa**

2b. Stems 3-7 mm thick; leaves 8-20 mm wide; spikelets with a single, hermaphrodite flower; glumes subdistichous; style shortly or hardly bilobed; style base longer than nut, conspicuously furrowed on both sides; nut 2.5-3.5 mm long .... **1. R. corymbosa**

1. **Rhynchospora corymbosa** (L.) Britt., Trans. New York Acad. Sci. 11: 84. 1892; Fischer in Gamble, Fl. Pres. Madras 1672. 1931 (3: 1160. 1957, repr.ed.); Kern in van Steenis, Fl. Males. 1,7: 713. 1974; Hooper in Saldanha & Nicolson, Fl. Hassan 695. 1976; Rao & Razi, Fl. Mysore 564. 1981; Rao & Verma, Cyp. NE India 50. 1982; Sharma *et al.*, Fl. Karnataka 315. 1984; Koyama in Dassanayake & Fosberg, Rev. Handb. Fl. Ceylon 5: 336. 1985; Karthik. *et al.*, Fl. Ind. Enum. Monocot. 67. 1989. *Scirpus corymbosus* L. in Torner, Cent. Pl. 2: 7. 1756. *Rhynchospora aurea* Vahl, Enum. Pl. 2: 229. 1805; Hook.f., Fl. Brit. India 6: 670. 1893; Cooke, Fl. Pres. Bombay 2: 901. 1908 (3: 416. 1958, repr.ed.).

Illus. : Matthew, Illus. Fl. Tamilnadu Carnatic Pls. 794 & 795. 1982.

Perennials with short rhizome, up to 120 (-190) cm tall. Stems solitary or a few together, stout, triquetrous, with several nodes, 3-7 mm thick, leafy up to top. Leaves radical as well as cauline; lowest 2-3 reduced to bladeless sheaths; cauline leaves up to top of the stem; blades broadly linear, long-acuminate at apex, up to 100 cm long, 8-20 mm wide, scabrous on the margins; sheaths of cauline leaves with a scarious appendage on orifice. Inflorescence 2-5 corymbiform anthelas subtended by leafy bracts. Anthelas compound or decompound, many-branched, 10-15 cm long, with numerous spikelets. Primary rays unequal, up to 12 cm long; secondary rays up to 3 cm long. Spikelets subterate, lanceolate, acute at apex, 6-8 mm long, rusty brown or chestnut brown, 2 or 3-flowered. Glumes 5-7, subdistichous, membranous, acute or mucronate at apex, 3-7 mm long; lower ones ovate; upper ones oblong-ovate. Lowest flower bisexual; upper one (or two) male. Bristles in bisexual flowers unequal, 4-5 mm long, exceeding the nut, antrorsely scabrous; 1-3 in male flowers, shorter. Stamens 3; anthers 2-2.5 mm long. Style shortly or hardly bilobed at apex; base elongated-conical, compressed, 4- 5 mm long, as wide as the nut, conspicuously furrowed on both sides. Nut compressed, obovate or oblong-obovate, truncate at top, 2.5-3.5 x *ca* 2 mm, finely transversely wrinkled in the middle portion, brownish.

Fls. & Frts.: July - April.

Habitat: Open swampy areas, river banks, rice fields, in shallow waters, in moist areas among tall weeds dominated by *Eupatorium* sp., *Mimosa* sp. etc.

Distrib.: Pantropical. INDIA: Throughout (except the North-West). KARNATAKA: Belgaum, Chikmagalur (Rao *et al.*, 2000), Hassan, Mandya, Mysore (Rao & Razi, *l.c.*), Shimoga, Uttara Kannada.

Specimens examined: Belgaum: Anmod, *Saldanha* & *Prakash* 3514, 25.10.1978 (JCB). Hassan: Mankanahalli, Bisle Ghat, *Ramamoorthy* 306, 8.7.1970 (JCB). Mandya: Ranganathittu, *Ramesh* 6234, 7.2.1979 (JCB); Srirangapatana, *Radha* 4, 6.1.1985 (MGH). Shimoga: Thirthahalli, *Raghavan* 74336, 24.6.1961 (BSI); Jog falls, *Prasad* 173792, 4.11.1995 (BSI). Uttara Kannada: Yellapur, *Talbot* 1045, 20.9. 1884 (BSI); Ekambi, *Gammie* 13834 & 13835, 22.4.1900 (BSI); Castle Rock, *Almeida* 943, 27.12.1968 (BLAT); Dusk river, Castle Rock, *Almeida* MRA-1178, 25.2.1981 (BNHS); Chanduguli, *Ramesh & Shivprakash* 13341, 20.7.1981 (JCB).

2. **Rhynchospora rugosa** (Vahl) Gale in Rhodora 46: 275, t. 835, f. 1A-B. 1944; Kern in Back & Bakh. f., Fl. Java 3: 484. 1968 *et* in van Steenis, Fl. Males. 1, 7: 720. 1974; Rao & Verma, Cyp. NE India 52. 1982; Sharma *et al.*, Fl. Karnataka 315. 1984; Karthik. *et al.*, Fl. Ind. Enum. Monocot. 67. 1989. *Schoenus rugosus* Vahl, Eclog. Amer. 5. 1798. *Rhynchospora glauca* Vahl, Enum. Pl. 2: 233. 1806, *nom. illegit.* ; Clarke in Hook.f., Fl. Brit. India 6: 671. 1893; Fischer in Gamble, Fl. Pres. Madras 1672. 1931 (3: 1160. 1957, repr.ed.). *R. laxa* R. Br., Prodr. Fl. Nov. Holl. 230. 1810, *non* Vahl, 1806. *R. brownii* Roem. & Schult., Syst. Veg. 2: 86. 1817. *R. glauca* Vahl var. *condensata* Kukenth. in Bot. Jahrb. Syst. 69: 259. 1938.

Type: America Meridionali, *Von Rohr* .

Illus. : Clarke, Illus. Cyp. t. 73, f. 7. 1909; Matthew, Furth. Illus. Fl. Tamilnadu Carnatic Pl. 699. 1988.

Perennials with short rhizome. Stems densely tufted, slender, trigonous, smooth or slightly scaberulous at top, many- leaved at base and with some distant cauline leaves, 30-75 (-100) cm high, 1-1.7 mm thick. Leaves shorter than stem, rigid, flat or canaliculate, long-acuminate, with strong midrib and scaberulous margins, 2-3 (-5) mm wide; basal sheaths brown. Inflorescence paniculate, narrow, with 3-4 (-6) distant corymbiform anthelas; lateral panicles solitary or 2 together, compressed, often long-exserted from the sheath; branches very unequal. Bracts erect, about as long as the anthelas. Spikelets solitary or in small clusters, short-peduncled, terete, ovate or ovate-lanceolate, (3-) 4-6 mm long, 2-4-flowered. Glumes 5-8, spiral, membranous, short-mucronate. Flowers bisexual; upper ones tabescent. Bristles 5-6, somewhat shorter or longer than nut, antrorsely scabrous. Stamens (1-) 2(-3); anthers 1-2.5 mm long. Style bifid halfway; base conical, glabrous, half to as long as the nut. Nut biconvex, dorsiventrally compressed, obovate to broadly obovate, finely transversely wrinkled, light brown to castaneous, 1.5-2 x 1-1.75 mm; epidermal cells longitudinally oblong.

Fls. & Frts.: Not known from Karnataka.

Habitat: Swampy areas, open grasslands, river banks.

Distrib. : Widely distributed in tropical and subtropical regions of the world. INDIA: Peninsular India, Central & North-East India and Eastern Himalayas. KARNATAKA: Without locality (Sharma *et al., l.c.*) .

Note: A highly polymorphic species. As specimens are not available in BSI, description given above is as given by Kern, *l.c.*.

3. **Rhynchospora wightiana** (Nees) Steud., Syn. 2: 148. 1855; Clarke in Hook.f., Fl. Brit. India 6: 669. 1893; Cooke, Fl. Pres. Bombay 3: 901. 1908 (3: 415. 1958, repr.ed.); Fischer in Gamble, Fl. Pres. Madras 1672. 1931 (3: 1160. 1957, repr.ed.); Satyanarayan & Shankaranarayan in Ann. Arid Zone 2: 146. 1964; Kern in van Steenis, Fl. Males. 1, 7: 718, f. 101 c-d. 1974; Rao & Razi, Fl. Mysore 564. 1981; Sharma *et al.*, Fl. Karnataka 315. 1984; Singh, Fl. E. Karnataka 2: 646. 1988; Karthik. *et al.*, Fl. Ind. Enum. Monocot. 68. 1989. *Haplostylis wightiana* Nees, Nov. Act. Ac. Caes. Leop.- Car. 19, Suppl. 2: 148. 1855. **Fig. 53.**

Illus. : Clarke, Illus. Cyp. t. 64, f. 8-11.1909; Kern, *l.c.*

Annuals with fibrous roots, 6-40 cm high. Stems solitary or tufted, slender, striate, glabrous, 0.5-1.2 mm thick. Leaves radical as well as cauline towards base of the stem, much shorter than stem, linear, gradually narrowed to apex, 2-3 mm wide. Inflorescence a terminal, subglobose to globose head, 1.3-1.8 cm across, yellowish-brown, with numerous densely arranged spikelets. Bracts 4-7, unequal, longest up to 6 cm long, much overtopping the inflorescence, ciliate on margins towards the dilated base. Spikelets linear-lanceolate, compressed, 2-flowered (lower female, upper male), 5-7 mm long. Glumes 6, distichous, acute at apex, keeled or with a prominent midvein; lowest broadly ovate, *ca* 1 mm long; 2nd ovate, *ca* 1.5 mm long; 3rd ovate, *ca* 2 mm long; 4th oblong-ellipsoid, boat-shaped, *ca* 4 mm long; 5th oblong- lanceolate, longer than all others, *ca* 6 mm long; 6th linear- lanceolate, hyaline, not very prominent, *ca* 5 mm long. Perianth

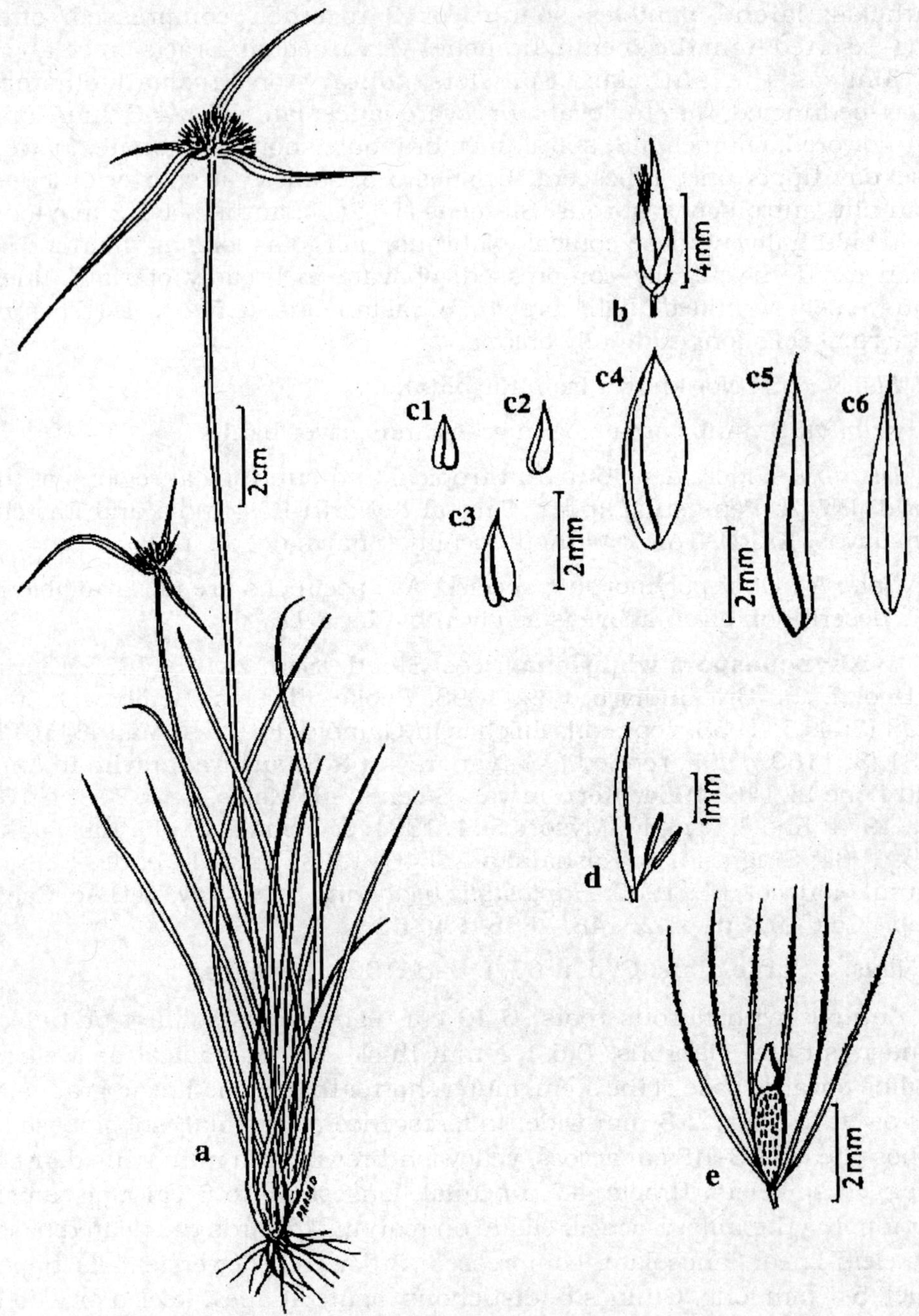

Fig. 53. *Rhynchospora wightiana* (Nees) Steud.
a. Habit, **b**. Spikelet, **c1.** Lowest glume, **c2**. 2nd glume, **c3.** 3rd glume, **c4**. 4th glume, **c5.** 5th glume, **c6.** 6th glume, **d.** 6th glume with male flower, **e**. Nut with bristles.

bristles in male flowers 0-4, short. Stamens 2; filaments elongate up to 5 mm; anthers linear, *ca* 1 mm long. Bristles in female flowers 6, antrorsely scabrous, much longer than nut, 5-6 mm long. Style very shortly bilobed; base compressed, *ca* 1 x 0.5 mm, persistent. Nut dorsiventrally compressed, oblong or obovate-oblong, tuberculate, 2-2.5 x 0.6 - 0.7 mm (excluding beak); epidermal cells isodiametric.

Fls. & Frts.: Aug.- March

Habitat: Moist areas, open grasslands, marshy areas, hill slopes.

Distrib. : Vietnam, Malesia. INDIA: Western Peninsula. KARNATAKA: Bellary (Satyanarayan & Shankaranarayan, *l.c.*), Dakshina Kannada, Hassan, Kodagu, Mysore, Shimoga, Uttara Kannada.

Specimens examined: Dakshina Kannada: Fahlsur, *Barber* 2441, 22.11.1900 (MH); Mangalore, *Faulker* 4857, Jan. 1902 (MH); Belve, *Saldanha* C.S.6005, 16.10.1960 (BLAT); Padubri-Belmanna Road, *Saldanha & Prakash* 4095, 12.11.1918 (JCB). Hassan: Shiradi, Just above the village, *Saldanha & Hooper* 2574, 24.11.1971 (JCB). Kodagu: Keelar, *Rao* 74818, 30.9.1961 (BSI). Mysore: Attikan, *Rao* 1222, 31.1.1971 (JCB) & (MGH). Shimoga: Karodi, Thirthahalli, *Raghavan* 82757, 28.9.1962 (BSI); Grassland above Jog falls, Sagar taluk, *Prasad* 173763, 3.11.1995 (BSI). Uttara Kannada: Yellapur, *Talbot s.n.* (Acc. No. 815), 10.8.1884 and 816, Sept. 1884 (BSI); Karwar, *Talbot s.n.* (Acc. No. 814), Aug. 1887 (BSI); Castle rock, *Gammie* 15699, 27.10.1902 (BSI) and *Bhide s.n.* (Acc. No. 2912), Oct. 1909 (BSI); Castle Rock, *Fernandez* J.F.480, 18.11.1949 (BLAT); Magod-Yellapur, *Ramesh & Udayakumar* 13623, 18.9.1981 (JCB).

21. RIKLIELLA

J. Raynal in Adansonia 13:154. 1973. *Scirpus* subgen. *Chloroscirpus* Cherm., Arch. Bot. Caen 7, Mem. 2: 2. 1936, *nom. nud. Scirpus* sect. *Squarrosi* Cherm., Arch. Bot. Caen 7, mem. 2: 2. 1936, *nom. nud.*

Type: Rikliella rehmanii (Ridl.) J. Raynal (=*Scirpus rehmanii* Redl.)

Small, tufted annuals with slender stems and setaceous radical leaves. Inflorescence a terminal or pseudolateral cluster of sessile spikelets. Involucral bracts 1-few, spreading or erect. Glumes caducous, oblanceolate or obovate, with a prominent midrib, excurrent into an awn. Flowers bisexual. Perianth absent. Stamen 1. Style very short; stigmas 2 or 3. Nut elliptical or obovate, puncticulate with isodiametric epidermal cells.

3 species, distributed in Southern Asia and tropical Africa. 2 in India; both are found in Karnataka.

Key to the species

1a. Inflorescence terminal (except when very young or with a single spikelet); stigmas 2; nut compressed. . . . . **1. R. kernii**

1b. Inflorescence always pseudolateral; stigmas 3; nut trigonous . . . . **2. R. squarrosa**

1. **Rikliella kernii** (Raymond) J. Raynal, Adansonia 2, 13: 155. 1973; Hooper in Saldanha & Nicolson, Fl. Hassan 696. 1976; Sharma *et al.*, Fl. Karnataka 315. 1985; Karthik. *et al.*, Fl. Ind. Enum. Monocot. 68. 1989. *Scirpus kernii* Raymond, Naturaliste Canad. 86: 230. 1959; J. Raynal, Adansonia 2, 8: 95, t. 1, f. 1-5.1968. **Fig. 54.**

Type : Senegal, *Berhaut* 4692-Isotype (P).

Illus.: J. Raynal, *l.c.*

Tufted annuals with fibrous roots, 4-15 cm high. Stems slender, compressed, *ca* 0.5 mm thick. Leaves shorter than stem, smooth, slender, canaliculate, narrowed at apex, *ca* 0.5 mm wide; ligule absent; sheaths purplish-red, striate, up to 1.5 cm long. Inflorescence a terminal head of 1-3 spikelets (pseudolateral when very young or with a single spikelet). Involucral bracts 1 or 2, spreading (except when the inflorescence is very young or with a single spikelet); lowest much longer, 1-3 cm. Spikelets sessile, squarrose, oblong-ovoid, obtuse at apex, 3-5 x 3-4 mm, densely many-flowered. Glumes membranous, obovate, with a long recurved mucro, 2-2.2 x *ca* 0.7 mm (including the *ca* 1 mm long mucro); midvein very prominent. Stamen 1; filament elongate up to 2 mm; anther oblong, *ca* 0.3 mm long. Style much short, inconspicuous; stigmas 2, *ca* 0.7 mm long. Nut compressed, obovate, obtuse at apex, slightly ridged on both surfaces towards the base, *ca* 0.8 x 0.4 mm, blackish when mature; epidermal cells in *ca* 20 vertical rows on both surfaces.

Fls. & Frts.: Nov.

Habitat : Wet sandy areas near tanks, in rice fields.

Distrib.: Tropical Africa. INDIA: Western peninsula, Central India. KARNATAKA: Hassan.

Specimen examined : Hassan: Tank near Dandiganahalli, *Hooper & Gandhi* 2402, 11.11.1971 (JCB).

Note : This species is over looked in the field, due to the small size. Even though, compared to *R.squarrosa* it is rare. It can be easily distinguished from the other species by the spreading bracts and by the presence of 2 stigmas.

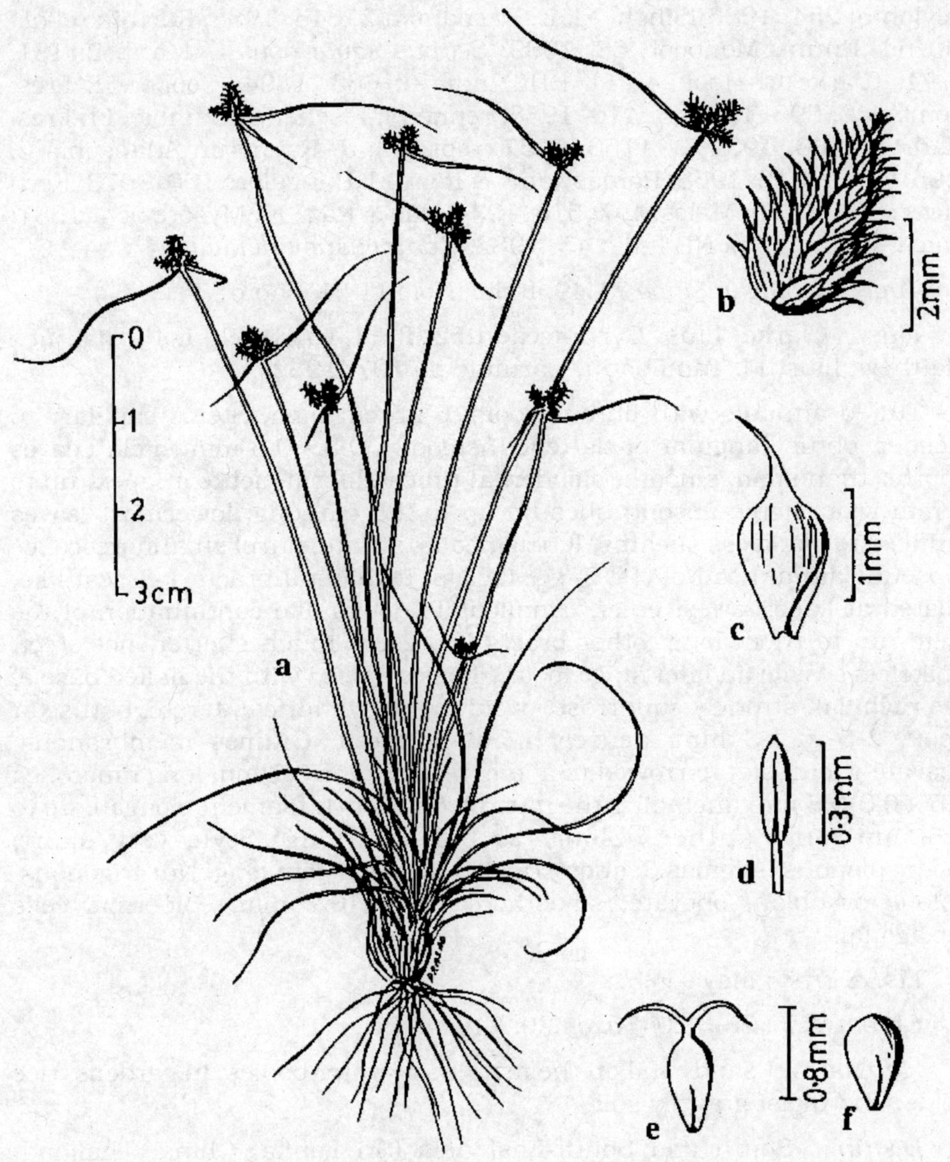

Fig. 54. *Rikliella kernii* (Raymond) J. Raynal
a. Habit, **b.** Spikelet, **c.** Glume, **d.** Stamen, **e.** Nut with style (early stage), **f.** Nut

2. **Rikliella squarrosa** (L.) J. Raynal in Adansonia 2, 13:154. 1973; Hooper in Saldanha & Nicolson, Fl. Hassan 696. 1976; Sharma *et al.*, Fl. Karnataka 215. 1984; Koyama in Dassanayake & Fosberg, Rev. Handb. Fl. Ceylon 5: 254. 1985; Singh, Fl. E. Karnataka 2: 645. 1988; Karthik. *et al.*, Fl. Ind. Enum. Monocot. 68. 1989. *Scirpus squarrosus* L. Mant. 2: 181. 1771; Clarke in Hook. f., Fl. Brit. India 6: 663. 1894; Cooke, Fl. Pres. Bombay 2: 896. 1908 (3: 410. 1958, repr.ed.); Fischer in Gamble, Fl. Pres. Madras 1166. 1931 (3: 1156. 1957, repr.ed.); J. Raynal in Adansonia 2, 8:95, t. 1, f. 9-12. 1968; Ramaswamy & Razi, Fl. Bangalore 100. 1973; Kern in van Steenis, Fl. Males. 1, 7: 516. 1974; Rao & Razi, Fl. Mysore 564. 1981; Rao & Verma, Cyp. NE India 45. 1982. *Kadu vusumani hullu.*

Type : *Koenig in Linne* 71.49, Inde. (Holo-LINN. Photo - P).

Illus. : Clarke, Illus. Cyperaceae t.52, f. 11-13. 1909; J. Raynal, *l.c.*; Matthew, Illus. Fl. Tamilnadu, Carnatic pl. 797. 1982.

Tufted annuals with fibrous roots, 6-25 cm high. Stems capillary to slender, obtuse-angular or flattened, smooth, 0.3 - 0.6 mm thick. Leaves shorter than stem, smooth, slender, at times filiform, acute at apex, up to 1 mm wide; ligule absent; sheaths up to 2.5 cm long; lowermost leaves reduced to bladeless sheaths, ferrugineous, striate, purplish. Inflorescence a pseudolateral head of (1-) 2-4 spikelets. Involucral bracts 1-3, leaf-like, dilated at base; lowest erect, canaliculate, looks like continuation of the stem, up to 6 cm long; other bracts if present much shorter, not erect. Spikelets sessile (in later stage looks like peduncled with the naked base of the rachilla), strongly squarrose, ovoid to subcylindrical, terete, obtuse at apex, 2-5 x 2-3 mm, densely many-flowered. Glumes membranous, obovate-rhomboid, narrowed to a long recurved *ca* 0.7 mm long mucro, *ca* 1.5 x 0.3-0.5 mm (including the mucro). Stamen 1; filament elongate up to 0.6 mm long; anther oblong, *ca* 0.3 mm long. Style very short, inconspicuous; stigmas 3, short, recurved, *ca* 0.1 mm long. Nut trigonous, oblong or oblong-obovate, smooth, *ca* 0.6 x 0.2 mm; epidermal cells hexagonal.

Fls. & Frts. : May - Feb.

Chrom. No. : 2n = 20 (*Taxon* 20: 216. 1971).

Habitat: Wet sandy soil on the margins of water bodies, in gardens, rice fields, and moist gravelly soil.

Distrib. : South and South-East Asia (Sri Lanka, China, Thailand, Malesia). INDIA: Throughout. KARNATAKA: Bangalore, Bidar, Dakshina Kannada, Gulbarga, Hassan, Kodagu, Kolar, Mandya, Mysore, Raichur, Shimoga, Uttara Kannada.

Specimens examined : Bangalore: Bannergatta, *Saldanha & Hooper* 18046, 7.11.1971 (JCB). Bidar: Bidar-Chincholi Road 33rd mile, *Singh* 128610, 9.2.1975 (BSI). Dakshina Kannada: Kadlu, without coll. name

16756, 2.9.1920 (MH); Someshwar, Ullal, *Hooper & Saldanha* 2515, 25.11.1971 (JCB); Kunjargiri, *Bhat* 623, 21.7.1980 (MGH). Hassan: Bourdalboore, *Nicolson et al.* 2352, 26.10.1971 (JCB). Kodagu: Kushalnagar, *Bhat* 1045, 21.9.1981 (MGH). Kolar: Chickballapur-Bogepalli Road, *Singh* 134549, 27.2.1975 (BSI). Mandya: Sreerangapattana, *Rao* 35, 10.11.1964 (MGH); Paschimavahini, *Padma Rani* 20, 8.8.1970 (MGH); Sangam, *Murthy & Prakash* 4582, 25.11.1978 (JCB); Narayana Surya, *Dinesh* 611, 14.10.1983 (MGH). Mysore: Talkad, *Rao* 694, 28.5.1971 (JCB). Raichur: Kushtagi-Ron Road, *Singh* 141566, 11.11.1975 (BSI). Shimoga: Agumbe, *Bhat* 700, 26.9.1980 (MGH). Uttara Kannada: Karwar, *Talbot* 1511, Aug. 1885 (BSI); without exact locality, *Talbot*, 867 (BSI).

Note: This species is also usually overlooked due to its small size.

22. SCHOENOPLECTUS

(Reichb.) Palla in Sitzber, Zool. Bot. Ges. Wien. 38: 49. 1888 *et* in Bot. Jahrb. Syst. 10: 298. 1888. *Scirpus* L. sect. *Schoenoplectus* Reichb., Ic. Fl. Germ. 8: 40. 1846.

Type: Schoenoplectus lacustris (L.) Palla

Annuals or perennials, some times with decumbent or long- creeping rhizome. Stems trigonous or terete, naked, rarely with 1 or 2 nodes above base, solid or rarely transversely septate. Leaves as a rule reduced to bladeless sheaths surrounding the stem base; at times with a short blade. Inflorescence pseudolateral, anthelate or crowded in a head, bearing few to many spikelets. Involucral bract usually single, looks like continuation of the stem, at times reflexed, more or less sheathing at base. Spikelets usually ovoid, oblong-ovoid or oblong-ellipsoid, many-flowered. Glumes spiral, mostly membranous or thinly herbaceous. Flowers bisexual, with or without hypogynous bristles. Bristles up to 6, spinulose-scabrous. Stamens 3; anther connective crested. Stigmas generally 3. Nut trigonous or biconvex, often transversely wrinkled, brown to black when mature.

Above 130 species throughout the world; *ca* 15 in India; 7 in Karnataka.

Key to the species

1a. Stems and involucral bracts transversely septate; bract usually much longer than stem proper 2

1b. Stems and involucral bracts not septate; bract shorter than stem.. 3

2a. Stems stout, 2-10 mm thick, glumes acute at apex, 4-7 mm long; nut 1.8 - 2.2 mm long . . . . **1. S. articulatus**

2b. Stems slender, up to 1.5 mm thick; glumes obtuse at apex, 2.5 - 3 mm long; nut *ca* 1.3 mm long. . . . . **7. S. senegalensis**

3a. Inflorescence capitate 4

3b. Inflorescence anthelate 5

4a. Perennials; stems rather stout, often in a row from a short or elongated rhizome, shortly 3-angled, 3-7 mm thick; glumes 3.5-5 mm long; stigmas 3 .... **6. S. mucronatus**

4b. Annuals, or short-lived perennials; stems rather slender, tufted, subterete or obtusely many-angled, 1-2 mm thick; glumes *ca* 3 mm long; stigmas 2 (at times with a minute inconspicous extra stigma) .... **3. S. juncoides**

5a. Annuals; glumes 2-2.2 mm long; staminal filaments elongate up to 2.5 mm; anthers oblong, *ca* 0.6 mm long; style *ca* 0.8 mm long; nut *ca* 1.2 mm long, conspicously transversely wavy-ridged .... **4. S. lateriflorus**

5b. Perennials; glumes 3.5 - 4 mm long; staminal filaments elongate up to 3 or 4 mm; anthers linear-oblong, 1.75-2 mm long; style *ca* 2 mm long; nut 1.8-2 mm long, not transversely wavy- ridged 6

6a. Rhizome creeping; stems usually in a row from the horizontal rhizome; perianth bristles usually absent, if present not plumosely fringed in the upper portion; anthers without fimbriate connective appendage .... **2. S. corymbosus**

6b. Rhizome not creeping; stems tufted; perianth bristles always present, plumosely fringed in the upper portion; anthers with fimbriate connective appendage .... **5. S. litoralis**

1. **Schoenoplectus articulatus** (L.) Palla in Bot. Jahrb. Syst. 10: 299. 1888; Hooper in Saldanha & Nicolson, Fl. Hassan 697. 1976; Sharma *et al.*, Fl. Karnataka 315. 1984; Koyama in Dassanayake & Fosberg, Rev. Handb. Fl. Ceylon 5: 163, f.7. 1985; Singh, Fl. E. Karnataka 2: 646. 1988; Karthik. *et al.*, Fl. Ind. Enum. Monocot. 68. 1989; Keshava Murthy & Yoganarasimhan, Fl. Coorg. 516. 1990. *Scirpus articulatus* L., Sp. pl. 1, 47. 1753; Clarke in Hook.f., Fl. Brit. India 6: 656. 1894; Cooke, Fl. Pres. Bombay 2: 891. 1908 (3: 406. 1958, repr.ed.); Fischer in Gamble, Fl. Pres. Madras 1666. 1931 (3: 1156. 1957, repr.ed.); Satyanarayan & Shankaranarayan in Ann. Arid Zone 2 (2): 146. 1964; Ramaswamy & Razi, Fl. Bangalore 98. 1973; Kern in van Steenis, Fl. Males. 1,7: 513. 1974; Rao & Razi, Fl. Mysore 564. 1981; Rao & Verma, Cyp. NE India 43. 1982.

Type : India.

Illus.: Koyama, *l.c.;* Matthew, Furth. Illus., Fl. Tamilnadu Carnatic pl. 700. 1988.

Annuals or short-lived perennials with fibrous roots, 15-80 cm high. Rhizome inconspicuous. Stems highly tufted, terete, smooth, glabrous, 2-10 mm thick. Leaves reduced to 2-3 basal sheaths; lower 1 or 2 very small or scale-like, brownish; upper sheath 4-15 cm long, cylindrical, obliquely truncate at apex, with membranous margins. Inflorescence a pseudolateral, dense head of several to numerous spikelets, 1-4.5 cm across. Involucral bract stem-like, usually much longer than stem proper, erect or weakly recurved, terete, fistulose, narrowed at terminal portion to a subacute apex, septate, smooth, glabrous, 2-11 mm thick, hardly widened at base. Spikelets sessile, oblong-ovoid to ovoid, terete, acute at apex, 7-20 x 4-6 mm, densely many-flowered. Glumes membranous, spiral, broadly ovate, acute and mucronate at apex, with a less prominent keel, 4-7 x 3.5-6 mm, many-nerved. Perianth bristles absent. Stamens 3; filaments elongate up to 4 mm; anthers oblong, apiculate at apex, *ca* 1.2 mm long. Ovary deeply triangular, obovate, *ca* 1 mm long; style *ca* 2 mm long, slightly narrowed to base; stigmas 3, much shorter than style, *ca* 0.5 mm long. Nut triquetrous, obovate, apiculate at apex, 1.8-2.2 x 1.5 mm, creamish and faintly transversely wavy- ridged in early stage, blackish-brown and smooth later.

Fls. & Frts.: July - April.

Chrom. No.: 2n = 32 (*Phyton* 6: 7-22. 1956).

Habitat: Rice fields, in and around shallow waters of tanks, ponds and other marshy areas; usually at low altitude.

Distrib.: Mediterranian region and paleotropics (South Asia to Australia, Africa). INDIA: Throughout. KARNATAKA: Bangalore, Bellary (Satyanarayan & Shankaranarayan, *l.c.*), Chikmagalur (Sharma *et al., l.c.*), Dakshina Kannada, Dharwar, Hassan, Kodagu, Mysore, Shimoga, Tumkur, Uttara Kannada.

Specimens examined : Bangalore: Bangalore, *Camaron* 623, Oct. 1891 (MH); Magadi, *Manohar & Murthy* 2870, 23.9.1978 (JCB). Dakshina Kannada: Udupi, *Bhat* 307, 12.10.1976 (MGH); Near Faringipet, *Saldanha* 10608, 28.12.1979 (JCB). Dharwar: Kalgeri tank, *Bhide s.n.* (Acc. Nos. 2789, 2790, 2791), 6.9.1911 (BSI); Kalgeri, *Bhide s.n.* (Acc. No. 86792). 6.9.1911 (BLAT). Hassan: Bagi, *Saldanha* 12190, 15.1.1969 (JCB). Kodagu: Ponnampet, *Bhat* 911, 26.1.1981 (MGH). Mysore: Pond near Chamundi hills, *Rao* 837, 21.8.1970 (JCB). Shimoga: Ghat Road, Agumbe, *Raghavan* 62669, 18.5.1960 (BSI); Tirthahalli, *Raghavan* 96903, 16.3.1963 (BSI); Gajanur, *Ramesh & Prakash* 5228, 11.12.1978 (JCB). Tumkur: Devarayanadurga, *Ramesh & Prakash* 6702, 14.4.1979 (JCB). Uttara Kannada: Kumta, *Talbot s.n.* (Acc. No. 780), 6.12.1883 (BSI); Mundgad without Coll. name, No. and date (BSI).

Note : Size of the plant and position of the inflorescence usually depends on water depth. In dried up marshy areas short plants with thin stems (2-3 mm thick) are found with the inflorescence very near to the base. Plants growing in shallow waters are much larger with broad stems (up to 10 mm thick) and inflorescence is found just above the water level. It is a polymorphic species especially in the size of the plant.

2. **Schoenoplectus corymbosus** (Roth ex Roem. & Schult.) J. Raynal in J.B. Lebrun *et al.*, Cat. Niger Plants 343. 1976; Hooper in Saldanha & Nicolson, Fl. Hassan 697. 1976; Sharma *et al.*, Fl. Karnataka 315. 1984; Karthik. *et al.*, Fl. Ind. Enum. Monocot. 68. 1989. *Isolepis corymbosus* Roth ex Roem. & Schult., Syst. Veg. 2: 110. 1817. *Scirpus corymbosus* auct. non L. 1756, nec Forssk. 1775; Clarke in Hook.f., Fl. Brit. India 6:657. 1893; Cooke, Fl. Pres. Bombay 2: 892. 1908 (3:407. 1958, repr.ed.); Fischer in Gamble, Fl. Pres. Madras 1667. 1931 (3: 1156. 1957, repr.ed.). **Fig. 55.**

Perennials with stout creeping rhizome and long thick fibrous roots, up to 180 cm high. Stems usually in a row from the horizontal rhizome, stout, erect, terete, gradually narrowed to apex, smooth, glabrous, 5-10 mm thick below the middle portion. Leaves reduced to 2-3 bladeless, papery sheaths; lowest short; upper one up to 31 cm long, with an oblique mouth, membranous on margins of the orifice, brownish. Inflorescence pseudolateral, with 8-14 rays bearing clusters of spikelets. Bract stem-like, much smaller than stem proper, terete, erect or slightly reflexed inward, narrowed to an acute apex, 2-10 cm long. Primary rays up to 11 cm long, compressed, striate. Spikelets clustered on secondary or primary rays, ovoid or ellipsoid-ovoid, faintly angled, 5-6 x 2-3 mm, brownish, many-flowered. Glumes membranous, somewhat loosely arranged, oblong-ovate to oblong-lanceolate, boat-shaped, acute-mucronate at apex, not strongly keeled except towards apex, 3.5 - 4 x 1-1.5 mm; midrib very prominent. Perianth bristles usually absent (but few specimens were found with short bristles). Bristles when present, usually 3, quite unequal, antrorsely scabrous, longest much smaller than nut; at times reduced to a single, rudimentary bristle. Stamens 3; filaments elongate up to 4 mm; anthers linear-oblong, apiculate, often slightly curved, *ca* 2 mm long. Ovary oblong, *ca* 7 mm long; style *ca* 2 mm long, slightly narrowed towards base, purplish-brown; stigmas 3, as long as style. Nut unequally triquetrous, ellipsoid to obovoid, apiculate at apex, 1.8-2 x *ca* 1 mm, shining blackish-brown, smooth.

Fls. & Frts.: Sept.-Nov.

Habitat: Standing water in lakes, ponds and tanks; up to 1000 m.

Distrib.: Paleotropics. INDIA: Throughout the plains. KARNATAKA: Bangalore, Hassan, Mysore (Fischer, *l.c.*) .

Specimens examined: Bangalore: Kulanahalli, on Bangalore-Tumkur Road, *Hooper & Gandhi* 2368, 10.11.1971 (JCB). Hassan: Near Railway station, Hassan, *Hooper & Gandhi* 2374, 10.11.1971 (JCB).

Note: Normally perianth bristles are absent in this species. But few specimens were found with short bristles which are much shorter than the nut. Also nut is somewhat obovate in these specimens, but usually ellipsoid in normal case (*Prasad & Singh*, 1997b).

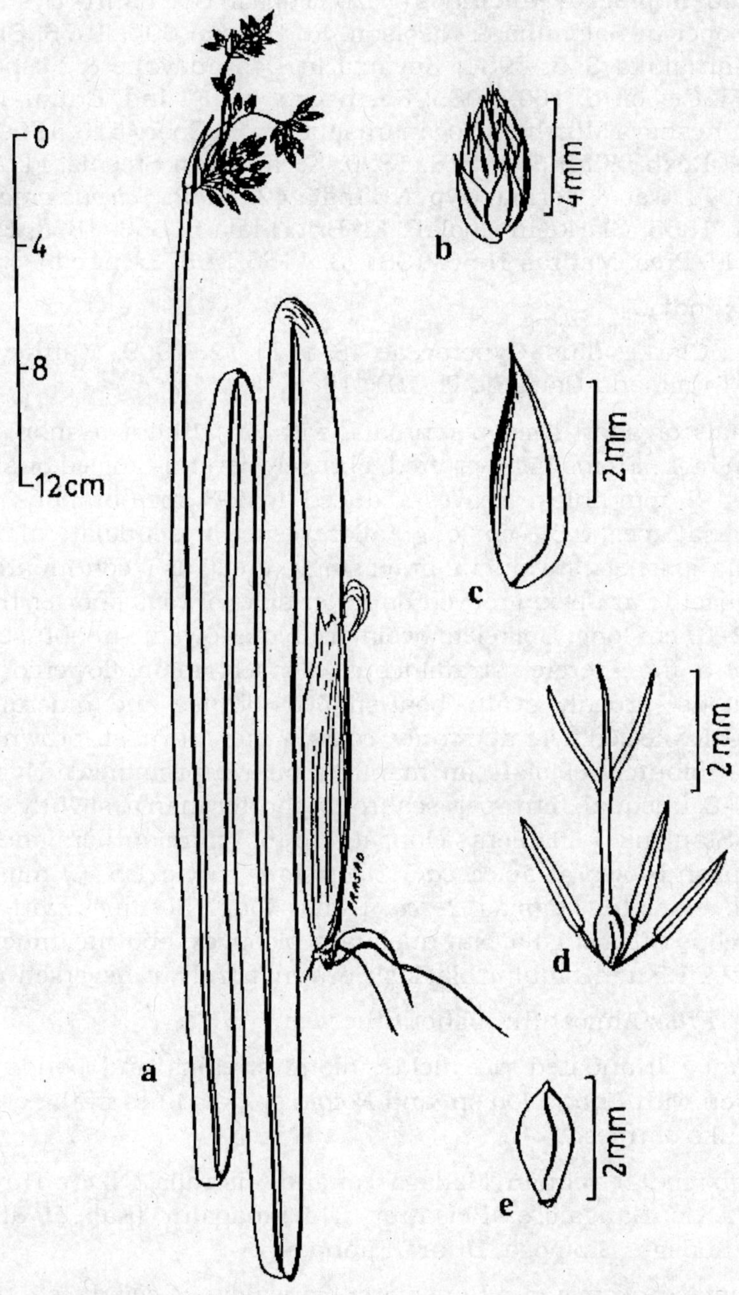

Fig. 55. *Schoenoplectus corymbosus* (Roth ex Roem. & Schult.) J. Raynal
a. Habit, **b.** Spikelet, **c.** Glume, **d.** Flower, **e.** Nut

3. **Schoenoplectus juncoides** (Roxb.) Palla in Bot. Jahrb. Syst. 10: 299. 1888; Hooper in Saldanha & Nicolson, Fl. Hassan 697. 1976; Sharma *et al.*, Fl. Karnataka 315. 1984; Koyama in Dassanayake & Fosberg, Rev. Handb. Fl. Ceylon 5: 160. 1985; Karthik. *et al.*, Fl. Ind. Enum. Monocot. 69. 1989; Keshava Murthy & Yoganarasimhan, Fl. Coorg 516. 1990. *Scirpus juncoides* Roxb., Fl. Ind. 1: 218. 1820; Kern in van Steenis, Fl. Males. 1, 7: 512. 1974; Rao & Verma, Cyp. NE India 42. 1982. *Scirpus erectus* auct. non Poir. 1805; Clarke in Hook.f., Fl. Brit. India 6: 656. 1893; Fischer in Gamble, Fl. Pres. Madras 1666. 1931 (3: 1156. 1957, repr.ed.).

Type : India.

Illus. : Clarke, Illus. Cyperaceae 48, f. 11-12. 1909; Matthew, Furth. Illus. Fl. Tamilnadu Carnatic Pl. 701. 1988.

Annuals or short-lived perennials, with or without a short rhizome, 15-60 cm high. Stems highly tufted, obtusely several- angled or subterete, smooth, 1-2 mm thick. Leaves reduced to 2-3 membranous sheaths, mucronate at apex, 0.5-8 cm long. Inflorescence a pseudolateral head with 2-6 sessile spikelets. Involucral bract single, erect, as if continuation of the stem, subacute at apex, grooved on inner side, always shorter than stem proper, 2-10 cm long. Spikelets ovoid or oblong-ovoid, subobtuse at apex, 5-12 x 4-5 mm, terete, stramineous, densely many-flowered. Glumes membranous, broadly ovate, boat-shaped, obtuse and mucronulate at apex, slightly keeled near apex only, *ca* 3 x 2 mm, purplish-brown lineolate on sides, minutely ciliolate on margins; midrib prominent. Hypogynous bristles 4-6, unequal, retrorsely scabrous, shorter than to slightly exceeding the nut. Stamens 3; filaments elongate up to 3 mm; anthers linear, acute and purplish-brown at apex, *ca* 1.3 mm long. Style 1.5 - 2 mm, slightly flattened, ciliolate; stigmas 2, *ca* 1 mm long, at times with a short inconspicous extra stigma. Nut unequally biconvex, obovate, mucronate at apex, *ca* 2 x 1.5 mm, shining blackish-brown, faintly transversely wrinkled.

Fls. & Frts.: Almost throughout the year.

Habitat : Inundated rice fields, along streams and ponds often in association with *Eriocaulon* sp. and *Rotala* sp., at times in the crevices on large trunks of trees.

Distrib.: India to Japan, Madagascar and Australia. INDIA: Throughout. KARNATAKA: Bangalore, Belgaum, Chickmagalur (Rao *et al.*, 2000), Hassan, Kodagu, Shimoga, Uttara Kannada.

Specimens examined : Bangalore: Bangalore, *Govindu s.n.* (Acc. No. 95996), 10.1.1953 (MH); Ghati, *Ramesh* 617, 29.3.1978 (JCB). Belgaum: Jog falls, *Vartak s.n.* (S. No. 10244), Nov. 1971 (MACS); Anmod, *Saldanha & Prakash* 3518, 25.11.1978 (JCB). Hassan: Bagi, *Saldanha* 12773, 25.2.1969 (JCB). Kodagu: Hardur, Sunticoppa, *Rao* 75100, 2.3.1963 (BSI);

Kamble, Sunticoppa, *Rao* 86113, 5.3.1963 (BSI); Mercara, *Vartak s.n.* (S. No. 10243), Nov. 1971 (MACS); Kotakeri, *Bhat* 727, 18.12.1980 (MGH); Bhagamandala, *Bhat* 800, 19.12.1980 (MGH); Kirugoor, *Bhat* 930, 26.1.1981 (MGH); Abbe falls, Mercara, *Yoganarasimhan et al.* 3838, 19.3.1983 (RRCBI). Shimoga: Hosur, near Yedur, *Raghavan* 83002, 5.10.1962 (BSI). Uttara Kannada: Yellapur, *Talbot* 934, 1884 (BSI); Castle Rock, *Vartak* 23074, 1.1.1993 (MACS).

4. **Schoenoplectus lateriflorus** (Gmel.) Lye, Bot. Notis. 124: 120. 1971; Hooper in Saldanha & Nicolson, Fl. Hassan 698. 1976; Sharma *et al.*, Fl. Karnataka 315. 1984; Singh, Fl. E. Karnataka 2: 645. 1988; Karthik. *et al.*, Fl. Ind. Enum. Monocot. 69. 1989; Keshava Murthy & Yoganarasimhan, Fl. Coorg 516. 1990. *Scirpus lateriflorus* Gmel., Syst. Veg. 1:127. 1791; Kern in van Steenis, Fl. Males. 1, 7: 514. 1974; Rao & Razi, Fl. Mysore 564. 1981; Rao & Verma Cyp. NE India 42. 1982; *S. supinus* auct. non L. 1753; Clarke in Hook.f., Fl. Brit. India 6: 655. 1894; Cooke, Fl. Pres. Bombay 2: 892. 1908. (3: 406. 1958, repr.ed.); Fischer in Gamble, Fl. Pres. Madras 1666. 1931 (3: 1156. 1957, repr.ed.); Ramaswamy & Razi, Fl. Bangalore 100. 1973. *S. supinus* var. *uninoides* Clarke in Hook. f., Fl. Brit. India 6: 656. 1894. *Schoenoplectus supinus* (L.) Palla subsp. *lateriflorus* (Gmel.) Koyama in Hooper *et al.*, Enum. Pl. Nepal 1: 119. 1978 & in Dassanayake & Fosberg, Rev. Handb. Fl. Ceylon 5: 158. 1985.

Type: Sri Lanka.

Illus. : Clarke, Illus. Cyperaceae 48, f. 13-14. 1909; Matthew, Furth. Illus. Fl. Tamilnadu Carnatic Pl. 702. 1988.

Annuals, 10-40 (-65) cm high. Stems highly tufted, narrow to slender, obtusely triangular, smooth, glabrous, with 1 or 2 nodes above the base, 1-2 (-3) mm thick. Leaves reduced to 1 or 2 sheaths, mucronate or with a short blade at apex, 1-14 cm long, often with a solitary female flower from the axil with a very long trifid style. Inflorescence pseudolateral, usually corymbose, some times head-like, with 2-10 (-22) spikelets. Rays up to 4, longest up to 3 cm long. Involucral bract 1 or 2; the lower erect, continuous to and looks like the stem, ventrally sulcate, always shorter than stem, 2-15 (-19) cm long; second bract, if present much shorter, up to 4 cm long. Spikelets clustered on rays or on the stem, oblong-ovoid, acute at apex, angular, 4-8 x 2-3 mm, densely many-flowered. Glumes membranous, broadly ovate, mucronate at apex, strongly keeled, 2-2.2 x *ca* 1.5 mm, brownish variegated on sides, very minutely ciliolate at margins; keel 3-nerved. Perianth bristles absent. Stamens 3; filaments elongate up to 2.5 mm; anthers oblong, with a prominent bristly connective appendage at apex, *ca* 0.6 mm long. Style *ca* 0.8 mm long; stigmas 3, as long as the style. Nut trigonous, broadly obovoid, shortly apiculate, *ca* 1.2 x 1 mm conspicously transversely wavy-ridged, straw-coloured, ultimately blackish.

As already mentioned above, sometimes a solitary female flower is found in the axil of leaf-sheaths with an obovate ovary; style trifid (not from the same point), *ca* 1.5 cm long (including the stigmas). Nut developed from this basal flower is planoconvex, obtuse or hemispherical with a long mucro at apex, double the size of normal nuts, *ca* 2 x 1.5 mm, not prominently wavy on the faces, stramineous; mucro *ca* 0.5 mm long, blackish at tip.

Fls. & Frts. : July - Sept.; Jan. - Feb.

Chrom. No. : 2n = 28 (*Phyton* 6: 7-22. 1956).

Habitat: Muddy soil along the river banks, ponds and canals, edges of rice fields, swamps and open wet places, usually in association with other sedges and grasses.

Distrib.: South and South-East Asia (India to China, Malesia & Formosa) and Australia. INDIA: Throughout (except North-West). KARNATAKA: Bangalore, Bellary, Bijapur, Chikmagalur (Sharma *et al., l.c.*), Dakshina Kannada, Hassan, Kodagu, Mandya, Mysore, Shimoga, Tumkur, Uttara Kannada.

Specimens examined: Bangalore: Bannergatta, *Saldanha* 122161, 17.9.1980 (JCB). Bellary: Near Hospet, *Manohar & Ramesh* 5919, 19.1.1979 (JCB). Bijapur: Mamdapur R.F., *Singh* 142931, 14.8.1976 (BSI). Dakshina Kannada: Rice Research Station, Mangalore, *Mudaliar* 93902, 12.3.1948 (MH); Udyavara, *Bhat* 273, 8.8.1976 (MGH). Hassan: Channarayapattana, *Saldanha* 15952, 6.1.1970 (JCB). Kodagu: Kushalnagar, *Bhat* 943, 16.2.1981 (MGH). Mandya: Karehalli, *Murthy* 5617, 7.1.1979 (JCB). Mysore: Hesaraghatta, *Govindu* CoA 165, Feb. 1933 (MH); Bilikeri, *Bhat* 4, 8.7.1970 (MGH); Chamundi Hills, *Padma Rani* 28, 23.8.1970 (MGH). Shimoga: Nalur, Agumbe, *Raghavan* 90433, 3.9.1963 (BSI). Tumkur: Devarayanadurga, *Ramesh & Murthy* 2729, 15.9.1978 (JCB). Uttara Kannada: Karwar, *Talbot* 1515, 10.8.1885 & 1307, 29.8.1885 (BSI); Wandre- Kolluru, *Raghavan* 145285, 23.9.1976 (BSI).

Note : Koyama (*l.c.*) treated this species as a subspecies of *S.supinus* (L.) Palla, a Eurasian species, which can be separated by its less slender habit, nodeless stems, absence of basal flowers, always capitate inflorescence and smooth connective appendage of the anther.

5. **Schoenoplectus litoralis** (Schrad.) Palla subsp. **subulatus** (Vahl) Koyama in Dassanayake & Forsberg, Rev. Handb. Fl. Ceylon 5: 157. 1985; Karthik. *et al.*, Fl. Ind. Enum. Monocot. 69. 1989. *Scirpus subulatus* Vahl, Enum. pl. 2: 268. 1806. *S. litoralis* sensu Clarke in Hook.f., Fl. Brit. India 6: 659. 1894; Cooke, Fl. Pres. Bombay 2: 894. 1908 (3: 408. 1958, repr.ed.); Fischer in Gamble, Fl. Pres. Madras. 1667. 1931 (3: 1156. 1957, repr.ed.); Kern in van Steenis, Fl. Males. 1, 7: 510. 1974; Singh, Fl. E. Karnataka 646. 1988. *Schoenoplectus litoralis* subsp. *thermalis* (Trabut) Murbeck, Contrib. Fl. Maroc. 1: 18. 1922; Sharma *et al.*, Fl. Karnataka 315. 1984. *S. litoralis* var. *subulatus* (Vahl) Chiov. Pl. Nov. 16. 1928; Rao & Razi, Fl. Mysore 564. 1981. **Fig. 56.**

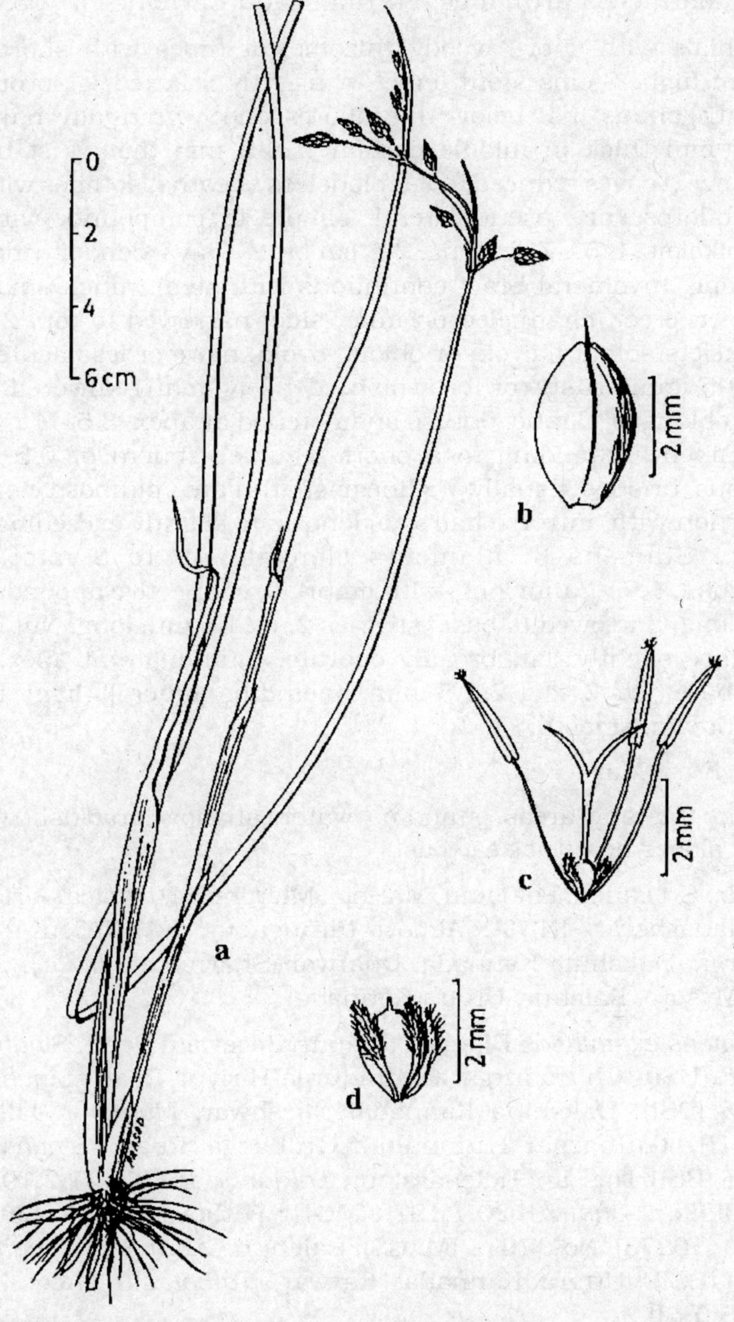

Fig. 56. *Schoenoplectus litoralis* subsp. *subulatus* (Vahl) Koyama
a. Habit, **b.** Glume, **c.** Flower, **d.** Nut with hypogynous bristles.

Illus. : Matthew, Furth. Illus. Fl. Tamilnadu Carnatic Pl. 703. 1988.

Perennials with short, woody rhizome, at times with short stolons, 60-140 cm high. Stems stout, erect or slightly reflexed, glabrous, terete (obtusely trigonous just below the inflorescence), gradually narrowed to apex, 3-7 mm thick in middle portion, 1.5-2 mm thick just below the inflorescence. Leaves reduced to 2-3 bladeless sheaths, at times with a short lamina. Inflorescence pseudolateral, simple to compound, with few to several spikelets, 3.5 - 7 cm long, 2-6 cm broad; rays slender, unequal, up to 4 cm long. Involucral bract continuous with stem, much smaller than stem proper, erect, channeled on inner side, narrowed to top, 2.5-10 cm long. Spikelets solitary, ovoid or oblong-ovoid, more or less acute at apex, 5-13 x 3-5 mm, rusty or brownish, densely many-flowered. Glumes scarious, oblong or elliptic, obtuse and notched at apex, 3.5 - 4 x *ca* 2 mm; keel prominent, extending to a short recurved mucro of 0.5 - 1 mm. Hypogynous bristles usually 4, linear-spathulate, plumosely fringed in upper portion with antrorse hairs, as long as to slightly exceeding the nut, brownish. Stamens 3; filaments elongate up to 3 mm; anthers linear-oblong, 1.75-2 mm long, with fimbriate connective appendage. Style *ca* 2 mm long, narrowed to base; stigmas 2, *ca* 1.5 mm long. Nut biconvex, but one face slightly flat, broadly obovate, mucronate at apex, cuneate towards base, *ca* 2 x 1.2-1.5 mm (including mucro), light brown to blackish-brown, smooth.

Fls. & Frts.: May - Feb.

Habitat: Marshy areas, standing water, shallow brackish water and other wet places near the sea coast.

Distrib.: Sri Lanka, Pakistan, Malesia, Micronesia, tropical Africa, South Africa, Madagascar. INDIA: Almost throughout. KARNATAKA: Bijapur, Chitradurga, Dakshina Kannada, Dharwar (Sharma *et al., l.c.*), Gulbarga, Hassan, Mysore, Raichur, Uttara Kannada.

Specimens examined : Bijapur: Bijapur-Bogewadi Road, *Singh* 142985, 16.8.1976 (BSI). Chitradurga: Chitradurga-Hiriyur Road, *Singh* 132605, 24.2.1975 (BSI). Dakshina Kannada: Nileshwar, *Mudaliar* 94279, Aug. 1940 (MH). Gulbarga: Humnabad-Gulbarga Road, *Singh* 142861, 11.8.1976 (BSI). Hassan: Holenarsipur, *Saldanha* 16302, 21.2.1970 (JCB). Mysore: Bilikeri, *Bhat* 29, 20.7.1970 (MGH); Without exact locality, *Vartak s.n.* (S. No. 10273), Nov. 1971 (MACS). Raichur: Gunthgola, *Singh* 132777, 8.9.1974 (BSI). Uttara Kannada: Karwar, *Talbot s.n.* (Acc. No. 784), 30.8.1885 (BSI).

Note: In the typical subspecies, disrtibuted in the mediterranean region, Middle East, China and Mangolia, the stems are shortly traingular throughout.

6. **Schoenoplectus mucronatus** (L.) Palla, Bot. Jahrb. Syst. 10: 299. 1889; Hooper in Saldanha & Nicolson, Fl. Hassan 699. 1976; Sharma *et al.*, Fl. Karnataka 316. 1984; Koyama in Dassanayake & Fosberg, Rev. Handb. Fl. Ceylon 5: 160. 1985; Karthik. *et al.*, Fl. Ind. Enum. Monocot. 69. 1989. *Scirpus mucronatus* L., Sp. Pl. 1, 50. 1753; Clarke in Hook.f., Fl. Brit. India 6: 657. 1893; Fischer in Gamble, Fl. Pres. Madras 1666. 1931 (3:1156. 1957, repr.ed.); Ramaswamy & Razi, Fl. Bangalore 99. 1973; Kern in van Steenis, Fl. Males. 1, 7: 510, f. 30. 1974; Rao & Verma, Cyp. NE India 43. 1982. *Hommagali hullu.*

Type: Europe.

Illus.: Kern, *l.c.;* Matthew, Furth. Illus. Fl. Tamilnadu Carnatic Pl. 704. 1988.

Perennials with a short horizontal rhizome, 50-90 cm high. Stems often in a row from the short or slightly elongated rhizome, stout, erect, sharpely 3-angled, smooth, 3-7 mm thick. Leaves reduced to 2-3 membranous sheaths; uppermost 12-20 cm long, obtuse-acute at apex, stramineous-brown. Inflorescence pseudolateral, a hemispherical cluster of few to many spikelets, 1-4 cm across. Involucral bract looks like continuation of the stem, much shorter than stem proper, triangular, tapering to an acute apex, erect or reflexed in later stage, 1.5-11.5 cm long. Spikelets sessile, ovoid or oblong-ovoid, acute at apex, terete, 10-20 x 4-6 mm, stramineous-brownish, densely many-flowered. Glumes membranous, ovate, boat-shaped, acute and mucronate at apex, 3.5 - 5 x 2-2.5 mm, with a prominent midrib and faintly keeled apex, minutely ciliolate on margins, many-nerved on both sides. Perianth bristles 5-6, unequal in size, slightly longer than nut, retrorsely scabrid. Stamens 3; filaments elongate up to 4 mm; anthers oblong, apiculate and purplish at apex, *ca* 1.2 mm long. Ovary obovate, *ca* 1 mm long; style *ca* 2 mm long, minutely papillose; stigmas 3, as long as the style or slightly shorter. Nut planoconvex with a distinct dorsal angle, broadly obovate, mucronate at apex, *ca* 2 x 1.8 mm, blackish-brown when mature.

Fls. & Frts.: April-Jan.

Habitat: Shallow water in ponds, tanks, lakes and ditches; also in open wet places and margins of standing water.

Distrib.: Warmer parts of old world (South Europe to Japan, South Asia to Australia and tropical Africa). INDIA: Throughout. Karnataka: Bangalore (Ramaswamy & Razi, *l.c.*), Belgaum, Chikmagalur (Rao *et al.*, 2000), Hassan, Kodagu, Mysore (Fischer, *l.c.*), Shimoga, Uttara Kannada.

Uses: In Malesia dried and flattened stems are often used for making cheap and durable mats, bags and strings. In Sumatra it is cultivated for this purpose (Kern, *l.c.*).

Specimens examined : Belgaum: Londa, *Gammie*, 14072, 21.5.1900 (BSI); Khanapur, *Ahuja* 47753, 5.1.1959 (BSI). Hassan: Tank near Railway station, Hassan, *Hooper & Gandhi* 2380, 10.11.1971 (JCB). Kodagu: Nagarhole, *Sreenath & Singh* 10362, 10.11.1979 (JCB). Shimoga: Kanagalgudda, Tirthahalli, *Raghavan* 81076, 28.5.1962 (BSI); Honnati, *Sreenath & Ramesh* 7306, 22.4.1979 (JCB). Uttara Kannada: Yellapur, *Talbot s.n.*, 1883 (BSI) and 588, 25.5.1883 (BSI).

7. **Schoenoplectus senegalensis** (Hochst. ex Steud.) Palla ex J. Raynal in P.de Fabregues & J. -P. Lebrun, Cat. Pl. Vasc. Niger 3: 344. 1976 *et* in Adansonia 2, 16: 149. 1976; Hooper in Saldanha & Nicolson, Fl. Hassan 699. 1976; Sharma *et al.*, Fl. Karnataka 316. 1984; Karthik. *et al.*, Fl. Ind. Enum. Monocot. 70. 1989. *Isolepis senegalensis* Hochst. ex Steud., Syn. Pl. Gl. 2: 96. 1855, *non Scirpus senegalensis* Lamk. 1791. *Scirpus jacobi* Fischer in Bull. Misc. Inf. 1931: 103. 1931 *et* in Gamble, Fl. Pres. Madras 1666. 1931 (3: 1156. 1957, repr.ed.); Ramaswamy & Razi, Fl. Bangalore 99. 1973. *Schoenoplectus jacobi* (Fischer) Lye, Bot. Notis. 124: 290. 1971.

Illus.: Matthew, Illus. Fl. Tamilnadu Carnatic Pl. 796. 1982.

Tufted annuals. Stems terete, slender 7.5 - 30 cm high, *ca* 1.5 mm thick, transversely septate. Leaves reduced to sheaths, with oblique mouth. Inflorescence a dense, lateral head of few to many spikelets, arising immediately above the mouth of the sheath. Bract much longer than stem proper. Spikelets sessile, ovoid, 4-8 mm long. Glumes suborbicular, obtuse at apex, at times minutely apiculate, slightly concave, not keeled, 2.5 - 3 mm long, slightly broader than long, pale brown. Perianth bristles absent. Stigmas 3. Nut obtusely trigonous, obovoid, not narrowed at apex, apiculate, *ca* 1.3 mm long, with a vertical ridge along each angle, with 8-10 strong transverse undulate ridges on the convex faces, black.

Fls. & Frts. : Dec. - Feb.

Chrom.No.: 2n = 28 (*Taxon* 21: 683. 1972).

Habitat: Wet sandy soil near open water and marshy areas.

Distrib.: Africa. INDIA: Peninsular India. KARNATAKA: Bangalore (Ramaswamy & Razi, *l.c.*), Hassan, Mysore.

Specimens examined: Hassan: Nagpuri, *Saldanha* 16092, 21.1.1970 (JCB). Mysore: Mysore, *Geetha* 1, 12.2.1982 (MGH).

Note : Very similar to slender plants of *S. articulatus* (L.) Palla. But can be distinquished by the distinctly undulate nuts.

23. SCLERIA

Berg., Kongl. Vetensk. Acad. Handl. 26: 142, t. 4, 5. 1765; Boeck. in Linnaea 38: 436-542. 1874. *Hypoporum* Nees in Edinburgh New Philos. Journ. 17: 266. 1834.

Type: Scleria flagellum-nigrorum Berg.

Perennials with short, creeping, often nodose rhizome, or annuals with fibrous, reddish roots. Stems solitary or more or less tufted, mostly erect, rarely straggling, triquetrous or trigonous, many-noded, smooth or scabrous, leafy. Leaves 3-ranked, linear, sheathing the stem, smooth to scabrous on the margins and main nerves, lower ones reduced to bladeless sheaths or with a very short lamina; sheaths closed, often 3-winged, apex of the ventral side truncate or produced into a tongue-like contraligule. Inflorescence paniculate, terminal and usually a few lateral ones from the apical nodes subtended by leafy bracts, some times reduced to glomerate-spiciform with glume-like bracts. Spikelets all bisexual or bisexual and male, or female and male; bisexual spikelets with 1 terminal female flower and 1 to several lateral male flowers; female spikelets with 1 female flower and often with 1-2 lateral empty glumes which are reduced male flowers; male spikelets with many to several flowers. Glumes distichous except the upper ones of male spikelets and the reduced male parts of bisexual flowers. Flowers unisexual. Male ones with 1-3 stamens only; anthers oblong to linear, with more or less produced connective. Female flowers with a 3-carpellate pistil; style continuous with the ovary, base often persistent on the nut; stigmas 3. Nut terete or trigonous, globose, ovoid, ellipsoid or pyramidal, bony, smooth or variously sculptured, glabrous or hairy, white or rarely bluish, borne on a gynophore. Gynophore persistent on the nut, cup-like or disk-like, usually trilobate, at times much reduced and obsolete.

A large pantropic genus of *ca* 200 species, some extending to subtropical and warm temperate regions. Though many species are widely distributed in Asia, Australia and also in Africa, only *Scleria lithosperma* is really pantropic. Many species have restricted distribution. In India *ca* 27 species, 11 in Karnataka.

Literature: **KERN, J.H.** (1961) Florae Malesianae Precursores - XXX. The genus *Scleria* in Malaysia, in *Blumea* 11: 140-218.

Key to the species

1a. Annuals with fibrous roots (rarely short-lived perennials with a short rhizome in *S. stocksiana*) 2

1b. Perennials, usually with short, creeping, horizontal rhizome. 7

2a. Disk obsolete or not lobed 3

2b. Disk always 3-lobed 4

3a. Plants up to 1 m high; leaves 3-5 mm wide; inflorescence consisting of one terminal and 2-3 lateral panicles; bracts well developed, overtopping the panicles; nut *ca* 2 x 1.5 mm, very smooth . . . . **1. S. annularis**

3b. Plants 25-50 cm high; leaves 0.5-2 mm wide; inflorescence spiciform, linear, unbranched, consisting of 5-25 clusters of spikelets; bracts inconspicuous; nut 1-1.3 mm across, lacuno- rugose and trabeculate especially towards the top .... **7. S. pergracilis**

4a. Disk-lobes triangular-lanceolate, gradually narrowed to an acute apex, up to half the length of nut; nut globose, 1.5- 1.8 mm across, cancellate with regular vertical rows of pits .... **2. S. biflora**

4b. Disk-lobes broadly ovate, broadly oblong or orbicular-ovate, obtuse to truncate at apex, never up to half the length of nut; nut broadly ovoid or cylindric-ellipsoid, 3 mm or more long, rugulose, tubercled or longitudinally lacunose 5

5a. Nut cylindric-ellipsoid, longitudinally lacunose; disk- lobes broadly oblong, truncate at apex .... **11. S. tessellata**

5b. Nut broadly ovoid, rugulose, tubercled or longitudinally verrucose-lacunose; disk-lobes broadly ovate, orbicular-ovate or subobtuse 6

6a. Lateral panicles pendulous in later stage; nut usually longitudinally verrucose-lacunose throughout .... **4. S. foliosa**

6b. Lateral panicles always erect; nut somewhat rugulose or tubercled especially near the base, smooth near the apex. .... **9. S. stocksiana**

7a. Inflorescence a single, terminal, much branched panicle (very rarely with a lateral panicle); bract small, setaceous, never overtopping the panicle, sometimes absent .... **8. S. poaeformis**

7b. Inflorescence consisting of a terminal panicle and 1-several lateral ones; bracts well developed, leaf-like, overtopping the panicle. 8

8a. Plants comparatively short and slender, up to 1 m or slightly more high; stems 1-3 mm thick; leaves 1-6 (-8) mm wide 9

8b. Plants tall and stout, up to 2 m or more tall (often straggling over shrubs, then with very long stems); stems 6-10 mm thick; leaves 6-20 mm wide . . 10

9a. Leaf sheaths prominently winged (especially towards upper portion of the stem); contraligule suborbicular; spikelets unisexual; nut almost globose or globose-ovoid, *ca* 3 x 2 mm, without depressions at base; disk well developed, with 3 ovate- lanceolate lobes reaching almost half the length of the nut .... **5. S. levis**

9b. Leaf sheaths not winged; contraligule triangular-ovate, spikelets bisexual; nut broadly obovoid or oblong-obovoid, *ca* 2.5 x 2 mm, with 3 depressions at the base; disk reduced to a minute, brown ring adnate to base of the nut .... **6. S. lithosperma**

10a. Leaves abruptly narrowed to tip; leaf sheaths never winged; contraligule broadly ovate-triangular; spikelets bisexual and males; nut ovoid, acute at apex, with 3 shallow depressions at base, smooth, glabrous; disk very much reduced, triangular .... **3. S. corymbosa**

10b. Leaves gradually narrowed to tip; leaf sheaths often 3- winged; contraligule obtuse; spikelets always unisexual; nut broadly ovoid to subglobose, umbonulate at apex, without depressions at base, reticulate-cancellate, minutely hairy, ultimately glabrescent; disk well developed, with 3 obtuse lobes .... **10. S. terrestris**

1. **Scleria annularis** (Kunth, Enum. Pl. 2: 359. 1837 *nom.nud.*) Nees ex Steud., Syn. Pl. Glum. 2: 176. 1855, *ex descr.*; Clarke in Hook.f., Fl. Brit. India 6: 687. 1894; Cooke, Fl. Pres. Bombay 2:903. 1908 (3:418. 1958, repr.ed.); Kern in Blumea 11: 200. 1961 *et* in van Steenis, Fl. Males. 1, 7: 744, f. 106 (25). 1974; Sharma *et al.*, Fl. Karnataka 316. 1984; Karthik. *et al.*, Fl. Ind. Enum. Monocot. 71. 1989. *Hypoporum annulare* Nees in Linnaea 9: 303. 1835, *nom.nud.*

Illus.: Kern, *l.c.*

Annuals, up to 100 cm high. Stems slender, sharply triquetrous, retrorsely scabrous on the angles, 1.5-2.5 mm thick. Leaves linear, narrowed to the obtuse tip, 3-5 mm wide, antrorsely scabrous on the margins, especially towards upper portion; sheaths usually retrorsely scabrous on the angles, glabrous or pubescent on anterior side; contraligule ovate or triangular, 1-2 mm long, glabrous or ciliate. Inflorescence consisting of a terminal panicle and 2-3 lateral ones; terminal panicle up to 5 cm long, up to 1.5 cm broad; lateral panicles usually solitary, shorter than terminal one. Main bract erect, overtopping the inflorescence; secondary bracts smaller, but exserting from panicles. Spikelets subsessile to short-pedicelled, bisexual or unisexual. Male spikelets lanceolate, *ca* 3 mm long; pedicels *ca* 1 mm long. Lowest 2 glumes empty, long-awned, much longer than fertile glumes, up to 4 mm long (including *ca* 2 mm long awn); fertile glumes oblong-lanceolate, mucronulate at apex, 2.5 - 3 mm long. Stamens 2 (3 in bisexual spikelets); filaments up to 2.7 mm long; anthers linear-oblong, apiculate, *ca* 1.2 mm long. Nut-bearing spikelets *ca* 4 mm long. Glumes broadly ovate-lanceolate, mucronulate at apex, *ca* 4 mm long. Disk not lobed, triangular with obtuse angles. Nut ovoid, slightly laterally compressed, obtuse at apex, not apiculate, *ca* 2 x 1.5 mm, glabrous, very smooth, shining white.

Fls. & Frts.: Not known from Karnataka.

Habitat : Grasslands.

Distrib. : Central China, Malesia. INDIA: South, Central, North-West and Eastern India. KARNATAKA: Kanara (Cooke, *l.c.*) .

Note: Though reported from Karnataka, no specimen is available in BSI from this State. It can be easily distinguished by the very smooth, shining white nuts. The description given above is based on specimens available from Maharashtra and Dadra.

2. **Scleria biflora** Roxb., Fl. Ind. ed. 2, 3: 573. 1832; Clarke in Hook. f., Fl. Brit. India 6: 687. 1894; Cooke, Fl. Pres. Bombay 2: 904. 1908 (3: 419. 1958, repr.ed.); Kern in Blumea 11: 197, f. 7c.1961 *et* in van Steenis, Fl. Males. 1, 7: 743, f. 113 a-g. 1974; Koyama in Dassanayake & Fosberg, Rev. Handb. Fl. Ceylon 5: 362. 1985; Karthik. *et al.*, Fl. Ind. Enum. Monocot. 71. 1989. **Fig. 57.**

Illus.: Clarke, Illus. Cyp. t. 127, f. 1-2. 1909; Kern, *l.c.*

Annual herbs, 15-30 (-45) cm high. Stems slender, smooth, 1-2 mm thick. Leaves linear, narrowed to an obtuse tip, 6-16 cm long, 2-4 mm wide, minutely scabrous on the margins towards apex; sheaths trigonous, narrowly winged; contraligule very short, much broader than long, obtuse. Inflorescence of 2-4 panicles, terminal one larger. Primary bract usually overtopping the panicle. Spikelets solitary or paired. Male spikelet lanceolate, short-pedicellate, *ca* 3 mm long; pedicels 2-2.5 mm long. Glumes lanceolate, 2-3 mm long. Stamens 2-3 to each glume; filaments *ca* 2 mm long; anthers linear-oblong, apiculate, *ca* 1 mm long. Nut-bearing spikelet obovoid, 3-4 mm long. Glumes broadly ovate-lanceolate, acuminate at apex, 3-4 mm long. Nut globose or slightly depressed-globose, with a dark brown beak of style base at apex, 1.5 - 1.8 mm across, cancellate with regular vertical rows of pits on the surface, ferrugineous-pubescent on ridges between pits, whitish. Disk deeply 3-lobed; lobes triangular-lanceolate, gradually narrowed upwards, acute at apex, up to half the length of the nut.

Fls. & Frts.: Aug. - Nov.

Habitat: Wet grasslands, grassy roadsides and as a weed in rice fields.

Distrib.: Sri Lanka, South China, Philippines and West Malesia. INDIA: Eastern India and Western peninsula. KARNATAKA: Dakshina Kannada, Hassan, Uttara Kannada; not common.

Specimens examined: Dakshina Kannada: Charmadi ghat, *Saldanha* 6245, 27.10.1960 (JCB); Venoor, *Bhat* 715, 16.11.1980 (MGH). Hassan: Bannuhalla, *Saldanha* 15501, 30.10.1969 (JCB). Uttara Kannada: Without exact locality, *Talbot* 1506, without date (BSI); Karwar, *Talbot s.n.*, 25-8-1885 (BSI); Tinai, *Talbot* 2563, 14.10.1890 (BSI); Karwar, *Hallberg & Mc Cann s.n.* (Acc. No. 80985), Oct. 1919 (BLAT).

3. **Scleria corymbosa** Roxb., Fl. Ind. 2, 3: 574. 1832; Clarke in Hook. f., Fl. Brit. India 6: 686. 1894; Fischer in Gamble, Fl. Pres. Madras 1677.1931 (3: 1163. 1957, repr.ed.); Kern in Blumea 11: 189. 1961 *et* in van Steenis, Fl. Males. 1, 7: 740, f. 105. 1974; Arora *et al.*, Bot. S. Kanara 62. 1981; Rao & Verma, Cyp. NE India 57. 1982; Sharma *et al.*, Fl. Karnataka 316. 1984; Koyama in Dassanayake & Fosberg, Rev. Handb. Fl. Ceylon 5: 351. 1985; Karthik. *et al.*, Fl. Ind. Enum. Monocot. 71. 1989. *S. ridleyi* Clarke in Hook. f., Fl. Brit. India 6: 686. 1894.

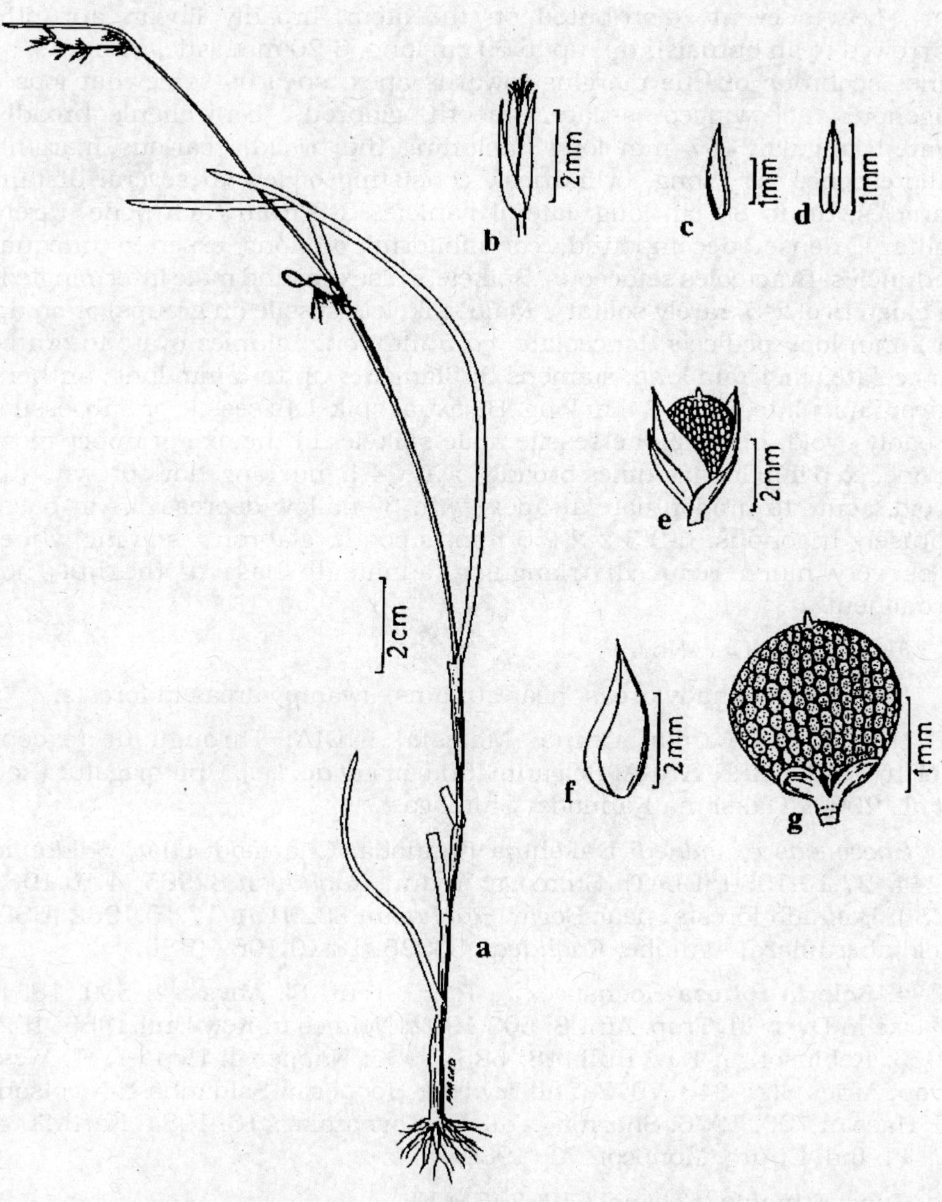

Fig. 57. *Scleria biflora* Roxb.
a. Habit, **b**. Male spikelet, **c**. Male glume, **d**. Anther, **e**. Nut-bearing spikelet, **f.** Female glume, **g.** Nut

Illus.: Clarke, Illus. Cyp. t. 124, f-1-4. 1909; Kern, *l.c.*

Perennials with woody rhizome, up to 2 m tall. Stems solitary, usually robust, trigonous, *ca* 1 cm thick, smooth, rarely scabrous towards upper part. Leaves evenly distributed on the stem, broadly linear, abruptly narrowed to an obtusish tip, up to 60 cm long, 6-20 mm wide, glabrous, at times scabrous on the margins towards apex; sheaths somewhat loose, trigonous, not winged, striate, smooth, glabrous; contraligule broadly ovate-triangular, 3-7 mm long (including the broad, scarious margin). Inflorescence very long, cylindrical, consisting of few to several distant panicles, up to 80 cm long; lateral panicles 2-3 from each node, rarely solitary, dense, decompound, corymbiform, on long exserted unequal peduncles. Bracteoles setaceous. Spikelets bisexual and male intermingled, in clusters of 2-5, rarely solitary. Male spikelets sessile (in groups) or on up to 2 mm long pedicels, lanceolate, *ca* 5 mm long; glumes ovate to ovate-lanceolate, *ca* 4 mm long; stamens 3; filaments up to 5 mm long; anthers linear, apiculate, up to 2 mm long. Bisexual spikelets sessile or subsessile, broadly ovoid, bearing one sessile male spikelet in the axil of upper most glume, *ca* 5 mm long; glumes broadly ovate, 4-5 mm long, dark brown. Nut ovoid, acute to umbonulate at apex, with 3 shallow depressions at base, obtusely trigonous, 3-3.5 x 2-2.5 mm, smooth, glabrous, shining white. Disk very much reduced, triangular, adnate to base of the nut, not prominent.

Fls. & Frts.: Oct.-Nov.

Habitat: Wet shady areas, near streams, swampy areas in forests.

Distrib.: Sri Lanka, China, Malesia. INDIA: Throughout (except North-West). KARNATAKA: Belgaum (Sharma *et al., l.c.*), Chikmagalur (Rao *et al.*, 2000), Dakshina Kannada, Shimoga.

Specimens examined: Dakshina Kannada: Charmadi ghat, *Saldanha* 6244, 27.10.1960 (BLAT). Shimoga: Yedur, *Raghavan* 82965, 4.10.1962 (BSI); Bakhalli forests, near Bogar, *Raghavan* 83291 A, 17.10.1962 (BSI); Kokkaberumara, Agumbe, *Raghavan* 83325, 18.10.1962 (BSI).

4. **Scleria foliosa** Hochst. ex A. Rich., Tent. Fl. Abyss. 2: 509. 1851; Clarke in Dyer, Fl. Trop. Afri. 8: 503 1902; Nelmes in Kew Bull.1956: 102. 1956; Robinson in Kew Bull. 18: 525. 1966; Napper in Hepper, Fl. West Trop. Africa 3(2): 343. 1972 (2nd rev.ed.); Hooper in Saldanha & Nicolson, Fl. Hassan 700. 1976; Sharma *et al.*, Fl. Karnataka 316. 1984; Karthik. *et al.*, Fl. Ind. Enum. Monocot. 71. 1989.

Type: Ethiopia, *Schimper* 1232 (BM,K).

Medium sized, tufted annuals, nearly glabrous. Stems 15- 150 cm high, 1-3 mm thick, usually scabrous on the acute angles. Leaves basal as well as at longer intervals in the upper part of the stem, 2-7 mm wide, glabrous, scabrid on the margins above and sometimes on the midrib beneath and on 2 prominent nerves on the upper surface towards apex; lower leaves reduced to bladeless sheaths; sheaths usually glabrous, at times pubescent; contraligule glabrous to densely villous, usually with blackish-red margin which sometimes has white edge. Inflorescence consisting of 2-4 panicles; terminal panicle sessile; lateral ones single at nodes, distinctly spaced, scarcely to shortly exserted from the sheaths on peduncles which become pendulous when mature. Bracts overtopping the panicles. Male spikelets 4-5 mm long, subsessile or pedicel shorter than spikelet. Female glumes ovate-lanceolate, 3-5 mm long, glabrous, light brown to blackish-red. Nut broadly ovoid, shortly tapering beak-like at apex, 3-4 x 2-2.5 mm, mostly longitudinally verrucose-lacunose, glabrous, white to grey. Disk 3-lobed; lobes orbicular-ovate, less than 1 mm long.

Fls. & Frts.: Oct.-Nov.(Hooper, *l.c.*) .

Habitat : Seasonal wet areas, shady areas near water in plains and ghats.

Distrib.: Tropical and Southern Africa, Madagascar. INDIA: South India. KARNATAKA: Hassan (Hooper, *l.c.*) .

Note: This species is included on the authority of Hooper (*l.c*) . As no specimen is available in BSI, description given above is adopted from Robinson, *l.c.* and Nelmes, *l.c.* Probably rare in Karnataka.

5. **Scleria levis** Retz., Observ. 4, 13. 1786; Kern in Blumea 11: 164, f. 34. 1961 *et* in van Steenis, Fl. Males. 1, 7: 732, f. 104 (6) 1974; Rao & Verma, Cyp. NE India 58. 1982; Sharma *et al.*, Fl. Karnataka 316. 1984; Koyama in Dassanayake & Fosberg, Rev. Handb. Fl. Ceylon 5: 345. 1985; Singh, Fl. E. Karnataka 2: 646. 1988; Karthik. *et al.*, Fl. Ind. Enum. Monocot. 71. 1989. *S. hebecarpa* Nees in Wight, Contr. Bot. India 117. 1834; Clarke in Hook. f., Fl. Brit. India 6: 689. 1894; Cooke, Fl. Pres. Bombay 2: 904. 1908 (3: 419. 1958, repr.ed.) ; Fischer in Gamble, Fl. Pres. Madras 1677. 1931 (3: 1163. 1957, repr.ed.) ; Rao & Razi, Fl. Mysore, 565. 1981. **Fig. 58.**

Type: India.

Illus.: Kern, *l.c.*

Key to the varieties

1a. Plants 40-90 cm high; stems 2-3 mm thick in the middle portion, glabrous or minutely puberulous on the faces; leaves 3-6 (-8) mm wide in the middle portion; nut subglobose or globose-ovoid, *ca* 3 mm long, pubescent to glabrescent . . . . var. **levis**

1b. Plants up to 125 cm high; stems *ca* 3.5 mm thick in the middle portion, puberulous on the faces; leaves *ca* 10 mm in the middle portion; nut broadly oblong, *ca* 4 mm long, with stellate hairs in the basal half (ultimately glabrous) . . . . var. **pubescens**

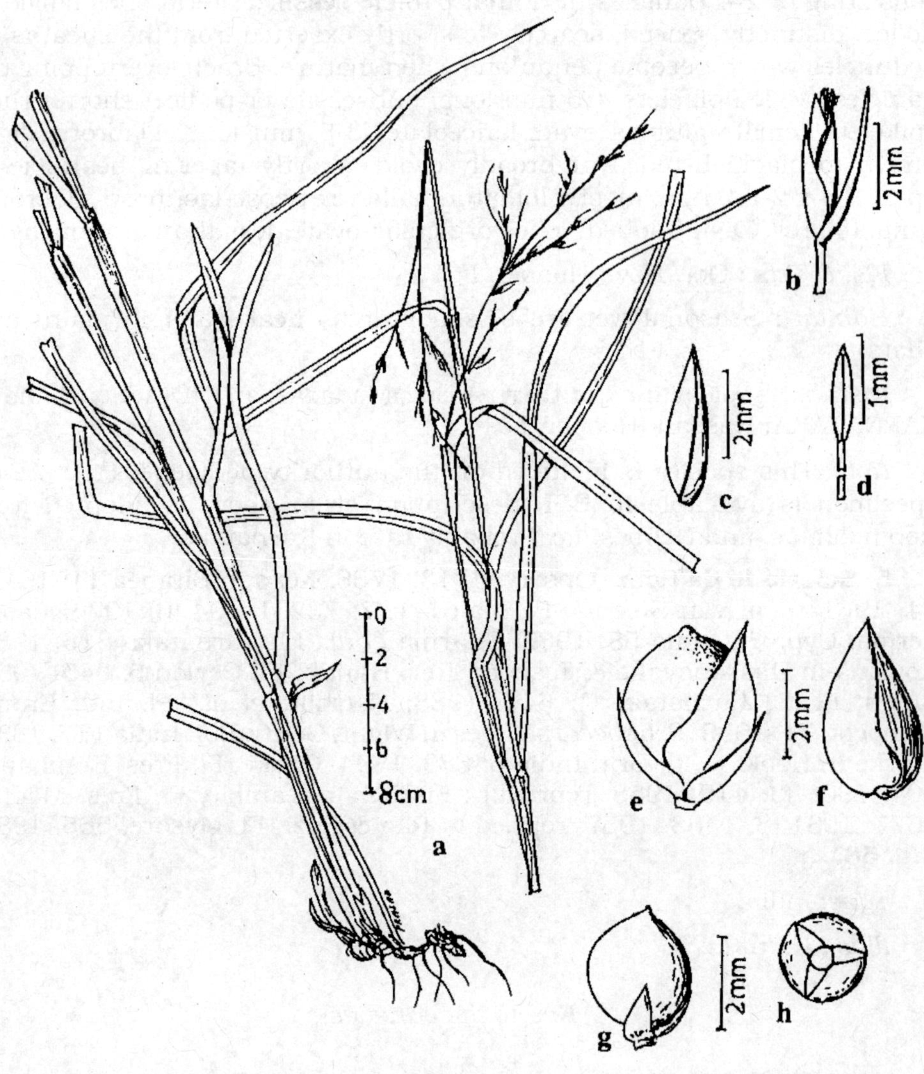

Fig. 58. *Scleria levis* Retz. var. *levis*
a. Habit, **b**. Male spikelet, **c.** Male glume, **d.** Stamen, **e.** Nut-bearing spikelet, **f.** Female glume, **g.** Nut, **h**. Basal view of nut showing the disk

var. **levis**

Perennials, 40-90 cm high. Rhizome stout, short-creeping, covered with scales. Stems somewhat slender, trigonous, glabrous or minutely puberulous on concave faces, at times retrorsely scabrous on angles, 2-3 mm thick in the middle portion. Leaves evenly distributed on the stem, linear, narrowed to an obtusish tip, 10-45 cm long, 3-6 (-8) mm wide, glabrous, often pubescent with long white hairs below, usually antrorsely scabrous on the margins and the midrib below towards apex. Upper sheaths usually prominently winged and retrorsely scabrous on angles; lower ones hardly or not winged; contraligule suborbicular, 1-2 mm long, densely hirsute. Inflorescence consisting of a terminal panicle and 1 or 2 smaller axillary ones; terminal panicle 7-13 cm long, with loose branches; lateral panicles usually not branched, spike-like, on exserted peduncles. Primary bracts overtopping the inflorescence. Spikelets unisexual, in clusters of 2-4, at times mostly males. Male spikelets oblong-lanceolate, 3.5-4mm long; pedicels up to 5 mm long. Glumes lanceolate, acute to mucronate at apex, 2.5 - 3.5 mm long. Stamens 2 or 3, filaments up to 3 mm long; anthers linear-oblong, apiculate, *ca* 1 mm long. Nut-bearing spikelets sessile, obovoid, 4-6 mm long; glumes broadly ovate, acute to mucronate at apex, 3-4.5 mm long. Nut almost globose or globose-ovoid, almost terete or very broadly trigonous, apiculate at apex, *ca* 3 x 2 mm (including the disk), smooth or faintly transversely rugulose, pubescent to glabrescent, whitish. Disk deeply 3-lobed; lobes ovate-lanceolate, acute at apex, appressed to nut, almost half as long as the nut.

Fls. & Frts.: July-May.

Habitat : Open moist grasslands, moist shady places, among grasses in deciduous forests; sea level to 800 m.

Distrib.: Sri Lanka, South China, Malesia, Japan, Northern Australia. INDIA: Throughout (except North-West arid region). KARNATAKA: Belgaum, Chikmagalur (Sharma *et al.*, *l.c.*), Chitradurga (Singh, *l.c.*), Dakshina Kannada, Kodagu, Mysore (Rao & Razi, *l.c.*), Shimoga, Uttara Kannada.

Specimens examined : Belgaum, Dudwawada R.F., Londa, *Prasad* 172894, 17.12.1994 (BSI). Dakshina Kannada: Varavi R.F., without coll. name, 018088, 18.11.1927 (MH); Kushalnagar, *Bhat* 1046, 21.9.1981 (MGH). Shimoga: Sharavati river bed, above Jog falls, *Prasad* 173770, 3.11.1995 (BSI). Uttara Kannada: Birchy, *Talbot* 1694, 10.5.1888 (BSI); Tinai Ghat, *Gammie* 15818, 1.11.1902 (BSI); Arabail ghat, *Ramesh & Shivprakash* 13357, 20.7.1981 (JCB).

var. **pubescens** (Steud.) C.Z. Zheng, Fl. Zhejiang 7: 286. 1986. *Scleria pubescens* Steud., Syn. Pl. Glum. 2: 168. 1855. *S.hebecarpa* var. *pubescens* (Steud.) Clarke in Hook. f., Fl. Brit. India 6: 689. 1894. **Fig. 59.**

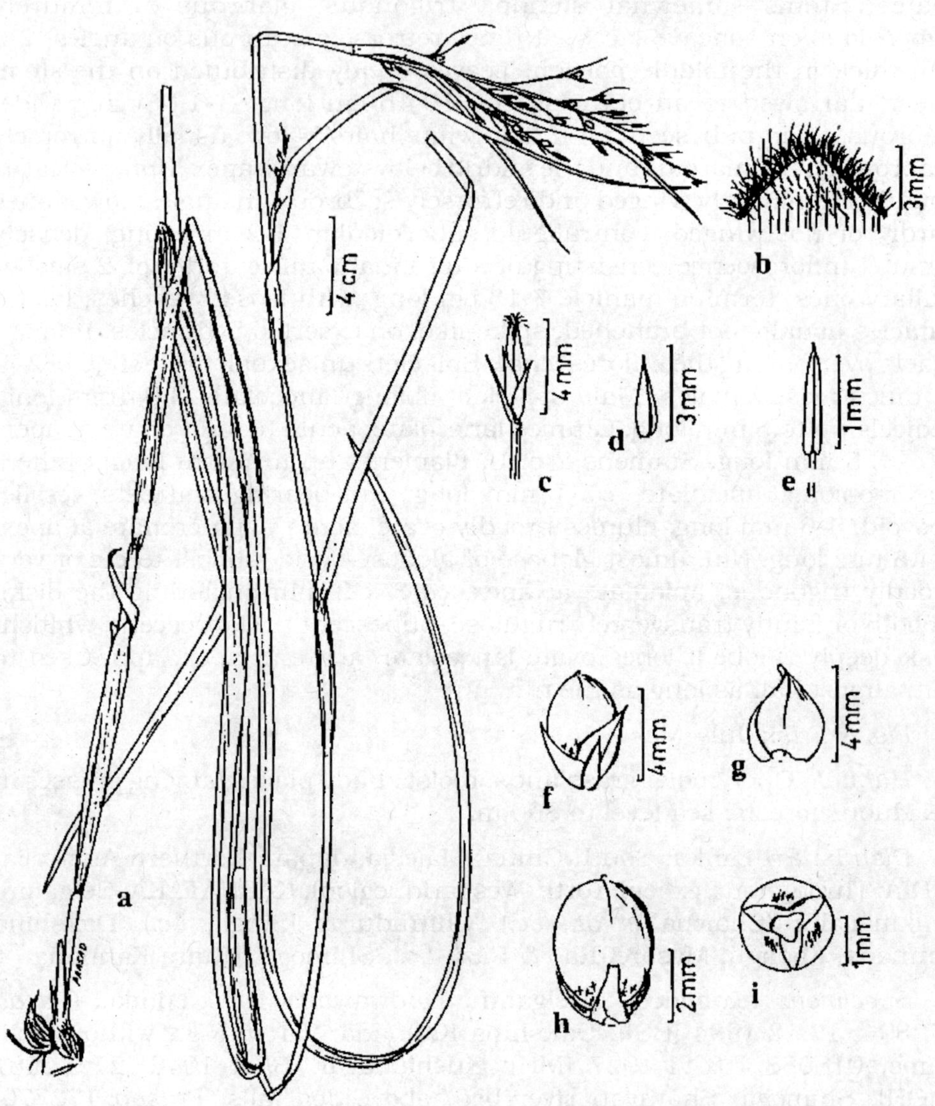

Fig. 59. *Scleria levis* var. *pubescens* (Steud.) C.Z. Zheng
a. Habit, **b.** Contraligule, **c.** Male spikelet, **d.** Male glume, **e.** Stamen, **f.** Nut-bearing spikelet, **g.** Female glume, **h.** Nut, **i.** Bsal view of nut showing the disk.

Stems up to 110 cm long, *ca* 3.5 mm thick in the middle portion, puberulous on the faces. Leaves *ca* 10 mm wide in the middle portion; contraligule broadly ovate or triangular, 3-4 mm long. Nut broadly oblong, *ca* 4 x 2 mm, with stellate hairs in the basal half (but ultimately glabrous), with 3 faint lines from the apex gradually fading towards the middle of the nut.

Fls. & Frts.: Sept.

Habitat: On the bank of nallahs at high altitude.

Distrib.: China, Malaya. INDIA: North-East. KARNATAKA: Mysore.

Specimen examined : Mysore: Basavangudi, Biligirirangan hill range, *Rao* 73767, 7.9.1961 (BSI).

Note : A perusal of literature revealed that the combination *Scleria levis* var. *pubescens* was published twice by different authors viz., C.Z. Zheng (1986) and Tjitrosedidjo (1995). In both the cases the basionym is the same, i.e. *Scleria pubescens* Steud. It seems Tjitrosedidjo was unaware of the existence of the above combination made by C.Z. Zheng (1986), and thereby rendered a superfluous combination which resulted in an isonym, which according to the rules of botanical nomenclature has to be rejected. Therefore, C.Z. Zheng is the actual authority of the name.

This is the first report of this variety from peninsular India. So far it was reported fom North-Eastern States only (Prasad & Singh, 2001).

6. **Scleria lithosperma** (L.) Sw., Prodr. 18. 1788; Clarke in Hook.f., Fl. Brit. India 6: 685. 1894; Cooke, Fl. Pres. Bombay 2: 903. 1908 (3:418. 1958, repr.ed.); Fischer in Gamble, Fl. Pres. Madras 1677. 1931 (3: 1163. 1957, repr.ed.); Ramaswamy & Razi, Fl. Bangalore 111. 1973; Kern in Blumea 11: 191, f. 6 h. 1961 *et* in van Steenis, Fl. Males. 1, 7: 740, f.105 (19).1974; Hooper in Saldanha & Nicolson, Fl. Hassan 700. 1976; Rao & Razi, Fl. Mysore 565. 1981; Rao & Verma, Cyp. NE India 56. 1982; Sharma *et al.*, Fl. Karnataka 316. 1984; Koyama in Dassanayake & Fosberg, Rev. Handb. Fl. Ceylon 5: 352. 1985; Karthik. *et al.*, Fl. Ind. Enum. Monocot. 71. 1989; Keshava Murthy & Yoganarasimhan, Fl. Coorg 517. 1990. *Scirpus lithospermus* L., Sp. Pl. 51. 1753.

Type : India.

Illus.: Clarke, Illus. Cyp. f. 123, f. 1-4. 1909; Kern, *l.c.*

Perennials with slender stems, up to 110 cm high; stems smooth, trigonous, 1-2 mm thick. Leaves linear, gradually narrowed to apex, 1-4 mm wide; sheaths not winged, usually pubescent on the faces; contraligule short, triangular or ovate, obtuse, hirsute or ciliate. Inflorescence loose, terminal panicle along with 2-3 axillary ones with few spikelets; terminal one up to 25 cm long; bracts usually overtopping the panicles. Spikelets bisexual, solitary or 2-3 together, with one female flower and a few to several male flowers, 3-4 mm long. Glumes ovate to lanceolate, acuminate, cuspidate or mucronulate at apex, 2-3 mm long, ferrugineous; stamen 1(-2); filament *ca* 3.5 mm long; anther linear, 0.75-1.5 mm long. Style short; stigmas *ca* 2.2 mm long. Nut obtusely trigonous, broadly obovoid or oblong-obovoid, obtuse and minutely umbonulate at apex, cuneate and with a depression on each face at base, *ca* 2.5 x 2 mm, smooth, shining white. Disk reduced to a minute, brown, glandular ring adnate to nut base.

Fls. & Frts. : June-Dec.

Habitat: Edges of forests, moist shady areas, moist evergreen forests of hilly areas, shady areas in sholas and as an undergrowth in dry forests; sea level to 1300 m.

Distrib.: The most widely distributed species of *Scleria*, pantropic. INDIA: Throughout (except in the Western arid zone). KARNATAKA: Bangalore (Ramaswamy & Razi, *l.c.*), Dakshina Kannada, Hassan, Kodagu, Mandya, Mysore (Rao & Razi, *l.c.*), Shimoga, Uttara Kannada.

Specimens examined: Dakshina Kannada: Mangalore, *Barber* 4827, June 1902 (MH); Shiradi, without coll. name, 15604; 15.12.1918 (MH); Varavi, without coll. name 018142, 21.11.1929 (MH); Belve, *Saldanha* 5836, 14.10.1960 (BLAT); Kunjargiri, Udupi, *Bhat* 622, 21.7.1980 (MGH); Karinji, *Bhat* 706, 11.10.1980 (MGH). Hassan: Kempuhole, Shiradi, *Saldanha* 14632, 22.8.1969 (JCB). Kodagu: Kushalnagar, *Bhat* 1048, 21.9.1981 (MGH); Kerati R.F., Makut, *Srinivasan* 72306, 4.11.1981 (MH). Mandya: Narayana Surya, *Dinesh* 982, 18.6.1984 (MGH). Mysore: Bandipur R.F., *Naithani* 21184, 25.8.1964 (MH). Shimoga: Sharavati river bed, above Jog falls, *Prasad* 173771, 3.11.1995 (BSI). Uttara Kannada: Karwar, *Talbot* 536, 10.7.1883 (BSI); Guddhalli peak, Karwar, *Bell* 212, 9.12.1920 (BLAT).

7. **Scleria pergracilis** (Nees) Kunth, Enum. Pl. 2: 354. 1837; Clarke in Hook. f., Fl. Brit. India 6: 685. 1894; Kern in Blumea 11: 196. 1961 *et* in van Steenis, Fl. Males. 1, 7: 743, f. 105 (22). 1974; Hooper in Saldanha & Nicolson, Fl. Hassan 700. 1976; Rao & Verma, Cyp. NE India 56. 1982; Sharma *et al.*, Fl. Karnataka 316. 1984; Koyama in Dassanayake & Fosberg, Rev. Handb. Fl. Ceylon 365. 1985; Karthik. *et al.*, Fl. Ind. Enum. Monocot. 72. 1989. *Hypoporum pergracile* Nees, Edinburgh New Philos. J. 17 (34): 267. 1834.

Type : India, Silhet, *Wallich* 3406 (K).

Illus.: Clarke, Illus. Cyp. t. 121, f. 1-5. 1909; Kern, *l.c.*

Annuals, 25-50 cm high. Stems very slender, 0.5- 1 mm thick, smooth, glabrous. Leaves narrowly linear, acutish at apex, 0.5 - 2 mm wide, glabrous, scabrid towards apex; sheaths narrow, not winged, smooth, glabrous or sparsely pilose, truncate at mouth or with a very short membranous appendage. Inflorescence spiciform, linear, unbranched, consisting of 5-25 clusters of spikelets; each cluster small, almost sessile, with 2-5 spikelets, lower ones with 1-1.5 cm gap, upper ones subcontiguous. Bracts inconspicuous, not or hardly longer than spikelet-clusters. Spikelets bisexual, small, obovate, 2.5-3 mm long; glumes ovate-lanceolate, acute and muticous at apex, glabrous, densely lineolate with reddish glandular streaks; glumes bearing male flowers thinly membranous. Stamens 2; anthers linear, *ca* 1 mm long. Nut much shorter than glumes, obtusely trigonous, slightly depressed, apiculate, 1-1.3 mm long and wide, lacuno-rugose, trabeculate especially towards the top, glabrous, shining white. Cupula very small, triangular, *ca* 0.5 mm long. Disk obsolete, concrete with the nut, forming a brown triquetrous stipe, 0.3-0.5 mm long.

Fls. & Frts. : Oct.- Dec. (Hooper, *l.c.*) .

Habitat : Marshy areas.

Distrib. : Sri Lanka, Indo-China, Thailand, Malesia and tropical Africa. INDIA: Throughout (probably not in North-West). KARNATAKA: Hassan (Hooper, *l.c.*) .

Note: This species is included on the authority of Hooper (*l.c.*) . As no specimens available in BSI, description is adopted from Kern (*l.c.*) . Probably rare in Karnataka.

8. **Scleria poaeformis** Retz., Obs. 4: 13. 1786; Fischer in Kew Bull. 1931: 265. 1931 *et* in Gamble, Fl. Pres. Madras 1678. 1931 (3: 1163. 1957, repr.ed.); Kern in Blumea 11: 178, f. 5 g. 1961 *et* in van Steenis, Fl. Males. 1, 7: 736, f. 105 & 108. 1974; Rao & Verma, Cyp. NE India 61. 1982; Sharma *et al.*, Fl. Karnataka 316. 1984; Koyama in Dassanayake & Fosberg, Rev. Handb. Fl. Ceylon 5: 349. 1985; Karthik. *et al.*, Fl. Ind. Enum. Monocot. 72. 1989. *S. oryzoides* Presl, Reliq. Haenk. 1: 201. 1828; Clarke in Hook. f., Fl. Brit. India 6: 691. 1894.

Type: India, *Koenig s.n.* (L Z).

Illus.: Kern, *l.c.*

Perennials with thick, horizontally creeping rhizome, 1-2 m tall. Stems robust, smooth or scabrid on angles at top, up to 2 cm thick, often rooting from lower nodes. Leaves mostly basal or subbasal, 1-3 on the stem, abruptly narrowed to an obtusish, somewhat cucullate tip, up to 25 mm wide, smooth or scabrid on the margins and the 3 prominent nerves; upper sheaths acutely triquetrous or narrowly winged, smooth or scabrid on angles, mouth concave or truncate, with narrow scarious margin; lower sheaths spongy, purplish-red. Inflorescence as a rule a single, compound, long-peduncled panicle with a small setaceous bract at the base, or ebracteate, ovate or elliptic in outline, 10-20 x 5-10 cm, very rarely with a lateral panicle from the axil of leafy bract; ultimate branches obliquely erect, spiciform. Spikelets solitary, sessile, evenly distributed along the spiciform branches, usually unisexual. Male spikelets numerous, 4-5 mm long;

stamens 3; anthers linear, *ca* 1.5 mm long. Nut-bearing spikelets few, mostly restricted to base of the branches, 4-5 mm long, their male part reduced to a sessile glume or 1 or 2 flowers. Glumes ovate, acute and muticous at apex. Nut obtusely trigonous to almost terete, ovoid, ellipsoid or subglobose, not or scarcely apiculate at apex, 2.6-3 x 2.5-3 mm, with three depressions at base, smooth, glabrous, very shining white. Disk small, much narrow than base of the nut, thick, triangular-cordate, appressed to the nut.

Fls. & Frts.: Not known from Karnataka.

Habitat: Swampy areas.

Distrib.: Sri Lanka, Southern China, Indo-China, Hainan Island, Malesia, Africa and tropical Australia. INDIA: South and North- East India. KARNATAKA: Uttara Kannada (N. Kanara, Sharma *et al.,l.c.*) .

Note : This species is included based on the authority of Sharma *et al.(l.c.).* As no specimens available in BSI, description provided is as given by Kern (*l.c.*) .

9. **Scleria stocksiana** Boeck. in Linnaea 38: 474. 1874; Clarke in Hook. f., Fl. Brit. India 6: 687. 1894; Cooke, Fl. Pres. Bombay 2: 905. 1908 (3: 419. 1958, repr.ed.); Sharma *et al.*, Fl. Karnataka 316. 1984; Karthik. *et al.*, Fl. Ind. Enum. Monocot. 72. 1989. **Fig. 60.**

Annuals with fibrous roots or short-lived perennials with short rhizome, 20-55 cm high. Stems usually tufted, triquetrous, striate, glabrous, 1-3 mm thick, at times retrorsely scabrous on the angles. Leaves linear, narrowed to obtusish tip, 10-27 cm long, 3-7 mm wide, antrorsely scabrous on the margins and midrib below towards the apex, usually scabrous on two prominent lateral veins above; sheaths usually winged, smooth or retrorsely scabrous on angles; usually pubescent on anterior side below the mouth; contraligule much broader than long, glabrous or ciliate. Inflorescence consisting of a terminal panicle and 1 or 2 axillay ones; terminal panicle longer, 2-4.5 cm long; lateral ones smaller, on erect peduncles. Primary bracts overtopping the inflorescence. Spikelets usually unisexual and in clusters of 2- 3 females and one male. Male spikelets linear, *ca* 4 mm long; pedicels 2-3 mm long, winged. Glumes lanceolate, acute to mucronate at apex, 3.5 - 4 mm long. Stamens 2 or 3; filaments up to 3.5 mm long; anthers linear-oblong, apiculate with minutely hairy connective appendage, *ca* 1 mm long. Female glumes broadly ovate, acute to mucronate at apex, 4-4.5 mm long. Nut broadly ovoid, obtusely trigonous, shortly umbonate at apex, *ca* 3 x 2 mm (including the disk), glabrous, somewhat rugulose or tubercled, especially near the base, smooth near apex. Disk appears to be biseriate; outer layer very short, hardly or very shortly 3-lobed, usually dark brown with whitish margin; inner layer prominently 3-lobed, with broadly ovate

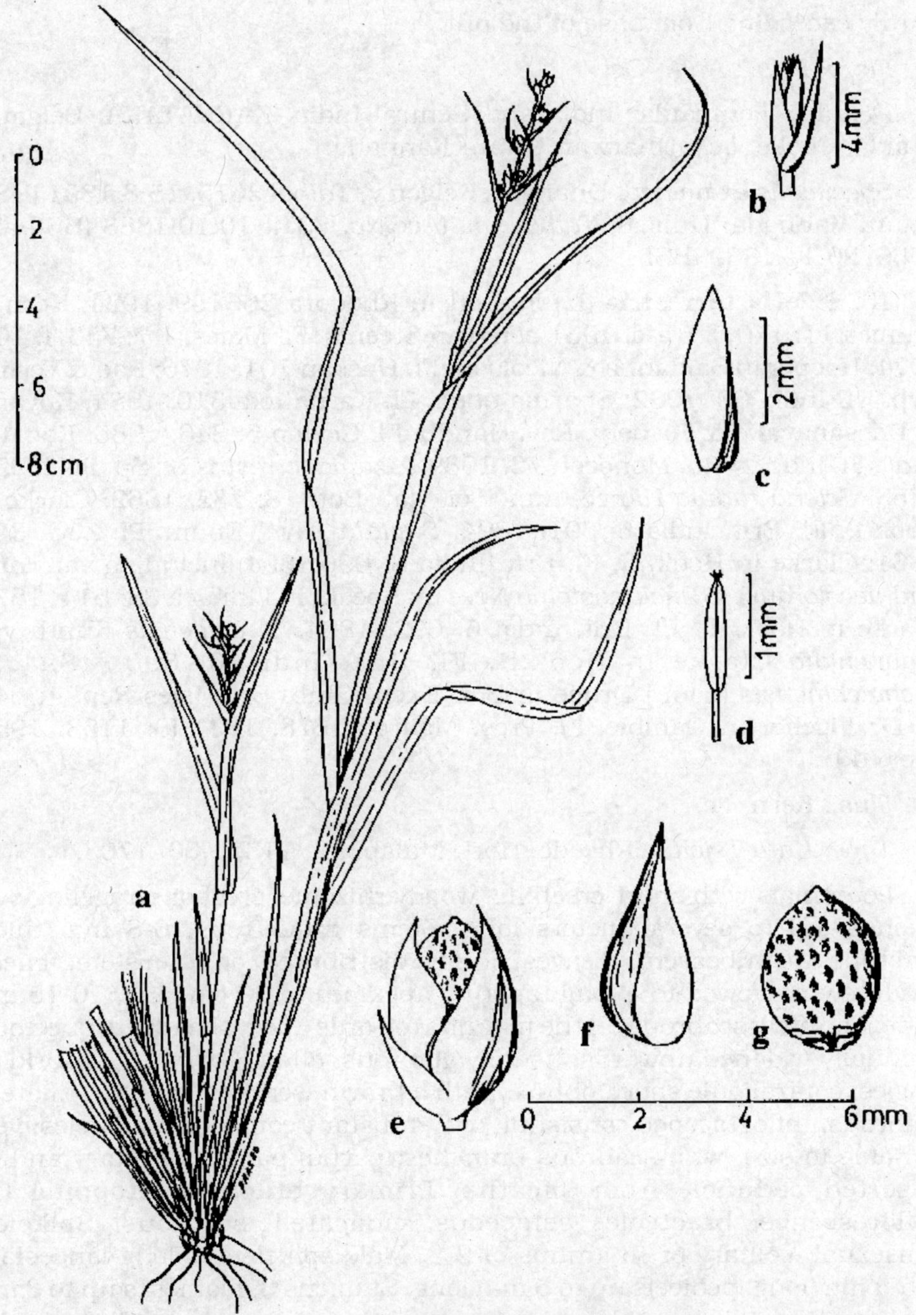

Fig. 60. *Scleria stocksiana* Boeck.
a. Habit, **b.** Male spikelet, **c.** Male glume, **d.** Stamen,
e. Nut-bearing spikelet, **f.** Female glume, **g.** Nut

or subobtuse lobes, with obtuse apex, closely appressed to base of the nut, hardly exceeding from base of the nut.

Fls. & Frts. : Aug.- Oct.

Distrib. : Peninsular India and Central India. KARNATAKA: Belgaum (Londa, Cooke, *l.c.*), Dharwar, Uttara Kannada.

Specimens examined: Dharwar: Kelgerry, *Talbot* 2615, 15.8.1891 (BSI). Uttara Kannada: Hulical, *Talbot s.n.* (Acc.No. 831), 10.10.1888 (BSI) and 2066, 20.11.1889 (BSI).

10. **Scleria terrestris** (L.) Fassett in Rhodora 26: 159. 1924; Kern in Blumea 11: 170, f. 5 a-c. 1961 *et* in van Steenis, Fl. Males, 1,7: 733, f. 104. 1974; Hooper in Saldanha & Nicolson, Fl. Hassan 701. 1976; Rao & Verma, Cyp. NE India 60. 1982; Sharma *et al.*, Fl. Karnataka 316. 1984; Koyama in Dassanayake & Fosberg, Rev. Handb. Fl. Ceylon 5: 346. 1985; Karthik. *et al.*, Fl. Ind. Enum. Monocot. 72. 1989. *Zizania terrestris* L., Sp. Pl. 2: 991. 1753. *Scleria radula* Hance, Ann. Sci. Nat. Bot. 18: 232, 1862; Clarke in Hook f., Fl. Brit. India 6: 691. 1894. *S. elata* Thw., Enum. Pl. Zeyl. 353. 1864; Clarke in Hook. f., Fl. Brit. India 6: 690.1894 (including var. *latior* and *decolorans*). *S. melanostoma* Nees ex Boeck. in Linnaea 38: 514. 1874; Clarke in Hook. f., Fl. Brit. India 6: 692. 1894. *S. chinensis* Kunth var. *biauriculata* Clarke in Hook.f., Fl. Brit. India 6: 690. 1894. *S. cochinchinensis* (Lour.) Druce in Bot. Exch. Club. Brit. Isles Rep. 4: 646. 1917; Fischer in Gamble, Fl. Pres. Madras 1678. 1931 (3: 1163. 1957, repr.ed.).

Illus.: Kern, *l.c.*

Type: *Katu-Tsjolum* Rheede, Hort. Malab. 12: 113, t. 60. 1703.

Perennials with short-creeping, woody rhizome, erect or straggling over bushes, up to several meters long. Stems triquetrous, 6-8 mm thick, glabrous to pubescent. Leaves evenly distributed on the stem, linear, gradually narrowed to a long tapering apex, up to 60 cm long, 10-15 mm wide, minutely scabrous on the margins towards apex; sheaths triquetrous, gradually widened upwards, usually glabrous, often 3- winged, scabrid on angles; contraligule short, obtuse, with a brown, scarious margin, ciliate or glabrous. Inflorescence consisting of 2-4 distant compound panicles, very variable in size, with scabrous branches; partial panicles solitary and on exserted peduncle from sheaths. Primary bracts overtopping the inflorescence; bracteoles setaceous, elongated, scabrous. Spikelets unisexual, solitary or in groups of 2-3. Male spikelets oblong-lanceolate, 3.5-4 mm long; pedicels up to 5 mm long. Stamens 3; filaments up to 3 mm long; anthers linear-oblong, apiculate, *ca* 1.8 mm long. Nut-bearing spikelets sessile, broadly obovate, 3.5-4.5 mm long. Glumes broadly ovate, 3.5-4.5 mm long. Disk shortly 3-lobed; lobes obtuse, appressed to the base of the nut, finely denticulate at apex. Nut broadly ovoid to subglobose, terete or obtusely trigonous, umbonulate, *ca* 3 x 2.3 mm (including the disk) reticulate-cancellate, minutely hairy, glabrescent ultimately, white to greyish.

Fls. & Frts.: Aug. - Dec.

Habitat: Shady areas in semi-evergreen forests as an undergrowth, swampy areas, often among the thick patch of shruby weeds.

Distrib.: Sri Lanka, China, Formosa, Malesia, Japan and Australia. INDIA: Throughout. KARNATAKA: Hassan, Kodagu, Shimoga, Uttara Kannada.

Specimens examined: Hassan: Mankarahalli, *Saldanha* 15779, 9.12.1967 (JCB). Kodagu: Abbe falls, Mercara, *Bhat* 1050, 25.9.1981 (MGH). Shimoga: Jog falls, *Prasad* 173790, 4.11.1995 (BSI). Uttara Kannada: Falls of Gersoppa, *Talbot* 2669, 28.12.1891 (BSI); Castle Rock, *Gammie* 15639, 25.10.1902 (BSI).

Note: A very variable species especially in the presence or absence of wings on leaf-sheaths, nature of nut surface and size of disk-lobes which resulted in the creation of many names for this species.

11. **Scleria tessellata** Willd., Sp. Pl.4: 315. 1805; Clarke in Hook.f., Fl. Brit. India 6: 686. 1894; Cooke, Fl. Pres. Bombay 2: 903. 1908 (3: 418. 1958, repr.ed.); Fischer in Gamble, Fl. Pres. Madras 1677. 1931 (3: 1163. 1957, repr.ed.); Nelmes in Kew Bull. 1956 (1): 107. 1956; Robinson in Kew Bull.18: 526. 1966; Sharma *et al.*, Fl. Karnataka 317. 1984; Karthik. *et al.*, Fl. Ind. Enum. Monocot. 73. 1989. *S. glandiformis* Boeck. in Linnaea 38: 458. 1874.

Type : India, Herb. Willdenow 17323 (B-W).

Illus.: Kern in van Steenis, Fl. Males. 1,7: 726, t. 106, f. 24 a. 1974 (nut).

Annuals with fibrous roots, *ca* 60 cm high. Stems usually tufted, deeply trigonous, 1.5-2 mm thick, smooth throughout, often sparsely and retrorsely scabrous on the angles, internodes rather long. Leaves mostly from near the base, linear, gradually narrowed to subobtuse tip, rather long (up to 27 cm or more), *ca* 5 mm wide, glabrous, often scabrous on 2 side-veins above and on the midnerve below towards apex; basal leaves usually reduced to bladeless sheaths or with a short lamina; sheaths usually glabrous, often pubescent below the mouth downwards; contraligule short; membranous, ciliate. Inflorescence consisting of a terminal panicle and 1 or 2 short, axillary ones; terminal panicle sessile, longer, *ca* 4 x 1.5 cm; lateral ones erect, short-peduncled, 1-2 cm long. Bracts usually overtopping the panicles. Spikelets mostly unisexual, at times bisexual, in clusters of 2-3, at times solitary. Male spikelets subsessile to short-pedicelled, lanceolate, *ca* 5 mm long; pedicels up to 1 mm long. Glumes lanceolate, acuminate at apex, 5-6 mm long. Stamens 2; filaments up to 4.5 mm long, somewhat coiled; anthers linear or linear-oblong, 1-1.8 mm long, with minutely fimbriate connective appendage. Female glumes broadly ovate to ovate-lanceolate, acute at apex, 4-6 mm long. Nut cylindric-ellipsoid, minutely or hardly beaked at apex, *ca* 3.2 x 2 mm (including disk), finely longitudinally lacunose with longitudinal walls of the lacunae more prominent. Disk deeply 3-lobed, thick, cream-coloured; lobes broadly oblong, almost truncate at apex, *ca* 1 mm long.

Fls. & Frts.: Not known from Karnataka.

Habitat: Swampy areas and wet grasslands, up to 1000 m.

Distrib. : Tropical Africa and Madagascar. INDIA: Throughout (except Western arid region). KARNATAKA: Belgaum (Sharma *et al., l.c.*), Mysore (without exact locality, Fischer, *l.c.*) .

Note : Most of the specimens labelled as *S. tessellata* Willd. in BSI are either *S. stocksiana* Boeck. or *S. parvula* Steud. Of the many specimens gone through, only a single specimen of this species was found in BSI, collected from Maharashtra. This specimen was used for making the above description. This species is included in this Flora based on earlier records by Fischer, *l.c.* and Sharma *et al., l.c.*

REFERENCES

Ahmedullah, M. & M.P. Nayar, 1986. *Endemic plants of the Indian Region.* Vol. 1. Botanical Survey of India, Calcutta.

Akiyama, S. 1935. Carices of Taisetsuzan Mts., in *Biogeographica* 1:45-53.

Akiyama, S. 1943. On the classification of the Japanese Carices, especially of the subgenus *Eucarex* [Preliminary report], in *Act. Phyotax. Geobot.* 13: 242-246.

Akiyama, S. 1949. On the classification of Carices indigenous to the Far Eastern region of Asia, in *Bot. Mag. Tokyo* 62: 23.

Akiyama, S. 1955. *Carices of the Far Eastern region of Asia* 1-257 & Pls. 1-248.

Anonymous, 1950. *Wealth of India.* Raw materials vol. 2. p. 423. CSIR, New Delhi.

Arora, R.K. 1961-1964. The forests of North Kanara district I - Scrub, in *J. Indian Bot. Soc.* 40: 187-200. 1961; II- Deciduous type. *Ibid.* 43: 75-86. 1964; III-Evergreen type. *Ibid.* 42: 38-60. 1963; IV- Successional trends and synthesis of vegetation. *Ibid.* 42: 629-636. 1963.

Arora, R.K., B.M. Wadhwa & M.B. Raizada. 1981. *The Botany of South Kanara District.* Dehra Dun.

Ascherson, P. & P. Graebner, 1902 *Synopsis der Mitteleuropaischen Flora* .

Baijnath, H. 1975. A study of *Cyperus alternifolius* L., sens. lat. (Cyperaceae), in *Kew Bull.* 30 : 521-526.

Barnes, E. 1944. Flowering plants of the Biligirirangan hills, in *J. Bombay nat. Hist. Soc.* 44 (3): 436-60.

Beetle, A.A. 1940-1946. Studies in the genus *Scirpus* L. in *Amer. J. Bot.* 27: 63-64. 1940; II. *Ibid.* 28: 469-476. 1941; III. *Ibid.* 28: 691-700. 1941; IV. *Ibid.* 29: 82-88. 1942; V. 29:653-656. 1942; VI. 30: 395-401. 1943; VII. 31: 261- 265. 1944; VIII. 33: 660-666. 1946.

Beetle, A.A. 1943. A key to the North American species of the genus *Scirpus* based on achene characters. *Amer. Midl. Nat.* 29: 533-538.

Beetle, A.A. 1945. The genus *Isolepis* R. Br., in *Amer. Midl. Nat.* 34: 723-734.

Beetle, A.A. 1950. Bulrushes and their multiple uses, in *Economic Botany* 4 (2): 132-138.

Bentham, G. 1881. Notes on Cyperaceae with special reference to Lestiboudois 'Essai' on Beauvois Genera., in *J. Linn. Soc.* (Bot.) 18: 360-368.

Bentham, G. 1883. Cyperaceae in Bentham & Hooker 1883. *Genera plantarum* 3: 1037-1073. London.

Bhat, K.G. 1971. *The Cyperaceae of Mysore City and surrounding areas.* M.Sc. dissertation, Univ. Mysore (in ed.)

Bhat, K.G. 1983. *Grasses and Sedges of Coorg and South Kanara districts of Karnataka - A Taxonomic study.* Ph.D. Thesis, 2 Vols. Univ. Mysore (in ed.)

Bhat, K.G. & B.A. Razi, 1971. Contribution from the Herbarium Manasa Gangotri, Mysore. III. The Cyperaceae of Mysore city and surrounding areas, in *J. Mysore Univ.* 24: 81-105.

Blake, S.T. 1954. The Cyperaceae collected in New Guinea by L.J. Brass, in *J. Arn. Arb.* 35: 217.

Blatter, E. & C.McCann, 1934. Revision of the Flora of the Bombay Presidency, in *J. Bombay nat. Hist. Soc.* 37: 255-277 & 532-548.

Boeckeler, O. 1868-1877. *Die Cyperaceen des Koeniglichen Herbariums* Zu Berlin, in *Linnaea* 35: 397-612. 1868; *Ibid.* 36: 271-512. 1870; *Ibid.* 36: 691-768. 1870; *Ibid.* 37: 1-128. 1871; *Ibid.* 37: 129-142. 1872; *Ibid.* 37: 520-544. 1872; *Ibid.* 37: 545-647. 1873; *Ibid.* 38: 223-544. 1874; *Ibid.* 37: 1-152. 1875; *Ibid.* 40: 327-452. 1876; *Ibid.* 41: 145-356. 1877.

Boott, F. 1858-1862. *Illustrations of the genus Carex.* Pt. I. tt. 1-200. 1858; Pt. II. tt. 201-310. 1860; Pt. III. tt. 311- 411. 1862. London.

Brown, R. 1810. Novae Cyperaceae. *Prodrumus Florae Novae Hollandiae et Insulae van-Diemen.* London.

Bruhl, J.J. 1995. Sedge genera of the world: relationships and a new classification of the Cyperaceae, in *Austr. Syst. Bot.* 8: 125-305.

Buchanan Hamilton, F. 1807. *A journey from Madras through the countries of Mysore, Canara and Malabar etc.* I & II (3 Vols. reprinted in 2 Vols. 1870). London

Burman, N. L.B. 1768. *Flora Indica.* Amsterdam.

Caius, J.F. 1935. The medicinal and poisonous sedges of India, in *J. Bombay nat. Hist. Soc.* 38: 163-170.

Cameron, J. 1894. *The Forest Flora of Mysore & Coorg.* Bangalore.

Champion, H.G. & S.K. Seth, 1968. *A revised Survey of the forest types of India.* Delhi.

Chopra, R.N., S.L. Nayar & I.C. Chopra, 1956. *Glossary of Indian Medicinal Plants.* C S I R, New Delhi.

Clarke, C.B. 1893 & 1894. Cyperaceae in: J.D. Hooker, *Flora of British India* 6: 585-672. 1893 *et* 6: 673-748. 1894. London.

Clarke, C.B. 1894. On certain authentic Cyperaceae of Linnaeus, in *J. Linn. Soc. (Bot.)* 30: 312.

Clarke, C.B. 1898. On the Sub-subareas of British India, Illustrated by the detailed distribution of the Cyperaceae in that Empire, in *J. Linn. Soc.* (Bot.) 34: 1-146.

Clarke, C.B. 1901-1902. Cyperaceae in: Thiselton-Dyer, W.T., *Fl. Trop. Africa* 8: 266-524.

Cooke, T. 1908. Cyperaceae in: *The Flora of the Presidency of Bombay* 2: 851-906 (3: 364-421. 1958, repr.ed.) London.

Datar, R. & V.D. Vartak, 1976. Taxonomic study of the genera *Lipocarpha* R. Br. and *Bulbostylis* Kunth from India, in *J. Univ. Poona* (Sci. & Tech. Sect.) 48: 19-32.

Datta, S.C. & A.K. Banerjee, 1978. Useful weeds of West Bengal rice fields, in *Economic Botany* 32 (3): 297-310.

Dinesh, M.S., S.N. Ramaswamy & C.R. Nagendran, 1990. Studies on the flowering plants of Ranganathittu Bird sanctuary, Mandya district, Karnataka state, in *Geobios New Reports*. 9: 3-7.

Eners, D.V. 1907. The evergreen forests of Manjrabad forest range, in *Indian For.* 33: 324-328.

Fischer, C.E.C. 1931. Cyperaceae in: Gamble, *Flora of the Presidency of Madras* Pt. 9: 1620-1687. (vol. 3: 1128-1169. 1957, repr.ed.). London.

Gamble, J.S. 1915-1936. *Flora of the Presidency of Madras*. London. (11 parts, of which 8-11 by C.E.C. Fischer) (Repr. ed. 3 Vols. 1957. Calcutta).

Govindarajalu, E. 1966-1978. Systematic anatomy of South Indian Cyperaceae: *Bulbostylis* Kunth, in *J. Linn. Soc. (Bot.)* 59: 289-304. 1966; *Fuirena* Rottb. *Ibid.* 62: 27-40. 1968; *Cyperus* L., subgen. *Kyllinga* (Rottb.) Suring. *Ibid.* 62: 41-58. 1968; *Cyperus* L. subgen. *Juncellus, Cyperus* subgen. *Mariscus* and *Lipocarpha* R.Br. *Ibid.* 68: 235-266. 1974; *Eleocharis* R. Br., *Rhynchospora* Vahl and *Scleria* Bergius. *Adansonia* 2, 14: 581-632. 1975; *Cyperus* subgen. *Pycreus* (P.Beauv.) C.B. Clarke. *Ibid.* 18: 95-128. 1978.

Govindarajalu, E. 1972-1985. Studies in Cyperaceae: VIII. Novelties in *Fimbristylis* (L.) Vahl, in *Proc. Ind. Acad. Sci.* 76B: 181-193. 1972; IX. *Ibid.* 78B: 45-58. 1973; XI. *Ibid.* 79B: 160-172. 1974. XII. *Ibid.* 80B: 41- 50. 1975; XVII. *Ibid.* 91(1): 43-53. 1982. XIX. *ibid.* 94 (1): 11-31. 1985.

Govindarajalu, E. 1975. Studies in Cyperaceae - XIII. Novelties in *Cyperus* L. subgen. *Pycreus* (Beauv.) C.B. Clarke, in *Proc. Ind. Acad. Sci.* 81 B: 187-196.

Govindarajalu, E. 1990-1996. Cyperaceae Indie Australis Precursores - A novelty in *Eleocharis* R. Br. and its vegetative anatomy, in *Proc. Ind. Acad. Sci.* 100:71-75. 1990; - Nova species in *Fimbristyle* (L.) Vahl and their vegetative anatomy. *Ibid.* 100: 77-84. 1990; -A novelty in *Cyperus* Linn. and its vegetative anatomy. *Ibid.* 100: 409- 413. 1990; - New species and Scanning electron microscopic observations in *Pycreus* sect. muricati. *Ibid.* 100: 415- 422. 1990; -New combinations in *Pycreus* Beauv. *Ibid.* 100: 423-433. 1990; -Novelties in *Pycreus* Beauv. in *J. Econ. Tax. Bot.* 20 (2):299-304. 1996.

Govindarajalu, E. 1996- 1998. Monographia Indicorum *Fimbristylium* precursores-Novelties in *Fimbristylis* Vahl- I, in *Rheedea* 6(2): 59-64. 1996. II. *Ibid.* 7(2): 115-126. 1997; III. *Ibid.* 8(1): 87-95. 1998.

Govindu, H.C. & M.J. Thirumalachar, 1960. The Cyperaceae of Mysore, in *Inf. Tech. Bull.* 1: 1-17. Bangalore.

Gowda, B., M.R. Gurudeva & G.C. Vijaya Kumar, 1986. Flora of Tumkur district, Karnataka, Part 1 - Kunigal taluk, in *J. Econ. Taxon. Bot.* 8 : 321-327.

Hooper, S.S. 1976. Cyperaceae in Saldanha, C.J. & D.H. Nicolson, *Flora of Hassan District, Karnataka, India* p.p. 655-701. New Delhi.

Hutchinson, J. 1959. Cyperaceae in the *Families of Flowering Plants* - II. Monocotyledons ed. 2. 704-710. Oxford.

Jain, S.K. & A. De Filipps, 1991. *Medicinal plants of India*, 2 Vols. Michigan. Kadambi, K. 1939. The montane evergreen forest, Bisale region, in *Indian For.* 65: 189-204.

Kadambi, K. 1941. The evergreen ghat rain - forest, Agumbe - Kilandus zone, in *Indian For.* 67: 184-203.

Kadambi, K. 1942. The evergreen ghat rain - forest of Tunga and Bhadra river sources, Kadur district. I, in *Indian For.* 68: 305-312.

Kadambi, K. 1944-45. The ghat rain-forests of Hassan district, Mysore State, in *Quart. J. Mysore For. Dept.* 26: 1-32 & 61-85.

Kadambi, K. 1950. The ghat rain-forests of the western ghats of Hassan district, Mysore state, in *Indian For.* 76: 18-30, 69-82, 121-132.

Karne, S.S. 1998. *Ethnobotany of Dakshina Kannada and Chikmagalur districts, Karnataka State.* Ph.D. Thesis. Univ. Poona (in ed.).

Kern, J.H., 1952-1954. Notes on Malaysian *Cyperaceae*, I, in *Reinwardtia* 2: 97-130. 1952; II. *Ibid.* 3: 27- 66.1954.

Kern, J.H. 1955-1960. Notes on Malaysian and some S.E. Asian Cyperaceae III, in *Blumea* 8: 110-169. 1955; IV. *Reinwardtia* 4: 89-97. 1956; V. *Blumea* 9: 215-236. 1958; VI. *Ibid.* Suppl. 4: 163-169. 1958; VII. *Act. Bot. Neerl.* 7: 786-800. 1958; VIII. *Blumea* 10: 635-651. 1960.

Kern, J.H. 1958. *Remirea* or *Cyperus*? Florae Malesianae Precursores XXI, in *Acta Bot. Neerl.* 7: 795-798.

Kern, J.H. 1961a. Cyperaceae of Siam (excl.Carex), in *Reinwardtia* 6: 25-83.

Kern, J.H. 1961b. Florae Malesianae Procursores XXX. The genus *Scleria* in Malaysia, in *Blumea* 11: 140-218.

Kern, J.H. 1962a. Cyperaceae of Thailand (excl. Carex), in *Reinwardtia* 6: 25-83.

Kern, J.H. 1962b. A new look at some Cyperaceae mainly from the tropical stand point, in *Advancement of Science* 19: 141- 148.

Kern, J.H. 1962c. Cyperaceae of Thailand II, in *Reinwardtia* 6: 145-154.

Kern, J.H. 1967. A new combination in *Fimbristylis* (Cyperaceae), in *Blumea* 15: 126.

Kern, J.H. 1974a. Cyperaceae in: van Steenis, *Flora Malesiana* 1, 7: Part 3: 435-753.

Kern, J.H. 1974b. Cyperaceae in: van Steenis, *Flora Malesiana* 1, 9: Part 1: 107-187.

Keshava Murthy, K.R. & S.N. Yoganarasimhan, 1990. *Flora of Coorg(Kodagu), Karnataka.* Bangalore.

Kirtikar, K.R. & B.D. Basu 1918. *Indian Medicinal Plants* Vol. 4: pp. 2632-2647, (Rev. ed. 1933). Allahabad.

Koyama, T. 1955-1958. Taxonomic studies of Cyperaceae in Asia. (3) in *Acta phytotax. Geobot.* 16: 5-12. 1955; (4) 16: 33- 41. 1955; (5) *Bot. Mag. Tokyo* 69:

59-67. 1956; (6) *J. Jap. Bot.* 32:147-153.1957; (7) *Bot. Mag. Tokyo* 70: 347-357. 1957; (8) *Act. Phytotax. Geobot.* 17: 97-102. 1958.

Koyama, T. 1958. Taxonomic study of the genus *Scirpus* L., in *J. Fac. Sci. Univ. Tokyo*, 3, 7: 271-366.

Koyama, T. 1961a. Taxonomic studies of Cypereaceae-12, in *Natur. Canad.* 88: 237-252.

Koyama, T. 1961b & 1962. Classification of family Cyperaceae-I, in *J. Fac. Sci. Univ. Tokyo* 3, 8: 37-148. 1961; II. *Ibid.* 8: 149-278. 1962; III. *Quart. J. Taiwan Mus.* 14: 159-194. 1961.

Koyama, T. 1963. The genus *Scirpus* L. Critical species of the section *Pterolepis*, in *Canad. J. Bot.* 41: 1107- 1131.

Koyama, T. 1971. Systematic interrelationships among Sclerieae, Lagenocarpeae and Mapanieae (Cyperaceae), in *Mitt. Bot. Staatssmml. Munchen* 10: 604-617.

Koyama, T. 1972. Cyperaceae - Rhynchosporeae & Cladieae, in *Mem. New York Bot. Gard.* 23: 23-89.

Koyama, T. 1973. New Cyperaceae from Nepal and adjoining Tibet, in *Bot. Mag. Tokyo* 86 (10004): 275-283.

Koyama, T. 1978a. Cyperaceae in: Hara *et al.*, *An enumeration of the flowering plants of Nepal.* 1: 96-120. London.

Koyama, T. 1978b. Studies in the Cyperaceae of Thailand. I. New and otherwise noteworthy species of *Fimbristylis*, in *Brittonia* 30: 102-108.

Koyama, T. 1979. Studies in the Cyperaceae of Thailand. II. Miscellaneous taxa of Fimbristylideae, Rhynchosporeae, Scirpeae and Sclerieae, in *Brittonia* 31: 284-293.

Koyama, T. 1985. Cyperaceae in: Dassanayake & Fosberg, *A Revised Handbook of Flora of Ceylon* 5: 125-405. New Delhi.

Koyama, T. 1988. Variations of *Fimbristylis dichotoma* (L.) Vahl (Cyperaceae) in Japan, in *J. Jap. Bot.* 63: 86-95.

Kreczetowicz, V.I. 1935. *Carex* in: Komarov, *Fl. URSS.* 3: 111-464.

Kshirsagar, R.D. 1998. *Ethnobotany of Mysore and Coorg districts, Karnataka state.* Ph.D. Thesis, Univ. Poona (in ed.).

Kukenthal, G. 1909. Cyperaceae-Caricoideae in: Engler, *Pflanzenreich* 4(20), Heft 38. Weinheim.

Kukenthal, G. 1935-1936. Cyperaceae-Scirpoideae, Cypereae in: Engler, *Pflanzenreich* 4(20), Heft 101. Weinheim.

Kukenthal, G. 1949, 1950, 1951. A monograph of the genus *Rhynchospora*, in *Bot. Jahrb.* 74: 375-509. 1949; *Ibid.* 75: 90-195-1950. *Ibid.* 75: 273-314. 1951.

Kunth, C.S. 1837. *Enumeratio plantarum* (Cyperaceae) - II. Stuttgart.

Linnaeus, C. 1753 & 1762-1763. *Species plantarum* ed. 1. (1753) *et* ed. 2. (1762-1763). Stockholm.

Linnaeus, C. 1754. *Genera plantarum* ed. 5. Stockholm.

Linnaeus, C. 1767 & 1771. *Mantissa plantarum.* Stockholm.

Lovery, E.P. 1888. *Catalogue of forest trees growing in Shimoga district.* Bangalore.

Mackenzie, K.K. 1931-1935. Cyperaceae tribe Cariceae, in *North Amer. Fl.* 18: 1-478.

Mackenzie, K.K. 1940. *North American Cariceae.* 2 vols. New York.

Maheshwari, P. & U. Singh, 1965. *Dictionary of Economic plants in India* p. 54. I C A R, New Delhi.

Mattfeld, J. 1936. Cyperaceae, in: Engler-Diels, *Syllabus der Pflanzenf.* 11 Aufl. 142-143.

McAtee, W.L. 1939. *Wildfowl food plants* p. 141.

Metcalf, F.P. 1931. Wild duck foods of North Dakota lakes, in *U.S. Dept. Agri. Tech. Bull.* 221.

Micheli, P.T. 1729. *Nova plantarum genera juxta Tournafortii deposita.* Florence.

Miquel, F.A.W. 1855-1859. *Flora van Nederlandsch Indie* - alternative title, *Flora Indiae Botavae* Vols. 1-3. Amsterdam.

Muller, T. 1915. The Industrial fibre plants of the Philippines, in *J. New York Bot. Gard.* 16:75.

Nayar, M.P. 1977. Changing patterns of the Indian Flora, in *Bull. Bot. Surv. India* 19: 145-155.

Nayar, M.P. 1980. Endemic flora of Peninsular India and its significance, in *Bull. Bot. Surv. India* 22: 12-13.

Nayar, M.P., K. Ramamurthy & V.S. Agarwal, 1989 & 1994. *Economic plants of India.* Vol. 1 & 2. Botanical Survey of India, Calcutta.

Nees Von Esenbeck, 1834. Cyperaceae Indicae praecipue juxta herbaria Wightii, Wallichii, Roylei, *et* Lindleyi in: Wight, R., *Contributions to the Botany of India.* London.

Nees Von Esenbeck, 1835. Ubersicht der Cyperaceen-Gattungen, in *Linnaea* 9: 273-308.

Nelmes, E. 1946. Key to the Carices of Malaysia and Polynesia, in *Kew Bull.* 1946: 5-29.

Nelmes, E. 1951. The genus Carex in Malaysia, in *Reinwardtia* 1: 221-450.

Nelmes, E. 1955. The genus Carex in Indo-China including Thailand and lower Burma, in *Mem. Mus. Hist. nat. Paris n.s.* Bot. 4: 83-182.

Ohwi, J. 1936. Cyperaceae Japonicae. I. A synopsis of the Caricoideae of Japan, including Sahalien, Kuriles, Korea and Formosa, in *Mem. Coll. Sci. Kyoto Imp. Univ.* B, 11: 229-530.

Ohwi, J. 1938. Notes on Japanese *Fimbristylis*, in *J. Jap. Bot.* 14: 64-579.

Ohwi, J. 1944. Cyperaceae Japonicae. II, in *Mem. Coll. Sci. Kyoto Imp. Univ.* B, 18: 1-182.

Pathak, G.P. 1920. Some famine foods in Ahmedabad, in *Agri. J. India* 15: 40-45.

Pax, F. 1888. Cyperaceae in: Engler & Prantl, *Die Naturlichen* Pflanzen-familien 11, 2: 98-126. Berlin.

Prasad, V.P. 1988. Wetland angiosperms of Keoladeo National Park, Bharatpur-Rajasthan, in *J. Econ. Taxon. Bot.* 12: 457-466.

Prasad, V.P. 1989. *Ecological studies on weed flora in cultivated fields around Jodhpur.* M.Sc. Dissertation, Univ. Jodhpur (in ed.).

Prasad, V.P. & N.P. Singh, 1996a. Notes on Cyperaceae of Karnataka, South India - 1: A few additions and exclusions, in *J. Econ. Taxon. Bot.* 20: 673-675.

Prasad, V.P. & N.P. Singh, 1996b. Polymorphism in the nuts of *Hypolytrum nemorum* (Vahl) Spreng. in *J. Econ. Taxon. Bot.* 20: 693-695.

Prasad, V.P. & N.P. Singh, 1997a. Notes on the Distribution and Endemism of Indian *Fimbristylis*, in *J. Bombay nat. Hist. Soc.* 94 (1): 22-26.

Prasad, V.P. & N.P. Singh, 1997b. On the perianth bristles in *Schoenoplectus corymbosus* (Roth ex Roem. & Schult.) J. Raynal, in *J. Bombay nat. Hist. Soc.* 94 (2): 441-443.

Prasad, V. P. & N.P. Singh, (1997c) 1998. Notes on Cyperaceae of Karnataka, South India-II. A new species, one new report and an exclusion, in *J. Econ. Taxon. Bot.* 21(3): 667-670.

Prasad, V.P. & N.P. Singh, (1997d) 1998. Two new species of *Fimbristylis* (Cyperaceae) from Maharashtra, India, in *J. Econ. Taxon. Bot.* 21(3): 671-675.

Prasad, V.P. & N.P. Singh, 1999. Two new species of *Fimbristylis* (Cyperaceae) from Western peninsular India, in *J. Bombay nat. Hist. Soc.* 96(3): 454-458.

Prasad, V.P. & N.P. Singh, 2001. Notes on Cyperaceae of Karnataka, South India: Four new reports, A new combination and three exclusions, in *Phytotaxonomy* 1 : 63-65.

Raizada, M.B., R.K. Arora & B.M. Wadhwa, 1979. A contribution to the Botany of Chikmagalur district, Karnátaka, in *Indian J. For.* 2 : 336-349.

Ramaswamy, S.V. & B.A. Razi, 1973. *Flora of Bangalore District.* Mysore.

Rao, A.S. & D.M. Verma, 1981. Cyperaceae (Morphology and salient taxonomic characters), in *Rec. Bot. Surv. India* 21(2): 207-220.

Rao, A.S. & D.M. Verma, 1982. Notes on Cyperaceae of Assam, in *Bull. Bot. Surv. India* 22: 80-90.

Rao, R.R. & B.A. Razi, 1981. *A synoptic Flora of Mysore District.* New Delhi.

Rao, Usha *et al.*, 2000. A contribution to Flora of Chikmagalur district, Karnataka, in *J. Econ. Taxon. Bot.* 24 : 349-357.

Raymond, M. 1951. Sedges as materials for phytogeographical studies, in *Mem. Jard. Bot. Montreal* 20: 1-24.

Razi, B.A. 1950a. An account of the forests of Mysore, in *J. Mysore Univ.* B 10: 47-58.

Razi, B.A. 1950b. An index to collections of flowering plants of Mysore as represented in Herbaria.I & II, in *J. Mysore Univ.* B 11 (1): 1-20 & 21-55.

Razi, B.A. 1955. An account of vegetation of the semi-arid tracts of Mysore State and Karnataka, in *J. Mysore Univ.* B15 (2): 1-6.

Razi, B.A. 1955. Some observations on plants of the South Indian hilltops and their distribution, in *Proc. nat. Inst. Sci. India* 21B : 79-89.

Razi, B.A. 1972. Vegetation and floristic composition of the Deccan Trap areas of Mysore state, in *Bull. Indian nat. Sci. Acad.* 45: 168-177.

Razi, B.A. 1977 (1979). Floristic studies in Karnataka - Past, Present and Future, in *Bull. Bot. Surv. India* 19 (1-4): 112-21.

Retzius, A.J. 1786-1791. *Fasciculus Observationum Botanicum.* fasc. 4. 1786/87; fasc. 5. 1788; fasc. 6. 1791.

Ridley, H.N. 1930. *The dispersal of plants throughout the world,* p. 744.

Rottboell, C.F. 1773. *Descriptionum et Iconum rariores.* Koebenhavn.

Roxburgh, W. 1820-1824. *Flora Indica* ed. 1. 1820-1824; ed. 2. 1832. Serampore.

Rubsy, H.H. 1909. Report of the Botanical work on the pacific coast, in *J. New York Bot. Gard.* 10: 229.

Saldanha, C.J. 1984. *Flora of Karnataka* Vol. 1. New Delhi.

Saldanha, C.J. 1996. *Flora of Karnataka* Vol. 2. New Delhi.

Sampathkumaran, M.A., 1924. *Botany Indian Science Congress Handbook, Bangalore session* 35-44.

Satyanarayan, Y. 1958. Ecological studies of Bombay coastline strand vegetation, in *J. Biol. Sci.* 1: 53-55.

Satyanarayan, Y. 1961. Ecological studies of the evergreen vegetation of the western ghats, in *Proc. Symp. Humid. Trop. Veg. Tjiawi.* Indonesia. 196-211.

Satyanarayan, Y & K.A. Shankaranarayan. 1964. Vegetation of Bellary District, Mysore State, in *Ann. Arid. Zone* 2: 123- 149.

Sen, D.N. 1981. *Ecological Approaches to Indian weeds.* Geobios International, Jodhpur.

Shah, C.K. 1963a. *Cyperus papyrus* L. - A new record for India, in *Curr. Sci.* 32: 424-425.

Shah, C.K. 1963b. The sedges of Sabarmati, in *J. Gujarat Univ.* 6: 179-186.

Shah, C.K. 1964. Embryo development in *Eleocharis palustris* L. in *nat. Canad.*, 111: 40-49.

Shah, C.K. 1965. Embryogeny in some *Cyperaceae,* in *Phytomorphology* 15: 1-9.

Shah, C.K. 1967. Sedges as material for ecological and phytogeographical studies - an appraisal, in *J. Ind. Bot. Soc.* 46: 428-456.

Shah, G.L. & B. Suryanarayana, 1967. A note on the distribution of sexes in *Rhynchospora wightiana* (Nees) Steud. in *Curr. Sci.* 36: 157-158.

Shah, G.L. & M.B. Deshpande, 1968. A new species of *Fuirena* from Gujarat. *Fuirena tuwensis* Deshpande & Shah, in *Bull. Bot. Surv. India* 10:239-240.

Shah, G.L. & R.G. Bhatt 1974. Some interesting rare Cyperaceae of Gujarat as new record for erstwhile Bombay Presidency, in *Indian For.* 100(2):121-122.

Sharma, B.D., N.P. Singh, R.S. Raghavan & U.R. Deshpande 1984. Cyperaceae in *Flora of Karnataka - Analysis.* Botanical Survey of India, Calcutta.

Singh, N.P. 1983. New records for Karnataka state, in *J. Econ. Taxon. Bot.* 4: 177-180.

Singh, N.P. 1988. *Flora of Eastern Karnataka.* Vol. 1 & 2. Mittal Publications, Delhi.

Someran, C.J. van, 1879. *A list of Forest trees of Mysore and Coorg.* Bangalore.

Steudel, E.G. 1854-1855. *Synopsis plantarum glumacerum.* vol. 2. *Synopsis plantarum Cyperacearum.* fasc. 7: 1-80. 1954 *et* fasc. 8-10: 81-348. 1855.

Subramanyam, K. & M.P. Nayar, 1974. Vegetation and phytogeography of the Western ghats, in: *Ecology Biogrography in India.* (ed.) M.S. Mani. 178-196. Hague.

Svenson, H.K. 1929-1939. Monographic studies in the genus *Eleocharis.* I, in *Rhodora* 31: 121-135, 152-163, 167-191, 199-219, 224-242. 1929; II. *Ibid.* 34: 193-203, 215-227. 1932; III. *Ibid.* 377-389. 1934; IV. *Ibid.* 39: 210-231, 236-273. 1937; V. *Ibid.* 41: 1-19, 43-77, 90- 110. 1939.

Swamy, H. Ramachandra, *et al.*, 1992. Additions to Flora of Chikmagalur district, Karnataka, in *J. Econ. Taxon. Bot.* 16 : 661-694.

Talbot, W.A. 1909 & 1911. *Forest Flora of the Bombay Presidency & Sind.* 2 Vols. Poona.

Tirumalachar, M.J., B.A. Razi & B.G.L. Swamy 1949. Contribution to the flora of Nandi hills. II. Additions to the common flowering plants of Nandi hills, in *J. Mysore univ.* 9(5): 73-83.

Tournfort, J.P. de. 1719. *Institutiones rei herbariae* ed. 3. Paris.

Turril, W.B. 1964. Plant Taxonomy, Phytogeography and Plant Ecology. in: *Vistas in Botany* Vol. 4, Pergamon Press, London.

Vahl, M. 1805-1806. *Enumeratio plantarum* vol. 2. Copenhagon.

Vartak, V.D. 1971. Taxonomic study of the genus *Fuirena* Rottb. from India, in *J. Univ. Poona* (Sci. & Tech. Sect.) 40: 185-203.

Vartak, V.D. 1982. Observations on multifarious uses of sedges from Western India, in *J. Econ. Taxon. Bot.* 3: 193-196.

Verma, D.M. & Veena Chandra, (1978) 1979. Distributional notes on Cyperaceae, in *Bull. Bot. Surv. India* 20: 140-142.

Verma, D.M. & Veena Chandra, 1981. Cyperaceae of Madhya Pradesh, in *Rec. Bot. Surv. India* 21 (2) 221-275.

Verma, D.M. & Veena Chandra, 1982. The genus *Rhynchospora* Vahl in India, in *Bull. Bot. Surv. India* 22: 126-135.

Willdenow, C.L. 1797-1830. *Species plantarum* ed. 4. Vols. 1-6. 1797-1830.

Yoganarasimhan, S.N., K. Subramanyam & B.A. Razi, 1981. *Flora of Chikmagalur District, Karnataka.* Dehradun.

*Some of the references given above were not seen in original

ADDENDA

When the present work was in press SEDGES AND GRASSES (*Dakshina Kannada and Udupi districts*) by Dr. K. Gopalakrishna Bhat & Dr. C.R. Nagendran was published which includes 74 species of sedges under 16 genera. *Fimbristylis monticola* Hochst. ex Steud. reported in it was found to be an addition to our work and hence incorporated here. Including this, the total number of *Fimbristylis* in the state has gone up to 40 and the total number of sedge species 164.

Fimbristylis monticola Hochst. ex Steud., Syn. Pl. Glum. 2 : 111. 1855; Clarke in Hook. f., Fl. Brit. India 6 : 642. 1893; Cooke, Fl. Pres. Bombay 2 : 885. 1908 (3 : 399. 1958, repr. ed.); Koyama in Dassanayake & Foseberg, Rev. Handb. Fl. Ceylon 5 : 291, f. 22. 1985; Karthik. *et al.*, Fl. Ind. Enum. Monocot. 53. 1989; Bhat & Nagendran, Sedges and Grasses 54. 2001.

Type: India, Tamilnadu, Nilgiri Mt., *Hochenacker* 940.

Perennials with slender tufted stems, up to 30 cm high. Leaves capillary, shorter or nearly as long as the stem; sheaths ciliolate on margin. Inflorescence simple or subcompound, bearing 2-6 spikelets. Bracts 2 or 3, setaceous. Rays 1-4, up to 15 mm long, each bearing 1 or 2 spikelets. Spikelets solitary, elliptic oblong, acute at apex, 4.5-9 x 1.5-3 mm, subloosely 9-25-flowered. Glumes spiral, broadly ovate to deltoid-ovate, mucronate at apex, 2-2.75 x 1.5-2 mm, sanguineous-brown or dark brown; margins hyaline; keel distinctly 3-nerved. Stamens 3; anthers *ca* 1 mm long. Style up to 1.5 mm long, pyramidal at base; stigmas 3, up to 1.2 mm long. Nut trigonous, broadly obovate, obtuse at apex, 0.7-0.8 x 0.6-0.7 mm, yellow-brownish, turning grey, minutely tuberculate.

Fls. & Frts. : Sept.-Oct.

Habitat : Moist shady places.

Distrib. : Sri Lanka. INDIA : Southern peninsular India. KARNATAKA : Dakshina Kannada [Surathkal, presently in Udupi district (Bhat & Nagendran, *l.c.*)].

Note : This species is included on the authority of Bhat & Nagendran, *l.c.*

INDEX TO BOTANICAL NAMES

INDEX TO VERNACULAR NAMES